D1210129

Laboratory Experiments in

# ORGANIC CHEMISTRY

# Laboratory Experiments in

# ORGANIC

MACMILLAN PUBLISHING CO., INC.
NEW YORK

SIXTH EDITION

# CHEMISTRY

**Roger Adams**
UNIVERSITY OF ILLINOIS

**John R. Johnson**
CORNELL UNIVERSITY

**Charles F. Wilcox, Jr.**
CORNELL UNIVERSITY

COLLIER MACMILLAN PUBLISHERS
LONDON

Library of Congress catalog card number: 70-87890.

MACMILLAN PUBLISHING CO., INC.
866 Third Avenue, New York, New York 10022

COLLIER-MACMILLAN CANADA, LTD.

*Printed in the United States of America*

Printing 9 10 Year 7 8 9

# Preface to the Sixth Edition

In this edition of the manual directions for assembling equipment have been modified to allow for use of either standard taper ground glass apparatus or conventional glassware with corks or rubber stoppers. Advantages of standard taper glassware are substantial saving of time in setting up an assembly of apparatus and the neatness of its appearance. A student's first impression of organic laboratory work and his attitude toward experimentation are likely to be affected favorably by these factors.

Some changes have been made in the organization of the subject matter and new material on spectroscopic and chromatographic methods has been added. A simple outline of systematic procedures for classification and identification of organic compounds has been introduced. This is to be used in conjunction with the small-scale tests and semimacro preparations included in various experiments.

The procedures for basic laboratory operations such as distillation and crystallization are now designated as *Exercises* to set them apart from *Experiments* on preparations and reactions. The current tendency to shorten the time available for laboratory work often makes it desirable to consider the principles and practice of a particular operation in conjunction with an experiment in which it is actually used, rather than as a separate exercise.

Several new experiments have been introduced to broaden the selection and extend the coverage of important reactions. Examples are: 17d, Friedel-Crafts acylation of a solid hydrocarbon, using an inert solvent (dichloroethane); 26, N-Phenylsydnone, a mesoionic heterocycle; 32, Meerwein arylation to give a stilbene derivative; 33, Modified Wittig synthesis of an alkene; 40, Reduction of a ketone by sodium borohydride; 46, Stork enamine synthesis of a diketone; 55b, Benzyne synthesis of triptycene. Supplementary preparations have been added to a few experiments: Dulcin, from *p*-phenetidine and isocyanic acid; naphthylcarbamates of alcohols; α-benzoylcinnamic azlactone from hippuric acid. Several new synthetic sequences are now available.

An effort has been made to stress the principle that the method and procedure used for a specific compound usually illustrate a broad general method applicable to other compounds of the same type. To this end the molar relationships of reactants have been emphasized. This also leads the student to consider the reasons for using one reactant in excess, or in equimolar amount.

Favorable response to the chapter on *Literature of Organic Chemistry* and on using Beilstein's *Handbuch* and the indexes of *Chemical Abstracts* has encouraged us to enlarge and improve this section.

The authors welcome suggestions and critical comments from colleagues and from students who use this book. Grateful acknowledgment is made for many suggestions that have helped in the preparation of this new edition.

R. A.
J. R. J.
C. F. W., Jr.

# General Preface

This laboratory manual is intended for beginning students in organic chemistry who may be specializing in chemistry and chemical engineering, or studying the subject because of its relationship to engineering, medicine, agriculture, food science, biology, the graphic arts, and other fields.

The more important manipulations and procedures encountered in organic chemical laboratory work, with details of correct manipulation and figures for typical assemblies of apparatus, are set forth in the first section of the manual. The intent is to emphasize the important principles involved in the laboratory procedures and not present merely the technique. Depending upon the time available for laboratory work, the instructor may wish to have students carry out exercises in manipulation before beginning preparative work, or defer these until actually encountered in a specific experiment.

The primary objective in selecting preparative experiments has been to choose syntheses that illustrate general reactions and furnish pure products in satisfactory yields. Laboratory work should encourage a student to be realistic—to consider the merits and faults of different synthetic routes to a particular goal.

Preference has been given to experiments using relatively inexpensive reagents and chemicals, whenever this could be done without sacrificing other desirable features. Many sequences of preparations are available so that a student can gain experience with stepwise syntheses that are so important in organic laboratory work. Examples are: *n*-butyl bromide—*n*-valeronitrile—*n*-valeric acid; aniline—acetanilide—*p*-nitroacetanilide—*p*-nitroaniline—Para Red (or *p*-nitrostilbene); *m*-dinitrobenzene—*m*-nitroaniline—*m*-nitrophenol. Other sequences are suggested in the introduction to the section on *Preparation and Reactions of Typical Organic Compounds*.

The scope of experiments is broad enough to provide generously for a full year's course. A list of *Suggestions for Supplementary Experiments* selected from well-known laboratory manuals, with references, has been provided to encourage students who are qualified for further synthetic work.

In many instances two or more experiments involving the same type reaction are presented. This permits students in a given class to use different examples to illustrate the same general reaction and emphasizes that aspect of the method used. Moreover, this allows certain changes in class assignments that may be desirable from one year to another.

Experimental procedures are described in great detail, particularly in the first half of the manual, even to specifying the amount of washing or drying reagent to be used. This approach is believed to be a good one, especially if the experiments are accompanied by questions concerning the procedures used and the principles involved. In later work the student is prepared to do without such details and will be able to exercise good judgment in assembling apparatus and using reagents. Laboratory conferences covering general instructions and some discussion of reactions and procedures, with occasional demonstrations, are effective in developing an intelligent attitude toward experimental work.

Questions given at the end of each experiment are intended to bring out the significance of the reactions studied and the general principles involved. Such questions can contribute in an important way to laboratory instruction. The teacher should indicate to a class which of the questions should be answered, since this will depend upon the extent of the theoretical organic chemistry presented.

The sequence in which the subject matter of organic chemistry is presented in the lecture course will differ according to a lecturer's choice and the objective of the course. It is a simple matter for the instructor to select a sequence of experiments to suit the pattern he favors. A good correlation of the theoretical aspects with the laboratory work is highly desirable.

In this manual aliphatic compounds are emphasized in the earlier experiments but aliphatic and aromatic compounds are usually included in the experiments illustrating characteristic reactions of the functional groups. For practical reasons the principal experiments on aromatic compounds may well be deferred until the student has acquired some degree of manipulative skill and experience. Experiments are available to permit the student to acquire experience in the purification and characterization of organic liquids as well as solids.

A judicious combination of syntheses and test reactions is desirable, since small-scale reactions in which a crystalline product is isolated serve to illustrate an important phase of organic laboratory work. Moreover, if the time available for the laboratory course is quite limited, a broader coverage of topics can be obtained in this way. The test reactions and *optional* preparations of solid derivatives enhance the student's familiarity with characteristic reactions of the functional groups and acquaint him with some of the special reagents developed for identification purposes.

Identification of simple unknowns, within limited classes of compounds, is suggested in the experiments on alcohols, carbonyl compounds, amines, carboxylic acids, and phenols. A brief outline of systematic procedures for the classification and identification of simple organic compounds, by solubility and chemical tests, has been included. It is intended that this outline be used in conjunction with the test reactions and small-scale preparation of derivatives mentioned above. This brief treatment is quite incomplete and one of the good texts in this field should be consulted for further information (see the chapter on *Literature of Organic Chemistry*). Recent advances in instrumental methods make possible a fruitful integration of structural information derived from chemical reactions with that obtained from spectroscopical methods. A few experiments with restricted groups of compounds serve to introduce the subject but cannot give the student an appreciation of the ingenuity and skill needed in the task of separating and identifying organic compounds.

A convenient procedure for students to follow in keeping notes and in recording experimental results is described but other methods may be devised to suit various individual situations. Copy of a convenient card to be used for reporting experiments and for entering the instructor's grade on the experiment has been included.

The concentrations of reagent solutions commonly placed on the side shelf in the laboratory are given in a table; other physical data are given in tabular form in the *Appendix*. For the instructor a list of equipment for the student's laboratory desk and a list of chemicals needed for each experiment are given in the *Appendix*.

The authors wish to acknowledge gratefully their indebtedness—to many colleagues and instructors in organic chemistry throughout the country for suggestions and friendly criticism—and to *Organic Syntheses* and other well-known laboratory manuals, for ideas and procedures.

R. A.
J. R. J.
C. F. W., Jr.

# Contents

## Exercises on Separation and Purification



## Identification of Organic Compounds

## The Preparation and Reactions of Typical Organic Compounds

# Appendix

# Exercises on Separation and Purification

# Introduction

Laboratory work that accompanies an introductory course in organic chemistry is intended primarily to acquaint the student with the principles and practice of laboratory operations used in this field as well as to reinforce the theoretical aspects of the subject. The experience of working with a variety of typical organic materials and observing their characteristic properties and transformations gives a sense of reality to the structural formulas and scientific names of organic molecules. Laboratory experiments should stimulate the student's intellectual curiosity and develop his powers of observation, in addition to giving him training in careful and skillful manipulation.

Two factors contributing to excellence in laboratory work are accuracy and neatness. Accuracy involves performance of experiments in a careful manner and in accordance with the laboratory directions, making careful observations and thoroughly understanding the experiment. Promptness in recording experimental observations in the notebook is necessary for accuracy. Neatness applies to the manipulation of reagents and apparatus, the cleaning of equipment, and the manner of recording notes.

Although each student is required to perform a certain number of experiments, all of the scientific training of laboratory work is lost if experiments are performed in an unenlightened routine fashion of reading one sentence and proceeding to the next without having beforehand a general knowledge of the whole sequence of laboratory operations and of the underlying principles.

## Laboratory Notes[1]

It is essential to have a suitable notebook in which to record directly the observations made during experiments and to assemble information that will aid in the performance of the experiment. For this purpose the student should procure a stiff-covered bound notebook, about 8 × 10 inches, provided preferably with cross-ruled paper (for facilitating the preparation of tables of physical constants required in the later experiments). The use of spiral or loose-leaf notebooks for laboratory records is not satisfactory, and the recording of experimental observations on loose sheets or scraps of paper is not permissible. It is preferable to record notes in ink, and if corrections are necessary these should be made by additional notes rather than by erasures.

For the exercises on separation and purification the student should proceed in the following way:

1. Read the descriptive pages concerning the laboratory operation to be carried out (these are found immediately preceding each experiment). In the notebook, write a title and general statement of the process to be studied.
2. Read the laboratory directions for the *entire exercise*, note particularly the cautions for handling materials, and consider the reasons for the procedure to be followed. In the notebook jot down any points that require special observation or reminders of specific details.

   Write the names and formulas of the compounds to be used, and in experiments where chemical tests are made, write equations for the reactions.

   To gain time for laboratory work, the student should carry out the instructions in the foregoing paragraphs before coming to the laboratory. The schedule of experiments will be announced beforehand, so that the student will have an opportunity to prepare his notebook and be ready to start laboratory work at the beginning of the laboratory period.

   Careful planning of laboratory work is essential. Effective use of laboratory time requires that the student know in advance just what he is going to do in the laboratory. Instead of watching idly while a liquid is being heated for an hour or more, one can use periods *when full attention is not required*, to conduct another experiment, to clean apparatus or prepare for subsequent work.
3. After completing the instructions in the preceding paragraphs, arrange the apparatus for the experiment and secure the approval of the laboratory instructor for the setup.

[1] The following specific directions for the preparation of notebooks and the general laboratory procedure are based upon those that have been used in the elementary courses in organic chemistry at Cornell University. For the particular conditions that obtain in other laboratories, the instructor may wish to alter these directions or substitute others.

4. Perform the experiment according to the laboratory directions and record observations directly in the notebook. When the experiment has been completed, dismantle the setup and clean the glassware and apparatus at once (see page 9).
5. Write answers to the questions given at the end of the experiment. Make complete statements in answering the questions.

Submit the completed notes and the report card (see page 160), together with any substance prepared, to the laboratory instructor for approval as soon as possible after the completion of the experiment. At this time the instructor will examine the notes and may ask questions pertaining to both the theoretical and practical parts of the experiment.

The grade for the experiment is based upon the student's neatness and skill in carrying out the experiment, and upon his knowledge of the underlying principles and generalizations. Credit is not given for the performance of experiments if the notes and report card are not submitted.

**TABLE 1**
**Desk Reagents**

| REAGENT | DENSITY *g/ml* | GRAMS OF REAGENT *Per 100 g* | GRAMS OF REAGENT *Per 100 ml* | MOLES PER LITER (APPROX.) |
|---|---|---|---|---|
| Acetic acid (glacial) | 1.06 | 99.5 | 105.5 | 17.5 |
| Hydrochloric acid (concd) | 1.18 | 35.4 | 42 | 12 |
| Nitric acid (concd) | 1.42 | 70 | 100 | 17 |
| Sulfuric acid (concd) | 1.84 | 96 | 176 | 18 |
| Sodium Hydroxide solution (dil) | 1.11 | 10 | 11.1 | 3 |
| Ammonia solution (concd) | 0.90 | 29 | 26 | 15 |

## General Suggestions

**Weighing and Measuring Reagents.** In performing laboratory experiments it is important to weigh or measure the amounts of materials carefully and to use the *exact* quantities called for in the directions. When reagents need not be accurately measured (as in certain tests), the laboratory directions indicate approximate quantities. When an approximate quantity is indicated, as 1–2 ml, the student should use a quantity within the specified limits. At first it will be advisable actually to measure the quantity used, in order to learn to judge such quantities, but after some experience the student should be able to estimate approximate quantities.

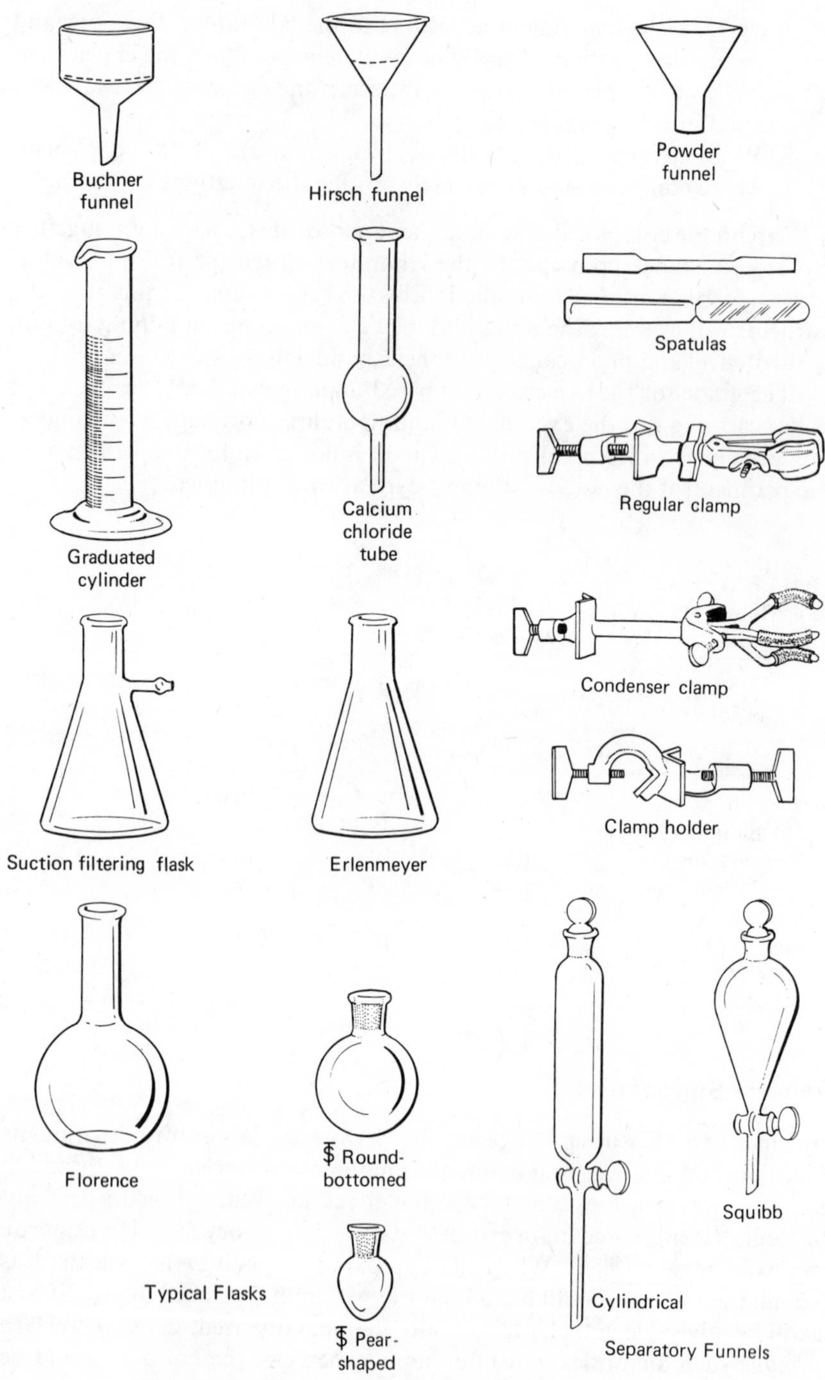


**Figure 1.** Common Organic Laboratory Apparatus

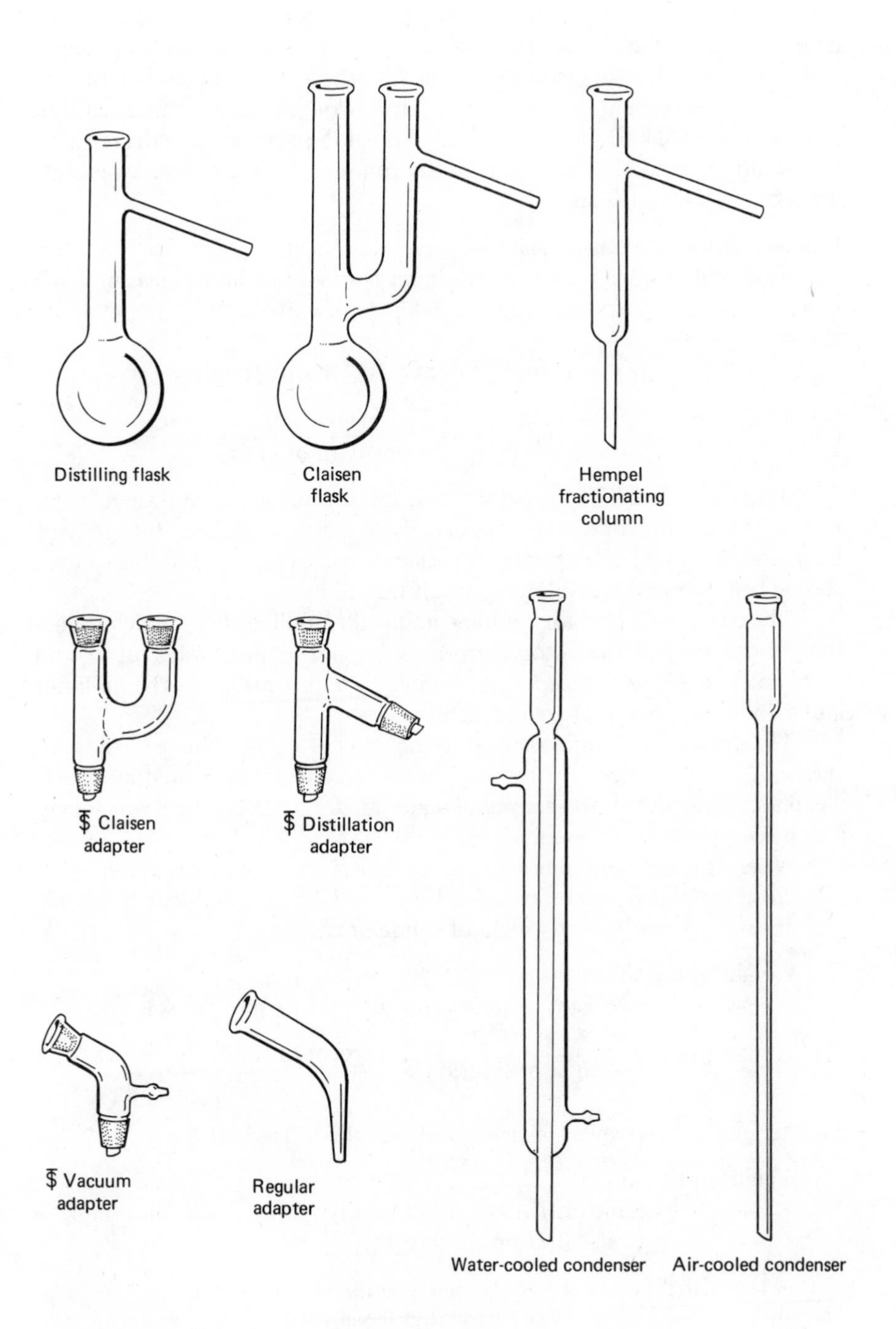


**Figure 1.** *(continued)*

In many cases the success of an operation depends on using the starting materials and certain reagents in definite amounts. It is usually advantageous and sometimes necessary to know at least fairly accurately the amount of each reagent that is present. A careful laboratory worker will acquire a habit of using solutions of known strength, and of weighing or measuring the reagents and solutions used. The strengths of the common laboratory desk reagents are listed in Table 1 (page 5).

**Interconversion of Weights and Volumes.** In laboratory practice it is often necessary or desirable to convert weight measures into volume measures, and *vice versa*. These conversions may be made by use of the following relationships:

$$\text{Weight (g)} = \text{Volume (ml) at } t^\circ \times \text{Density (g/ml) at } t^\circ$$

$$\text{Volume (ml) at } t^\circ = \frac{\text{Weight (g)}}{\text{Density (g/ml) at } t^\circ}$$

The numerical values of the density and the specific gravity of a particular liquid (at a given temperature) are usually so nearly equal that they may be used interchangeably for approximate calculations. Nevertheless, the student should bear in mind the following accurate definitions:

The common laboratory volume unit is the milliliter (ml), which is one-thousandth part of the volume of one kilogram of pure water at 4° and 760 mm. For all work except highest precision measurements the milliliter and cubic centimeter (cc) are identical.[2]

The density of a liquid is equal to the mass of a unit volume of the substance. An accurate density value includes a statement of the temperature and the units; for example, the density of water at 20° is 0.9982 g/ml, which may also be expressed by the notation $d_4^{20} = 0.9982$.

When aqueous solutions are used, or substances containing water, such as hydrochloric acid, concentrated sulfuric acid, 95 percent ethanol, etc., it is necessary to calculate the weight of solute present.

$$\begin{aligned}\text{Weight of the solute (g)} &\\ &= \text{Weight of solution} \times \text{g of solute/g of the solution}\\ &= \text{Weight of solution} \times \frac{\text{percent of solute by weight}}{100}\end{aligned}$$

Another general statement of the equation is the following:

$$\begin{aligned}\text{Weight of the solute} &\\ &= \text{Volume (ml) at } t\,(^\circ\text{C}) \times \text{Density (g/ml) of the solution}\\ &\quad \times \text{Concentration (by weight)}\end{aligned}$$

[2] In setting up the metric system it was intended that the volume of the liter should be exactly 1000 cc. Owing to very slight discrepancies in the interrelation of the fundamental units of mass and length, the actual result is that a volume of one cubic centimeter equals 0.999973 ml.

For convenience, the physical constants of solutions of the common acids and of ethanol are included in the appendix. More complete tables may be found in the chemical handbooks and in reference works.[3]

The use of these physical constants is shown by the following example. Let us suppose that we wish to find the weight of pure ethanol present in 30 ml of ordinary ethanol (95 percent by volume, or 92.5 percent by weight). By reference to an "alcohol table" the density of 92.5 percent (by weight) ethanol at 20° is found to be 0.8112 ($d_4^{20}$). Using the preceding equations:

$$\text{Weight of 30 ml of 92.5 percent ethanol} = 30\text{ ml} \times 0.8112\text{ g/ml} = 24.34\text{ g}$$

$$\text{Weight of pure ethanol (100 percent)} = 24.34\text{ g} \times \frac{92.5}{100} = 22.5\text{ g}$$

**Cleaning and Drying Glassware.** It is advantageous to clean laboratory glassware immediately after use, since tars and gummy matter are most easily removed before they harden. Much time is saved by having glassware clean and dry, ready for use at once. Many water-insoluble organic compounds and gums can be removed quickly and economically by use of scouring powder, a brush, and warm water. The use of strong acids such as concentrated sulfuric or sulfuric-chromic acid cleaning solution, is dangerous and messy. *Nitric acid is particularly dangerous* as a cleaning agent because it reacts explosively with many organic compounds.

To remove resins and gummy material from glassware, first pour or scrape out as much material as possible, directly into a waste crock; *never put organic tars, paper, or other solid wastes into the sink.* Next try to remove or loosen the resin by using a *small amount* of acetone or benzene (10–20 ml) and allowing the solvent to stand in contact with the material for 5 or 10 min. The solvent action may be hastened by warming on a steam bath (not over a flame) with care to avoid accidental ignition of the flammable solvent vapor. Do not expect tars and gums to dissolve quickly; allow ample time for the organic solvent to act. Benzene or acetone is usually a good solvent for tars; ethanol is generally not effective.

To remove the remaining small amounts of tars and dirt use scouring powder and a large test tube brush. By proper bending of the brush it will reach the inner surfaces of flasks. The use of a little washing powder or liquid detergent ("Lakeseal Laboratory Glassware Cleanser," "Alconox," "Lux," etc.) followed by a good water rinse will give glassware a clean brilliant sparkle when it dries. The best way to dry apparatus is to allow it to stand overnight in the laboratory desk. Beakers and flasks should be inverted to permit drainage; test tubes and small funnels may be inverted over crumpled paper placed in the bottom of a large beaker.

[3] Such as the Chemical Tables from the *Handbook of Chemistry and Physics*, Chemical Rubber Publishing Co., Cleveland, Ohio; *Physical Properties of Chemical Compounds*, Volumes I–III (cumulative name index in Volume **III**), American Chemical Society Advances in Chemistry Series, Volumes **15**, **22**, **29**.

If wet glassware must be dried quickly for immediate use it may be rinsed with one or two small portions, *not over 10 ml*, of acetone, allowed to drain, and the last traces of acetone removed by drawing or passing a current of dry air through the apparatus. Methanol or ethanol may be used instead of acetone but they evaporate less quickly. Ordinary compressed air is not suitable for drying purposes unless a good drying train is used, since the air is apt to be nearly saturated with water and may even contain suspended droplets of water or oil. It is more convenient to draw a stream of air through the apparatus by means of a glass tube connected to a suction pump.

**Apparatus with Interchangeable Ground Glass Joints.** In advanced laboratory work it is common practice to use apparatus having interchangeable ground glass joints (standard taper joints). The principal advantage of ground-jointed apparatus is that the joints are not affected by corrosive liquids and vapors that attack corks and rubber stoppers (chlorosulfonic acid, phosphorus trichloride, bromine, nitric acid, etc.). Reaction mixtures containing such corrosive materials may be distilled or refluxed without contamination of the product or loss of material through leakage at the joints. Dimensions of the joints have been standardized (₮ 12/18, 14/20, etc.) so that a variety of assemblies can be set up from a small stock of standard taper flasks, condensers and adapters.

The initial cost of ground-jointed apparatus is much greater than regular glassware, but ground-jointed apparatus is sturdier so the breakage cost is significantly less. Relatively inexpensive kits of basic pieces of apparatus (costing about $60) are available in convenient storage boxes. Pieces for special purposes may be added.

The foremost rule for assembling ground-jointed apparatus is that the ground surfaces be free of any gritty material that might score them when they are mated. It is good practice to wipe each surface gently with a lint-free cloth or tissue before assembly.

Because of the highly precise grinding of standard taper apparatus it is not necessary in most ordinary laboratory work to apply grease to the ground surfaces before assembly (this does *not* apply to stopcocks or other apparatus with ground surfaces that must be rotated during use). Grease *should* be applied in the following situations:

1. When the apparatus will be heated above 150°
2. When there is any possibility that the joint will come in contact with strongly alkaline solutions
3. When the apparatus will be required to hold a vacuum
4. When the surfaces will be rotated during use.

A good method for applying grease is to place several small dabs on one of the two surfaces to be mated. The joint members are then placed together and rotated back and forth gently until the grease has formed a thin, continuous film between the two ground surfaces. Care should be taken to avoid excess grease since the excess will gradually flow out of the bottom of the joint and contaminate any material with which it comes in contact. It should be remembered also that if a liquid is to be poured out of a flask having a greased joint that the surface must first be wiped clean.

Greased joints should always be cleaned thoroughly when the apparatus is disassembled. Hydrocarbon greases are readily removed by acetone, carbon tetrachloride, and many other organic solvents. Except for special situations, the use of silicone greases is to be discouraged. Although these have the advantage over hydrocarbon greases of lower vapor pressure and lower solubility in most organic solvents, they present a formidable cleaning problem.

Another limitation of ground-jointed apparatus is that, unlike glassware assembled with corks or rubber stoppers, it has no mechanical flexibility. Special care must be exercised when a clamp is tightened onto a portion of an assembly that is jointed to another clamped member.

**Bending Glass Tubing.** Glass tubing of small bore can be cut cleanly by making a short mark with a sharp triangular file and then exerting a *gentle* pressure with the thumbs on the side opposite the file mark. Risk of cutting the thumbs or fingers is avoided by holding the tube in a piece of cloth or toweling. Soft glass tubing may also be cut by drawing out a piece of waste tubing to a thin capillary, breaking off a 6-cm length of the fine tubing, heating one end in a hot flame, and pressing the resulting small globule of molten glass quickly against the file mark. Glass rods require a somewhat deeper file mark. The rough edges of glass tubing and stirring rods should be smoothed by fire-polishing. This is done by heating the rough end in a moderate flame until the glass barely begins to soften. Overheating the ends of glass tubing should be avoided as this will reduce the size of the opening at the end of the tube.

Good smooth bends are made by heating the glass tubing in the slightly luminous flame of a burner provided with a wing top (fishtail). In an ordinary burner flame only a short length of tubing is softened, with the result that bending causes a constriction in the diameter of the tube and a thinning of the glass on the outer curve of the bend. To avoid burning the fingers and to facilitate handling the hot tubing, it is advisable to make bends with a convenient length of tubing and to cut to the desired length after bending.

**Boring Corks and Rubber Stoppers.** For use with organic compounds, cork is generally much more satisfactory than rubber. Contact with hot organic liquids or vapors causes rubber to soften and swell, and often the organic liquids extract troublesome sulfur-containing compounds from rubber stoppers and tubing. Corks have many advantages over rubber stoppers: corks can be bored much more easily and usually soften only slightly in contact with organic liquids; generally, the small amounts of coloring matter extracted from corks by organic liquids are not troublesome and can be removed easily; and corks are much less expensive than rubber stoppers. Corks have some limitations: they are attacked and eventually disintegrated by halogens, halogen acids, nitric and sulfuric acid, alkalies, and by prolonged contact with vapors of high-boiling organic compounds. Corks that have become soiled must be discarded.

Select a good cork that will at first fit only a short distance into the opening of the flask. Before boring the hole, soften the cork by rolling underfoot on the floor or in a cork roller. After rolling, the cork is reduced slightly in diameter and will extend one-third to one-half its length into the opening. This is the proper fit for a cork—it is bad practice to use a cork that is too small and fits more than halfway or one that is too large and barely enters the opening, as either type of misfit leads to an unstable assembly.

To bore the hole, select a sharp cork borer *a trifle smaller* than the final diameter desired. A good method is to hold the borer in the right hand and cork in the left, and work the borer gently at right angles with small twists into the larger end of the cork. Do not bore a hole by pressing down against a hard surface or desk top. To secure a clean perpendicular hole the cork itself should be twisted with the left hand after each twist of the borer; it is unnecessary to remove the borer from the right hand. During the operation observe and correct the perpendicular alignment of the borer at frequent intervals and do not hurry. After the hole has been bored halfway through, remove the borer and push out the plug with a short metal rod. Bore the remainder of the hole from the opposite end of the cork, watching the perpendicular alignment so that the holes will meet neatly. The hole may now be smoothed by reaming out *slightly* by means of a small round file; the cork is rotated during the filing so that the opening will remain circular and not become elliptical. The hole should fit the tube or thermometer snugly but not so tightly that forcing is required.

Moisten the outside of the tube to be inserted, grasp it close to the cork, and work it gently into the hole with a twisting motion. Severe cuts and damage to apparatus often result from attempting to force tubes or thermometers into holes that are too small or from grasping the tube at a point too far from the cork. After the tube has been inserted, the cork is pushed firmly into the neck of the flask or other apparatus with a twisting motion. Time and care spent in making a neat and correct assembly of apparatus is a good investment that yields dividends over and above the personal satisfaction of good craftsmanship.

It is more difficult to bore rubber stoppers properly than to bore corks. The borer must be freshly sharpened and the boring done slowly and patiently. The borer is moistened with isopropyl alcohol or glycerol as lubricant and only slight pressure exerted; haste and excessive pressure will result in a hole of diminishing diameter as the boring progresses. To facilitate the insertion of a tube into a rubber stopper or into a rubber tube, the outside of the glass tube and the inside of the hole should be dusted lightly with talcum powder or moistened with a liquid lubricant. Rubber stoppers and rubber tubing should be removed from glass tubes and thermometers directly after use, to avoid stiffening and sticking fast to the glass, and stoppers may be stored under water to avoid hardening while not in use.

Special care is needed to loosen rubber that has become stuck fast to glass. It is helpful to loosen the rubber at the point of contact by gentle prying with a knife blade and working a little isopropyl alcohol into the opening. After standing a short time, the stopper is pulled gently and worked loose from the glass, always with the glass tube grasped close to the stopper. Sometimes it is necessary to cut off hardened stoppers or tubing with a knife. Rubber stoppers in good condition are retained for repeated use.

**Wash Bottles, Dropping Tubes, etc.** The student will find it helpful to assemble or construct a few pieces of apparatus that are used frequently throughout the laboratory work, and some that are needed occasionally for special purposes. It is convenient to have a 500-ml wash bottle for water and one or two smaller wash bottles of 125-ml or 250-ml capacity for organic solvents such as acetone, ethanol, or benzene. Wash bottles fabricated from polyethylene are available commercially and are suitable for water and most common organic solvents. For powerful solvents such as dimethyl sulfoxide or dimethylformamide a wash bottle can be assembled from a Florence flask, glass tubing and a drilled cork.

Small pipettes and dropping tubes may be made easily by drawing out ordinary glass tubing, cutting off the tip and fire-polishing carefully. For use in measuring roughly quantities of 1 and 2 ml, the pipette can be calibrated by drawing up water from a small graduated cylinder and making a shallow file mark. For use as a dropping tube, the size of the tip should be adjusted so that 1 ml of water will give 20–30 drops per ml. Ordinary medicine droppers may also be used but it is necessary to keep the rubber bulb clean and out of contact with organic chemicals and reagents.

A number of glass stirring rods of different lengths should be made to fit beakers of various sizes; a few rods with disk ends and a few with flattened ends. Disk ends are made by heating the extreme tip of the rod to redness and pressing against a piece of asbestos board or wood; flattened and curved ends are obtained by pressing the hot softened glass with crucible tongs or pliers. The ends of stirring rods must always be fire-polished.

One or two small metal spatulas made from metal rodding (see footnote, page 73) will prove useful in handling organic solids.

**Stirring.** In the organic laboratory, stirring is needed often to hasten solution or reaction by bringing a solid and liquid, or two immiscible liquids, into good contact. Stirring of a homogeneous solution is advantageous only when one desires to bring the liquid into good contact with the walls of the vessel to render external heating or cooling more effective and to insure uniform temperature throughout the liquid. Often the movement of a boiling liquid is sufficient to give good mixing.

For ordinary laboratory preparations, satisfactory mixing is generally accomplished by intermittent shaking by hand; a circular swirling motion is most effective and reduces the danger of splashing. Occasionally a rather violent shaking is required, when a heavy solid such as metallic zinc or iron must be brought into contact with an organic liquid. For material in a beaker sufficient mixing can usually be obtained by stirring by hand with a glass rod or a wooden paddle.[4]

Mechanical stirring is generally essential for large-scale preparations and in any operation where continuous stirring is required over a long period of time. Descriptions of stirring assemblies with motor driven stirrers may be found in laboratory manuals for advanced work (see also page 170).[5]

## General Precautions

**➤Safety Glasses.** The eyes are particularly vulnerable to injury by splashing droplets of corrosive chemicals and flying particles of glass or other solid fragments. It is good practice to wear safety glasses *at all times* in the laboratory. This is a required precaution in many college and industrial research laboratories. Safety glasses are made in a number of styles and with plain or ground lenses.[6]

**➤Fire Hazards.** One of the chief dangers of organic laboratory work is the fire hazard attendant upon the manipulation of volatile, flammable, organic liquids. With few exceptions, organic liquids and vapors catch fire readily and many organic vapors form explosive mixtures with air. Obviously, organic liquids must not be manipulated near an open flame, and precautions must be taken to avoid the escape of organic vapors into the laboratory. For general safety a student should form the habit of scanning the adjacent space for lighted burners before working with flammable solvents, and it is good practice to look around for fire hazard to yourself and to adjacent workers before lighting a match or a burner.

[4] The wooden tongue depressors used by physicians are convenient paddles for mixing viscous solutions or thick suspensions, and may be used also to transfer moist materials to a filter.

[5] *Organic Syntheses*, Collective Volumes I and II, John Wiley and Sons, New York; Fieser, *Experiments in Organic Chemistry*, D. C. Heath, Boston.

[6] Safety glasses of inexpensive type, costing from $1.50 to $3.00, may be purchased through the usual apparatus supply houses or directly from the manufacturer.

The degree of flammability of organic compounds varies widely. The vapor of diethyl ether, petroleum ether, benzene, acetone, and ethanol catches flame quite readily and the manipulation of these liquids requires careful attention at all times to fire hazards. Methylene chloride (bp 40°) is a safer solvent, and carbon tetrachloride is nonflammable. Carbon disulfide is so readily ignited (even by a hot steam pipe) that it should *never* be used by an inexperienced worker.

➤**Chemical Burns and Cuts.** Specific precautions for handling particularly dangerous chemicals are noted in the directions for the experiments where they are used, but any ordinary chemical or piece of apparatus can be dangerous if manipulated carelessly. The student does well to develop a general awareness of dangers and accidents that arise from carelessness in simple routine operations. For example, a severe cut or laceration may result from carelessness in inserting a glass tube into a cork, and a severe explosion and fire may result from attempting to distill a substance in a completely closed system.

Procedures to be followed in case of accidents are given in the appendix and also are printed on the inside back cover of this manual.

➤**Laboratory Apparel.** It is desirable to protect clothing from soiling and damage from chemicals and accidents of various sorts by wearing an inexpensive laboratory coat or a rubber apron. For freedom of arm movement tight sleeves should be avoided; loose and bulky sleeves may cause overturning of fragile apparatus. Unprotected wearing apparel of light-weight, flammable fabrics constitutes a serious fire hazard in the organic laboratory. Many synthetic fabrics are soluble in acetone and many other common organic solvents.

In experiments requiring transfer of corrosive chemicals it is desirable to wear some type of resistant glove. An inexpensive, disposable glove made from polyethylene is available.[7]

[7] Poly gloves, distributed by Cole-Parmer, come in large and medium size, $4.90/100; similar gloves are available from Will Scientific, Inc. in large, medium, and small sizes, $4.95/100.

# Simple Distillation

Since organic compounds do not usually occur in pure condition in nature, and are accompanied by impurities when synthesized, the purification of materials forms an important part of laboratory work in chemistry. Four general separation procedures are used frequently in organic work in the laboratory and in industry: distillation, chromatography, crystallization, and extraction. Sublimation is used occasionally, and various special techniques such as electrophoresis and zone-refining are available for advanced work. The process used in any particular case depends upon the characteristics of the substance to be purified and the impurities to be removed. In order to select the most appropriate process and to employ it effectively it is important that the student understand the principles involved as well as the correct methods of manipulation.

Simple distillation has been placed first because it depends in a clear way on the physical concept of vapor pressure which is needed to understand the other separation techniques to be described. Simple distillation also happens to be one of the most commonly used purification methods.

## Principles of Distillation

**Boiling.** In a liquid the molecules are in constant motion and tend to escape from the surface and become gaseous molecules, even at temperatures far below the boiling point. When a liquid is placed in an enclosed space, the pressure exerted by the gaseous molecules rises until it reaches the equilibrium value for that particular temperature. The equilibrium pressure is known as the vapor pressure and is a constant characteristic of the material at a specific temperature. Although vapor pressures vary widely with different materials, vapor pressure always increases as the temperature increases (Figure 2). The vapor pressure is commonly expressed as the height, in millimeters, of a mercury column that produces an equivalent pressure. The vapor pressure of a pure compound is altered by the addition of soluble substances.

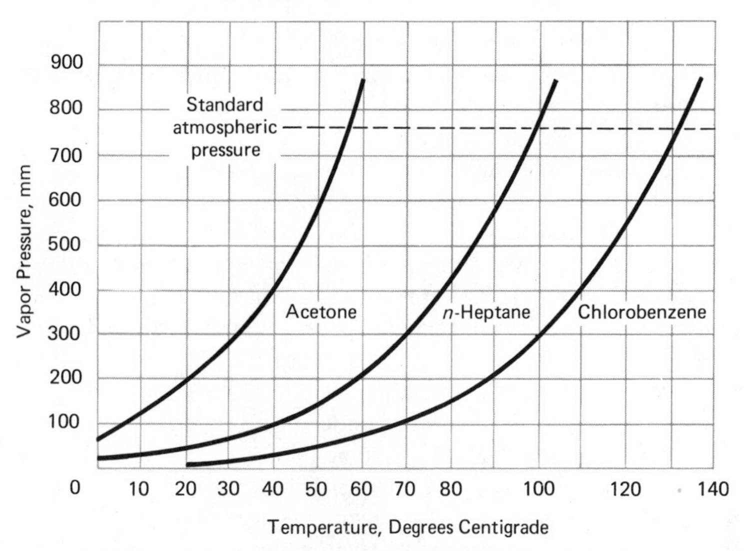


**Figure 2.** Temperature Variation of Vapor Pressure

In ordinary glass flasks, there are microscopic pockets of air trapped in the pores and crevices of the walls. With a liquid in the container the pockets are filled with vapor of the liquid at its equilibrium vapor pressure. When the temperature of the liquid is raised, the vapor remains compressed until the vapor pressure exceeds the applied pressure (the pressure at the liquid surface plus the hydrostatic pressure), whereupon the trapped vapor rapidly expands to form bubbles that rise to the surface and expel their vapor. The resulting agitation (boiling) churns more air bubbles into the liquid where they continue the process after receiving new charges of vapor. Liquids heated in containers that have been de-gassed do not boil although they vaporize explosively if heated to a sufficiently high temperature. To avoid the hazards associated with sudden irregular boiling (bumping) a dependable source of bubbles should always be introduced into a flask before its contents are heated to boiling. When a liquid is boiled at atmospheric pressure the bubble source is customarily a boiling chip (see page 28); with vacuum distillations a stream of air bubbles from a capillary tube is used (see page 55).

**Boiling Point and Boiling Temperature.** The boiling point of a liquid is defined as the temperature at which its vapor pressure equals the atmospheric pressure. By convention, boiling points are reported in scientific literature at a pressure of one atmosphere except when otherwise specified. The boiling temperature is the actual observed temperature when boiling occurs and is generally a few hundredths to a few degrees above the boiling point because of experimental difficulties involved in the measurement.

**Distillation of a Pure Compound.** Distillation consists of boiling a liquid and condensing the vapor in such a manner that the condensate (distillate) is collected in a separate container. Two simple apparatus assemblies appropriate for this operation are shown in Figure 9 (page 25) and Figure 10 (page 26).

The objective of distillation is to separate a mixture of two or more materials that differ in their ease of vaporization.

When a pure substance is distilled at constant pressure, the temperature of the distilling vapor will remain constant throughout the distillation, provided that sufficient heat is supplied to insure a uniform rate of distillation and superheating is avoided. In actual practice these ideal conditions are not realized: drafts in the laboratory can cause momentary condensation of vapors before they reach the thermometer. A certain amount of superheating of vapors occurs almost invariably under ordinary conditions. Because of these contrary effects a distillation range of 1° actually represents an essentially constant boiling point. With somewhat more refined apparatus and technique a distillation range of 0.1° can be observed for a pure compound.

The temperature reading of a thermometer *in the distilling vapor* represents the boiling point of that particular portion of the distillate. This temperature will be the same as the boiling point of the liquid in the distilling flask only if the distilling vapor and the boiling liquid are identical in composition. Since a pure liquid fulfills this condition, a constant thermometer reading is sometimes used as a criterion of purity of a liquid. It should be noted, however, that certain mixtures (such as azeotropes, page 22) give constant thermometer readings. Occasionally two liquids have such similar boiling points that no appreciable change in thermometer reading will be observed when a mixture of them is distilled.

**Ideal Solutions.** The pressure and composition of vapor above an ideal mixture of liquids at a given temperature can be calculated if its composition and the vapor pressures of the pure components are known. The total pressure is the sum of the partial vapor pressures of all components. The partial pressure of each component is given by Raoult's law:

$$P_A = P_A^\circ N_A \qquad (1)$$

where $P_A$ (the partial pressure of $A$) is the vapor pressure of $A$ above the mixture, $P_A^\circ$ is the vapor pressure of pure $A$, and $N_A$ is the mole-fraction of component $A$ in the mixture. Because there is a fixed number of molecules in a mole (gram-molecule), Raoult's law states in molecular terms that the vapor pressure of $A$ above a solution is proportional to the mole-fraction of the molecules of $A$ in the liquid. Application of Raoult's law to the two component mixture of carbon tetrachloride and toluene is illustrated graphically in Figure 3.

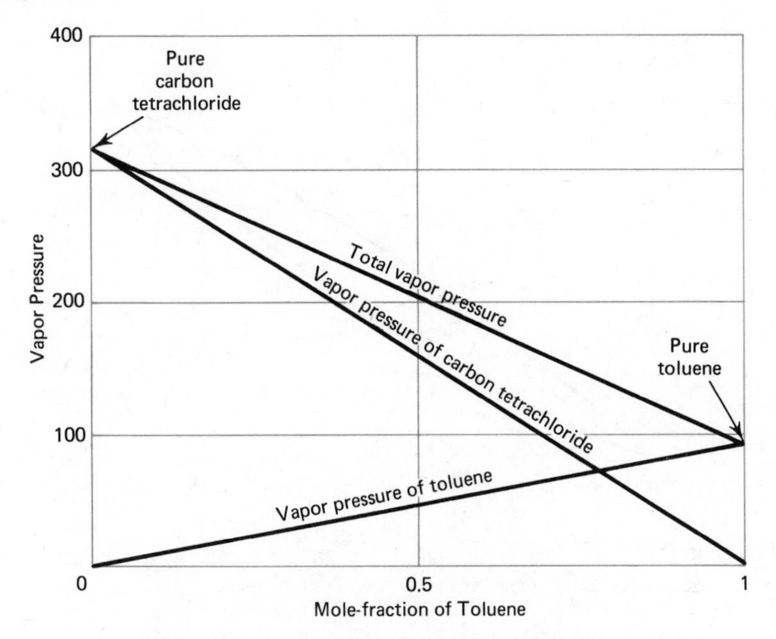


**Figure 3.** Graphical Application of Raoult's Law

The composition of the vapor with respect to each component can be calculated from Dalton's law:

$$Y_A = \frac{P_A}{\text{total vapor pressure}} = \frac{P_A}{P_A + P_B + P_C + \cdots} \tag{2}$$

where $Y_A$ is the mole-fraction of component $A$ in the vapor. Combination of Dalton's law and Raoult's law shows that for an ideal mixture at any temperature the most volatile component has a greater mole-fraction in the vapor than in the solution. In terms of the previously defined symbols, if $A$ is the most volatile component of the mixture, $Y_A$ is greater than $N_A$.

The boiling point of a *mixture* is defined as that temperature at which the *total* vapor pressure equals the pressure above the solution. From Raoult's law (also see Figure 2) it is apparent that the total vapor pressure of an ideal mixture is intermediate between the vapor pressures of the pure components. This means that the boiling points also will be intermediate between the boiling points of the pure substances. The general dependence of boiling point on composition of ideal binary mixtures resembles that depicted in Figure 4 for the specific system of carbon tetrachloride and toluene. The boiling point of any particular mixture is obtained by erecting a vertical line from the horizontal composition axis until it intersects the *liquid* curve. For example, from

Figure 4 it will be found that a 60 mole percent (0.6 mole-fraction in carbon tetrachloride) mixture of carbon tetrachloride in toluene boils at 87°.

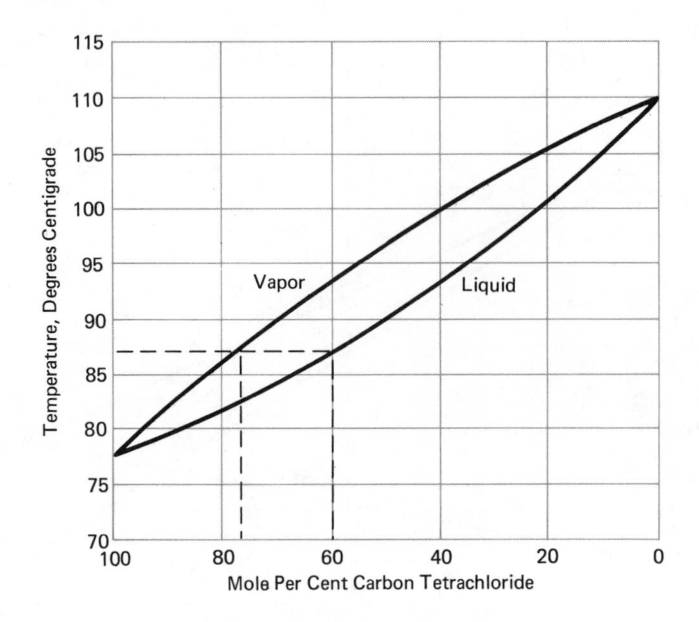


**Figure 4.** Boiling Point Diagram: Carbon Tetrachloride and Toluene

The composition of the vapor in equilibrium with any particular liquid composition is obtained from Figure 4 by projecting a horizontal line from the vertical intersection of the *liquid* curve over to the *vapor* curve and from that intersection back vertically to the composition axis. The vapor above a 60 mole percent solution of carbon tetrachloride in toluene contains 78 mole percent of carbon tetrachloride. Figure 4 demonstrates graphically the previous conclusion that the vapor is richer in the lower-boiling, more volatile component.

The boiling point and vapor composition calculated in this manner apply only to the initial state of a distillation. Because of the higher concentration of carbon tetrachloride in the vapor compared to the liquid remaining in the boiler, the composition of the liquid *gradually* shifts towards pure toluene as the distillation proceeds; the boiling point, reflecting this composition change, climbs gradually also. The actual rate of change of boiling point depends on how rapidly the mixture is distilled but a typical set of observations would resemble those of Figure 5, obtained with a 50:50 carbon tetrachloride-toluene mixture. The dotted line in Figure 5 shows the appearance of a perfect separation of this mixture. The technique of fractional distillation, to be discussed in the following section, is a method for more nearly approaching this perfect separation.

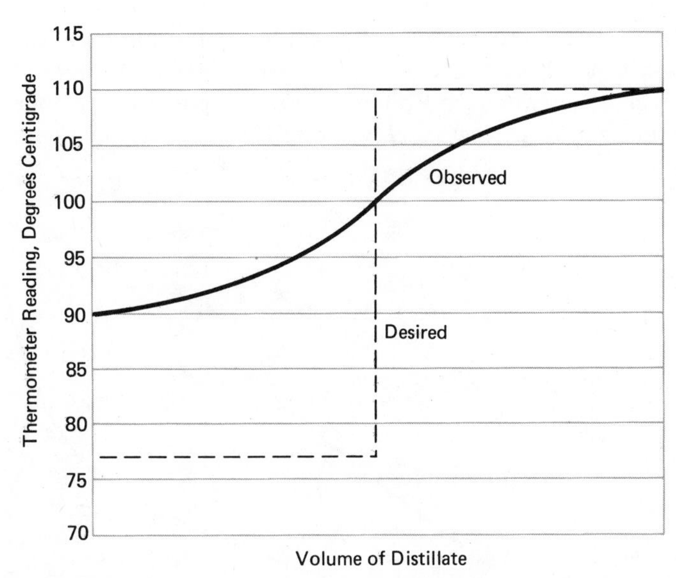


**Figure 5.** Distillation Curve for a 50:50 Carbon Tetrachloride–Toluene Mixture

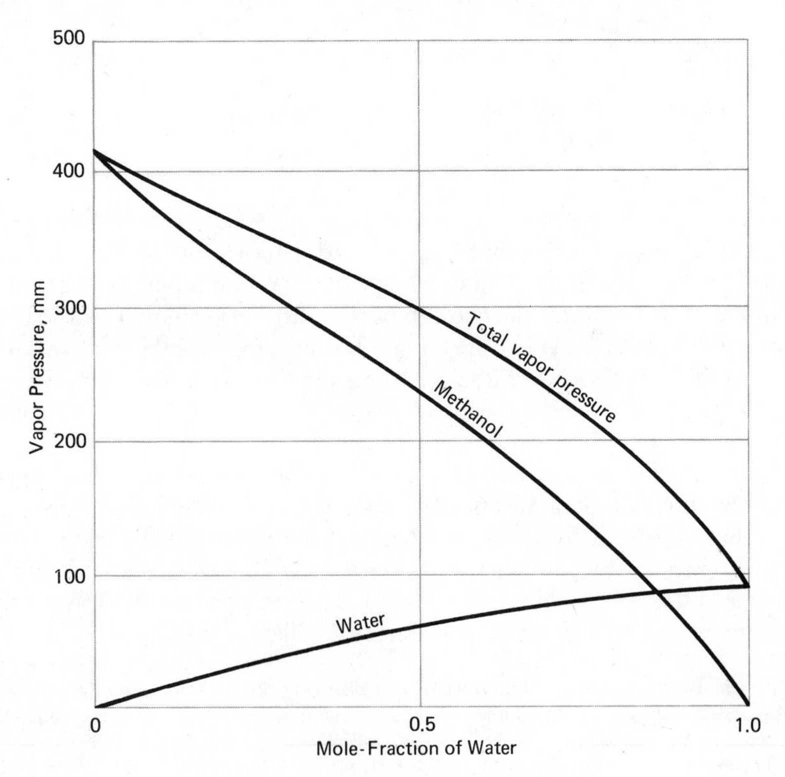


**Figure 6.** Deviations from Raoult's Law

**Non-Ideal Solutions and Azeotropes.** Many actual solutions depart widely from Raoult's ideal law. The methanol-water system is typical of those that show *positive* deviation from Raoult's law (see Figure 6). The boiling point composition curve for methanol-water mixtures shown in Figure 7 also

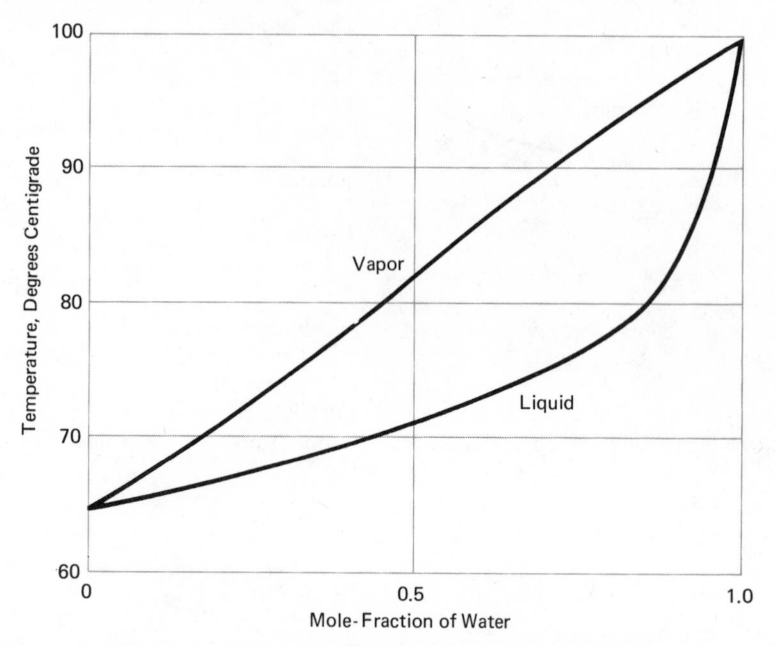


**Figure 7.** Boiling Point Diagram: Methanol–Water

reflects the non-ideal behavior by its distorted shape (compare with Figure 4). Although the molecules of liquids always attract one another, there are some binary mixtures in which the two components attract each other particularly strongly and cause the vapor pressure to be lower than ideal (negative deviations from Raoult's law).[1] Other mixtures show positive deviations because the molecules of one component are attracted to each other more strongly than they are to molecules of the other component.

Frequently, the deviations from ideality are so extreme that boiling point-composition diagrams have a maximum or a minimum (Figure 8). If a mixture showing this extreme behavior has the composition corresponding to the extreme boiling point (an *azeotropic mixture*), it will behave like a pure liquid and show a constant boiling point. The components of an azeotropic mixture cannot be separated by ordinary distillation processes because the

[1] Negative deviations from Raoult's law occur also whenever the components have different molecular volumes. For mixtures of low molecular weight compounds the effect is usually small but becomes appreciable for mixtures of high molecular weight polymers with low molecular weight compounds. This difference in size is largely responsible for the insolubility of polymers in many standard organic solvents.

vapor in equilibrium with the liquid has the same composition as the liquid itself. Table 2 gives the composition and boiling point of several examples of binary azeotropic mixtures. Azeotropic mixtures containing three components (ternary systems) are encountered also; for example, benzene-water-ethanol (see page 166) or ethanol-water-ethyl acetate give minimum-boiling point azeotropic mixtures.

The effect of even more extreme deviations from Raoult's law will be considered in connection with steam distillation.

**TABLE 2**
**Binary Azeotropic Mixtures**

| COMPONENT A | | COMPONENT B | | AZEOTROPIC MIXTURE | | |
|---|---|---|---|---|---|---|
| *Substance* | *Boiling Point, °C* | *Substance* | *Boiling Point, °C* | *Percentage of A (weight)* | *Percentage of B (weight)* | *Boiling Point, °C* |
| Acetone | 56.4 | Chloroform | 61.2 | 20 | 80 | 64.7 (maximum) |
| Nitric acid | 86.0 | Water | 100.0 | 68 | 32 | 120.5 (maximum) |
| Formic acid | 100.7 | Water | 100.0 | 77.5 | 22.5 | 107.3 (maximum) |
| *n*-Propyl alcohol | 97.2 | Water | 100.0 | 71.7 | 28.3 | 87.7 (minimum) |
| *t*-Butyl alcohol | 82.5 | Water | 100.0 | 88.2 | 11.8 | 79.9 (minimum) |
| Ethanol | 78.3 | Water | 100.0 | 95.6 | 4.4 | 78.15 (minimum) |
| Ethanol | 78.3 | Chloroform | 61.2 | 7 | 93 | 59.0 (minimum) |
| Ethanol | 78.3 | Toluene | 110.6 | 68 | 32 | 76.7 (minimum) |
| Acetic acid | 118.5 | Toluene | 110.6 | 28 | 72 | 105.4 (minimum) |

## Laboratory Practice

**Apparatus for Simple Distillation.** A simple distillation apparatus suitable for distillation of samples greater than 5 ml in volume is shown in Figure 9. This consists of a round-bottomed flask connected by means of a distillation adapter to a water-cooled condenser. A thermometer is held in place in the vertical arm of the distillation adapter by a special rubber connector[2] at a

[2] The thermometer can be mated to the vertical arm by means of a short length of soft rubber tubing. However, this practice is hazardous because if the thermometer is to be held snugly the tubing must be forced over the much larger adapter arm.

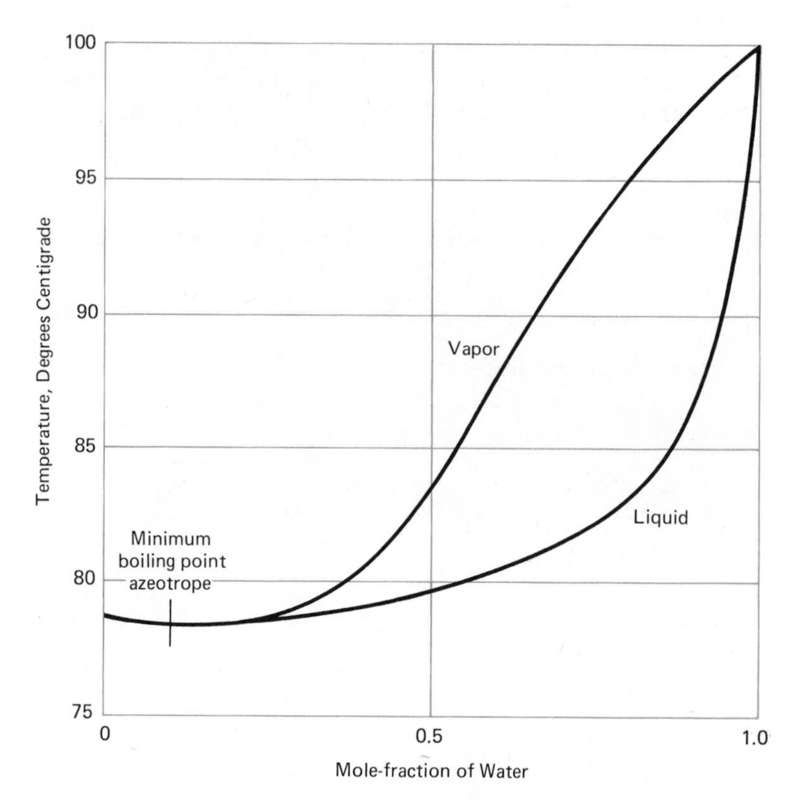


**Figure 8.** Boiling Point Diagram: Ethyl Alcohol–Water

height adjusted so that the top of the mercury bulb is 5–10 mm *below* the opening of the side-arm. A vacuum adapter is connected to the lower end of the condenser.

The distilled liquid is collected in a clean, dry receiver, commonly an Erlenmeyer flask or small-mouthed bottle. To reduce vapor losses and minimize fire hazards, it is desirable to insert the lower end of the adapter well into the mouth of the receiver. If the distillate is particularly flammable or hazardous in other ways a round-bottomed flask may be used as the receiver with its ground glass joint mated to the lower joint of the distillation adapter. *A distilling assembly must have an opening to the atmosphere* to avoid developing a dangerously high pressure within the system when heat is applied. When a mated round-bottomed flask is used as the receiver the side-arm on the distillation adapter becomes the opening and this arm must *not* be sealed.

The distilling flask rests on a piece of wire gauze (preferably one with an asbestos center),[3] which is supported on an iron ring. The main purpose

[3] Another arrangement is to place the wire gauze on a square of asbestos board having a circular hole about 15 mm in diameter bored in the center. There is more danger of cracking glass apparatus by heating on a *plain* wire gauze *with a hot flame* than by direct impingement of

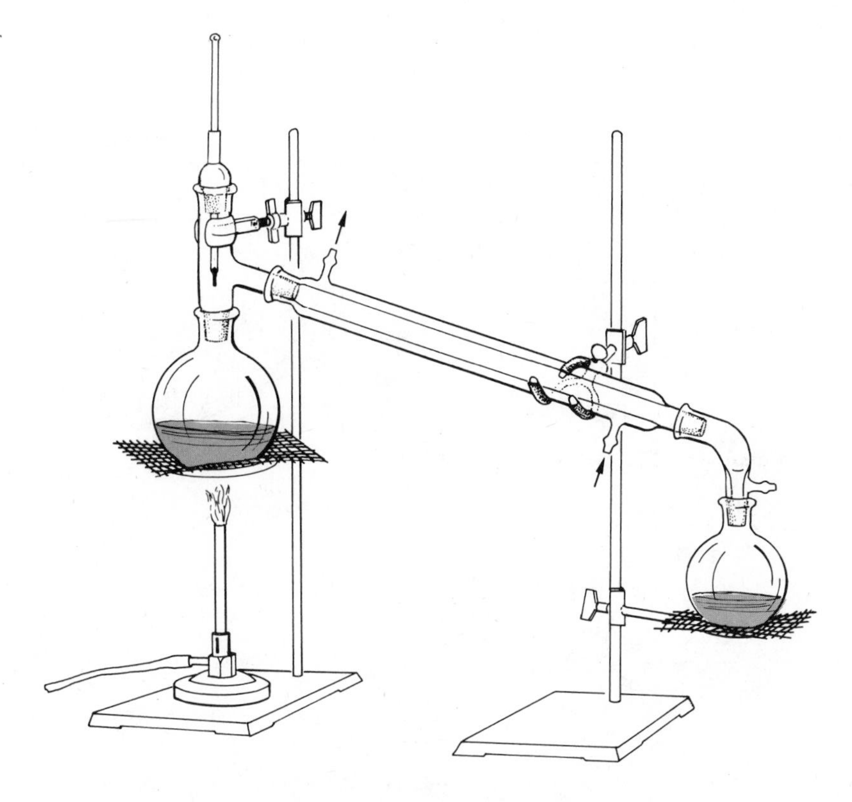

**Figure 9.** Apparatus for Simple Distillation: Ground Glass Joints

of the wire gauze is to avoid superheating and decomposition of the liquid or vapor that might result from heating the sides and upper portion of the flask. An alternative, used commonly in advanced work, is to heat the flask in a heating bath (see also page 166). Heating baths have the additional advantage of being a more even and easily regulated heat source.

If corks are used in place of ground-jointed glassware additional precautions must be observed. The connecting corks must be properly bored (see page 11). The side-arm of the distilling flask (see Figure 10) should be adjusted to project far enough (at least 25 mm) beyond the cork connecting with the condenser, so that the distilling liquid will not be contaminated or discolored by direct contact with the cork. The same precaution to avoid contact of the distillate with corks should be observed when attaching an adapter and receiving flask at the lower end of the condenser. It will be noted that the adapter available for use with apparatus joined by corks does not have a side-arm so that the receiver must *not* be corked tightly. A wad of

the flame on the glass, as local hot spots having a temperature of almost 1000° can be developed by the flame of a Bunsen burner directed against a plain metal gauze; see Wooster, *J. Chem. Educ.*, **18**, 196 (1941).

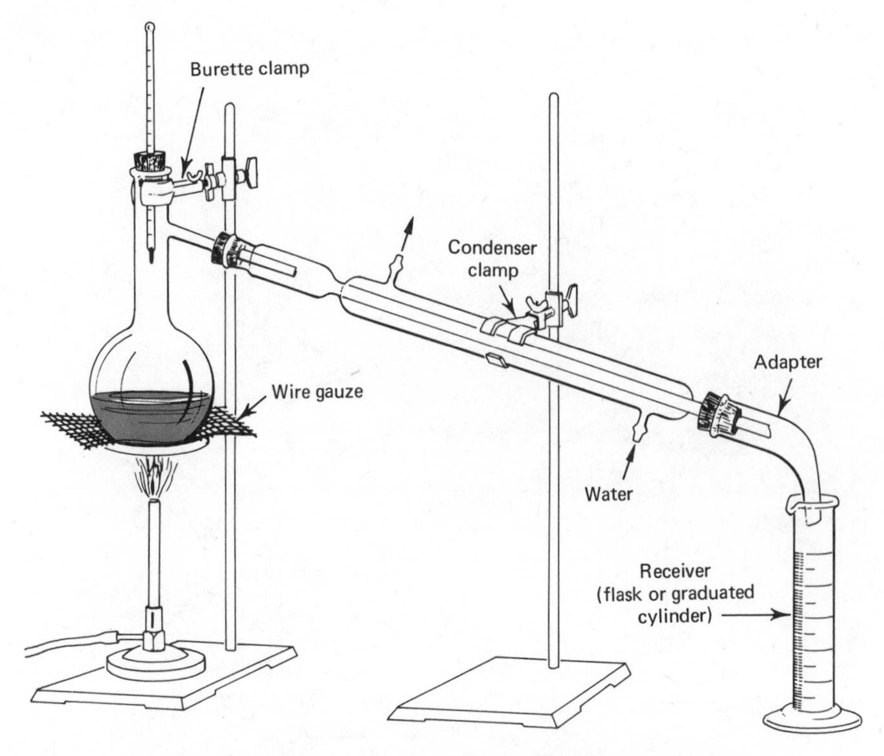


**Figure 10.** Apparatus for Simple Distillation: Cork Stoppers

cotton may be placed *loosely* around the mouth of the adapter but care must be taken to have it not come in contact with the distillate.

With either ground-jointed or corked apparatus the size of the distilling flask chosen should be such that the material to be distilled occupies between one-half to two-thirds of the bulb. If the bulb is more than two-thirds filled there is danger that some of the liquid may be carried over mechanically into the distillate. If the bulb is less than half-filled there will be an unnecessarily large loss resulting from the relatively large volume of vapor required to fill the flask. This loss is particularly serious with compounds of high molecular weight. Whether a pure compound or a mixture is distilled, a small portion of liquid will always be left in the flask at the end of the distillation as a result of condensation of the vapor within the flask upon cooling. The flask containing the material to be distilled should never be heated to dryness with a flame since there is a possibility that the glass vessel will crack.

**Small Scale Distillations.** When less than 5 ml of liquid is to be distilled neither the assembly in Figure 9 nor the one in Figure 10 is suitable. Surface tension causes the liquid to spread in a thin film over the inner surfaces of the distilling flask and the condenser and a large portion of the distillate would not be collected in the receiver.

An assembly is shown in Figure 11 that has minimal surface area and is suitable for distilling small samples. It consists of a 10-ml (or smaller) distilling flask connected by means of a well-bored cork (see page 11) to a side-arm test tube, which serves as both the receiver and the condenser. The test tube is immersed in a cooling bath if the distillation temperature is less than about 100°. For distillation temperatures in the range of 75° to 100° tap water is an adequate cooling bath liquid; for the range of 50° to 75° ice water can be used. For liquids distilling below 50° the bath temperature must be below 0° (see page 167). When the bath is cooler than room temperature it is essential that the side-arm be connected to a drying tube (see page 162), which dries the air entering the receiver so that moisture will not condense and contaminate the distillate. For distillation temperatures above 100° a cooling bath is not required since air convection will carry away enough heat to condense the small sample being vaporized. *A distilling assembly must have an opening to the atmosphere.* If during a distillation with the assembly shown

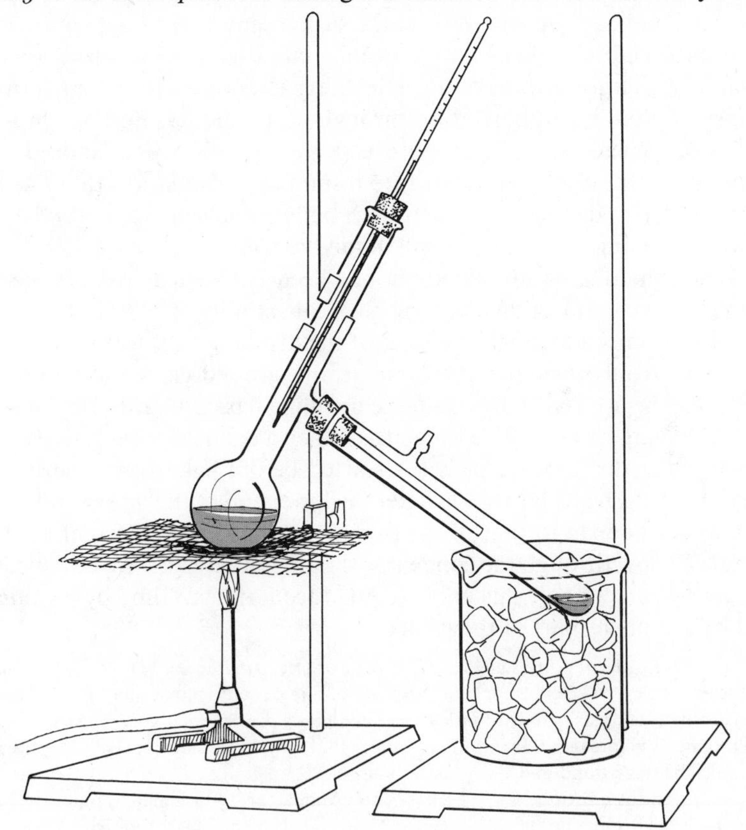

**Figure 11.** Apparatus for Distillation of Small Samples

in Figure 11 it is noticed that distillate vapors are being lost through the side-arm the proper remedy is to interrupt the distillation and replace the cooling bath by a colder one. *Under no circumstances must the side-arm be plugged.*

**Procedure.** The proper method of carrying out a distillation is to supply just enough heat at the distilling flask so that the liquid distills regularly at a uniform rate. Insufficient supply of heat will arrest the distillation temporarily and permit the bulb of the thermometer to cool below the distilling temperature, resulting in erratic temperature readings. Overheating and unsteady application of heat increase the opportunity for superheating the liquid and result in sudden irregular ebullition (bumping). Even under the proper conditions of heating it is necessary to introduce one or two tiny boiling chips of porous substance[4] or other antibump agent[5] into the liquid before heat is applied.

Superheating occurs because the transformation of a liquid into the vapor phase will not take place immediately, even at the boiling point, unless the liquid is in contact with a gaseous phase. Consequently, in a distilling flask the liquid can vaporize only at the surface unless gas bubbles are introduced into the body of the liquid. Boiling chips consist of porous material containing a large amount of air, which expands on heating and furnishes bubbles that initiate vaporization throughout the liquid. Boiling chips lose their effectiveness after a single use and must be discarded; indeed, fresh boiling chips should be added before resuming a distillation that has been interrupted. It is dangerous to introduce boiling chips into a liquid that is at or near its boiling point, as this will induce sudden violent ebullition.

When the distillation assembly has been completed it is checked for tightness of all connections and for physical stability. The liquid to be distilled is introduced through the neck of the distilling flask with the aid of a funnel, to avoid having it contaminate the ground-glass joint. A funnel should also be used with the one-piece distilling flask illustrated in Figure 11 to avoid having it run into the side-arm. When a condenser is being used, the flow of water through the jacket is started before lighting the burner and applying heat. The water should enter the lower end of the jacket and flow in a direction opposite to that of the organic vapor (countercurrent cooling). The rate of flow through the condenser should not be excessive but adequate to keep the jacket cool; this may be tested from time to time by placing the hand on the upper part of the jacket.

[4] Tiny fragments of porous unglazed clay plate or brick are frequently used for this purpose and it is convenient to keep a supply of these on hand in a small sample tube. The commercial material usually consists mainly of rather large and heavy pieces which must be broken up into small fragments before use. Carborundum chips, No. 12 mesh, are suitable also. They are available from The Carborundum Co., Niagara Falls, N.Y.

[5] A convenient substitute is a "Peerless Wood Applicator," which may be purchased in any drugstore. The effective surface of these wooden splints is greatly increased by breaking them and inserting the broken end into the liquid.

The rate of heating is controlled by regulating the size of the flame and the opening of the air entry of the burner. The flame must be adjusted so that the liquid boils gently and distills slowly at a uniform rate, generally between 30–60 drops (1–2 ml) per minute for simple distillation. A slower rate is used for fractional distillation or for vacuum distillation. Heating should be stopped just before the last traces of liquid have been vaporized in order to avoid decomposition and charring in the flask. The thermometer reading is recorded when the first drops of distillate appear at the end of the side-arm or on the walls of the condenser; this is called the "initial boiling point."[6] Thereafter the temperature and the volume of the distillate are recorded at frequent intervals. If the purpose of the distillation is to determine the composition of the liquid, many temperature-volume readings are required and it is convenient to collect the distillate directly in a graduated cylinder. The results should preferably be recorded in a neat tabular form such as that shown in Table 3 (page 30).

It is useful to plot a temperature–volume curve, from which the presence and amount of low-boiling impurities, the approximate distilling range of constant-boiling components of a mixture, etc. can be determined. When a substance containing small amounts of impurities is distilled, the first portion of distillate (called the forerun, or low-boiling fraction) will contain the more volatile impurity and a certain amount of the main liquid that is carried with it. As the temperature continues to rise, the bulk of the principal liquid will then distill over a short temperature range, usually 2–3° (called the principal fraction, or main fraction). After this fraction has distilled, the boiling point will rise, owing to the presence of the less volatile impurity. The next fraction (called the after-run, or high-boiling fraction) will consist of a mixture of the principal liquid and the less volatile impurity. The residual liquid in the distilling flask will contain the less volatile impurity along with some of the principal liquid which it holds back from distilling. However, even a pure substance will always leave a small amount of residual liquid (why?).

When the distillation is being carried out to purify a liquid it is better to use tared[7] flasks to collect the different fractions. If the distillation behavior is known or can be estimated (as when a liquid of known boiling point is being purified) it is a simple matter to use three receivers and to collect the forerun, the main fraction and after-run over the proper temperature ranges. When a liquid with unknown properties is being purified and sufficient sample is available it is a good strategy to determine first the temperature–

[6] A short time lag is necessary to permit the thermometer to warm up to the temperature of the distilling vapor. An incorrect value will be obtained if the temperature is read when the first drops of liquid appear on the thermometer bulb. During this lag the thermometer is coming into equilibrium with the distilling vapor and the mercury column is rising rapidly.

[7] A tared flask is a vessel which is weighed when it is clean and dry. The amount of liquid distilled is then easily calculated by subtracting the weight of the tared flask from the total weight of liquid plus flask.

**TABLE 3**
**Distillation of Toluene***
30 ml Sample; from "Rifco," Drum No. 48-2722

| VOLUME OF DISTILLATE (*ml*) | TEMP. OF VAPOR (°*C*) | TIME (MIN) *Actual* | *Elapsed* | OTHER OBSERVATIONS |
|---|---|---|---|---|
| | 95 | 10:17 | 0 | First drops cloudy (moisture!) |
| 2 | 107 | 10:18 | 1 | Distillate clear and colorless |
| 5 | 107.5 | 10:20 | 3 | |
| 10 | 108 | 10:23 | 6 | |
| 15 | 108 | 10:27 | 10 | Barometric pressure 742 mm |
| 20 | 108 | 10:30 | 13 | |
| 23 | 108 | 10:32 | 15 | |
| 25 | 109 | 10:34 | 17 | |
| 28 | 112 | 10:36 | 19 | Distillate clear and colorless |

Residue in distilling flask slightly yellow; volume 1.2 ml Thermometer No. 2: correction at 100° = +1.0°; calibrated against water in similar apparatus (see this Notebook, page 7, 21 Mar. 1963).

Distillate in the range of 108–113° (corrected) = 26 ml = 86.7% of 30 ml sample.

Material appears to conform to the manufacturer's No. 2 specification: "Not less than 90% distills between 105–115°C; not more than 5% below 100°C or more than 5% above 120°C."

* This chart is given merely to illustrate a neat and convenient form of recording distillation data in the notebook. The figures given are hypothetical and not the results of an actual distillation. For the introductory experiments the laboratory instructor may wish to issue specific instructions as to the method of recording the data; for example, to record the boiling point at volume intervals of 2 ml or 5 ml, *or* at time intervals of 1 or 2 min, etc. Such specific instructions should be jotted down in the notebook.

volume distillation curve. If the losses of two distillations cannot be tolerated, it is necessary to deduce the boiling behavior of the sample as the distillation proceeds. This requires close attention to the thermometer readings; it is desirable that several extra tared flasks be available in case the collection of the main fraction is begun or ended prematurely.

**Correction of Boiling Temperatures.** The thermometers usually used in chemical laboratories register the temperature in degrees Centigrade (°C), and all temperatures recorded in the manual are Centigrade temperatures. In industrial plant operations, thermometers registering temperatures in degrees Fahrenheit (°F) are frequently employed. Conversion of Centigrade to

Fahrenheit degrees or the reverse is accomplished by the following simple formulas.

$$(°C \times \tfrac{9}{5}) + 32 = °F$$

$$(°F - 32) \times \tfrac{5}{9} = °C$$

The temperature readings registered directly on an ordinary thermometer in the course of laboratory distillations (or determinations of melting points) are subject to several sources of error. Two rather important contributions are errors resulting from exposure of a portion of the mercury column to atmospheric cooling, and inaccurate or incorrect graduations of the thermometer scale.

When a typical thermometer with a long scale (250–300 mm) is used, the true boiling point (or melting point) is not registered because the mercury column is not entirely at the temperature of the mercury in the bulb of the thermometer. The portion of the mercury column that extends above the stopper of the distillation adapter (or the surface of a melting point bath) is cooled by the surrounding atmosphere, and the registered temperature is therefore below the true temperature of the vapor in the distilling flask. For temperatures below 100° this cooling effect does not cause any considerable error, but for high temperatures the observed reading may be several degrees below the true temperature. This error can be corrected by adding a *stem correction* calculated by the formula:

$$\text{Stem Correction (deg)} = 0.00154(t - t')N$$

0.000154 = coefficient of apparent expansion of mercury in glass
$N$ = number of degrees on the stem of the thermometer from the lower exposed level to the temperature read
$t$ = temperature read
$t'$ = average temperature of the exposed mercury column

In practice, this correction is subject to an error since $t'$ is not accurately known but it may be taken roughly to be one-half of the difference between room temperature and the observed temperature.

Some thermometers have graduated scales that already include a correction for an assumed 3-inch (76-mm) immersion of the stem and temperature readings taken with them should not be corrected. Such partial immersion thermometers are designated by having an engraved line circling the stem 76 mm above the bottom of the mercury bulb.

Many important errors of temperature readings in ordinary laboratory work may be due to incorrect graduation and calibration of the thermometer scale. To determine whether or not a thermometer registers correctly it may be tested by verification at several temperatures against the boiling points of pure liquids or the melting points of pure solids, or by comparison with previously standardized short-scale thermometers.

**TABLE 4**
**Reference Temperatures for Calibration**

| BOILING POINTS (°C) (°C AT 760 MM PRESSURE*) | | MELTING POINTS (°C) | |
|---|---|---|---|
| Acetone | 56.1 | Water-ice | 0.0 |
| Water | 100.0 | 1,3-Dinitrobenzene | 89.7 |
| Ethylene dibromide | 131.6 | Benzoic acid | 121.7 |
| Aniline | 184.4 | Benzilic acid | 150 |
| Nitrobenzene | 210.9 | Hippuric acid | 187.5 |
| 2-Bromonaphthalene | 281.1 | 3,5-Dinitrobenzoic acid | 204 |
| Benzophenone | 305.9 | Cinchonine | 264 |

* The boiling points of azeotropes may also be used for reference temperatures.

In a simple distillation the pressure upon a liquid is the atmospheric pressure. For ordinary work the variation in boiling point due to small deviations in pressure from one atmosphere (760 mm) may be neglected but for accurate work it is necessary to record the barometric pressure during distillation. Examples of the effect of pressure changes are shown in Table 5.

**TABLE 5**
**Effect of Pressure on Boiling Points**

| | BOILING POINTS (°C) | | | |
|---|---|---|---|---|
| PRESSURE (*mm*) | *Ethanol* | *Benzene* | *Water* | *Aniline* |
| 780 | 79.0 | 81.1 | 100.73 | 185.6 |
| 770 | 78.6 | 80.6 | 100.37 | 184.9 |
| 760 | 78.32 | 80.2 | 100.00 | 184.4 |
| 750 | 78.0 | 79.8 | 99.63 | 183.8 |
| 740 | 77.6 | 79.4 | 99.26 | 183.3 |
| 730 | 77.3 | 79.0 | 98.88 | 182.5 |
| 100 | 34.3 | 25.8 | 51.58 | 121.0 |
| 20 | 7.1 | −5.6 | 22.14 | 81.9 |

The boiling point of a reference liquid must be corrected if the atmospheric pressure during standardization is other than 760 mm. For water and several other liquids the changes of boiling point at pressures near 760 mm are given in Table 5. The boiling point at pressures in the region of 760 mm can be calculated with sufficient accuracy for most purposes by the Rule of Crafts, in the following convenient form:

$$\text{bp at } p \text{ mm} = \text{bp at 760 mm} - \frac{(273 + \text{bp at 760 mm})(760 - p)}{10{,}000}$$

No correction for variations from 760 mm is needed when standardizations are made by means of melting points, since the effect of small pressure changes on melting points is negligible.

The most satisfactory way to determine the true temperatures that correspond to observed temperature readings is to calibrate the thermometer using the same conditions under which the thermometer is to be employed. Thus, a thermometer may be calibrated for a fixed partial immersion of the stem: for example, a thermometer to be used for distillation may be calibrated for 3-inch (76-mm) immersion of the stem; one to be used for melting point determinations may be calibrated for 1-inch (25-mm) immersion of the stem.

# EXERCISE 1

# Simple Distillation

The purpose of this experiment is to give the student sufficient practice in purification of liquids by simple distillation so that he can carry out this operation in subsequent experiments skillfully and without reference to detailed directions. The laboratory instructor may wish to skip one or more parts of the experiment.

Arrange a distillation assembly similar to that shown in Figure 9 or 10 (page 25), using a 50-ml boiling flask. Follow the correct methods for supporting the apparatus and lubricating the joints as described on pages 10 and 11.

**(A) Distillation of a Pure Compound.** In the dry 50-ml boiling flask introduce 25 ml of pure, dry carbon tetrachloride by means of a clean, dry funnel. Add one or two tiny boiling chips, attach the boiling flask, and make certain that all connections are tight. Arrange a graduated cylinder to serve as receiver. Heat the flask gently on a wire gauze, preferably one having an asbestos center, using a small flame, and record the temperature when the first drops of distillate collect in the condenser. Continue to distill the liquid slowly (not over 2 ml per minute) and record the distilling temperature at regular intervals during the distillation—when the total distillate amounts to 5 ml, 10 ml, 15 ml, and 20 ml. Discontinue the distillation (and extinguish the burner flame) when all but 2–3 ml of the liquid has distilled. Record the temperature range from the beginning to the end of the distillation; this is the observed boiling point. If the boiling point differs from the literature value record the correction in your laboratory notebook for future reference.

Transfer the used carbon tetrachloride to a bottle provided for this purpose on the side shelf. From your data draw a rough distillation graph for pure carbon tetrachloride, plotting distilling temperature on the vertical axis against total volume of distillate on the horizontal axis.

➤**CAUTION:** Manipulate carbon tetrachloride carefully. Avoid inhaling the vapor; if the liquid is spilled on the skin, wash it off promptly with soap and water.

**(B) Distillation of a Mixture.** By means of a clean, dry funnel introduce 25 ml of a mixture of carbon tetrachloride and toluene into the distilling flask, add a few tiny boiling chips, and distill the mixture slowly. Follow the same procedure used for distilling pure carbon tetrachloride. Draw a rough distillation graph and compare it with that observed for pure carbon tetrachloride. From the graph estimate the composition of the liquid and record your analysis in your notebook.

Transfer the distillate to a bottle for this purpose on the side shelf, labeled "Recovered Carbon Tetrachloride-Toluene Mixture."

**(C) Purification of a Liquid.** From your instructor obtain a 25-ml sample of an impure unknown. Carry out a preliminary distillation to determine the distillation behavior of the mixture and its approximate composition. Redistill the liquid and collect in a tared receiver the main fraction boiling over a 4–5° range. Record the boiling range and weight of the main fraction.

## Questions

**1.** Define accurately the term boiling point. What effect does a reduction of the external pressure have upon the boiling point?

**2.** What effect on the boiling point is produced by: (a) a soluble nonvolatile impurity? (b) an *insoluble* admixed foreign substance such as sand, fragments of wood or cork, etc.?

**3.** Why should a distilling flask at the beginning of a distillation be (a) filled to not more than two-thirds of its capacity? (b) filled to not less than one-third its capacity?

**4.** Calculate the weight of vapor of benzene ($C_6H_6$) required to fill a 50-ml flask at the boiling point under normal atmospheric pressure. Make similar calculations for toluene ($C_7H_8$) and for carbon tetrachloride ($CCl_4$).

**5.** Why is the apparatus shown in Figure 10 not suitable for distillation of samples with a volume of 5 ml or less?

**6.** Why is it dangerous to heat an organic compound in a distilling assembly that is closed tightly at every joint and has no vent or opening to the atmosphere?

**7.** Calculate the stem correction for observed temperature readings of 125°, 175°, and 250°; assume that the thermometer scale is exposed above the 25° mark and the average temperature of the exposed portion is half the difference between the observed temperature and room temperature (20°).

**8.** In a distillation assembly why is it advantageous to have the cooling water enter the condenser jacket at the lower end and leave from the upper end, rather than to have it flow in the opposite direction?

# Fractional Distillation

The term fractional distillation has two meanings for chemists. The original use of the term refers to the multistep process of changing receivers several times during distillation of a mixture so that several portions (called fractions or cuts) of distillate boiling over successively higher temperature ranges are collected. The first fraction is enriched in the lower-boiling, more volatile component while the second and subsequent fractions contain successively higher amounts of the higher-boiling material. The process of separating the distillate into fractions of progressively higher boiling point and subjecting the fractions to redistillation results in a much sharper separation of the mixture into its components than would be obtained by a single simple distillation. The more times this process is repeated the more complete is the separation, until in the limit, after an infinite number of repetitions, the different receivers would contain only pure components.

The second and common use of the term fractional distillation refers to a distillation operation where a *fractionating column* has been inserted between the boiler and the vapor take off to the condenser (Figure 14). The effect of this column when properly operated is to give in a single distillation a separation equivalent to several successive simple distillations. This represents a considerable saving of time and makes the selection and proper operation of fractionating columns an important subject for chemists.

**Fractionating Columns.** The easiest approach to understanding the principles by which fractionating columns give their superior separations is to consider first a rather special type of column known as a *Bubble Plate Column*. The essential features of a bubble plate column are illustrated in Figure 12 and consist of (1) a series of horizontal plates, *A*, which support a layer of distillate, (2) capped risers, *B*, through which the distilling vapors ascend, and (3) overflow pipes, *C*, which return any excess distillate to the next lower plate. At the beginning of a distillation, the vapors coming up from the boiler pass through the first riser and are deflected downward by the cap onto the first plate, where they are condensed. As simple vaporization and condensation continue the rising vapors are forced to bubble through the liquid on the plate. The liquid level rises to the top of the overflow tube and then flows downward to the boiler. The liquid on the first plate corresponds to the first

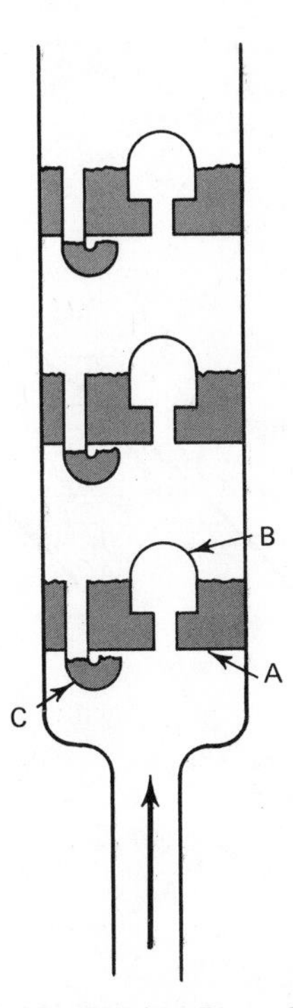


**Figure 12.** Bubble Plate Column

fraction in a simple distillation—it is enriched in the lower-boiling component. It follows that the temperature of the vapor bubbling through the liquid is *above* the boiling point of the liquid on the plate; through heat exchange the liquid is brought to its boiling point and its vapor rises to the second plate where the same processes are repeated. As the distillation continues, each plate becomes filled with a layer of liquid whose composition is that of the vapor rising from the next lower plate. Under ideal circumstances each plate achieves an increment of separation equivalent to one simple distillation.

The overflow tubes serve a more important function than just acting as returns for excess condensate. Since the vapor leaving any plate is richer in the lower-boiling component than the vapor entering the plate, the higher-boiling materials tend to accumulate on the plate. The overflow returns this higher-boiling material to the lower plate, so that an equilibrium balance of low-boiling to high-boiling components is maintained. In effect vapor and condensate are passing in opposite directions through the column: the more volatile component ascends the column in the vapor stream while the less volatile components descend. The *counterflow* is essential for effective separation in a fractionating column.

A diagram giving the relations between liquid and vapor composition for mixtures of carbon tetrachloride (bp 77°) and toluene (bp 111°), as determined by actual experiment, is shown in Figure 13. This plot is different from

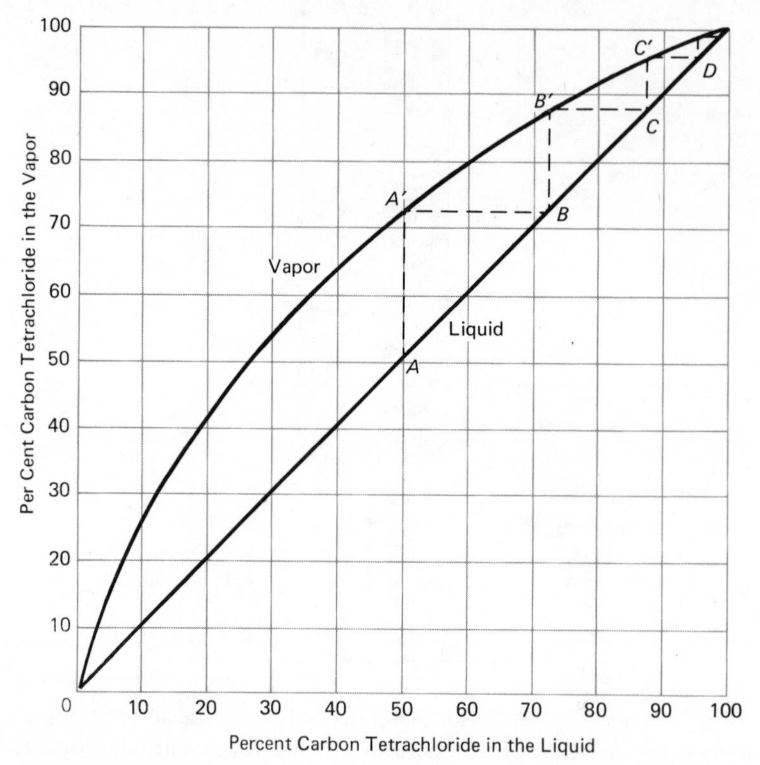


**Figure 13.** Liquid–Vapor Diagram: Carbon Tetrachloride–Toluene

the composition–boiling point diagram (Figure 4, page 20) and is a form that is particularly convenient for studies of fractional distillation. The mole percent of carbon tetrachloride in the liquid mixture is plotted on the $x$-axis, and the percentage of carbon tetrachloride in the vapor in equilibrium with a given liquid mixture is plotted on the $y$-axis; the percentage of toluene in the mixtures is obtained by subtracting the carbon tetrachloride content from 100 percent. The 45° diagonal (liquid line) can be used to show the effect of the bubble plates in concentrating the more volatile component. A liquid mixture containing 50 percent carbon tetrachloride (point $A$, on the liquid line) is in equilibrium with vapor containing 71 percent carbon tetrachloride (point $A'$, on the vapor curve). If the liquid $A$ is partially vaporized and the vapor $A'$ condensed completely on the first bubble plate, the condensate will be represented by $B$ (on the liquid line). Repetition of this process with liquid $B$ yields a new distillate $C$, containing 88 percent carbon tetrachloride, that condenses on the second bubble plate. Each successive bubble plate achieves an additional increment of separation.

Bubble plate columns have the drawback of requiring large samples for effective operation and a substantial portion of material is withheld on the plates (*holdup*). To overcome these disadvantages, small scale laboratory fractionations are usually done with cylindrical columns packed with materials having large surface area (glass beads or helices, small sections of twisted metal, carborundum chips, and the like). The principles of operation of *packed columns* are quite similar to those of the bubble plate column. The layers of packing material, like the bubble plates, serve as support for films of condensate; vapor passing through the layers is enriched in the lower-boiling component, and the higher-boiling components move downward to lower layers. The scrubbing action of the packing material effects the counter-flow of vapor and condensate that is essential for fractionating efficiency.

**Relative Efficiency of Fractionating Columns.** Since column packings differ widely in efficiency, it is desirable to have a means of comparing their effectiveness for separating mixtures. The enrichment factor ($\alpha$), relating the relative volatility of two components of a mixture, is expressed as the quotient of the ratio of the mole-fractions of the components in the vapor to the ratio of their mole-fractions in the liquid:

$$\alpha = \frac{y_1/y_2 \text{ (vapor)}}{x_1/x_2 \text{ (liquid)}} \tag{1}$$

A theoretical plate is the unit of separation corresponding to the difference in composition, $\alpha$, that exists as equilibrium between a liquid mixture and its vapor. This concept may be illustrated by considering a 50:50 mole percent mixture of carbon tetrachloride and toluene. The vapor in equilibrium with the liquid (bp 90°) contains 71 mole percent carbon tetrachloride and 29 mole percent toluene. This amount of enrichment corresponds to one theoretical plate.

$$\alpha = \frac{\frac{71}{29}}{\frac{50}{50}} = \sim 2.5$$

The length of packed column required to obtain this degree of separation in the mixture is known as the Height Equivalent to a Theoretical Plate (usually abbreviated HETP). The smaller the value of the HETP the more efficient is the column. While the exact HETP of any given packing depends on operating factors (diameter of the column, density of packing, rate of distillation, etc.) it is useful to have rough estimates of relative values. Table 6 records representative values of HETP for several packings as measured under normal working conditions using student apparatus to separate a benzene-toluene mixture. Also shown in Table 6 are representative values of the column holdup per unit volume of packing. These are dimensionless because they are the ratio of two volumes. Both the HETP and the holdup values will vary with the manner of packing and subsequent treatment of the column.

**TABLE 6**
**HETP for Common Packing Materials***

| PACKING | HETP (CM) | HOLDUP/PLATE (G) |
|---|---|---|
| Carborundum chips | 6 | 1.2 |
| Glass beads | 8–9 | 0.9 |
| Glass helices | 4–5 | 0.6 |
| Metal helices | 8–9 | 0.9 |
| Metal sponge | 30 | 1.6 |

* Obtained with 25-cm packed column using a benzene-toluene mixture.

In addition to packed columns, special columns are available that achieve mixing of the ascending vapor and the descending condensate by their special construction. One of the simplest, least expensive, and most widely used is the Vigreux column illustrated in Figure 15. Under normal working conditions the Vigreux column has a relatively low efficiency (high HETP of ~10 cm) but its low resistance to vapor flow permits a large *throughput* (volume of distillate per unit of time) that makes the column well suited to distillation of bulk solvents. Because of its small surface area the column has a low holdup and is sometimes used for preliminary purification of small samples.

The spiral wire column is also widely used. It consists of a wire wound spirally on a glass rod that is held concentrically within an outer glass tube. Spiral wire columns are slightly more efficient than columns packed with glass beads (HETP of ~2 cm) and have about half the holdup of a packed column capable of the same throughput. Its limitation is throughput, which is essentially fixed (about 0.5 ml/min maximum) whereas packed columns can be scaled up as needed. Because of their simple construction spiral wire columns are generally built in the laboratory rather than purchased.

A unique column deserving special mention is the "Auto Annular Still."[1] This unit, resembling a spiral wire column, contains an annular

[1] Nester Faust Mfg. Corp., Newark, Del. Another, though less efficient, column of related design is manufactured by Podbielniak, Franklin Park, Ill.

Teflon helix wrapped around a Teflon rod. A motor spins the Teflon helix at high speeds, which mixes the ascending vapors extremely effectively with the descending condensate. It is claimed that the 24-inch column (45-inch with accessories attached) produces more than 150 theoretical plates with a holdup of less than 0.5 ml and a throughput of 15–60 ml per hour. This remarkable unit unfortunately costs several thousand dollars.

**Separation Efficiency.** The total number of theoretical plates, $n$, present in a column is equal to the height of the packed portion of the column divided by the HETP of the packing material. The composition of the vapor at the top of the column, $(y_1/y_2)_T$, is related to the composition in the boiler, $(x_1/x_2)_B$, in the following way:

$$\left(\frac{y_1}{y_2}\right)_T = \alpha^{n+1}\left(\frac{x_1}{x_2}\right)_B \qquad (2)$$

The exponent of $\alpha$ is $n + 1$ rather than $n$ because of the vapor enrichment that occurs in vaporizing the mixture in the boiler. Although this equation has theoretical significance, it is more practical to have an expression for the number of theoretical plates required to separate a given mixture. An approximate expression (equation 3) has been derived for fractional distillation of 50:50 mixtures, such that the first 40 percent of the material distilled will have an average purity of 95 percent in the lower-boiling component. Equation 3 shows that as the relative volatility, $\alpha$, approaches unity the number

$$\begin{array}{l}\text{Number of theoretical plates required}\\ \text{to achieve a standard separation}\end{array} = \frac{2.85}{\log_{10}\alpha} \qquad (3)$$

of theoretical plates required to achieve 95 percent purity rises steeply. This relationship becomes still more useful (but also more approximate) if one substitutes for $\log_{10}\alpha$ an expression involving the difference in boiling points of the two components. Equation 4 results for ideal mixtures.[2] This equation makes it clear that tall high-platage fractionating columns are required to separate cleanly materials boiling a few degrees apart. When the required

$$\begin{array}{l}\text{Number of theoretical plates required}\\ \text{for standard separation}\end{array} = \frac{250}{T_B - T_A} \qquad (4)$$

number of theoretical plates is unavailable for practical reasons, it is necessary to collect a smaller portion of the low-boiling distillate.

**Reflux Ratio and Holdup.** Equations 2 and 3 were derived for an ideal fractional distillation where there is equilibrium between the rising and

[2] The approximate number of theoretical plates required to achieve a separation such that the first 40 percent of the material distilled will have an average purity greater than 99.5 percent in the lower-boiling component is given by $450/(T_B - T_A)$.

descending counterflowing streams of materials. For this equilibrium to be attained it is essential that vapor reaching the top of the column be condensed and liquid returned to the column (*reflux*). If a large portion of the vapor reaching the top of the column is removed as distillate (*takeoff*) the equilibrium is seriously disturbed and much lower separation efficiency results. The extreme modes of operation are known as *total reflux* and *total takeoff*. Since the first mode yields no distillate and the second gives distillate of much lower purity than is possible with the column, in practice some intermediate ratio of takeoff to reflux is employed. The best practical compromise is to adjust the reflux ratio so that it equals the number of theoretical plates of the column. Higher rates of collecting distillate (lower reflux ratios) give poorer

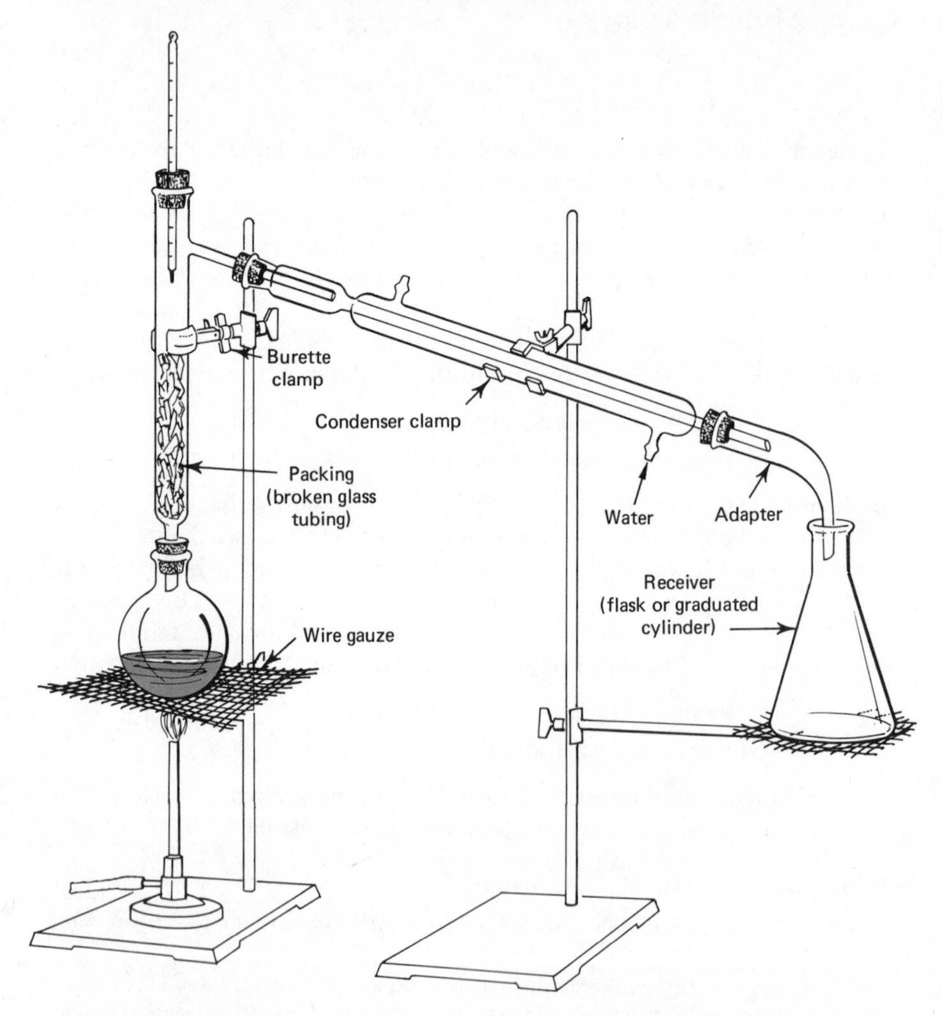


**Figure 14.** Apparatus for Fractional Distillation (with packed Hempel column)

separations; slower rates are overly time-consuming and do not provide significantly better separations.

Another factor that seriously affects separation efficiency is the total amount of liquid and vapor in the column at any instant (*holdup*). A great drop in separation efficiency occurs if the holdup is more than about 10 percent of the amount of sample to be distilled.

## Laboratory Practice

Successful fractional distillation demands a column with an adequate number of plates. As simplified as it is, equation 4 is a useful guide to the required number. Estimation of the desired number of plates requires knowledge of the composition and boiling behavior of the mixture to be separated. When this information is lacking it is desirable to run a preliminary simple distillation and to plot a graph showing the actual relation of distillation temperature to volume of distillate (see Figure 14).

It may happen that none of the fractionating columns available has an adequate number of plates. In this case it will be necessary to separate the mixture into a number of fractions of progressively higher boiling point, and to refractionate these separately in a systematic way until an acceptable separation of the components is achieved. Table 7 illustrates a representative distillation of a benzene-toluene mixture (bp difference of 31°) through a short Hempel distillation column packed with about 10 cm of broken glass

**TABLE 7**
**Fractional Distillation of Benzene-Toluene Mixture Containing 100 g of Each Component***

| FRACTION | TEMPERATURE INTERVAL IN °C | WEIGHT OF FRACTION IN GRAMS | | | |
|---|---|---|---|---|---|
| | | *1st Dist'n* | *2nd Dist'n* | *3rd Dist'n* | *4th Dist'n* |
| 1 | 80–85 | — | 37 | 59 | 70 |
| 2 | 85–91 | 60 | 42 | 26 | 17 |
| 3 | 91–99 | 40 | 26 | 16 | 10 |
| 4 | 99–105 | 34 | 18 | 12 | 8 |
| 5 | 105–111 | 53 | 50 | 48 | 43 |
| Residue | Above 111 | 11 | 23 | 32 | 42 |
| Total Weight of Distillate | | 198 g | 196 g | 193 g | 190 g |
| Total Losses during Manipulation | | 2 g | 4 g | 7 g | 10 g |

* The values given in this chart are such as can be obtained under ordinary laboratory conditions using a small Hempel column. The exact weights of various fractions, however, are so dependent upon rate of distillation, type and size of column, and on other physical conditions, that the chart is valuable merely as an indication of what may be expected.

tubing (approximately three theoretical plates). A diagram of the fractionating assembly used is shown in Figure 14.

In the simpler and less efficient columns (Figure 14) the returning liquid is provided merely by atmospheric cooling of the vapor in the upper portion of the column (but the main portion should be insulated to avoid excessive heat loss). In more effective columns an adequate quantity of refluxing liquid is obtained by placing a partial condenser (cold finger) at the head of the column, which is adjusted to give the desired ratio of distillate to reflux.

The upper portion of a relatively efficient and inexpensive packed column is shown in detail in Figure 15. The head of the column is equipped

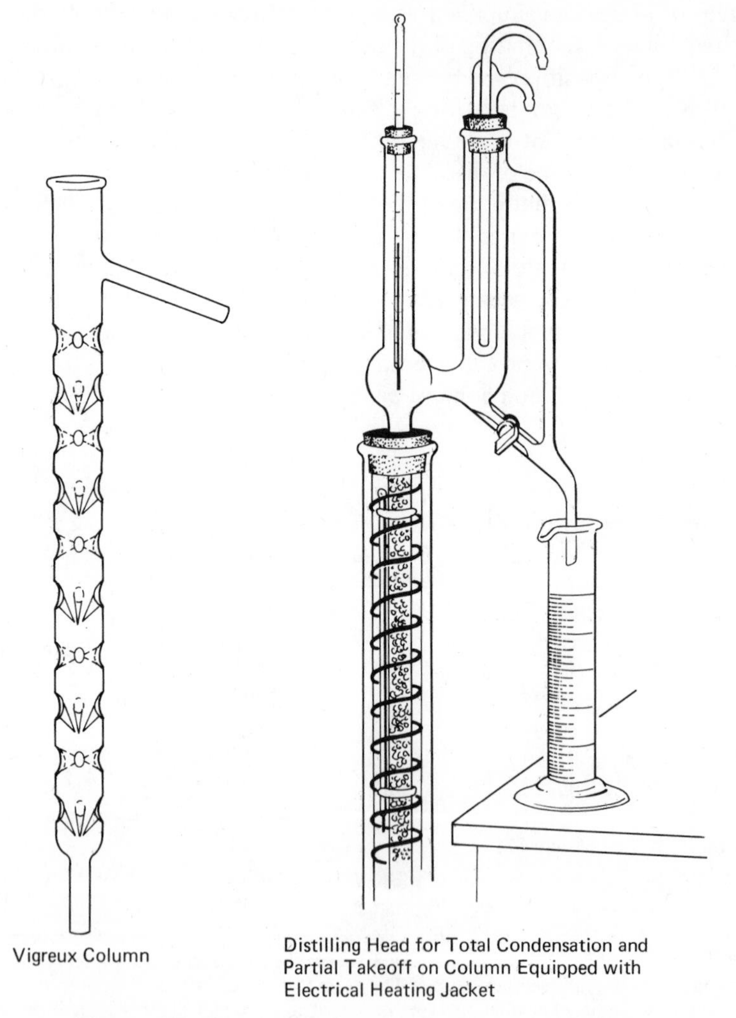

Vigreux Column

Distilling Head for Total Condensation and Partial Takeoff on Column Equipped with Electrical Heating Jacket

**Figure 15.** Special Devices for Fractional Distillation

with a reflux condenser and an adjustable stopcock for controlling the reflux ratio. The column is surrounded by a glass jacket on which is wound a spiral heating wire that is heated electrically so as to minimize the heat loss from the column. The temperature of the upper section of the column is indicated by a thermometer taped to the outside of the packed column. To further reduce and even out the heat loss from the column the heating jacket is surrounded by another glass tube.

Increasing the length of a column or the reflux ratio improves the efficiency of separation but care must be taken to avoid flooding the column with liquid. Flooding diminishes the contact area between vapor and liquid, and the pressure of ascending vapor may force the liquid upward in the column. To obtain good heat exchange between vapor and liquid, and to avoid flooding, a column should be well insulated. For liquids that distill below 100° a wrapping of asbestos paper is usually sufficient; for higher-boiling liquids or very long columns, an evacuated or electrically heated jacket may be used.

In a packed column it is essential to leave sufficient free space for the countercurrent flow of liquid and vapor. With packing materials like carborundum chips, glass beads or short lengths of glass tubing the column may be filled simply by pouring in the packing, but with glass or metal spirals the best results are obtained by dropping the spirals into the column singly.

EXERCISE

# 2

# Fractional Distillation

The purpose of this experiment is to give the student practice in purification of liquids by fractional distillation. It is not expected that he will master this technique after a single distillation but he should become aware of difficulties involved. The student is cautioned to carry out a fractional distillation methodically since haste will lead to sharply lowered separation efficiency.

If the student is using ground-jointed glassware the apparatus used for simple distillation (Figure 9, page 25) is assembled with a packed column placed between the boiling flask and the distillation adapter. A ground-jointed condenser makes an effective thermally insulated distillation column.

If the student is using apparatus with corks, arrange an assembly for fractional distillation as shown in Figure 10 (page 26). The Hempel column should be insulated by wrapping with wet asbestos paper and the wrapping allowed to dry before the column is packed.

The laboratory instructor will indicate what kind of packing is to be used and issue any special instructions for placing it in the column. Prepare five clean, dry receivers (50–150 ml Erlenmeyer flasks), provide each with a *clean* tightly fitting cork, and label them *A*, *B*, *C*, *D*, and *E* (residue). The boiling flask should be selected so that it is approximately 60 percent full at the beginning of the distillation.

*Before starting the distillation* check your apparatus carefully and have it approved by the laboratory instructor.

***Carbon Tetrachloride and Toluene.***[3] In the distilling flask place 120 ml of a mixture of carbon tetrachloride and toluene, 1:1 by volume, add two boiling chips, and fit the flask securely to the column. Heat the flask on a wire gauze, having an asbestos center, *using a small flame* that impinges directly

[3] If desired the composition of the distillate may be determined by gas chromatography using a 5-ft 5 percent SE 30/Chromosorb W column at 25°. With these data the number of theoretical plates can be calculated from equation 1.

below the flask. As soon as the mixture starts to boil, regulate the flame especially carefully so that the liquid distills slowly and regularly at the rate of about 2 ml (60 drops) per min. In the first distillation, collect the fraction (if any) which distills from 76–81° in flask *A*, from 81–88° in flask *B*, from 88–98° in flask *C*, from 98–108° in flask *D*.[4] After fraction *D* has distilled, extinguish the flame, cool the flask, allow the column to drain, disconnect the flask, and pour the residue into *E*. Measure the volume of each fraction and record the results in tabular form.[5]

**Fractional Distillation of Carbon Tetrachloride and Toluene Mixture**

| FRACTION | TEMPERATURE RANGE | VOLUME OF FRACTIONS IN ML | | |
|---|---|---|---|---|
| | | *1st Dist'n* | *2nd Dist'n* | *3rd Dist'n* |
| *A* | 76–81° | | | |
| *B* | 81–88° | | | |
| *C* | 88–98° | | | |
| *D* | 98–108° | | | |
| *E* | Residue | | | |
| Total Volume of Fractions | | | | |

If the separation efficiency of the column was not adequate it will be necessary to redistill the different fractions. In the subsequent distillations proceed in the following way: Pour the contents of flask *A* into the round-bottomed flask, add one or two tiny boiling chips,[6] and redistill, collecting the distillate from 76–81° in the same flask *A*. When the thermometer reaches 81°, stop the distillation and add the contents of flask *B*. Continue the distillation and collect the fraction from 76–81° in flask *A*, and from 81–88° in flask *B*. When the thermometer reaches 88°, stop the distillation and add the contents of flask *C*. Continue the distillation and collect the fraction from 76–81° in flask *A*, from 81–88° in flask *B*, and from 88–98° in flask *C*. When

[4] In the first distillation there may be no distillate in the range 76–81°, so that flask *A* will remain empty and flask *B* will be used. In the subsequent distillations, likewise, there may be little or no distillate in the intermediate ranges *B*, *C*, or *D*. The results vary widely depending on the type of column and the care used in operation.

[5] If one desires to record the weight of each fraction instead of the volume, it is convenient to weigh each receiver empty with its cork and record this weight (called the tare) on the label of the receiver. It is good practice to record the tare of each receiver also in the notebook; when the receivers with distillate are weighed after a distillation, the gross weight is recorded, the tare subtracted, and the net weight of the fraction entered in the tabular form. If all of the weights are recorded in the notebook, one can check the figures at a later date for arithmetic errors if a discrepancy shows up.

[6] It is desirable to add a *tiny* fresh boiling chip each time the distillation is stopped and a new fraction introduced. At the end of the distillation series the residual liquid in the still is poured off and the accumulated used chips are discarded.

the thermometer reaches 98°, stop the distillation and add the contents of flask *D*. Continue the distillation and collect the fractions *A*, *B*, *C*, and *D*. When the thermometer reaches 108°, stop the distillation and add the contents of flask *E*. Continue the distillation and collect the fractions *A*, *B*, *C*, and *D*. After fraction *D* has distilled, extinguish the flame, cool the flask, allow the column to drain, disconnect the flask, and pour the residue into *E*. Measure the volume of each fraction and record the results in tabular form.

If *B*, *C*, and *D* at this stage contain a total of more than 15–20 ml of liquid, carry out a third distillation in the same manner. If necessary, carry out a fourth or fifth distillation, so that the fraction *A* will contain almost all the carbon tetrachloride and the residue *E* almost all the toluene. To remove mechanical impurities and coloring matter, and to obtain almost pure toluene, *E* may be redistilled from a small distilling flask without a column.

Draw rough distillation graphs for each successive distillation, plotting the midpoint of the temperature range of the fractions against total volume of distillate.

In a research laboratory it is customary to follow the progress of a distillation by some convenient analytical procedure. Gas chromatography (Chapter 8, page 114) is used commonly.

➤**CAUTION:** Extinguish or remove the flame when transferring fresh fractions into the round-bottomed flask, since toluene has a high vapor pressure and is flammable. If the lower end of the condenser is fitted with an adapter, there is usually less loss of material by evaporation and less danger of fire. A *loose* plug of cotton between the adapter and the receiver will also diminish the evaporation losses. Care must be taken to prevent the cotton from coming in contact with the distillate and from plugging the opening tightly!

***Benzene and Toluene.***[3] A mixture of benzene and toluene may be separated by the procedure outlined, if, to increase the efficiency of the fractionation, the column is well packed and the rate of distillation reduced somewhat. The following ranges are satisfactory for the fractions: *A*, from 80–85°; *B*, from 85–91°; *C*, from 91–99°; *D*, from 99–106°; *E*, residue.

***Benzene and Acetic Acid.*** For the separation of a mixture of benzene and glacial acetic acid, the following temperature ranges are satisfactory for the fractions: *A*, from 80–85°; *B*, from 85–90°; *C*, from 90–100°; *D*, from 100–105°; *E*, residue. The progress of the separation of this mixture may be followed conveniently by titrating an aliquot portion (1 or 2 ml) of each fraction against standard alkali to determine the acetic acid content. The acetic acid content of the original mixture should be determined in the same way before the material is fractionated.

***Methanol and Water.*** For the separation of a mixture of methanol and water, the following temperature ranges are satisfactory for the fractions: *A*, from 64–70°; *B*, from 70–80°; *C*, from 80–90°; *D*, from 90–95°; *E*, residue.

## Questions

**1.** Why is it necessary to have liquid flowing back through the fractionating column in order to obtain efficient fractionation?

**2.** What is an azeotropic mixture, and why cannot its components be separated by fractional distillation?

**3.** What physical constants may be used to test the purity of the samples of purified material obtained after a fractional distillation?

**4.** What is meant by the temperature gradient of a column? Why is it desirable to maintain a uniform temperature gradient and how is this achieved?

**5.** It is common to use an oil-bath to heat the boiling flask during a fractional distillation. What is the advantage of this indirect heating?

# Vacuum Distillation

Since the boiling temperature of a material is decreased by diminishing the pressure upon its surface, it is possible to effect distillation at a lower temperature by using a closed system inside which the pressure has been reduced. This procedure is useful for purifying liquids (or low-melting solids) that are decomposed at elevated temperatures. For example, glycerol boils with some decomposition at 290° under 760 mm pressure but may be distilled without decomposition under 12 mm pressure, where its boiling point is 180°. A possible disadvantage of fractional distillation under reduced pressure is the reduction in separation efficiency of most fractionating columns.

In planning a vacuum distillation three aspects must be considered: the pressure needed to achieve the desired boiling point, the type of vacuum pump needed to lower the pressure to the required level, and the associated glassware, pressure measuring, and heating devices.

**Estimation of Boiling Point.**[1] One useful relationship between pressure and boiling point is given in equation 1.

$$\log_{10}\left(\frac{760}{P}\right) = 5.46\left(\frac{\text{normal boiling point}}{T} - 1\right) \tag{1}$$

in which $P$ is the pressure over the liquid and $T$ is the boiling point at this pressure. In this equation both boiling temperatures are expressed in degrees Kelvin (°K = °C + 273). The equation is fairly precise for most organic liquids but is in error for substances possessing unusually large attractions between molecules (water, alcohols, acids). More precise relationships have

[1] In equation 1 the value of the constant 5.46 has been changed from its usual theoretical value of 4.81 in order to correlate a wider range of data.

been developed but the extra work required to use them is not justified for preparative organic chemistry.

An example of the equations application is outlined below for nitrobenzene.

1. Normal boiling point of nitrobenzene = 211°C = 484°K
2. If the desired boiling point is 100°C = 373°K the equation becomes

$$\log_{10}\left(\frac{760}{P}\right) = 5.46\left(\frac{484}{373} - 1\right) = 1.62$$

3. The expression is solved for $P$

$$P_{\text{predicted}} = 18.0 \text{ mm}$$

4. At a pressure of 18.0 mm it is observed that nitrobenzene boils at 98°C instead of the desired 100°C.

In Table 8 are recorded the pressures required to achieve several ratios of normal boiling point to reduced pressure boiling point (both expressed in °K). Ratios are given for both hydrocarbons and for hydrogen-bonded liquids. Most other organic materials will have an intermediate ratio.

**TABLE 8**
**Pressure-Boiling Point Relationships**

| PRESSURE | RATIO OF NORMAL BOILING POINT, °K TO REDUCED PRESSURE BOILING POINT, °K | |
|---|---|---|
| *mm* | *Hydrocarbons* | *Hydrogen-bonded Liquids* |
| 0.001 | 2.089 | 1.905 |
| 0.01 | 1.904 | 1.751 |
| 0.1 | 1.719 | 1.597 |
| 1.0 | 1.533 | 1.443 |
| 5.0 | 1.404 | 1.336 |
| 10.0 | 1.348 | 1.289 |
| 50.0 | 1.219 | 1.182 |
| 100.0 | 1.163 | 1.135 |
| 250.0 | 1.089 | 1.074 |
| 760.0 | 1.000 | 1.000 |

**Vacuum Pumps.** The pump used to reduce the pressure in the system is selected according to the range of pressure required. A water pump is used for pressures above about 25 mm, a rotary oil pump is usually used for the range of 0.01 to 25 mm and a diffusion pump is used for pressures below about 0.01 mm. The user should be aware of the characteristic features of each type of pump.

A water pump in good condition can produce a vacuum almost down to the vapor pressure of the water flowing through it. With room temperature water, a pressure of about 25 mm can be produced but during the winter if the water is quite chilled pressures of 10 mm or less may be obtained. It is not feasible to vary the water flow through the pump in order to regulate the pressure. Instead one should turn the water flow fully on and control the pressure by adjusting a valve that leaks air into the system. An inexpensive valve can be obtained by using the valve at the base of an adjustable Bunsen burner. A hose is connected from the apparatus to the gas inlet tube of the burner and the reverse flow of air into the apparatus is regulated at the burner base. A troublesome characteristic of water pumps is their tendency to allow water to flow back into the system if the water pressure drops momentarily. Modern water pumps come with internal check valves to prevent this back flow but they should not be trusted; it is wise to include a safety trap between the water pump and the rest of the system. The safety trap also provides a convenient means of interconnecting the pump, the air leak, and the mano-

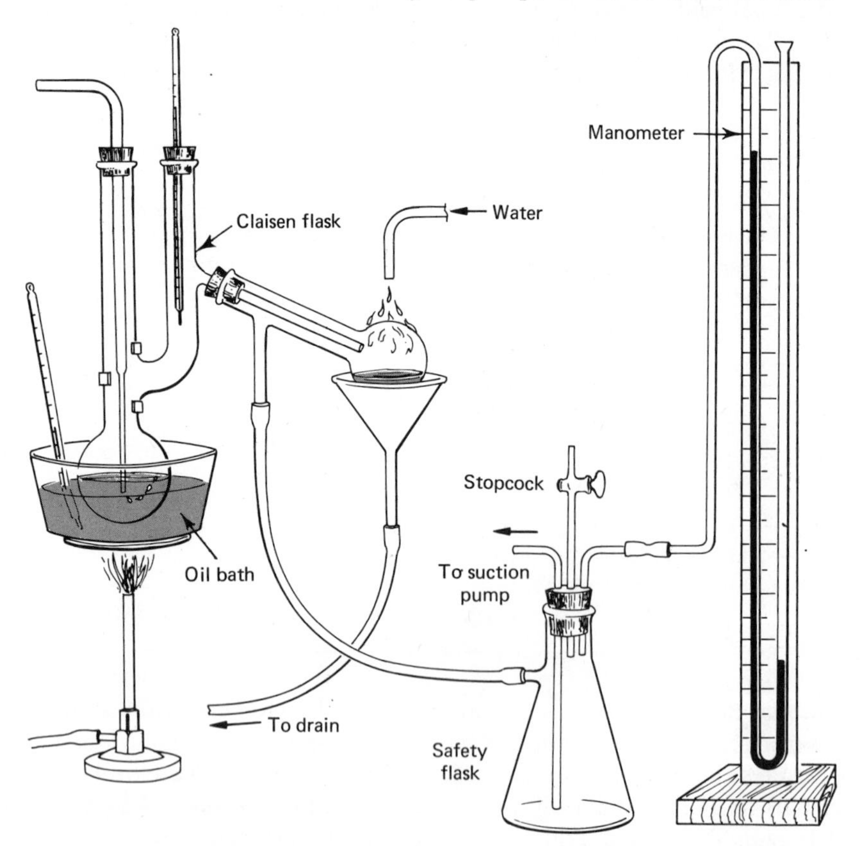


**Figure 16.** Assembly for Distillation under Diminished Pressure

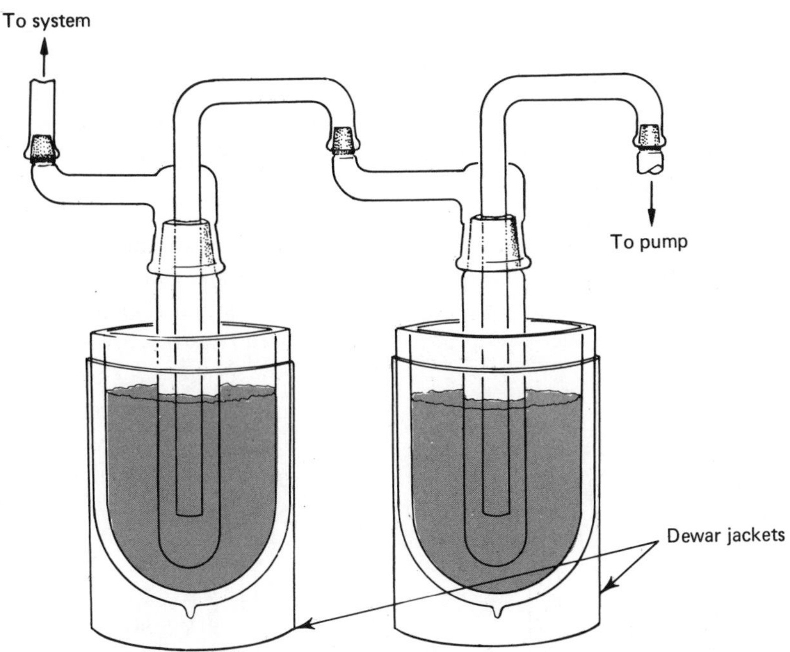


**Figure 17.** Trap Arrangement for Rotary Oil Pumps

meter used to measure the internal pressure. A practical arrangement is shown in Figure 16.

In using a rotary oil pump the pump oil has such a low vapor pressure that contamination of the distillate is improbable, however one must guard against contamination of oil by any uncondensed materials from the distillation. Oil pumps work by compressing small samples drawn from the vapor inside the distillation apparatus to a pressure above the atmospheric pressure and then releasing them from the pump. This pumping operation requires good seals between two reciprocating vanes and a rotating eccentric piston. If the oil becomes contaminated with acidic materials the movable vanes will corrode and the pumping capacity of the pump will be sharply diminished. Many other non-acidic organic materials may polymerize in the pump to give sludges that also will wear the movable vanes with a corresponding reduction in pump effectiveness. It is important that the pump be protected by at least one vapor trap that is maintained at the Dry Ice sublimation temperature (−78°) or lower. If the mixture being distilled evolves gases that would sweep vapors into the pump much more efficient trapping devices are required. Figure 17 displays one widely used trap arrangement for rotary oil pumps.

Diffusion pumps, like the water aspirator, work on the Bernoulli principle,[2] except that in place of a rapidly flowing stream of water they use a jet

[2] When liquids or gases flow through a pipe of variable cross section the pressure is smallest where the cross section is least and the velocity is greatest.

of mercury or oil vapor. Diffusion pumps produce pressures in the range of $10^{-2}$ to $10^{-6}$ mm. They require heaters to vaporize the mercury or oil and cooling devices for recondensing the vapor after it has passed through the jet. Unlike aspirators and mechanical pumps diffusion pumps require an additional fore pump ("backing pump") that reduces the pressure in the pump to a critical level (usually about 0.1 to 0.01 mm).

**Manometers.** There are many styles of manometers for measuring the pressure in the system, each designed for maximum precision over a small range of pressures. Two general purpose manometers that together cover a sufficient range with adequate precision for preparative work are the tilting

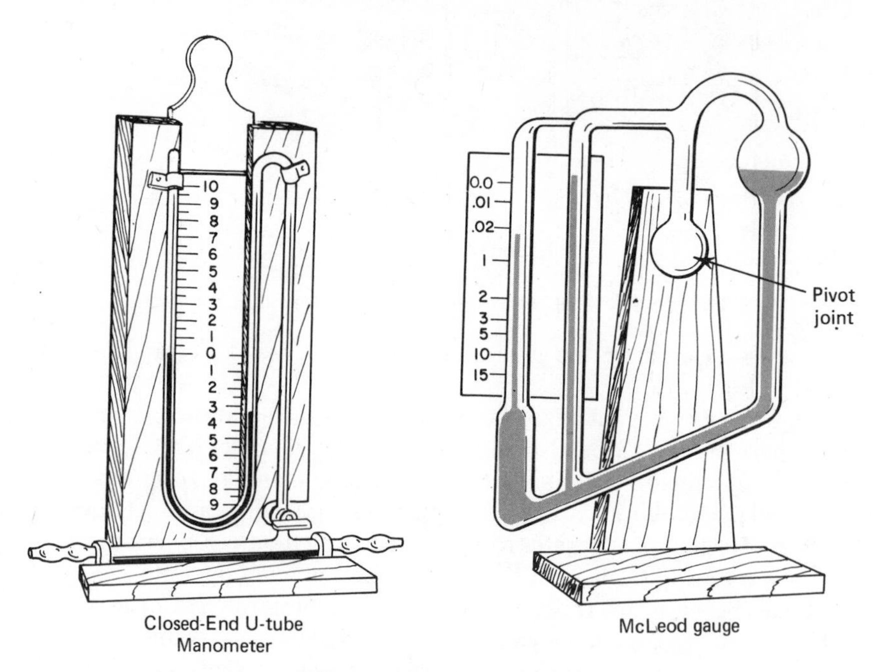


**Figure 18.** Manometers

McLeod gauge ($10^{-2}$ mm to 10 mm) and a closed end **U**-tube mercury manometer (5 mm to 200 mm), which are shown in Figure 18.

The McLeod gauge is operated by tilting the movable section of the gauge towards the upright position until the higher of the two mercury columns is level with the top of the bore of the lower capillary column. If the gauge has been calibrated properly the pressure can be read directly from the height of the lower capillary column against the gauge markings.

With the open tube manometer it is necessary to subtract the net height of the mercury column in the manometer from the barometric pressure.

## Laboratory Practice

An apparatus for vacuum distillation using corks or rubber stoppers is diagrammed in Figure 16. When standard taper ware is being used the special Claisen flask is replaced by a regular round-bottomed flask and the triply-jointed Claisen adapter; a regular condenser terminated by a vacuum adapter is used in place of the distillation flask receiver. For samples larger than about 10 ml the ground-jointed apparatus is easier to use, however for smaller samples the one piece Claisen flask is recommended in order to avoid excessive mechanical losses.

The principle purpose of the Claisen flask is to diminish the chance of contamination of the distillate from frothing or violent bumping. Both of these conditions are more troublesome in vacuum distillation than in ordinary distillation. The size of the Claisen flask should be such that it is not quite half-filled at the start of the distillation. The flask is usually heated in an oil bath or sand bath to insure regular heating. The bath temperature is usually 15 to 25° higher than that of the distilling vapor.

To promote regular boiling and to minimize bumping, a fine stream of air bubbles is introduced at the bottom of the flask through a thin flexible capillary tube.[3] Boiling chips are not effective in vacuum distillation. In distilling a substance easily oxidized by atmospheric oxygen it is advantageous to use an extremely fine capillary thread for the introduction of bubbles. For more refined work a fine stream of an inert gas (argon, nitrogen, carbon dioxide) may be used.

With distillations using corks or rubber stoppers, ordinary side-arm distilling flasks are used for receivers. Several of these, sufficient to receive all the fractions, are prepared before the distillation is begun. The necks of all the receiving flasks must be about the same dimensions so that they will all fit tightly on the stopper of the side tube of the Claisen flask. The delivery tube of the Claisen flask should reach just into the bulb of the receiver. In order to simplify the apparatus the ordinary condenser is omitted; instead the receiver is supported above a large funnel and is cooled directly by a jet of cold water. This type of cooling is satisfactory for liquids that distill above 50°; for liquids that distill below this temperature a more elaborate cooling device is necessary.

All connections in the apparatus must be tight, and special attention

[3] The capillary is prepared by drawing out a piece of ordinary glass tubing, 7–8 mm in diameter, to capillary dimensions and drawing out this first capillary, in a small luminous flame, to an extremely fine and flexible capillary thread. The capillary thread is tested by blowing into the tube while the thread is held under ethanol or ether; it should emit a fine stream of very minute bubbles. The top of the tube which bears the capillary should be bent at right angles in order to facilitate adjustment of the depth of the capillary tube so that it will reach exactly to the bottom of the distilling flask. In place of a capillary, long wooden splints may be inserted in the flask to serve the same purpose. It is convenient to use "Peerless Wood Applicators" which may be purchased in any drug store.

must be given to the quality and size of the tubing and stoppers. Rubber stoppers are often used because they simplify the preparation of tight joints, but in many cases they are likely to contaminate the distillate with impurities. Clean corks of good quality are preferable although their use requires more skill and care to obtain tight connections. The corks should be rolled and pressed before use and should be of such size that not more than one-half (nor less than one-quarter) of the cork projects into the neck of the opening. Their porosity can be diminished and tight joints may be obtained by applying a *thin* coating of collodion[4] after the system has been evacuated. The external pressure serves to force the collodion into the pores of the cork.

With distillations using ground-jointed apparatus regular round-bottomed boiling flasks are used in conjunction with a condenser and the vacuum adapter. It is imperative that the joints be properly lubricated to prevent leaks during the distillation and ease of separation of the joints afterwards.

In advanced work the single receiver is replaced by a device consisting of a flask with two or more arms to each of which is attached a receiving flask. By rotation of this "cow" the different receivers in turn can be brought in line with the drips of condensate without having to break any of the vacuum seals.

The following points should be observed in carrying out a vacuum distillation:

1. Never use Erlenmeyer flasks as receivers. Even with the small 50-ml Erlenmeyer flask the force acting on the flat bottom is about 50 lb. Remember that the force at any point is proportional to the difference between the internal and external pressure. A distillation at 100 mm places nearly as much stress on the apparatus as one at $10^{-6}$ mm.
2. Test the completely assembled apparatus before placing the liquid in the Claisen flask, to detect leaks and to make certain that all of the parts of the apparatus will withstand the external pressure. *Use safety glasses to protect the eyes.*
3. In using a water pump, turn on the water to the full pressure, otherwise water may be sucked back into the safety flask.
4. In releasing the vacuum in the distilling flask, open the stopcock on the safety flask and gradually allow the pressure to reach atmospheric pressure in the apparatus before shutting off the water pump.
5. In using a rotary pump be certain that the pump is adequately guarded with traps and that they are filled with coolant.
6. When changing receivers remove the flame from beneath the bath, and allow the distilling flask to cool slightly before releasing the vacuum (by lowering the bath or raising the flask from the bath).

[4] Collodion is a solution of cellulose nitrate in ether-alcohol. Similar material can be prepared by solution of polystyrene in benzene.

After the receiver has been changed the system should be evacuated again before the heating is continued.

7. If a closed tube manometer is used, close the stopcock of the manometer before releasing the vacuum. If this is not done the abrupt surge of the mercury column may break the glass tube.

EXERCISE

# 3

# Vacuum Distillation

**(A) Purification of Benzaldehyde.** Purify a 60-g (58-ml) sample of technical benzaldehyde[5] in the following way. Wash with two 20-ml portions of sodium carbonate solution (10 percent), then with water, and dry over 5–10 g of anhydrous magnesium sulfate. It is advantageous to add a few small crystals of hydroquinone or catechol (anti-oxidants) during the drying operation. Decant through a fluted filter into a Claisen flask of suitable size and distill under diminished pressure (preferably below 30 mm) in the manner described above. Place a few crystals of hydroquinone or catechol (anti-oxidants) in the receiving flask in which the purified benzaldehyde is collected. To determine the boiling point of benzaldehyde under the particular pressure in your apparatus, prepare a graph from the following boiling point data for benzaldehyde: 180°/760 mm, 95°/50 mm, 87°/35 mm, 80°/25 mm, 70°/15 mm, 62°/10 mm.

**Auto-oxidation.** In common with many oxidizable substances, benzaldehyde is capable of combining directly with oxygen of the air (auto-oxidation) and is converted eventually to benzoic acid.

Auto-oxidation is extremely sensitive to the effect of catalysts, which are considered to act upon an unstable intermediate complex of "peroxide" character that is the initial step of oxidation. Catalysts that accelerate auto-oxidation are called *pro-oxidants*; those that retard or inhibit auto-oxidation are called *anti-oxidants*. The latter find important technical application for the preservation of organic materials; for example, the deterioration of rubber is greatly retarded by the incorporation of anti-oxidants. Likewise, the auto-oxidation of benzaldehyde can be effectively inhibited by the addition of a trace (less than 0.1 percent is sufficient) of hydroquinone or catechol.

[5] Technical benzaldehyde usually contains a small amount of benzoic acid. Benzaldehyde made by hydrolysis of benzylidene chloride contains small amounts of chlorobenzaldehydes because of ring substituted impurities in the benzylidene chloride.

**(B) Purification of Ethyl Acetoacetate.** Place a 60-g (60-ml) sample of technical ethyl acetoacetate[6] in a Claisen flask of suitable size and distill under diminished pressure (preferably below 30 mm) in the manner described above.

During the distillation of the low-boiling fraction, containing ethyl acetate, the high vapor pressure of the latter may increase the pressure in the system above 30 mm. If this occurs, the distillation of the first fraction may be carried out at a higher pressure, but the pressure should be maintained below 30 mm for the distillation of the remaining fractions. To obtain the correct boiling point of ethyl acetoacetate under the particular pressure in your apparatus prepare a graph from the following boiling point data for ethyl acetoacetate: 180°/760 mm, 100°/80 mm, 97°/60 mm, 88°/30 mm, 78°/18 mm, 74°/14 mm, 71°/12 mm.

Collect the purified ethyl acetoacetate over an interval of 5–6°, determined from the graph, and calculate the percentage recovery from the crude product.

## Questions

**1.** Give the pressures under which bromobenzene, aniline, and ethyl benzoate could be vacuum distilled at 100°.

**2.** What are the principal advantages and disadvantages of vacuum distillation compared to simple distillation at atmospheric pressure?

**3.** Why is bumping more troublesome in vacuum distillation than in ordinary distillation?

**4.** What precautions must be observed in using a water pump for vacuum distillation?

[6] Technical ethyl acetoacetate may contain a little ethyl acetate, acetic acid and water. Since ethyl acetoacetate decomposes to some extent on heating to its boiling point at atmospheric pressure with the formation of dehydracetic acid, it is purified by distillation under diminished pressure. It may also be purified by *rapid* distillation at atmospheric pressure.

# Steam Distillation

Steam distillation consists of distilling a mixture of water and an immiscible or partly immiscible substance.[1] The practical advantage of steam distillation is that the mixture distills at a temperature below the boiling point of the lower-boiling component. Consequently, it is possible to effect steam distillation of a high-boiling organic compound at a temperature much below its boiling point without resorting to vacuum distillation. Steam distillation is useful also in separating mixtures when one component has an appreciable vapor pressure (at least 5 mm) in the vicinity of 100° and the other has a negligibly low vapor pressure. The process of steam distillation is employed widely in the laboratory and in industry; for example, for the isolation of aniline, nitrobenzene, and many natural essences and flavoring oils.

## Principles of Steam Distillation

Mixtures of two immiscible substances behave quite differently from homogeneous solutions and the description of their behavior requires a different physical law. The basis of the new law can be grasped by considering the consequence of increasingly positive deviations from Raoult's law (see pages 21–23). One symptom of small positive deviations is a skewed boiling point composition diagram as is found with methanol-water solutions (Figure 7, page 22). Greater positive deviations such as occur in ethanol-water solutions lead to maxima in the total vapor pressure curve and to low-boiling azeotropes (Figure 8, page 24). With still greater positive deviations the two components can separate into two immiscible phases. Typical are *n*-butanol-water mixtures with a vapor pressure composition diagram shown

[1] Occasionally the principle of steam distillation is extended to mixtures of two immiscible organic compounds such as ethylene glycol and hydrocarbons and is then called co-distillation.

in Figure 19 and a boiling point composition diagram shown in Figure 20. As indicated by Figure 20 two liquid phases coexist in the composition range of 8 to 69 percent (by weight) *n*-butanol; compositions outside this range form homogeneous solutions. In the limit of very large positive deviations from Raoult's law the two components are completely insoluble and each component vaporizes independently of the other to give a constant total vapor pressure that is the sum of the individual vapor pressures. Most water-insoluble organic compounds approximate this extreme behavior so that steam distillation calculations are normally based on the simple law: *the total vapor pressure equals the sum of the vapor pressures of each component.*

If the vapor pressures of the two components are known, the distillation temperature and the composition of the distilling vapor mixture can be determined easily. The temperature of steam distillation is found readily by plotting the vapor pressure curves of the individual components and making a third curve showing the sum of the vapor pressures at the various temperatures (Figure 21). The steam distillation temperature will be the ordinate of the point where the sum equals the atmospheric pressure. Knowing the distillation temperature of the mixture and the vapor pressures of the pure components at that temperature, one can calculate the composition of the distillate by means of Dalton's law of partial pressures.

According to Dalton's law the total pressure ($P$) in any mixture of gases is equal to the sum of the partial pressures of the individual gaseous components ($p_A$, $p_B$, etc.). The proportion *by volume* of the two components in the

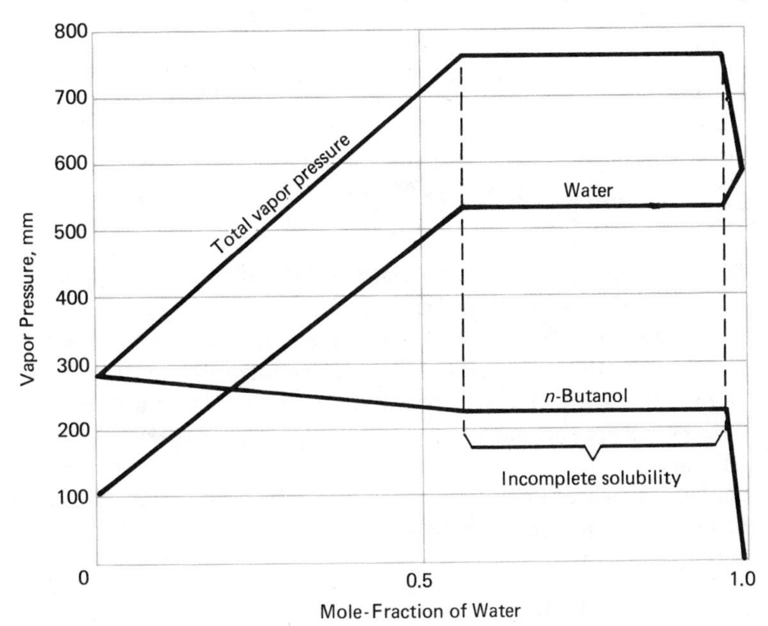


**Figure 19.** Vapor Pressure Diagram: *n*-Butanol–Water (92.7°)

distilling vapor will consequently be equal to the ratio of the partial pressures at that temperature; *the molar proportion* of the two components ($n_A$ and $n_B$) in steam distillation will be given by the relationship: $n_A/n_B = p_A/p_B$, where $p_A + p_B$ equals the atmospheric pressure. The weight proportion of the components is obtained by introducing the molecular weights ($M_A$ and $M_B$):

$$\frac{\text{Weight of } A}{\text{Weight of } B} = \frac{p_A \times M_A}{p_B \times M_B}$$

Application of this method is given in the next section for the steam distillation of bromobenzene.

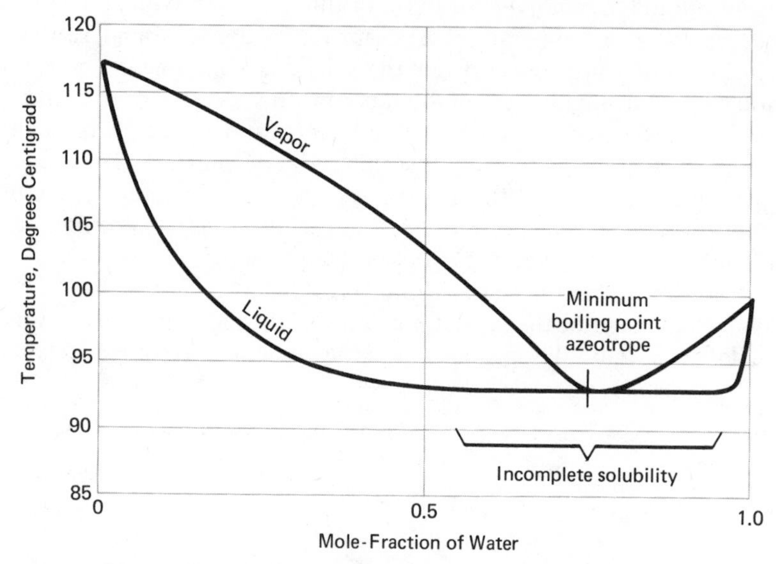


**Figure 20.** Boiling Point Diagram: *n*-Butanol–Water

**Distillation Temperature and Composition of Distillate.** The application of the general principles mentioned in the preceding text may be made clearer by considering a specific case, such as the steam distillation of bromobenzene and water. Since the sum of the individual vapor pressures (see Figure 21) attains 760 mm at 95°, the mixture will distill at this temperature. At 95° the vapor pressures are: bromobenzene, 120 mm; water, 640 mm. According to Dalton's law the vapor at 95° will be composed of molecules of bromobenzene and of water in the proportion of 120 to 640. The proportion by weight of the components can be obtained by introducing their molecular weights:

$$\frac{\text{Weight of bromobenzene}}{\text{Weight of water}} = \frac{120 \times 157}{640 \times 18} = \frac{1.63}{1.00}$$

The weight composition of the distillate will therefore be 62 percent bromobenzene and 38 percent water:

$$\text{Bromobenzene} = \left(\frac{1.63}{1.00 + 1.63} \times 100\right) = 62 \text{ percent}$$

$$\text{Water} = \left(\frac{1.00}{1.00 + 1.63} \times 100\right) = 38 \text{ percent}$$

This calculation gives the minimum amount of water in the distillate. In practice an excess of water or steam is used in the distilling flask to sweep out the vapor mixture and the system is not operated under equilibrium conditions.

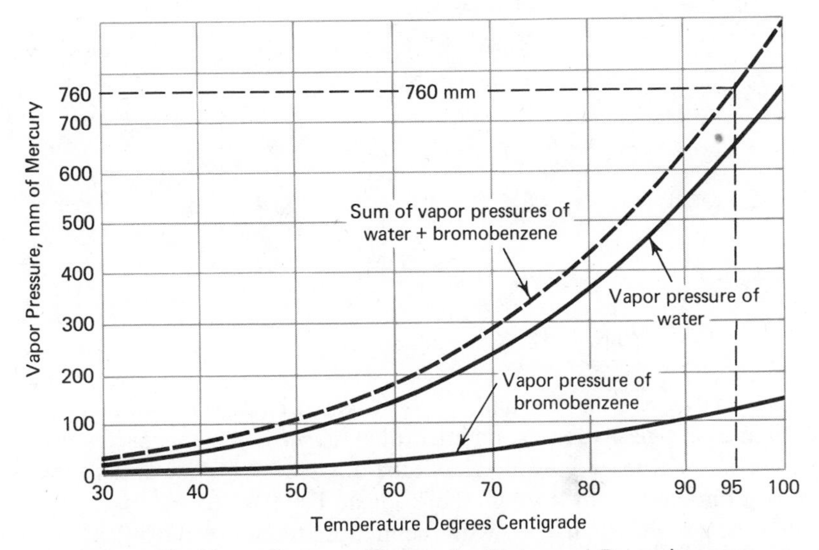


**Figure 21.** Vapor Pressure Curves for Water and Bromobenzene

It can be seen from calculations of the type illustrated (for bromobenzene) that there are several requirements for the actual practical use of steam distillation in the laboratory: the substance to be steam distilled must be insoluble, or only sparingly soluble in water; it must not be decomposed by prolonged contact with boiling water or steam; and it must have an appreciable vapor pressure (preferably, at least 5 mm) in the neighborhood of 100°. That water has a very low molecular weight (18) compared with that of typical organic compounds is a favorable circumstance for steam distillation, since this permits a substance to be steam distilled at a practical rate even though its vapor pressure is relatively small near 100°. To meet certain special conditions, laboratory methods have been devised for using steam distillation under diminished pressures and superheated steam under pressures somewhat above one atmosphere (above 15 lb per sq in).[2]

**Steam Distillation Apparatus.** A simple assembly for ordinary steam distillation in the laboratory is shown in Figure 22. A large round-bottomed flask,

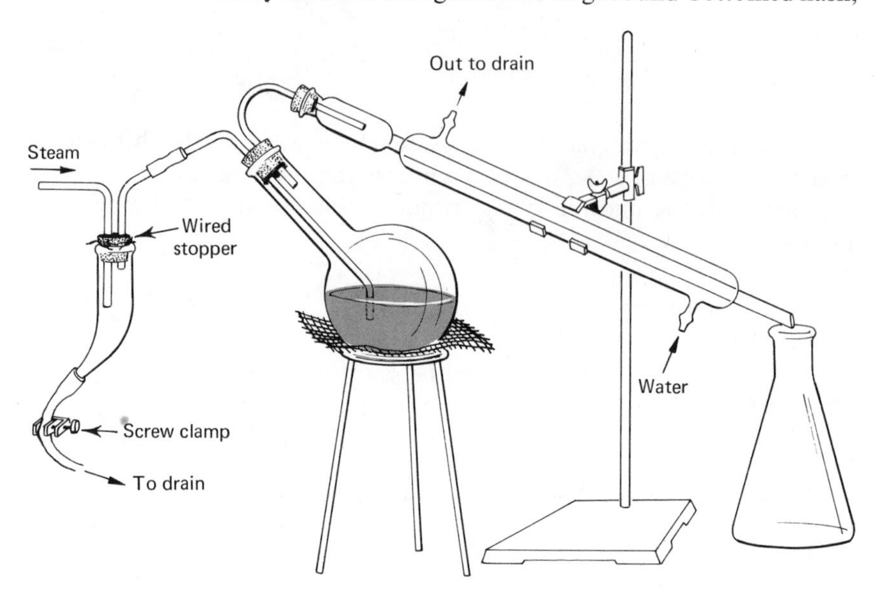


**Figure 22.** Assembly for Steam Distillation

which serves as the distillation flask, is fitted with a two-hole rubber stopper holding a bent glass tube (steam inlet tube) *reaching to within 5–10 mm of the bottom of the flask* and a wide glass tube of **U**-shape (vapor outlet tube) extending only 5–10 mm through the stopper. The flask for steam distillation should be less than half-filled with the mixture to be steam distilled as additional water will be condensed in it during the steam distillation. To reduce danger of carrying material over into the distillate mechanically as fine spray by the current of steam, the flask is tilted at an angle of 30–45° from the vertical. Owing to the mechanical action of the steam it is essential to hold the flask firmly in place with a clamp, and support it on a wire gauze on an iron ring. For steam distillation over an extended period of time, it is advantageous to warm the flask gently with a small flame at the outset and during the distillation. This avoids excessive condensation of steam in the distillation

[2] Distillation with superheated steam under moderate pressure may be used for steam distillation of a substance whose vapor pressure is very low in the vicinity of 100°. It is possible to effect steam distillation under diminished pressure but this usually offers no advantage, as it may decrease the proportion of organic component in the distillate (i.e., reduce the rate of steam distillation of the organic component).

flask from the steam used for heating the contents to boiling and from heat losses due to atmospheric cooling.

If ground-jointed apparatus is being used the assembly should consist of a round-bottomed boiling flask fitted with the Claisen adapter to which is attached the condenser set downward for distillation, and a thermometer (needed only to plug the small opening of the adapter). In the large central opening of the Claisen adapter is fitted a one-hole rubber stopper holding a straight glass tube (steam inlet tube) *reaching to within 5–10 mm of the bottom of the flask*. The flask should be less than half-filled with the mixture to be steam distilled. The Claisen adapter minimizes the probability that material will be carried over mechanically.

Steam for the distillation may be obtained by means of a simple laboratory generator[3] or from a laboratory supply line. Steam from a supply line must be passed through a trap to remove water than has condensed in the pipes and foreign matter (rust, grease, etc.) before it is permitted to enter the steam inlet tube of the distillation flask. A satisfactory trap may be made by fitting the large end of an adapter tube with a two-hole stopper bearing a steam inlet tube and an exit tube, and attaching a long rubber tube to the small end to serve as water outlet (see Figure 22). The water outlet tube is conducted to a drain or sink and is provided with a screw clamp; during the distillation the screw clamp is adjusted so that the rate at which the water is drained off is about equal to the rate at which it collects in the trap. The stopper bearing the steam inlet and exit tubes should be wired in place.

Another form of trap may be made from a 1-liter Florence flask by fitting it with a three-hole stopper holding a steam inlet tube extending 50–75 mm into the flask, a steam outlet tube extending only 5–10 mm through the stopper, and a water outlet tube reaching to within 10 mm of the bottom of the flask. The water outlet is fitted with a long rubber tube leading to a drain; a screw clamp is used to regulate the outlet during the distillation (see Figure 23).

The material to be steam distilled is placed in the distillation flask and, if water is not already present, a small quantity of water is mixed with it. A moderately rapid and steady current of steam is passed into the mixture and the steam distillate is collected. The distillation is stopped when a small test portion shows that no water-insoluble material (or only minute traces) is passing over. *Before stopping the current of steam it is essential to open wide*

[3] If steam is not available in the laboratory, it may be generated conveniently by boiling water in a 1-liter Florence flask (see Figure 23). In this case, provide the generating flask with a three-hole stopper which holds a steam outlet tube, a safety tube (50–60 cm in length) extending nearly to the bottom of the flask, and a cylindrical separatory funnel (to introduce more water if necessary). Support the generating flask on a wire gauze, introduce 500–600 ml of water, and add a few tiny boiling chips to produce regular ebullition. Heat the water rapidly to boiling and adjust the flame so as to obtain a steady current of steam. Pass the steam directly into the distillation flask without using a trap. It is advantageous to hold the stopper of the generator in place by means of soft wire.

*the screw clamp of the water outlet of the steam trap and disconnect the distillation flask*, or the contents of the latter will be sucked into the trap.

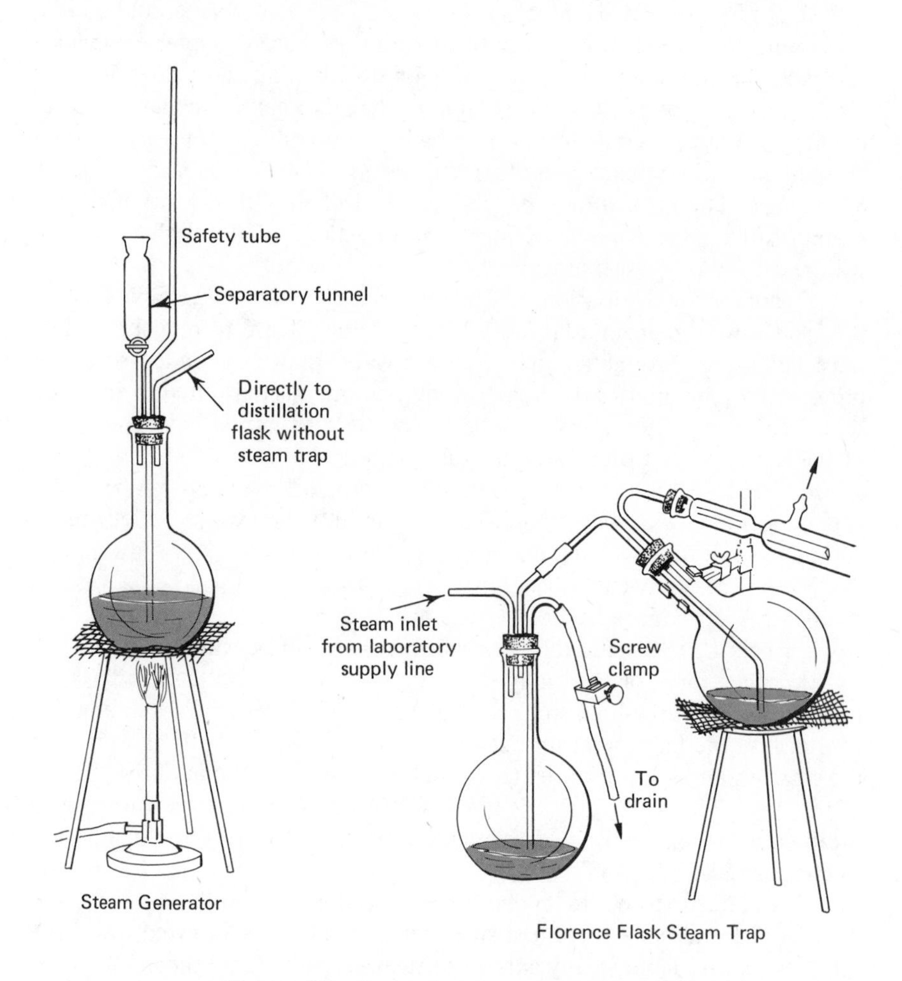


**Figure 23.** Steam Distillation Accessories

Many substances that are solids at room temperature may be steam distilled. With such materials, which may solidify in the condenser and form a mass that obstructs the tube, it is *essential* to watch the condenser tube carefully. If a mass of crystals forms, the flow of water through the jacket is stopped *temporarily* and the water allowed to drain out of the jacket. The heat from the hot distillate will cause the crystalline mass to melt and flow out into the receiver. When the tube is clear the flow of water through the jacket is started again.

# EXERCISE 4

# Steam Distillation

**(A) Steam Distillation of *o*-Chlorotoluene.**[4] Arrange an apparatus for simple distillation as shown in Figure 10, page 26, using a 125-ml distilling flask and a 100-ml graduated cylinder. In the flask, place 30 g (28 ml) of *o*-chlorotoluene (bp 159° at 760 mm) and 50 ml of water. Add two boiling chips and heat the flask on a wire gauze with a flame adjusted to give vigorous boiling. It is essential for the success of this experiment that the mixture boil rapidly with good mixing of the two phases. Discard the first 10 ml of distillate and collect the *next* 30 ml of distillate. Record the volumes of the water and *o*-chlorotoluene layers in this distillate. Compare the ratio of the volumes actually found with the ratio calculated from the ideal steam distillation law using the tabulated vapor pressures and densities.[5] Compare the observed distillation temperature with the calculated value.

**(B) Separation of a Mixture by Steam Distillation.** Arrange an apparatus for steam distillation as shown in Figure 22 using a 500-ml round-bottomed flask (preferably with a long neck)[6] as the distillation flask and a 250-ml Erlenmeyer flask as a receiver. If ground-jointed apparatus is being used, the Claisen adapter described on page 65 should be used. If steam from a laboratory supply line is to be used, arrange a suitable steam trap from an adapter (Figure 22) or a 1-liter Florence flask (Figure 23). Use a water-cooled condenser 60–70 cm in length if available.

[4] This experiment illustrates the level of agreement to be expected between predictions from the ideal steam distillation law and an actual distillation. The instructor may wish to substitute bromobenzene, *p*-chlorotoluene, or furfural; appropriate vapor pressure data have been included in the Appendix. If desired, the conventional steam distillation apparatus of Figure 22 may be used for this experiment. With ground-jointed apparatus, a 250-ml flask should be used.

[5] The vapor pressure of water and *o*-chlorotoluene are given in the Appendix. The density of *o*-chlorotoluene at 25° is 1.08.

[6] A long-necked flask such as *Kimax* boiling flask No. 25200 is suitable.

Have the setup approved by the laboratory instructor before starting the distillation. Make sure that the apparatus is supported firmly and that all stoppers and connections are tight.

Weigh out 5-g samples of *p*-dichlorobenzene and of salicylic acid, and determine the melting point of each sample.

In a clean porcelain mortar, thoroughly mix together the samples of *p*-dichlorobenzene and salicylic acid, and determine the melting point of the mixture. Transfer the mixture to the steam distillation flask, add about 50 ml of water, and pass a rapid, steady current of steam into the mixture. Continue to distill with steam until a test portion of the distillate shows that no more water-insoluble material is passing over. When the distillation is finished, open the screw clamp of the drain tube and disconnect the distillation flask *before* shutting off the steam.[7] Save the distillate and the residue in the round-bottomed flask for further examination.

➤**CAUTION:** Since the material which distills with steam may solidify in the condenser, watch carefully to avoid the formation of a crystalline mass which will completely obstruct the condenser tube. If a large crystalline mass collects in the tube, stop the flow of water through the condenser *momentarily* and drain the water from the condenser jacket. The heat from the vapors will then melt the crystals and the obstruction will be removed. As soon as this occurs, start the water again through the condenser jacket.

Before the residue in the flask has had a chance to cool, filter the solution through a fluted filter (page 87), and collect the filtrate in a clean beaker. Cool the solution rapidly to room temperature or below, and collect the crystals with suction in a Büchner funnel. Allow the crystals to dry and determine their melting point. What is this substance which did not distill with steam?

Separate the solid material in the distillate by filtering with suction, and press it as dry as possible with a clean cork or spatula. Allow the crystals to dry completely and determine their melting point. What is the substance that distilled with steam?

## Questions

**1.** Why is the steam distillation flask tipped at an angle? Why is a steam trap used?

**2.** What properties must a substance have in order that it may be practical to distill it with steam?

**3.** What are the advantages and disadvantages of steam distillation as a method of purification?

[7] If the system is not opened, condensation of steam in the trap will cause a partial vacuum and the contents of the distilling flask will be drawn into the trap.

**4.** If a mixture of bromobenzene and water were subjected to steam distillation under diminished pressure, at 100 mm pressure (see the Appendix for vapor pressure data), what would be the temperature of distillation and the weight composition of the distillate? Compare the results with the composition at 760 mm, given on page 63.

**5.** Calculate the distillation temperature and the theoretical weight composition of the distillate for the steam distillation of *p*-dibromobenzene at 760 mm. How would a mixture of bromobenzene and *p*-dibromobenzene behave when subjected to steam distillation? (See Experiment 19.)

# Melting Points

The true melting point of a crystalline substance is defined as the temperature at which the solid phase exists in equilibrium with the liquid phase. At the melting point the vapor pressure of the solid phase equals the vapor pressure of the liquid phase (*melt*). On account of the relatively large samples that are required for such determinations, the melting points usually observed in the organic laboratory (as well as most of the melting points recorded in handbooks and in the chemical literature) are not true but capillary melting points. These values are comparable among themselves but are usually slightly higher than true melting points.

For the determination of the melting point of a crystalline substance, a small amount of the finely powdered crystals is introduced into a thin-walled capillary tube, which is fastened to a thermometer, and the tube and thermometer are heated in a well-stirred bath of paraffin oil or glycerol (or other suitable liquid). The temperatures recorded as the observed melting point are the temperatures at which the substance *begins* to liquefy and that at which it becomes *completely* liquefied. The melting point range (the interval between the beginning of liquefaction and complete liquefaction) and the actual temperature of the melting point are valuable indications of the purity of the substance.

A pure crystalline organic substance usually possesses a sharp melting point; that is, melts completely over a very short temperature range (in practice, not more than 0.5–1.0°, provided good technique is used). The melting point range is influenced not merely by the purity of the material but also by factors such as the size of the crystals, the amount of material, rate of heating, etc. A finite time is necessary for the transfer of heat from the hot bath liquid, through the walls of the capillary tube and throughout the mass of the sample. If the bath liquid is heated too rapidly the temperature of the bath will rise several degrees during the time lag required for the melting process to occur and this results in an observed range much larger than the true one; likewise, too rapid heating will cause the thermometer reading to be lower than the actual bath temperature, on account of the time required for heat transfer to the mercury. Consequently, the observed temperature will differ from the actual temperature of melting as a result of differences between the time lag of melting and the time lag of the mercury column of the thermometer.[1]

To obtain satisfactory results it is essential in the vicinity of the melting point to *heat the melting point bath slowly at a uniform rate* (about 2° rise per minute for ordinary work). To insure a minimum lag in the melting process and heat transfer, it is necessary to use a small amount of material, powdered finely and packed densely, and to employ a fine capillary tube having a small diameter and very thin walls. In practice the minimum diameter of the melting point tube is limited by the difficulty of introducing the sample into it. The height of the column of tightly packed material in the tube should be just sufficient to permit adequate observation of the behavior on melting; usually a height of 3–5 mm is satisfactory.

The presence of even small amounts of miscible, or at least partially miscible, impurities will usually produce a marked increase in the melting point range and a depression of the actual temperature at which fusion begins to occur. These facts are made use of in establishing the identity of organic compounds by determining the melting points of mixtures, a method commonly known as "mixed melting points." The lowering of the melting point of a pure solid substance on admixture with an impurity soluble in the melt results from the lowering of the partial vapor pressure of the pure substance in the melt. For ideal mixtures the lowering of the melting point (or depression of the freezing point) is proportional to the amount of soluble impurity present. In practice, this relationship holds only for very small amounts of impurity, and many deviations are encountered (see next paragraph). Insoluble impurities do not affect the actual melting point but their presence will result in a melt that is not clear.

A wide melting point range does not necessarily indicate the presence of an impurity in the original sample, but may result from the circumstance that the pure substance undergoes some decomposition before the melting point is reached. In some instances the material undergoes a slight liquefaction and contraction (sintering) at a temperature below the true melting point; in others the material may decompose completely and become so badly discolored that a definite melting point cannot be observed.[2] To interpret the

[1] Two other sources of error in thermometer readings are atmospheric cooling of the mercury in the stem and incorrect calibration of the thermometer scale (see page 30).

[2] For substances that undergo decomposition on heating, devices such as the Maquenne block or Dennis-Shelton melting point bar [*J. Amer. Chem. Soc.*, **52**, 3128 (1930)] may be used. The finely powdered material is dusted in small portions onto a heated metal surface and the temperature noted at which the fine particles melt instantly on striking the hot metal.

results in a proper way, the behavior on melting should be observed closely and recorded in the notebook: for example, "melts sharply at 113.0–113.5°"; "mp 175–177°, with decomposition"; "discolors slightly at 85°, melts slowly at 87–89°"; "mp 75–75.5° (sharp), material resolidified quickly and did not remelt until bath was raised to 145–147°."[3]

A sharp melting point is usually an indication of high purity of a crystalline substance, but it is not necessarily so. Eutectic mixtures and other mixtures of organic substances, particularly of homologs and of isomers, possess rather sharp melting points which may remain practically constant after repeated crystallizations from a solvent. A more reliable indication is obtained by fractional crystallization from two or more widely different types of solvents, and determination of the melting points of the various fractions. In the case of a pure compound the melting points of all of the fractions will remain the same.

**Melting Point Apparatus.** A melting point tube (actual size), packed with a sample, and a typical laboratory assembly for determining melting points are illustrated in Figure 24. A small beaker (100-ml capacity) is fitted with a stirring rod prepared by bending a long glass rod into a circle that can be moved freely up and down between the wall of the beaker and the thermometer. The rod is bent at right angles to parallel the wall of the beaker and to pass over the top of the beaker. The handle projecting from the top should be long enough to avoid burning the fingers when the bath is being heated, and the ends should be fire-polished. The thermometer is inserted in a 20–25-mm cork, into a properly bored hole of the correct size, and supported by means of a burette clamp. The position of the thermometer is adjusted so that it is well centered vertically, *with the lower end of the mercury bulb about 1 cm above the bottom of the beaker.*

Capillary melting point tubes are prepared by drawing out a test tube, or a piece of soft glass tubing, about 10–15 mm in diameter. The tube is heated in a blast lamp or in the hottest flame of a Bunsen burner, and rotated slowly, without drawing it out, until the glass has softened thoroughly and the hot portion begins to droop. The tube is then withdrawn from the flame and the ends are drawn apart in a straight horizontal line, evenly and not too rapidly, until the hot section has resolidified and its diameter reduced to 1–2 mm. After a little practice, a length of thin tubing sufficient for ten or more melting point tubes will be secured in one operation. The thin tubing is cut into sections 60–80 mm in length, by making a small scratch with a file

[3] The last observation illustrates an unusual one that can arise with a pure individual that exhibits polymorphism, and one of the allotropic forms undergoes transition to a higher-melting form at its melting point. Other abnormal behavior may result from slow and gradual allotropic change without indication other than sintering, evolution of gas arising from decomposition, or evolution of a highly volatile solvent (water, ethanol, benzene, etc.) from a crystalline solvate or from traces of solvent held mechanically, and other phenomena.

and pressing gently. One end of the tube is sealed by touching it to the edge of a small hot flame; no attempt is made to fire-polish the other end as this is unnecessary and is extremely difficult to accomplish without constricting or sealing the opening. It is convenient to prepare a good supply of melting point tubes at one time and store them in a clean, dry test tube fitted with a clean cork.

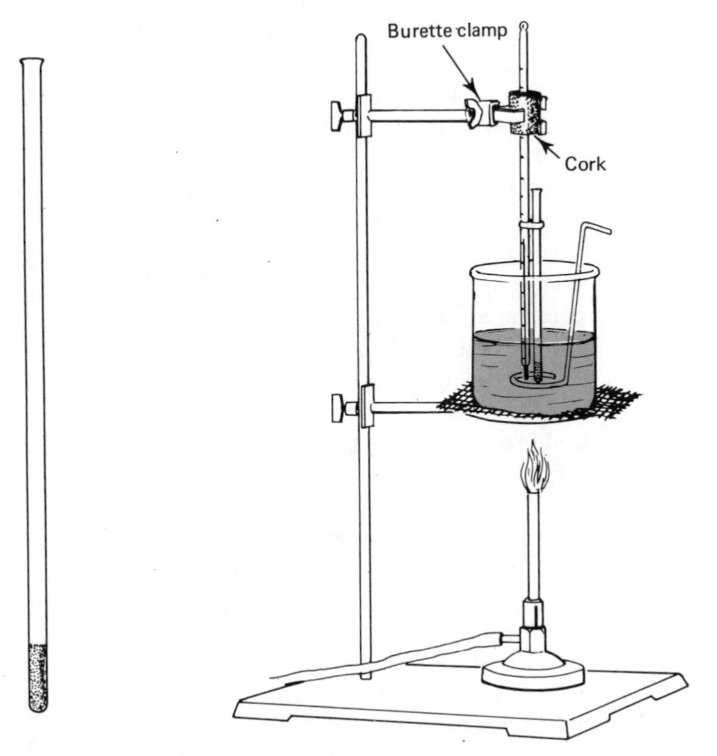


**Figure 24.** Melting Point Apparatus

A small amount of the material to be examined (0.1 g is ample) is pulverized finely by crushing with a clean spatula[4] or knife blade, on a piece of smooth hard paper or a watch glass. The crushed material is collected into a small mound and the open end of the melting point tube thrust into it. The solid is shaken down into the tube by drawing a triangular file *gently* along the upper portion of the tube and then tapping the lower end on the desk top; or, it may be forced down by dropping the tube (sealed end downward)

[4] For this and other manipulations of solid organic substances a small metal spatula is convenient. A satisfactory spatula can be made from a 12–15 cm length of thick wire of nickel or silver by flattening the ends with a hammer to a width of 5–6 mm. The ends are shaped by means of a file or buffer, and one end may be bent to a 45° angle about 12–15 mm from the tip. Stainless steel wire makes an excellent spatula but is much more difficult to flatten.

through a length of ordinary glass tubing onto the desk top. Further increments of the sample are introduced in the same way, until the material forms a compact column 3–5 mm high at the bottom of the tube after repeated tapping. It is essential that the material be pulverized finely and packed tightly.

Although the capillary tube will usually adhere to the thermometer by capillary action of a thin film of bath liquid, it is advisable to attach it more firmly by means of a thin slice of rubber tubing or a small rubber band. The tube should be adjusted so that the sample is placed just alongside the mercury bulb of the thermometer and the rubber fastening should be above the level of the bath liquid (to avoid softening of the rubber and discoloration of the bath).

When the apparatus has been arranged properly the beaker is removed temporarily and filled to *not more than two-thirds its capacity* with a suitable bath liquid.[5] A larger amount of liquid must not be used as the expansion on heating may lead to overflowing or to spattering during stirring.

The beaker is replaced on the wire gauze (or asbestos board with small circular opening) and heated with a small flame, while the bath liquid is stirred with the curved stirring rod. One may heat at a fairly rapid rate until the bath temperature approaches to within about 15–20° of the melting point (roughly determined in a preliminary trial, if necessary). Heating is then continued with a very small flame adjusted so that the temperature rises *slowly and regularly*, at a rate of about 2° per minute. Stirring is continued throughout, and the sample and temperature are observed carefully. The observed melting point is reported as the temperature range beginning with the thermometer reading when the substance starts to liquefy and ending with the reading when the melt becomes clear. The temperature readings and accessory observations are recorded at once in the notebook.

The thermometer used for melting points should be one that has been standardized by one of the methods described earlier (page 31). The observed readings are subject also to a correction for the exposure of the mercury column to atmospheric cooling. Often the stem correction is not applied; one may indicate whether or not this has been done by a notation such as, "mp 172–173° (uncor)" or "mp 172–173° (cor)."

The bath liquid is not discarded after each use; the bath and stirrer are stored in the desk (after cooling and covering with a watch glass) for subsequent use, since melting points are determined frequently during the course

[5] Liquids suitable for bath temperatures up to about 250° are: medicinal paraffin oil ("Nujol," etc.), anhydrous glycerol, di-*n*-butyl phthalate, medicinal castor oil, Monsanto "Arochlor #1242" (technical mixtures of chlorinated biphenyl), Dow-Corning "Silicone fluid #500" (mixtures of organosilicon compounds). Commercial tristearin or hydrogenated fatty oils ("Crisco," "Spry," etc.) are quite satisfactory, although the bath material forms a soft greasy solid at room temperature. Concentrated sulfuric acid is unsafe for use in an open type of bath. For temperatures below 100°, water may be used for the bath liquid.

of ordinary laboratory work. Melting point tubes are discarded after a single use.

Other convenient forms of melting point baths and other types of melting point apparatus have been devised for ordinary work and also for specific purposes. Mechanical agitation of the bath liquid may be provided by means of a small electrical or air-driven stirrer, i.e., the Hershberg apparatus. The Thiele apparatus is designed so that the bath liquid is caused to circulate by convection currents. At temperatures above 220–250° the usual bath liquids are unsuitable because of smoking or decomposition at the high temperatures. For the higher temperature ranges a metal block with a vertical hole for the thermometer and capillary tube, and a small window for observation of the sample may be used.

A number of commercial devices are available for the determination of melting points. They are electrically heated and the heat input is controlled by adjustment of a rheostat or Variac. Some are built around a silicone oil

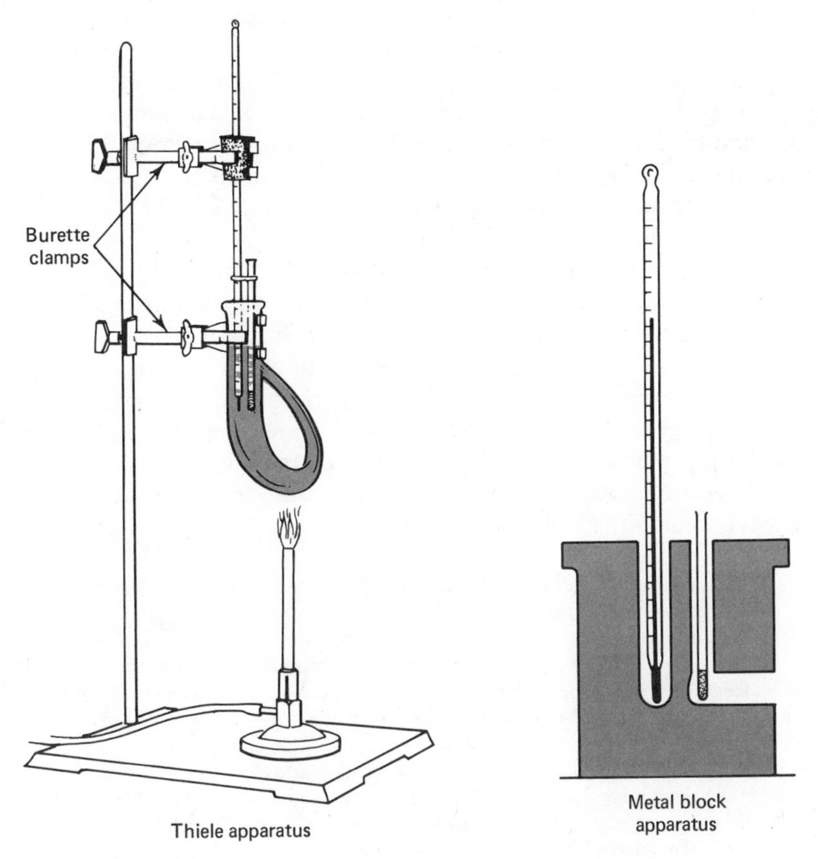


**Figure 25.** Various Forms of Melting Point Apparatus

bath; others use some variation of the heated metal block. Most are equipped with some optical magnifying device for easier viewing of the sample. These commercial instruments are quite expensive (about $100 and up) and do not give a more precise melting point than the simple apparatus illustrated in Figure 24 (some are *less* precise).

## Melting Points of Mixtures

The melting points of mixtures (mixed melting points) are used frequently to verify the identity of the product of a known reaction, to assist in the identification of an unknown substance, to establish the identity of two samples of material isolated from different sources or prepared by different methods, and to deal with other problems connected with characterization of solid organic compounds.

A typical illustration of this principle is afforded by the situation where a substance to be identified, designated by *X*, is suspected of being identical with one or the other of two known substances, *A* or *B*, and where the three pure compounds, *A* and *B* and *X*, have approximately the same melting point. Mixtures of about equal amounts of *A* and *X*, and of *B* and *X*, are prepared and the melting points of these mixtures are determined.[6] If *A* and *X* are identical, the melting point of a mixture of *A* and *X* in any proportions will always be the same as that of *A* and *X*, apart from any slight differences resulting from different degrees of purity of the samples of *A* and *X*. If *A* and *X* are different from one another, the melting point of the mixture of *A* and *X* will usually, but not invariably, be lower than that of either *A* or *X*. Similar reasoning is applied to the mixture of *B* and *X*. If the melting points of the mixtures of *A* and *X*, and of *B* and *X*, are both below the original melting points of the pure substances, one can safely conclude that *X* is not identical with either *A* or *B*. There are some mixtures that are exceptions to the general rule that the melting point of a pure compound will be lowered by the presence of a soluble impurity or by admixture with a different substance (see the next paragraph). Consequently, if the melting point of *X* is not depressed

[6] In practice it is highly desirable to determine the melting points of the mixtures and of a control sample of pure *X*, in a single operation, so that the rate of heating will be the same for the three samples. To avoid confusion the individual tubes are given some means of identification (by making one, two, and three small file scratches at the upper end, or cutting the tubes to three different lengths, etc.) and arranged in a definite manner on the thermometer. The symbols and arrangement should be jotted down in the notebook: for example, "*X* (medium tube, center)"; "*A* + *X* (long tube, at left)"; "*B* + *X* (short tube, at right)"; etc. It is difficult to observe carefully the melting point behavior of more than three samples and to avoid mistakes in the identity of the melting point tubes. In careful work a small pocket magnifying glass is used to observe closely the behavior of the samples on melting and to read the thermometer more accurately. Differences in the behavior of the samples are more significant than the actual temperature readings.

by admixture with *A* (or with *B*), one can conclude that *X* is probably identical with *A* (or with *B*) but one cannot assert with absolute certainty that *X* is identical with *A* (or with *B*). Comparisons of other physical constants of the two materials would be necessary to establish an identity beyond any reasonable doubt.

In practice it is often advisable to prepare a number of mixtures which contain varying proportions of the substances to be identified and the known substance; for example, a mixture containing 10 percent of *A* and 90 percent of *X*, another containing 50 percent of each, and a third mixture containing 90 percent of *A* and 10 percent of *X*. The results of the melting point determinations of these mixtures may then be plotted and a rough melting point curve may be drawn. For this purpose the temperature plotted is that at which the mixture liquefies completely. The nature of the melting point curves which are thus obtained is shown in Figure 26.

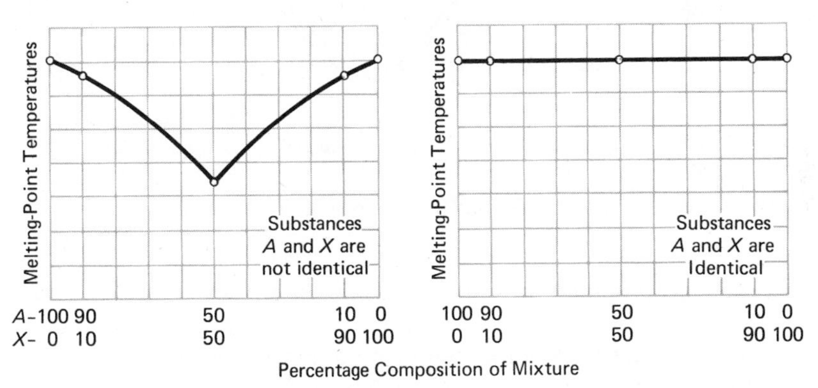


**Figure 26.** Melting Point Behavior of Mixtures

Actually, the melting points of only three mixtures would not give sufficient data to reveal the details of the whole curve. Complete melting point curves (also called freezing point diagrams) of binary mixtures illustrating the varied types of curve that may be encountered in practice are shown in Figure 27. When the molecular weights of the components are known it is advantageous to plot the compositions in mole percentages[7] rather than by weight.

**Formation of Eutectic Mixtures.** The behavior of naphthalene-biphenyl mixtures is typical of many organic compounds. The two components form

[7] Conversion of weight compositions to a molar basis is made as follows:

$$\text{Mole fraction of } A = \frac{\text{Wt percent of } A \div \text{mol wt of } A}{(\text{Wt percent of } A \div \text{mol wt of } A) + (\text{Wt percent of } B \div \text{mol wt of } B)}$$

$$\text{Mole percent of } A = \text{Mole fraction of } A \times 100 \text{ percent}$$

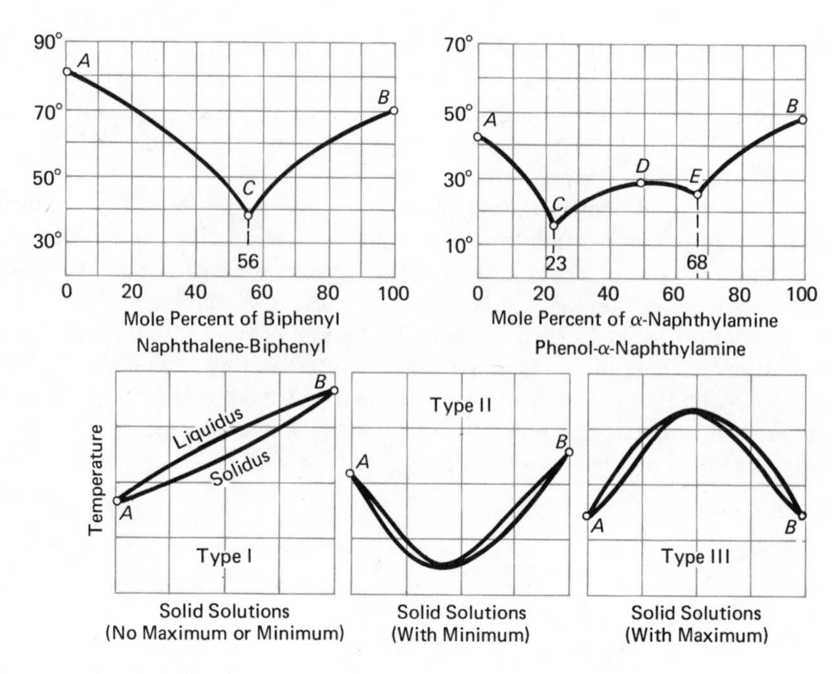


**Figure 27.** Melting Point–Composition Diagrams for Mixtures of Various Types

practically ideal solutions over the entire range (i.e., the freezing point lowering is proportional to the mole-fraction of the solute present). The curve *A–C* shows the lowering of the melting point of naphthalene (*A*, mp 80°) by admixture with biphenyl, and *B–C* the lowering of the melting point of biphenyl (*B*, mp 69°) by naphthalene. The intersection *C* is a eutectic point (39.4°) at which temperature *both* crystalline solids *A* and *B* can exist in equilibrium with a melt of fixed composition (44 mole percent naphthalene + 56 mole percent biphenyl).

**Compound Formation.** The phenol-α-naphthylamine curve illustrates a case in which the components *A* and *B* form a compound, of the composition *AB*. The addition of α-naphthylamine to phenol (*A*, mp 41°) lowers the melting point until the point *C* is reached; further addition then raises the melting point until *D* is reached. A similar phenomenon occurs upon addition of phenol to α-naphthylamine (*B*, mp 49°) until the point *D* is reached. This indicates the formation of a compound of the composition *AB* (mp 28.4°). Point *C* represents a eutectic of *A* and the compound *AB*, and point *E* a eutectic of *B* and the compound *AB*. It should be noted that the point *D* (mp of the compound formed) may be higher than *A* or *B*. The abscissa of *D* gives the composition of the compound formed; occasionally several compounds of *A* and *B* are formed, having different compositions ($A_2B$, $AB$, $AB_2$, etc.). The compound formed between phenol and α-naphthylamine is an example of proton-bonding between amines and phenols.

**Formation of Solid Solutions.** In the preceding examples the individual components (*A* or *B* or *AB*) form separate crystals in solidifying from the melt. In other cases a homogeneous solid solution (mixed crystals) of *A* and *B* (or *A* and *AB*, etc.) is formed. Three different types of curve are known for solid solutions, when the components are miscible in all proportions. Each of these diagrams is made up of two parts, one for the composition of the liquid solutions (liquidus curve) and another for the solid solutions (solidus).

In one type all mixtures of *A* and *B* melt at temperatures intermediate between the melting points of pure *A* and pure *B*, and there is neither a minimum nor maximum in the curve (Type I, Figure 27). This situation illustrates a case where the melting point of pure *A* is *raised* by admixture with *B*, but that of pure *B* is lowered by addition of *A*. The second type of curve (Type II, Figure 27) has a minimum, and the melting point of either *A* or *B* is lowered by admixture with the other. The third type of curve (Type III, Figure 27) passes through a maximum, and here the melting point of either *A* or *B* is *raised* by adding the other to it. This unusual form of curve is encountered occasionally with mixtures of optical isomers.

EXERCISE

# 5

# Determination of Melting Points

Arrange an assembly for determining melting points as illustrated in Figure 24 (page 73). For the 100-ml beaker that will serve to hold the bath liquid, provide a properly bent stirring rod having a long handle. Place the beaker on a wire gauze supported on a ring fastened to an iron stand. Mount a previously standardized thermometer (see page 31) in a 20–25-mm cork and support it by means of a burette clamp so that the thermometer is held vertically above the center of the beaker, with the lower end of the mercury bulb about 1 cm above the bottom of the beaker. Prepare a supply of at least 10–15 small capillary tubes sealed at one end (melting point tubes) by the method described on page 72.

When the apparatus has been arranged properly introduce about 60 ml of glycerol or paraffin oil (or other suitable bath liquid)[8] into the beaker. *The beaker must not be more than two-thirds filled with the bath liquid* at room temperature, as the liquid will expand appreciably in volume when heated. Have the setup and melting point tubes approved by the laboratory instructor before proceeding.

**(A) Melting Points of Pure Substances.**[9] Fill the lower end of a melting point tube with a tightly packed column (3–5 mm high) of finely pulverized benzoic acid, by the procedure outlined on page 73. Fill a second tube in a similar way with 2-naphthol. It is essential that the material be packed firmly and densely into the end of each tube. Mark the tubes with one and two small file scratches, respectively, so that they can be distinguished from one another (see footnote 6, page 76), attach the tubes to the thermometer by means of

[8] For a list of various bath liquids, see footnote 5, page 74.

[9] Urea and cinnamic acid may be used instead of benzoic acid and 2-naphthol in parts (A) and (B) of this experiment. The latter pair gives a more typical behavior in part (B) and appears to be more satisfactory in the hands of beginning students.

a thin ring of rubber tubing or small rubber band, and determine the melting points of the substances according to the procedure outlined on page 74.

Heat the beaker with a small flame and stir the bath liquid with the curved stirring rod. Apply heat at a moderately rapid rate until the bath liquid is within 15–20° of the melting point (in this instance to about 100°; when necessary, make a rough preliminary determination). Continue the heating with a *very small flame* adjusted so that the temperature rises slowly and at a uniform rate (about 2° per minute). Continue to stir the bath liquid, and observe carefully the samples in the melting point tubes and the thermometer reading. Record as the observed melting point the range between the thermometer reading when the sample starts to liquefy and that when the melt is clear. Note also whether the sample undergoes preliminary sintering or discoloration, melts sharply or slowly over a wide range, etc.

After the samples have melted, extinguish the flame and allow the bath to cool. The beaker with stirrer and bath liquid are retained for subsequent use in this experiment and throughout the term's work. Melting point tubes are discarded into the waste crock (not into the sink) after a single use.

**(B) Melting Points of Mixtures.** Prepare three different mixtures of benzoic acid and 2-naphthol having roughly the following proportions of the two components: (1) about 90 percent benzoic acid and 10 percent 2-naphthol; (2) about 50 percent of each; (3) about 10 percent benzoic acid and 90 percent 2-naphthol. For the purpose of this experiment it is permissible to judge the proportions very roughly by the relative lengths of long thin piles of the powdered crystals (e.g., 9 parts benzoic acid and 1 part 2-naphthol for mixture No. 1, etc.). The components must be powdered finely and each mixture mixed *very thoroughly*, on a piece of smooth hard paper or a watch glass, by means of a clean spatula or knife blade.

Introduce the mixture into melting point tubes marked appropriately for identifying them (see footnote, page 76) and determine the melting points of the mixtures by the same procedure used in part (A). Record the observed melting points directly in the notebook. From your results draw a rough melting point curve for mixtures of benzoic acid and 2-naphthol, plotting compositions (preferably in mole percentages)[10] on the $x$-axis and melting points on the $y$-axis. For this purpose the temperature taken as ordinate is that at which the mixture liquefies completely.

## Questions

**1.** Define accurately the term melting point.

**2.** What general conclusions can be drawn from the determination of melting points of mixtures?

[10] For conversion of weight compositions to mole percentages see footnote on page 77.

**3.** Suppose that two different organic compounds, $M$ and $N$, have about the same melting point, and that you are given an unknown organic compound, $X$, which also has the same melting point and is suspected to be identical with either $M$ or $N$. Describe a procedure for identifying $X$ and state the results you would obtain, in each of the following situations: $X$ is identical with $M$; $X$ is identical with $N$; $X$ is not identical with either $M$ or $N$.

**4.** What physical constants other than melting point and melting points of mixtures may be used to aid in establishing the identity of an organic solid? What constant other than boiling point may be used for an organic liquid?

**5.** Why is it essential: (a) to pack the sample densely and tightly in the melting point tube? (b) to heat the bath slowly and steadily in the vicinity of the melting point?

# Crystallization and Sublimation

In a typical laboratory preparation, a crystalline solid separating from a reaction mixture is usually contaminated with small amounts of impurities. Purification is generally accomplished by crystallization from an appropriate solvent. The procedure consists essentially of the following steps: (a) dissolving the substance in the solvent at the boiling point; (b) filtering the hot solution to remove insoluble impurities; (c) allowing the hot solution to cool and deposit crystals of the substance; (d) separating the crystals from the supernatant solution (the mother liquor); (e) washing the crystals to remove adhering mother liquor; (f) drying the crystals to remove the last traces of solvent.

**TABLE 9**
**Common Solvents for Crystallization**

| SOLVENT | BP, °C | SOLVENT | BP, °C |
|---|---|---|---|
| Water | 100 | Methylene chloride | 40 |
| *Methanol | 64 | Chloroform | 61 |
| *Ethanol | 78 | Carbon tetrachloride | 76 |
| **n*-Butyl alcohol | 117 | *Benzene | 80 |
| *Diethyl ether | 34 | *Petroleum ether | 20–90 |
| *Acetone | 56 | *Ligroin | 90–150 |
| *Ethyl acetate | 78 | *Acetic acid (glacial) | 118 |

* Flammable liquid—requires caution against fire hazards.

The selection of a suitable solvent for crystallization is of great practical importance. A good solvent for crystallization is one that will dissolve a moderate quantity of the substance at an elevated temperature but only a small quantity at low temperatures. The solvent should dissolve the impurities readily, even at low temperatures, and should be capable of being removed easily from the crystals of the purified substance. It is essential that the solvent shall not react in any way with the substance to be purified. Other factors such as ease of manipulation, flammability, cost, etc., should also be taken into consideration.

In selecting a solvent for the purification of a given substance, one should consider its effectiveness for removal of the particular impurities that are likely to be present. The following general categories of impurities may be encountered.

*Mechanical impurities* consisting of dust, grit, particles of paper, cork, etc., are readily removed by filtering the hot solution since they are insoluble in all of the common solvents. Admixed inorganic salts may often be separated in this way by using an organic solvent in which they are insoluble; an alternative method is to wash the crystals before recrystallization with a solvent such as water, in which the inorganic salts are soluble and the organic compound is insoluble.

*Traces of coloring matter and resinous impurities* may often be removed by warming the solution with a *small* amount of decolorizing carbon or other adsorbent (Norit, Darco, Nuchar, etc.) before filtering the hot solution. The action of decolorizing agents varies widely and the effectiveness in removing a particular impurity may differ markedly from one solvent to another. An excessive amount of decolorizing agent must be avoided since it will frequently also adsorb the compound which is being purified.

*Impurities more soluble in the solvent* are readily removed by crystallization since they will be retained in the mother liquor. Likewise, impurities having about the same solubility as the substance being purified, when present in moderately small amounts, are readily eliminated in the mother liquor.

*Impurities less soluble in the solvent* are very difficult to remove if they are present in considerable amount, since the hot solvent will dissolve an appreciable amount of the impurity and, on cooling, the impurity will crystallize out and contaminate the product. It is for this reason that one tries to select a solvent which will dissolve the impurities readily, even at room temperature.

In many instances information is available concerning the solvents suitable for crystallizing the substance which is to be purified. If such information is lacking, the common solvents may be tested experimentally, on a small scale, to select a satisfactory one.[1] This is done conveniently by using a series of *small* test tubes, and placing in each tube a small quantity of the substance to be purified and a small quantity of the solvent to be tested. The action of the solvent is tested in the cold and at the boiling point, and one notes whether well-formed crystals are produced abundantly on allowing the hot solution

[1] For hints on general solubility relationships, see page 104.

to cool. If two or more solvents appear to be promising from the preliminary tests, each of these may be tested more thoroughly with larger, weighed samples of material, to determine the loss in the crystallization process and to compare the purity of the recrystallized samples.

To secure a satisfactory recovery of the purified material it is essential to avoid using an unnecessarily large amount of solvent. The quantity of the substance lost through retention in the mother liquor will be minimized if the sample is dissolved in the smallest possible amount of the hot solvent. In practice it is desirable to employ *slightly more* (2–5 percent more) *than the minimum quantity of hot solvent* required to dissolve the sample, so that the hot solution will be not quite saturated with the solute. This aids in avoiding separation of crystals, as a result of slight cooling during filtration of the hot solution, which may clog the filter and funnel. With a substance that melts at a temperature below the boiling point of the solvent, enough of the solvent should be used to allow the solution to cool to a temperature below the melting point of the dissolved substance before the latter separates out; otherwise, the material will separate at first in oily droplets instead of well-formed crystals.

During the preparation of the solution in the hot solvent the liquid should be stirred or shaken to aid the solution process, as many organic substances dissolve quite slowly; an adequate time and good mixing are needed, and it is advantageous to crush any large crystals or lumps of the sample beforehand. One should avoid the use of an excessive amount of solvent through being hasty or through attempting to bring insoluble foreign matter into solution. It is better to err on the side of an insufficient amount of solvent—the undissolved residue remaining after filtration or decantation of the hot solution may then be tested with a fresh portion of solvent to see if more of it will dissolve.

Occasionally a mixture of two solvents (solvent-pair) is more satisfactory than a single solvent. Such solvent-pairs are made up of two mutually soluble liquids, one of which dissolves the substance readily and another which dissolves it very sparingly. Examples of solvent-pairs are ethanol and water, glacial acetic acid and water, diethyl ether and petroleum ether, and benzene and ethyl acetate. A typical procedure in using a solvent-pair, such as ethanol and water, is to dissolve the substance in ethanol at the boiling point and add water dropwise, with shaking or stirring, until a faint turbidity persists in the hot solution. A few drops of ethanol are then added, slightly more than the minimum amount required to form a clear solution, so that the hot solvent mixture will not be quite saturated with the solute. The hot solution is treated subsequently in the usual way (clarified with carbon, filtered, etc.).

Filtration of the hot solution to remove insoluble impurities, decolorizing carbon, etc., must be carried out rapidly and efficiently in order to avoid crystallization of the dissolved substance in the filter and funnel. For this purpose, a large fluted filter paper (see Figure 28) may be used to obtain rapid filtration, and a large funnel with a very short wide stem (or a stemless funnel) is employed to avoid clogging due to separation of crystals in the stem. Just before the filtration is started the funnel should be preheated by warming gently over a small flame (or in a jet of steam) or by placing it in the mouth of the flask in which the hot solution is being prepared. Suction filtration by means of a Büchner funnel[2] may be used for hot filtrations, particularly if a large volume of liquid is to be filtered or a large quantity of insoluble foreign material is present. Suction filtration of hot solutions requires careful manipulation to avoid clogging the pores of the filter plate and stem of the funnel as a result of rapid evaporation of the solvent under the diminished pressure in the apparatus. The necessary manipulative skill can be acquired after a little practice and experience.

The following procedure is suitable for suction filtration of hot solutions: The Büchner funnel is checked first to see that all parts are perfectly clean (the inner portion can be reached with a bent test tube brush) and that the holes in the perforated plate are not plugged (these may be cleaned with a pin or stiff wire). The funnel is preheated by pouring portions of the boiling solvent through it, dried quickly with a clean towel, and fitted with a filter paper of the proper size. For aqueous solutions the funnel may be preheated in a jet of steam; if steam is used to preheat the funnel for filtration of a non-aqueous solution, the inner portion must be dried by washing with a little ethanol or acetone. The hot funnel is fitted to a suction filtering flask and the filter paper moistened with a small portion of the hot solvent. Suction is applied by connecting the side tube to a water pump operating at the full water pressure, and the hot solution is poured quickly onto the filter. When filtration has started, the amount of suction applied is moderated by pinching the rubber tube leading to the water pump. As filtration progresses more of the hot solution is poured into the funnel, which must not be permitted to go dry until the operation is completed. During the filtration the suction is controlled by intermittently pinching the rubber tube with a rubber tubing clamp (*not* by changing the flow of water through the pump) so that rapid filtration is maintained without allowing the solvent to boil off due to the diminished pressure in the filtering flask (see page 32). If filtration is not rapid, cooling takes place and the material crystallizes in the inner portion of the funnel, causing clogging and loss of product. After the filtration has been completed the filtrate is transferred to a flask or covered beaker and allowed to cool; the Büchner funnel is cleaned thoroughly and readied for use in filtering the

[2] The usual Büchner funnel is a porcelain funnel provided with a fixed perforated plate, on which a circle of filter paper is placed flat without folding. The funnel is fitted by means of a rubber stopper into a suction filtering flask having a side tube for attachment to a water suction pump (Figure 30). Another type of Büchner funnel is made of glass and fitted with a porous plate of sintered glass.

The style of funnel was invented by Ernst Büchner (1888). He is often confused with Eduard Buchner who is well known for his work with yeast enzymes (Nobel Prize 1907).

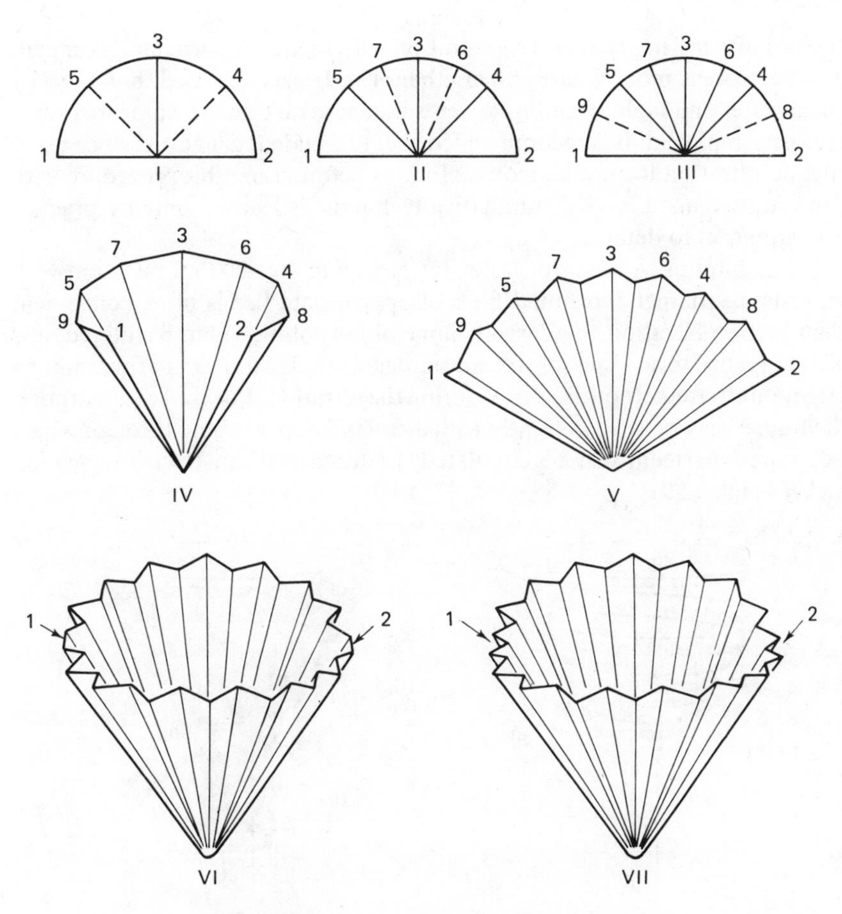


**Figure 28.** Preparation of a Fluted Filter

Select a filter paper and glass funnel of such dimensions that the top of the folded paper is 5–10 mm *below* the upper rim of the funnel (120 mm diam. paper for a *65*-mm funnel, 185 mm paper for a 100–120 mm funnel, etc.). Fold the paper in half, and again into quarters (part I, above). Bring each edge *into* the center fold and crease again, producing new folds at *4* and *5*. Do not crease the folds tightly at the center—this would weaken the central portion so that it might break during filtration. Continue the series of *inward* folds, bringing *2* into *5*, *1* into *4*, *2* into *4*, *1* into *9* (II and III).

Grasp the folded filter cone (IV) in the left hand and make a new fold in each segment, between *2* and *8*, *8* and *4*, etc., *in a direction opposite to the first series*. The result is a bellows or fan arrangement (V). Open the filter paper completely and note the two places *1* and *2* (VI) where the paper would lie flat against the side of the funnel. Fold each of these sections in half, forming two smaller flutings, only half as deep as the others. Strengthen all of the flutings by creasing lightly a second time, and the paper is ready for use.

crystals of purified product. Hot filtration with suction is particularly difficult when a solvent more volatile than ethanol or benzene is used, but is easier when water and higher-boiling solvents (acetic acid, toluene, chlorobenzene, etc.) are employed. It is recommended that the student gain experience using suction filtration for hot aqueous solutions before using this procedure with organic solvents. Good manipulative technique is learned only by practice and attention to details.

For filtering volumes of hot solutions up to 400–600 ml filtration with an ordinary funnel and fluted filter of appropriate size is more convenient than suction filtration. If a large volume of hot solution is to be filtered, it is advantageous to use suction filtration, or to provide a means of warming a large conical funnel continuously during the filtration. For the latter purpose, the funnel may be wound tightly with a coil of copper tubing through which hot water or steam can be circulated, or fitted with an electrical heating jacket (Figure 29).

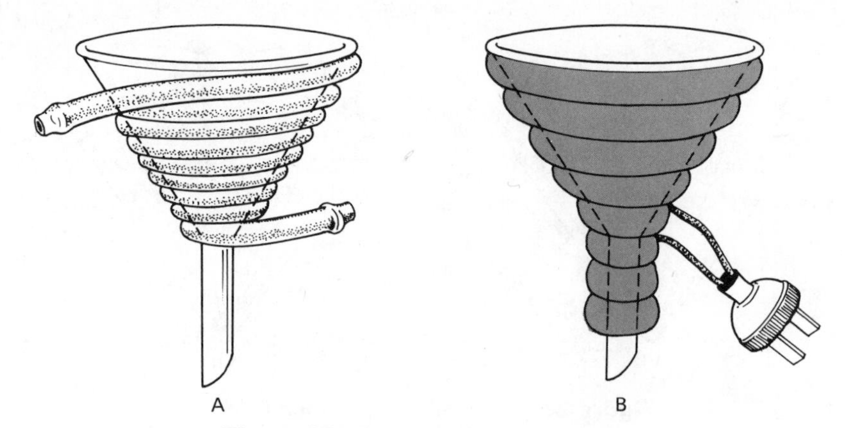


**Figure 29.** Funnels for Hot Filtration

It is desirable usually to produce small, uniform, well-formed crystals of the purified material. To accomplish this result the hot filtered solution should be reheated to redissolve any crystals that have deposited during filtration; crystals that have formed in the funnel may be recovered by placing it in the mouth of the flask, where condensed vapor of the solvent will redissolve the crystals. The clear solution is covered and set aside to cool undisturbed. With a substance that contains only traces of impurity and crystallizes quite readily, it may be advantageous to cool the hot solution rapidly with vigorous stirring, so that small uniform crystals are obtained. Since organic substances differ greatly in their rates of crystallization and materials of varying degrees of impurity will be encountered, it is necessary to adapt the crystallization procedure to the specific material in hand.

Not infrequently, one may meet with a crystalline organic compound that exhibits a marked tendency to separate from solution, even at temperatures well below the melting point of the pure substance, in the form of an oily liquid which cannot easily be induced to crystallize. This situation arises especially with low-melting solids and with mixtures of closely related compounds. In such cases satisfactory crystallization may be obtained by allowing the solution to cool slowly and to remain undisturbed for a considerable period of time. Often it is advantageous to inoculate the solution carefully with a few *tiny* crystals of the desired product (seed crystals), which will serve as crystal nuclei, and then to allow the solution to remain undisturbed. The necessary seed crystals may be secured by inducing crystallization in a test portion by vigorous scratching with a glass rod, or by reserving a small portion of the original sample of material. If a solution is already supersaturated, cooling to a very low temperature may not hasten crystallization since the rate of crystallization is apt to be reduced by lowering the temperature.

Occasionally it may be necessary to prepare a dilute solution and allow the solvent to evaporate slowly to secure crystals, but this practice should be used only as a last resort, since it is difficult to obtain a pure product in such a manner (why?).

For collecting the purified crystals and separating them from the mother liquor, filtration is carried out by means of a Büchner funnel and a suction filtering flask of heavy glass, having a side tube for attachment to a water pump to furnish suction (see Figure 30). All parts of the Büchner funnel,

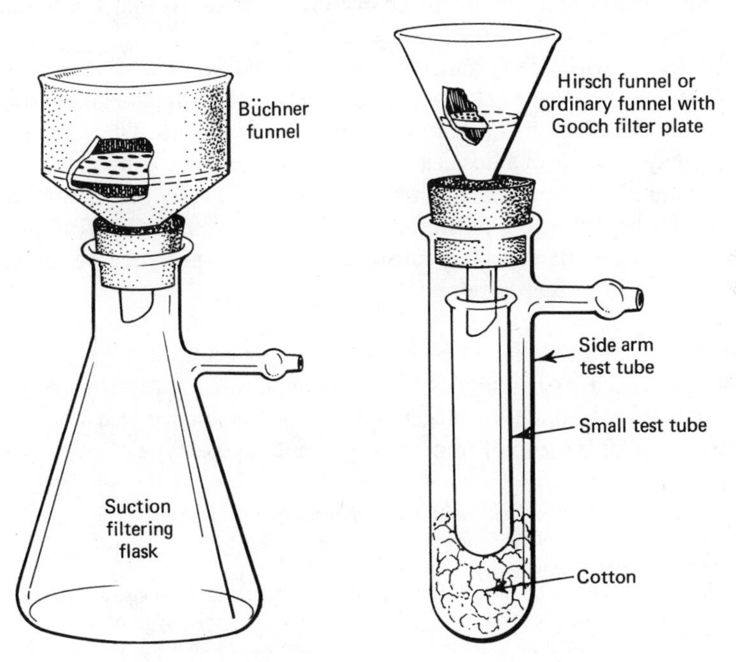


**Figure 30.** Apparatus for Suction Filtration

including the inner inaccessible portion, should be perfectly clean and all holes in the perforated plate should be open (clean with a pin if necessary). The filter paper for the Büchner funnel should be selected (or trimmed if necessary) so that it lies flat on the perforated plate and does not fold up against the side. It is desirable to moisten the filter paper with the solvent and apply suction before the filtration is started. The suspension of crystals is then poured onto the filter in such a way that a layer of uniform thickness is collected. Often it is convenient to use some of the filtrate to wash out crystals that adhere to the walls of the flask in which crystallization was carried out.

An objective of paramount importance in collecting a purified product is the complete *separation of the crystals from the mother liquor containing the dissolved impurities*. When carried out properly, suction filtration is very effective for this purpose.[3] The crystals are first pressed down firmly on the filter with a clean cork or spatula[4] (or a flat glass stopper) and sucked as dry as possible. Before washing the crystals the suction tube is disconnected and the crystalline cake is loosened carefully with a spatula. The fresh cold solvent for washing is added in small portions and the material is stirred into a smooth paste. When this has been accomplished suction is applied again and the crystals are pressed down firmly as before, to remove the wash liquid as completely as possible. Washing the crystals with two or three small portions of solvent is more effective than a single washing with the same amount of solvent. It is particularly important in the washing operations to stop the suction and break up the crystalline cake, so that the whole mass may come into contact with the fresh solvent.

After the washed crystals have been pressed firmly and sucked as dry as possible, the crystalline cake is removed and spread out on a large filter paper to permit evaporation of remaining traces of solvent. The final stages of drying may be done in a desiccator over an appropriate solid drying agent or in a drying oven regulated to a temperature well below the melting point of the substance. All traces of the solvent should be removed before taking the melting point of the purified substance, since the presence of solvent may lower the melting point appreciably.

**Fractional Crystallization of Mixtures.** Simple crystallization (or recrystallization) is used principally for the purification of a reaction product, or similar material, made up largely of a single substance and having only a relatively small portion of impurities present. In this type of purification the

[3] Crystals may also be freed of adhering mother liquor and washed, in a very efficient way, by means of a centrifuge. Small amounts of material (0.5–1 g) may be filtered in a Gooch crucible and centrifuged in an appropriate holder.

[4] For the manipulation of solid organic substances a small metal spatula is convenient. A satisfactory spatula can be made from a 12–15 cm length of thick wire of nickel or silver, by flattening the ends with a hammer to a width of 5–6 mm. The ends are shaped and smoothed by means of a file or buffer, and one end may be bent at a 45° angle about 12–15 mm from the tip.

mother liquor containing the other components of the mixture is eventually discarded. When two or more components of a mixture are to be separated and isolated in a pure state, it is usually necessary to subject the mixture to a series of preliminary partial purifications and separations in order to obtain the individual components in a sufficient degree of purity so that the small remainder of impurities can be removed by simple recrystallization. Sometimes an organic compound will separate in an amorphous condition and not form well defined crystals until a rather high degree of purity is attained. Methods used for preliminary purification include:[5] steam distillation, vacuum distillation, sublimation, simple and countercurrent extraction with solvents, partial crystallization from the melt, and chromatography. Several of these operations are illustrated in later experiments.

The basic procedure for fractional crystallization is not complex but the selection of appropriate solvents and experimental conditions for a specific mixture is usually difficult. Fractional crystallization may be carried out in many different ways. One method involves separation of the mixture into a series of fractions by allowing a hot solution to cool slowly to a selected temperature (chosen so that only a portion of the material will crystallize out) and filtering at the same temperature to collect a first crop of crystals; the mother liquor is allowed to cool to a lower temperature to obtain a second fraction of crystals, and so on. The last fractions of the series may be isolated by concentrating the mother liquor to successively smaller volumes, or by evaporating the final mother liquor and fractionating the residue from a different solvent.

Another method of obtaining a series of fractions is to bring the material into solution in one solvent and effect fractional precipitation by stepwise addition of a second solvent in which one of the components is sparingly soluble.

The series of fractions obtained from the first fractionation (designated by appropriate symbols, such as *A*-1, *A*-2, *A*-3, etc.) may be subjected to a second fractionation in the following way: the first fraction (*A*-1) is crystallized from a fresh portion of solvent to furnish the first fraction of the second series (*B*-1), and the mother-liquor from these crystals is used to recrystallize the second fraction from the first series (*A*-2). By continuing in this manner a second series of fractions (*B*-1, *B*-2, *B*-3, etc.) is obtained. The course of the separation and the identity of adjacent fractions may be followed by various means, such as: melting points of the successive fractions (and of mixtures with end fractions, etc.), thin-layer or gas chromatography, biological testing, colorimetric methods, acidimetry and alkalimetry, quantitative analysis for a characteristic element (N, S, Cl, etc.), etc.

[5] For information on special methods and apparatus, consult Morton, *Laboratory Technique in Organic Chemistry*, McGraw-Hill Book Co, New York (1938); Weissberger and Perry, *Physical Methods of Organic Chemistry*, Interscience Publishers, Inc., New York (ten volumes); Wiberg, *Laboratory Technique in Organic Chemistry*, McGraw-Hill Book Co, New York (1960).

The composition gradient of the fractions will be affected by the solubility relationship and the relative proportions of the components. If present in relatively small amount, the least soluble component may remain in solution in the mother liquor during the early stages of the fractionation, while a more soluble component present in relatively large amount is being collected in the earlier fractions. When testing indicates that one component has been concentrated sufficiently, the appropriate fraction (or fractions) may be recrystallized in the ordinary manner, preferably by changing to a different solvent.

**Sublimation.** The sublimation of a solid substance is an unusual type of distillation process in which the solid undergoes direct vaporization and the vapor is condensed directly to the solid state, *without the intermediate appearance of a liquid state*. For sublimation to occur it is necessary that the solid have a relatively high vapor pressure at a temperature below its melting point. As relatively few organic compounds fulfill this condition, this method is not used frequently in ordinary laboratory work. When applicable, sublimation often affords a method of obtaining a product of high purity; it is particularly

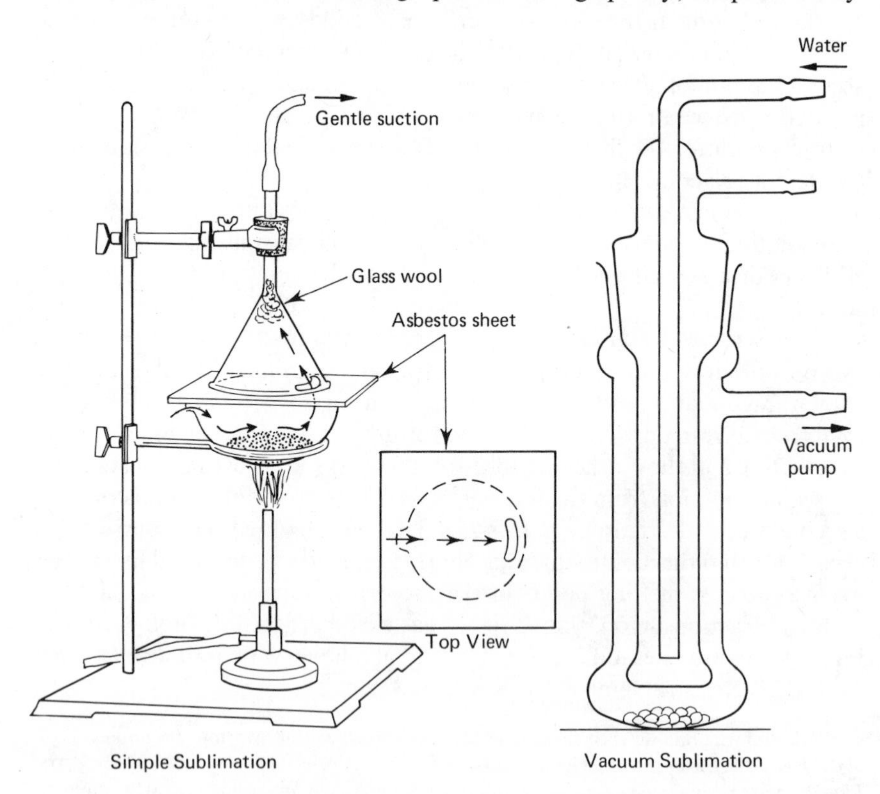


**Figure 31.** Apparatus for Sublimation

valuable for the isolation of a volatile solid from colored gums and tars, and is best adapted to use with small amounts of material.

Sublimation is a slow operation and the handling of the solid condensate offers some difficulty. For ordinary work involving the purification of 1–2 g of solid material by sublimation (at atmospheric pressure), a simple sublimation apparatus may be set up readily (Figure 31). The material to be sublimed is placed in a small porcelain or Pyrex dish and covered with a square of thin asbestos sheeting (or cardboard) having a series of small holes or an elliptical slot to serve for the exit of vapor. The cover is placed so that the openings for the vapor are above the inside of the rim of the dish and opposite the lip. A conical funnel with a loose plug of glass wool (or cotton) in the stem is inverted over the dish to condense the hot vapor. The dish is heated with a very small flame and the vapor is drawn into the condensing chamber by applying a *very gentle* suction.

A simple apparatus suitable for sublimation of samples of high vapor pressure is pictured in Figure 32. The sample is sublimed from a beaker and

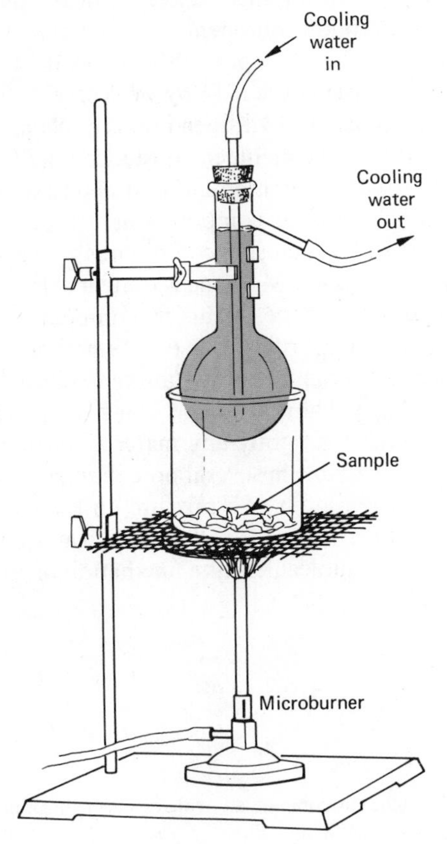


**Figure 32.** Sublimation of Samples with High Vapor Pressure

collected on the bottom of a distilling flask that has been converted into a condenser by passing cooling water in through the neck of the flask and out the side-arm (or a small Florence flask provided with water inlet and outlet).

Sublimation may be carried out more effectively under diminished pressure, and the range of applicability of the process is extended to a larger number of compounds by the use of vacuum sublimation. A form of apparatus suitable for this purpose is shown in Figure 31.

Examples of substances that can be purified by sublimation at atmospheric pressure are: camphor, hexachloroethane, anthraquinone, and many other quinones.

## Inclusion Compounds (Clathrate Complexes)

Many materials have the property of forming crystals that have open spaces between the molecules in the crystal lattice. Crystallization of such a material in the presence of another substance of the proper size and shape may incorporate the second component into the holes in the crystal lattice to form an *inclusion compound*.[6] In spite of the name, the enclosed molecules of inclusion compounds usually are held by weak intermolecular forces rather than stronger interactions like hydrogen bonds or chemical bonds. Since it is largely the dimensions of the included substance relative to the geometry of the crystal lattice of the host compound that determines the stability of the complex, some rather unusual combinations are observed. For example, hydroquinone forms stable inclusion compounds with argon and many other small chemically inert species. Methane is quite insoluble in water but forms an inclusion compound with ice. Inclusion compounds formed by urea are useful for separations and purifications (see Experiment 14).

Molecular sieves, synthetic zeolites (alumino-silicates) manufactured by Union Carbide, possess networks of molecular sized pores of uniform dimension. The pores will adsorb tightly any material small enough to fit inside. Molecular sieves are available in several pore diameters (3 Å to 10 Å) and by proper selection the separation of a mixture on the basis of molecular size is possible. For example, size 3 Å will adsorb water and ammonia tightly while excluding larger molecules like methanol or ethanol (see Drying Agents, page 162).

[6] For a survey of inclusion compounds, with excellent diagrams, see Brown, *Scientific American*, July 1962, pages 82–92; also Fieser, *J. Chem. Educ.*, **40**, 457 (1963). For a general discussion of Molecular Sieves see Breck, *J. Chem. Educ.*, **41**, 673 (1964).

EXERCISE

# 6

# Crystallization and Sublimation

**(A) Crystallization from Water.** Weigh out 10 g of "Impure Acetanilide for Recrystallization"[7] and transfer it to a 500-ml Erlenmeyer flask. Add about 100 ml of hot water and bring the mixture to the boiling point (note that the acetanilide melts to a heavy oil in the hot solution) by heating on a wire gauze (preferably one with an asbestos center). Add successive small portions of hot water, while stirring the mixture and boiling gently, until the acetanilide has dissolved completely; then add 10–15 ml more of the hot solvent. The objective is to dissolve the acetanilide in slightly more than the minimum amount of hot solvent. Do not attempt to dissolve resins, mechanical impurities, etc.

To the hot solution add gradually, with care to avoid excessive foaming, 1–2 g of decolorizing carbon (Norit, Darco, etc.) and boil for a short while longer, to aid in removing small amounts of colored impurities. Meanwhile prepare a funnel, a fluted filter (see page 87), and a clean flask to receive the filtrate. The funnel should be at least 120 mm in diameter and have a very short stem (15–30 mm).[8]

Heat the funnel by inverting on a steam bath, dry with a clean towel, place the fluted filter in the funnel and arrange the receiving flask to collect the hot filtrate. Without allowing the funnel or the solution to cool, pour the solution into the filter. If the solution cannot be poured into the filter in a single portion, replace it on the wire gauze and continue to heat with a small flame to prevent cooling. As soon as all of the solution has been filtered, cover the mouth of the flask containing the hot filtrate with a watch glass and

[7] A method of preparing satisfactory material for this purpose is given in the Appendix under Materials and Reagents.

[8] If desired, the hot solution may be filtered by suction filtration in a Büchner funnel, by the procedure given on page 86. This affords a good opportunity to gain experience in the art of hot suction filtration.

allow it to cool and stand undisturbed. Do not close the mouth of the flask tightly with a stopper (why?).

If crystals have separated in the hot filtrate during the filtration, the filtrate should be heated to redissolve them. Crystals that have formed in the funnel may be recovered by placing it in the mouth of the receiver and allowing condensing vapor from the hot solution to redissolve them.

When the product has separated completely, filter the crystals with suction on a clean Büchner funnel. Wash the crystals twice with a little cold water (see page 90), then press them as dry as possible on the funnel with a clean spatula or cork. Spread the crystals on a large clean paper, *allow to dry completely*, and weigh. Determine the melting point (see Exercise 5); if the product is not sufficiently pure, as shown by melting completely within a temperature range of 1–2°, recrystallize the material again in a similar way. Calculate the percentage recovery of pure acetanilide.

**(B) Crystallization from a Flammable Solvent.** Weigh out 10 g of "Impure Naphthalene for Recrystallization"[9] and transfer it to a 125-ml Erlenmeyer flask. Add 40 ml of 95 percent ethanol and warm the mixture on a steam bath until the solvent boils. Add successive small portions of ethanol (not more than 10–12 ml total) and boil gently after each addition, until the naphthalene has dissolved; then add 1–2 ml more of the solvent. Do not attempt to dissolve admixed particles of sand, grit, etc. Meanwhile prepare a fluted filter (see page 87) and arrange a funnel (65 mm diam, 10–20 mm stem) and clean dry flask to receive the filtrate.

➤**CAUTION:** Flammable solvents such as ether, alcohols, benzene, etc., must never be heated in an open flask over a burner, nor manipulated near a flame. These solvents should be heated on a water or steam bath (or an electric hot plate having no exposed hot filament). If a burner is used, the flask must be fitted with an upright reflux condenser. Take particular care to insure that no lighted burner is nearby during the filtration of the hot solution!

Remove the boiling solution from the steam bath, add gradually 0.5 g of decolorizing carbon (*caution—frothing*), and swirl the solution gently. Reheat to boiling and pour the hot solution into the fluted filter.[10] Cover the mouth of the flask containing the hot filtrate with a watch glass[11] and allow

[9] A method of preparing satisfactory material for this purpose is given in the Appendix under Materials and Reagents.

[10] Occasionally a few fine particles of the decolorizing agent may pass through the pores of the filter paper. Often this may be remedied by heating the filtrate to boiling and filtering again through the same filter; or a small amount of a filter aid ("Filter-Cel," "Super-Cel," etc.) may be added. If the filter has been damaged in folding or is defective, a fresh one must be used.

[11] If crystals have separated in the hot filtrate during the filtration, the filtrate should be heated to redissolve them. Crystals that have formed in the funnel may be recovered by placing it in the mouth of the receiver and allowing condensing vapor from the hot solution to redissolve them.

to cool and stand undisturbed. When the product has separated completely, collect the crystals with suction in a clean Büchner funnel (see page 89); wash all the crystals into the funnel by rinsing the Erlenmeyer flask with part of the filtrate. Discontinue the suction and wash the crystals with two 3–5 ml portions of fresh ethanol (cold). Apply suction again and press the crystals firmly with a cork or flat glass stopper. Spread the crystals on a large clean paper, *allow to dry thoroughly*, and record the weight of purified naphthalene. Calculate the percentage recovery and determine the melting point of the purified product.

**(C) Sublimation.** Assemble the apparatus pictured in Figure 32 using a 100-ml beaker and a 25-ml distilling flask. Place 1 g of impure[12] *p*-dichlorobenzene in the beaker and tap the beaker to spread the compound in a uniform layer over the bottom. Insert the condensing flask so that its bottom is about 10–15 mm above the top of the compound and clamp the flask and beaker in this position. Warm the beaker gently over a wire gauze with a small flame. When no more material sublimes, remove the condenser and scrape the condensate into a tared bottle. Record the weight of purified material and determine its melting point range.

## Questions

**1.** Outline the successive steps in the crystallization of an organic solid from a solvent and state the purpose of each operation.

**2.** What properties are necessary and desirable for a solvent in order that it be well suited for recrystallizing a particular organic compound?

**3.** Apart from any considerations of solubility relations and effectiveness for removal of impurities, in what respects would *n*-hexyl alcohol be a less desirable crystallization solvent than methanol or ethanol? Why might carbon tetrachloride be preferable to ether, or to benzene? (See Table 9, page 83.)

**4.** Mention at least two reasons why suction filtration is preferable to ordinary gravity filtration for separating the purified crystals from the mother liquor. Why is it desirable to release the suction before washing the crystals with small portions of the fresh solvent?

**5.** What means other than crystallization from a solvent may be used to purify an organic solid or to effect preliminary separation of a solid mixture?

[12] The impurity should be relatively non-volatile. Methyl orange or charcoal is suitable.

# Extraction with Solvents

The process of extraction with solvents is used in organic chemistry for the separation and isolation of substances from mixtures which occur in nature, for the isolation of dissolved substances from solutions, and for the removal of soluble impurities from mixtures.

The extraction of alkaloids from leaves and bark, flavoring extracts from seeds, perfume essences from flowers, and sugar from sugar cane, are examples of extractions of the first type. Solvents commonly used for this purpose are: ether, methylene chloride, chloroform, carbon disulfide, acetone, various alcohols, and water.

A common form of apparatus for continuous extraction of solids by means of volatile solvents is called a Soxhlet extractor. A typical laboratory setup employing this extraction device is shown, mounted and ready for use, in Figure 33.

The isolation of dissolved organic substances from solution is often accomplished by extraction with an immiscible solvent. The general principle underlying this process is known as the distribution law. In dilute solutions a substance distributes itself between two immiscible solvents so that the ratio of the concentration in one solvent to the concentration in the second solvent always remains constant (at constant temperature). This constant ratio of concentrations for the distribution of a solute between two particular solvents is called the distribution coefficient or the partition coefficient for the substance between the two solvents.

Distribution Coefficient[1] of $X$ between solvents $A$ and $B$

$$= \frac{\text{Concentration of } X \text{ in } A}{\text{Concentration of } X \text{ in } B} = K \text{ (at constant temperature)}$$

[1] In the case of dissolved substances which are ionized or associated in one of the solvents, the distribution law holds true for the ratio of the concentrations of the simple molecules only. To obtain an expression which will hold for the total concentrations it is necessary to introduce an expression for the ionization or association equilibrium.

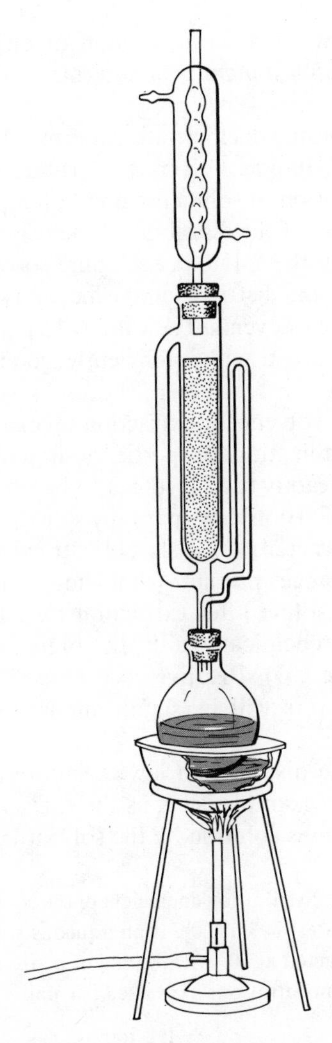

**Figure 33.** Soxhlet Extractor Assembly

where the concentrations are expressed in grams per milliliter of the solution.[2] Actually no two solvents are absolutely immiscible, but each solvent is at least slightly soluble in the other. In practice the most common application

[2] The values of the distribution coefficients found in the chemical literature are usually based on concentrations per volume *of solution*, but in making rough calculations of the type illustrated in the following paragraphs it is much simpler to use concentrations per unit volume *of the solvent*. For dilute solutions the error introduced in this way is negligible. It is important to observe the manner in which the ratio is expressed; for example, at 20° the distribution coefficient of *n*-butyric acid between ether and water is approximately 5 (Concn in ether/Concn in water = 5), but if the ratio is expressed as the distribution between water and ether the value is 0.2 (Concn in water/Concn in ether = $\frac{1}{5}$).

of the distribution law is to the extraction of dissolved substances from aqueous solutions by *almost immiscible* solvents such as ether,[3] benzene, and chloroform.

The actual distribution coefficient is determined by bringing the solvents and solute into equilibrium at a given temperature and determining experimentally the concentration of solute per unit volume of each separate phase. A rough approximation of the distribution coefficient can be made by determining the solubility of the solute in each pure solvent independently, since under ideal conditions the distribution coefficient is the same as the ratio of the solubilities in the two solvents. The values obtained in this way are subject to several errors but are usually sufficiently good for simple laboratory calculations.

The selection of a solvent for extraction involves considerations analogous to those for crystallization. Properties desired for a suitable solvent are: (a) it should dissolve readily the substance to be extracted (favorable distribution coefficient); (b) it should be sparingly soluble in the liquid from which the solute is to be extracted; (c) it should extract little or none of the impurities or other substances present; (d) it should be capable of being separated easily from the solute after extraction (usually by distillation); and (e) it should not react chemically with the solute in an undesired way (see reaction solvents, page 103). Relative cost, ease of manipulation, flammability, and similar factors will be significant in choosing among various possible solvents.

Application of the distribution law to laboratory extractions may be illustrated by a specific example, such as the extraction of an organic compound $A$ from an aqueous solution by the solvent benzene.

It is assumed that the distribution coefficient of the compound $A$ between benzene and water ($K$ benzene/water) is 3 at 20°. If an aqueous solution containing 6 g of $A$ in 100 ml of water is extracted at 20° with 100 ml of benzene, the following equations may be written (using concentrations expressed per unit volume of the solvent):

$$K = \frac{C_b}{C_w} = \frac{W_b/100 \text{ ml}}{W_w/100 \text{ ml}} = 3 \qquad (1)$$

$$W_b + W_w = 6 \text{ g} \qquad (2)$$

where $W_b$ = weight of $A$ in the benzene layer
$W_w$ = weight of $A$ in the water layer

[3] Ether, which is used frequently as an extraction solvent, is somewhat more soluble than other common solvents; for example, at 30° diethyl ether is soluble to the extent of 1 g in 18.8 g of water, and water is soluble to the extent of 1 g in 73 g of ether. Methylene chloride (bp 40°) has several advantages over ether and is to be preferred whenever it has satisfactory solvent properties; methylene chloride is less soluble in water (about 2 g per 100 ml at 20°) and is not flammable under ordinary conditions.

$C_b$ = concentration of $A$ in the benzene layer
$C_w$ = concentration of $A$ in the water layer

Since $W_w = 6 - W_b$ [from equation (2)], we may write

$$\frac{W_b/100 \text{ ml}}{(6 - W_b)/100 \text{ ml}} = 3$$

whence $W_b = 4.5$ g
$W_w = 1.5$ g

After one extraction with 100 ml of benzene, 4.5 g of $A$ (75 percent of the total) is removed by the benzene and 1.5 g (25 percent of the total) remains in the water layer.

**Multiple Extraction.** With a specified quantity of solvent for use in extraction, is it preferable to make a single extraction with the total quantity of solvent, or to make several successive extractions (multiple extraction) with portions of the solvent? As far as efficiency of extraction is concerned, the second method is usually preferable. The basis for this conclusion is shown by further consideration of the example given above.

If the same aqueous solution cited previously (6 g of $A$ in 100 ml of water) were extracted successively with two 50-ml portions of benzene instead of once with 100 ml of solvent, the following relationship will hold for the first 50-ml extraction:

$$\frac{W_b/50 \text{ ml}}{W_w/100 \text{ ml}} = \frac{W_b/50 \text{ ml}}{(6 - W_b)/100 \text{ ml}} = 3$$

whence $W_b = 3.6$ g
$W_w = 2.4$ g

The aqueous solution which now has 2.4 g of $A$ remaining in it, is extracted with the second 50-ml portion of benzene:

$$\frac{W_b/50 \text{ ml}}{(2.4 - W_b)/100 \text{ ml}} = 3$$

whence $W_b = 1.44$ g
$W_w = 0.96$ g

The two extractions with 50 ml of benzene have removed 5.04 g of $A$ (84 percent of the total) and 0.96 g of $A$ (16 percent) remains in the water layer. Obviously the multiple extraction is more efficient in this case, since a single extraction with 100 ml of solvent removed only 4.5 g of $A$ (75 percent) and left 1.5 g in the water.

As one would expect from this conclusion, multiple extraction with three portions of one-third of the total quantity of solvent would be even more efficient than two extractions with half portions. Thus, three extractions with $33\frac{1}{3}$-ml portions of benzene would remove 5.25 g of $A$ (87.5 percent) and leave only 0.75 g in the water. There is, of course, a practical limit to multiple extraction beyond which the small additional amount of material extracted does not justify the additional time and work involved in the larger number of extractions. If the distribution coefficient of the solute is very favorable for a solvent (for example, 10 or more), the number of extractions required for essentially complete removal of the solute will be relatively small.

**Problems:** (a) Calculate the amount of $A$ that would be extracted by using four 25-ml portions of benzene. Would this additional amount be sufficient to justify the additional extraction operation?

(b) Assume that a compound $B$ has a distribution coefficient of 15 between benzene and water, and that you have an aqueous solution of 6 g of $B$ in 100 ml of water. If 100 ml of benzene is to be used for extraction, how many extractions with portions of the solvent would be required to extract at least 95 percent of $B$ from the aqueous solution?

(c) Assume that a compound $C$ has a distribution coefficient of 0.5 between benzene and water, and other conditions similar to those in problem (b). Would multiple extraction be more or less efficient than a single extraction with 100 ml of solvent in this situation?

The general multiple extraction formula shown in equation 3 gives the fraction of compound $X$ remaining in volume $V_A$ of solvent $A$ after $n$ extractions with solvent $B$, each of volume $V_B/n$.

$$\text{Fraction remaining} = \frac{C_{\text{final}}}{C_{\text{initial}}} = \left(\frac{KV_A}{KV_A + V_B/n}\right)^n \tag{3}$$

where $C_{\text{final}}$ = the final concentration of $X$ in $A$
$C_{\text{initial}}$ = the initial concentration of $X$ in $A$
$K$ = distribution coefficient (concentration of $X$ in $A$ divided by the concentration of $X$ in $B$).

The removal of soluble impurities by extraction with an immiscible solvent is of great importance in the purification of organic substances. If a substance containing a soluble impurity is treated with an immiscible solvent, the impurity distributes itself (according to the distribution law) between the substance being purified and the immiscible solvent. Consequently a portion of the impurity will be removed by the immiscible solvent. If the distribution coefficient is known, calculations similar to those made above (for the extraction of a dissolved substance from solution) may be made for the extraction of the impurity. It is also true in this type of extraction that several extractions with portions of the solvent are more efficient for the removal of an impurity than a single extraction with the total amount of solvent.

**Effect of Salts on Solubility.** The solubility of organic substances in water is markedly affected by the presence of dissolved inorganic salts. For example, ethanol, which is perfectly miscible with pure water, is only slightly soluble in strong aqueous solutions of sodium chloride, potassium carbonate, and certain other inorganic salts. The same is true of acetone, pyridine, methanol, and many other water-soluble organic compounds. This phenomenon (*salting-out effect*) occurs commonly with salts having ions of small radius and concentrated charge. The opposite effect of enhanced solubility in salt solutions (*salting-in*) occurs frequently when the salt has ions of large radius and diffuse charge. Benzene, for example, is about 40 percent more soluble in 1 molar aqueous tetramethylammonium bromide than in pure water.

In the case of solutions of ionized organic substances, such as metallic salts of organic acids and salts of organic bases with mineral acids, it is possible that the *common-ion effect* may also act to decrease the solubility of the salt. Thus, sodium benzenesulfonate is quite soluble in water but is precipitated from an aqueous solution by adding sodium chloride; aniline hydrochloride is readily soluble in water but is only slightly soluble in strong hydrochloric acid solutions.

**Extraction by Chemically Active Solvents.** Extraction of an organic substance by means of a chemically inert, immiscible solvent depends merely upon simple solubility relations. Another type of extraction, of great importance, depends upon the use of a reagent that reacts chemically with the substance to be extracted. This method is used frequently to separate the components of a mixture, and also to remove small amounts of impurities in an organic compound. The principle of the method is to employ a reagent that acts selectively upon the organic substances that are present. Obviously this method requires a good knowledge of the chemical behavior of organic compounds toward various reagents, as well as a knowledge of solubility relationships.[4]

Commonly used *reaction solvents* are: dilute aqueous alkalies (5 percent sodium hydroxide or potassium hydroxide solutions); dilute aqueous mineral acids (5 percent hydrochloric acid, etc.); cold concentrated sulfuric acid.

Dilute sodium hydroxide solution (also sodium carbonate or bicarbonate) can be used to extract an organic acid from its solution in an organic solvent, or to remove traces of acid that are present as an impurity in an organic preparation. The use of aqueous alkali depends upon the conversion of the free acid to the corresponding sodium salt, which is soluble in water or dilute alkali. Thus, *n*-butyric acid may be extracted quantitatively from a benzene solution by dilute sodium hydroxide because this reagent converts the acid to sodium *n*-butyrate, which is very soluble in water or dilute alkali but insoluble in benzene. Likewise, an organic acid or mineral acid present as impurity in a water-insoluble liquid or solid, can be removed by washing with dilute alkali.

[4] For discussion of this method consult: Shriner, Fuson and Curtin, *The Systematic Identification of Organic Compounds*, John Wiley and Sons, Inc., New York, 5th edition (1964); Cheronis and Entrikin, *Semimicro Qualitative Organic Analysis*, Interscience Publishers, Inc., New York (1957).

Dilute hydrochloric acid can be used in a similar way to extract basic substances from mixtures, or to remove impurities. The use of dilute acids depends upon converting the base (amines, ammonia, etc.) into a water-soluble salt (amine hydrochloride, ammonium chloride, etc.). Thus in the preparation of benzanilide, unreacted aniline can be removed by washing with dilute hydrochloric acid and the benzoic acid side product can be removed by subsequent washing with sodium carbonate solution; the anilide does not react with either the acid or base, but the amine is converted into the water-soluble salt, aniline hydrochloride, and the benzoic acid is converted into the water-soluble salt, sodium benzoate.

Cold concentrated sulfuric acid can be used to remove unsaturated hydrocarbons, alcohols, ethers, esters, etc., present as impurities in inert compounds such as saturated hydrocarbons and alkyl halides. Olefins and alcohols are dissolved by chemical reaction to form alkyl hydrogen sulfates; ethers, esters, etc., form addition complexes that are soluble in concentrated sulfuric acid. Owing to its great reactivity this reagent has limited utility.

**Solubility Relationships.** The solubility of an organic compound in inert solvents (water, alcohols, ether, hydrocarbons, etc.) and in chemically active solvents is directly related to its molecular structure. Consequently one can predict solubility relations in a qualitative way for various classes of compounds by taking into account the structure of the solute and the physical and chemical characteristics of the solvent medium.

Solubility in a chemically active solvent depends upon chemical reaction to form a soluble salt, and is readily predicted from definite chemical knowledge of the solute and the solvent.

Solubility in an inert solvent may be predicted with fair success on the basis of empirical solubility rules:

1. A substance is most soluble in that solvent to which it is most closely related structurally. Thus, simple alcohols are soluble in water, esters are soluble in alcohol and ether, etc.
2. A compound having branched chains is more soluble in a given solvent than the straight chain isomer. Thus, isobutyl and *tert*-butyl alcohols are much more soluble than *n*-butyl alcohol in water.
3. In any homologous series the higher members become more and more like hydrocarbons in their physical properties. Thus, in most homologous series containing only one functional group, the solubility in water falls below 4–5 percent when the member attains five carbon atoms.
4. Compounds of high molecular weight (polymers, etc.) are usually sparingly soluble in inert solvents. However, polymers often form colloidal dispersions in certain solvents.
5. Liquids and low-melting solids are generally more soluble than high-melting solids in inert solvents.

These simple rules are a convenient guide to solubility relations but one must expect to encounter many exceptions in actual practice.

Solubility tests are often used as a preliminary step in the identification of organic compounds. The systematic outline developed by Kamm is shown in modified form on page 132. The reader is encouraged to compare the entries in this diagram with the previous discussion of extraction by chemically active solvents.

**Laboratory Extractions of Solutions.** Laboratory extractions are commonly carried out by shaking the solution to be extracted, together with the extraction solvent, in a glass separatory funnel. A long tapered funnel of the Squibb type, having a short stem cut off obliquely (Figure 1, page 6), is particularly convenient for this purpose. Cylindrical or pear-shaped funnels with a short stem may also be used. A long stem is disadvantageous because it holds a long column of the liquid that is being drawn off and makes for difficulties in manipulation.

The separatory funnel should be of such size that it is not more than three-fourths filled by the solution and solvent. The funnel is shaken to obtain good physical mixing of the insoluble liquids, and is then allowed to stand undisturbed until the layers have separated completely. Vigorous shaking is desirable provided it does not produce an emulsion[5] that interferes with subsequent separation of the layers; for such systems it is better to mix the layers by repeated gentle swirling.

During the shaking operation it is essential to grasp the funnel with both hands, one hand at the top and the other on the stopcock, in such a way that the stopper and stopcock are held firmly in place. The internal pressure should be released from time to time by inverting the funnel (so that the stem points upward) and opening the stopcock momentarily until the pressure is reduced. This is particularly important when a very volatile solvent such as ether is used.

➤**CAUTION:** Do not point the stem of the separatory funnel toward anyone when releasing the pressure. Drops of liquid caught in the stem may be ejected forcefully.

When the liquids have separated, the stopper is loosened or removed and the lower layer is drawn off carefully into an Erlenmeyer flask. The stopcock should be manipulated with both hands to avoid loosening it and consequent loss of material. If only one layer is to be retained, *it is a safe rule to save both layers* until you are certain which one contains the desired material.[6]

[5] Emulsions of water solutions with solvents such as ether and benzene may often be broken by adding 1–2 ml of ethanol or ethyl acetate, and swirling the contents of the funnel gently.

[6] An aqueous layer can be distinguished from a water-insoluble layer by adding a small test portion to a few milliliters of water in a test tube; the aqueous layer will form a homogeneous solution, whereas the nonaqueous layer will form a two-layer system.

As the interface of the two liquids approaches the stopcock, the liquid should be drawn off slowly. After the separation has been made, the funnel is rotated by a twisting motion with the stopcock closed, to assist in draining droplets of the heavier liquid from the walls. This small additional quantity is drawn off into the receiving flask. The upper layer is poured through the mouth of the funnel into a clean flask; it is not allowed to flow through the stopcock as this would lead to contamination with the liquid adhering to the stem. The organic solvent layer is usually dried by means of an appropriate solid drying agent (see Drying Agents, page 162) and the solvent is removed by distillation.

In any operation where a separatory funnel is used, the stopcock must be kept lubricated to avoid sticking or leakage during manipulations. After use the separatory funnel should be cleaned thoroughly and the stopcock freshly lubricated to inhibit "freezing" in a fixed position. If a silicone base lubricant is used, it should be removed before cleaning with an oxidizing mixture or a siliceous film will be formed on the glass surface. Stopcocks that have become "frozen" sometimes may be freed by warming the outer barrel with steam and applying gentle pressure to the stopcock plug (using a towel to protect the fingers). A vise-like stopcock plug remover is available from scientific supply firms.

For special purposes, laboratory methods and apparatus have been developed for continuous extraction of aqueous solutions with immiscible solvents, and for multiple counter-current extraction. In some instances an extraction solvent made up of a mixture of solvents is employed.

EXERCISE

# 7

# Simple and Multiple Extraction

In a small Erlenmeyer flask, place 110 ml of distilled water and add 5 ml of glacial acetic acid.[7] After mixing thoroughly remove a 10-ml portion and titrate the acidity against standard sodium hydroxide solution (about 0.3 *N*) using phenolphthalein as indicator. The acetic acid solution is then to be extracted in two different ways.

**(A) Simple Extraction.** Place 50 ml of the solution in a clean 250–300-ml separatory funnel and extract with 150 ml of ether in the following way. Stopper the separatory funnel, shake gently, and turn it upside down. While the separatory funnel is in this position, open the stopcock to release the internal pressure. Close the stopcock, shake vigorously, and again release the internal pressure. Repeat this procedure four or five times, then support the separatory funnel upright in a ring and let it stand undisturbed. When the liquids have separated completely, draw off the lower aqueous layer into a clean Erlenmeyer flask (following the procedure for laboratory extractions given on page 105).

➤ **CAUTION:** In the manipulation of organic solvents in a separatory funnel, it is important that the stopcock should be kept properly greased in order to avoid sticking, since the organic solvent removes the stopcock grease. After using an organic solvent in the separatory funnel, it is advisable to clean the separatory funnel thoroughly and to grease the stopcock properly before replacing the apparatus in the laboratory desk. If this is not done, the stopcock is likely to become "frozen" in a fixed position and the separatory funnel will be rendered useless.

[7] Propionic acid, if available, may be used advantageously instead of acetic acid in this experiment. The distribution coefficient of propionic acid is somewhat more favorable than acetic acid, and the difference between single and multiple extraction is somewhat greater than with acetic acid.

Titrate the aqueous solution with standard sodium hydroxide solution (about 0.3 *N*), using phenolphthalein as indicator. Calculate (a) the amount of acetic acid left in the water layer, (b) the amount extracted by the ether, (c) the percentage of the acetic acid left in the water, (d) the percentage extracted by the ether. Pour the ether extract into a bottle labeled "Ether from Extraction Experiments."

➤CAUTION: Ether is a readily flammable solvent and must be manipulated with careful attention to fire hazards. Use both hands in operating the stopcock of the separatory funnel to avoid dislodging the stopcock.

**(B) Multiple Extraction.** Place the second 50-ml portion of the original acetic acid solution in a clean 100–150-ml separatory funnel and extract with 50 ml of ether as described in part (A). Separate the aqueous layer and extract it a second time, with a fresh portion of 50 ml of ether. Separate the aqueous layer and extract it a third time, with a fresh portion of 50 ml of ether. After the third extraction, draw off the aqueous layer into a clean Erlenmeyer flask and titrate with the same standard sodium hydroxide solution (about 0.3 *N*) that was used for part (A). Calculate (a) the amount of acetic acid left in the water layer, (b) the amount extracted by the ether, (c) the percentage of the acetic acid left in the water, (d) the percentage extracted by the ether.

Compare the effectiveness of extraction with 150 ml of ether by the two different procedures.

Pour all the ether extracts into a bottle labeled "Ether from Extraction Experiments."[8]

## Extraction of Caffeine from Tea

The stimulating effect of aqueous infusions of coffee beans, tea and maté leaves, and cola nuts is due mainly to the presence of caffeine, a nitrogen heterocycle of the molecular formula $C_8H_{10}N_4O_2$. Its structure has been established by study of its degradation products and by synthesis to be

[8] The ether from these experiments may be recovered by washing with dilute sodium hydroxide solution, drying over solid flake sodium hydroxide (commercial flake lye) and distilling.

Diethyl ether and other ethers on standing in the air for some time form peroxides, which impart a pungent odor and may cause explosion of the still residue when a large volume of ether is distilled. The presence of peroxides may be tested by shaking a few milliliters of the ether with an aqueous solution of potassium iodide acidified with dilute sulfuric acid to which 1–2 ml of starch solution has been added. The peroxides liberate iodine, which colors the starch blue. Peroxides may be removed before distillation by washing the ether with aqueous ferrous sulfate solution slightly acidified with sulfuric acid.

1,3,7-trimethyl-2,6-dioxopurine. Tea leaves contain 3–5 percent of caffeine and a trace of theophylline, a lower homolog lacking the methyl group at

Caffeine

Guanine

position 7. These compounds are related structurally to the important purines, adenine and guanine, that are present in the ribonucleic acids (RNA and DNA).

In this experiment the caffeine is extracted from tea leaves by hot water, in which it is quite soluble (about 18 g/100 ml at 80°; 2.2 g/100 ml at 20°). Colored impurities such as tannic acids can be removed as calcium salts by addition of calcium carbonate. From the aqueous extract the caffeine is isolated conveniently by multiple extractions with small portions of chloroform, in which caffeine is very soluble (18 g/100 ml at 20°). Caffeine forms a monohydrate that loses water readily on warming to give the anhydrous form, mp 238°.

In a 500-ml Erlenmeyer flask place 30 g of ordinary dry tea, 300 ml of water, and 15 g of powdered calcium carbonate. After boiling the mixture gently with occasional swirling for 20 min, filter the hot mixture on a Büchner funnel and press the filter cake firmly with a large cork to obtain as much as possible of the liquid.[9]

Cool the aqueous extract to 15–20°, transfer it to a separatory funnel, and extract the caffeine with three successive 25-ml portions of chloroform (following the procedure for laboratory extractions given on page 105).[10]

➤**CAUTION:** Avoid inhaling the vapor of chloroform or contact of the liquid with the skin. The vapor is irritating and nauseous.

Transfer the extracts to a 125-ml distilling flask fitted with an efficient condenser (see apparatus for simple distillation, page 23), and distill off all

[9] This filtration is difficult because the gelatinous material that separates on cooling tends to plug the pores of the filter paper (review Crystallization and Sublimation). An alternate procedure is to cool the mixture to room temperature, add 5 g of Celite or other filter aid, and filter on a Büchner funnel.

[10] Aqueous extracts of plant materials tend to form stubborn emulsions when extracted with organic solvents. An effective means for breaking difficult emulsions is to press, with the aid of a glass rod, a small wad of glass wool into the bottom of the separatory funnel and to draw off the lower layer through the glass wool.

theine 1,3,7-Trimethyl-xanthine
MW 194.2

but about 10 ml of the solvent, using a water bath. Save the recovered chloroform.[11] Pour the warm concentrated solution of caffeine into a small beaker and rinse the flask with 5–10 ml of the recovered solvent. Evaporate the combined solutions to dryness on a steam bath in a hood. If hood space is not available the vapors may be drawn off by means of an inverted funnel connected to a water aspirator. To purify the crude product dissolve it in about 10 ml of hot benzene on a steam bath, add 15–20 ml of ligroin (bp 60–90°), and allow the product to crystallize. Collect the crystals on a small suction filter and wash them with a little ligroin. Do not attempt to determine the melting point.

## Questions

**1.** What conclusions can you draw in regard to the most efficient method of extracting acetic acid from an aqueous solution by means of an immiscible solvent?

**2.** What is meant by the term distribution coefficient?

**3.** What properties do you look for in a good solvent for extraction?

**4.** Explain the fact that acetic acid can be extracted quantitatively from an ether solution by dilute aqueous sodium hydroxide solution.

**5.** When the following solvents are used for extracting an organic compound from an aqueous solution, will the organic solvent form the upper or lower layer: chloroform? benzene? *n*-heptane? methylene chloride?

**6.** If ethyl chloride (density 0.92) were used to extract ethylene bromohydrin (density 2.41) from an aqueous solution, could you be certain that the organic solution would form the upper layer? How could you test to determine which is the nonaqueous layer?

**7.** Mention two examples of organic compounds that cannot be extracted effectively from an aqueous solution by means of an immiscible organic solvent such as ether or benzene.

[11] Add a drop of ethanol to the recovered chloroform to stabilize it against oxidation, and store it in a dark bottle. Ordinary commercial chloroform contains 0.75 percent of ethanol as stabilizer. Ethanol-free chloroform can be obtained by distillation, rejecting the first 10 percent of distillate, since chloroform and ethanol form an azeotrope (bp 59°) containing 7 percent of ethanol. The residual pure chloroform boils at 61.2°.

# Chromatography

The term chromatography is applied to numerous purification processes that share the principle of distributing a sample between a stationary phase and a mobile one. As with liquid-liquid extraction the degree of separation is determined by differences in distribution coefficients.

## Liquid-Solid Chromatography

Solid surfaces adsorb thin layers of foreign molecules as a result of forces identical in character with those operating between molecules of a liquid or of a gas. Since adsorption strengths differ with the character of the solid surface, a properly chosen solid may adsorb selectively one component of a mixture. An important example of selective adsorption is the use of charcoal in crystallization to remove colored impurities. The ideal limiting law governing adsorption from a dilute solution is:

$$\frac{[\text{Amount of Solute } A \text{ Adsorbed per Unit Surface Area}]}{[\text{Concentration of Solute } A \text{ in Solution}]} = K_A$$

The factors that determine the extent of adsorption of a molecule on a solid surface are closely related to the factors that enter into solubility considerations. An additional complication in adsorption is that the solvent and the solute are competing for the active sites on the surface. For molecules containing polar functional groups the value of $K_A$ (the adsorption coefficient) is determined principally by the relative polarities of the substance and the solvent. Highly polar solvents tend to be preferentially adsorbed so that a low $K_A$ results. For molecules containing hydroxyl groups their relative abilities to form hydrogen bonds (proton-bonding) to the solid or the solvent are significant.

Two solutes having different adsorption coefficients toward a certain solid, can be separated by the process of liquid-solid chromatography. One of the most practical methods involves preparation of a cylindrical column of the solid (*stationary phase*) and addition of a concentrated solution (*liquid phase*) at the top of the column. As the solution penetrates the column the solutes are adsorbed. At the moment the solution has completely penetrated the column, fresh solvent is added at the top. The solvent flows down the column and redissolves the solutes in amounts determined by the adsorption law and carries them to lower clean sections of the column where they are readsorbed (in amounts governed by the adsorption law). As more solvent percolates through the column the cycle of adsorption-solution continues and the solutes gradually move down the column in concentrated bands (*development*). With solutes having different adsorption coefficients the least tightly adsorbed material tends to move ahead more readily. If the coefficients are sufficiently different or the column is sufficiently long the faster moving component will form a separate band below the slower moving one. At the lower end the solutes are forced off the column (*elution*) and can be collected separately in successive fractions.

For satisfactory separation by liquid-solid column chromatography it is essential to choose an appropriate combination of solid adsorbent and eluent that is compatible with the compounds to be separated. Compounds that are

**TABLE 10**
**Generalizations for Liquid-Solid Chromatography**

| SOLIDS IN DECREASING ORDER OF ADSORPTION STRENGTH FOR POLAR MOLECULES |
|---|
| Activated Alumina,* Charcoal |
| Activated Magnesium Silicate* |
| Activated Silicic Acid* |
| Inorganic Carbonates |
| Sucrose, Starch |

| SOLVENTS IN INCREASING ORDER OF ELUTING ABILITY† |
|---|
| Saturated Hydrocarbons |
| Aromatic Hydrocarbons |
| Ethers |
| Halogenated Hydrocarbons |
| Ketones |
| Alcohols |
| Organic Acids |

* The adsorption strength can be diminished by addition of water. Under carefully controlled conditions this strength is reproducible.

† This approximate order only applies to alumina. With non-polar solids the order tends to be inverted.

adsorbed very tightly require an excessive volume of eluent for development. Compounds that are adsorbed weakly may move too rapidly to give separation before being eluted. Table 10 gives some generalizations that are useful as a guide in selecting appropriate solid-solvent combinations.

Another common variation of liquid-solid chromatography is the use of a thin film of solid (mixed with a binder such as Plaster-of-Paris) on a sheet of glass or plastic. The solution is added as a spot at the bottom of the plate and the plate dipped vertically into a shallow layer of solvent, which ascends by capillary action and moves the solutes with it. The particular advantage of this technique is that the solutes are exposed and can be isolated readily or treated on the plate at any moment. The method is widely used for qualitative identification of mixture components because of its exceptionally good resolution. For a fixed combination of solid, binder, and solvent each substance will travel along the thin-layer plate a characteristic fraction of the distance traveled by the solvent. It is customary to report thin-layer chromatography data as $R_f$ values, defined as the distance of the spot from the starting point divided by the distance of the solvent front from the starting point. Thin-layer chromatography is restricted to small samples. A method known as dry-column chromatography, which combines the high resolution of thin-layer chromatography with the large sample capacity of regular column chromatography, is described under Laboratory Practice.

## Ion-Exchange Chromatography

Ion-exchange chromatography is a special example of liquid-solid chromatography wherein strong ionic attractions replace relatively weak polar adsorption forces.

A column of solid acidic material (such as Amberlite IR-120, a resin of polystyrene beads containing free sulfonic acid groups) can donate protons to any bases present in the surrounding liquid phase to form cations and anions which strongly attract each other. The extent of proton transfer depends on the basicity of the solute and can be described by an equilibrium constant $K$ (analogous to the previously discussed adsorption coefficient).

$$\text{Solid}^{-}\text{—H}^{+} + \text{Base} \overset{K}{\rightleftarrows} [\text{Solid}]^{-} \cdots [\text{H-Base}]^{+}$$

$$K = \frac{[\text{Number of Donated Protons}]}{\begin{bmatrix}\text{Number of Available} \\ \text{Proton Donating Sites}\end{bmatrix}[\text{Concentration of Free Base}]}$$

Bases with larger values of $K$ are present in lower concentration as free base and descend the column more slowly. A mixture of bases having sufficiently different constants can be separated by this method.

When all components of a mixture are held tightly, as happens frequently, it is necessary to percolate dilute acid through the column in order to move the components down the column (*displacement development*).

Basic columns also are available (such as Dowex 3, a resin of polystyrene beads containing free amino groups); these accept protons and can be used to separate mixtures of organic acids of different acid strengths. Special column materials that form ionic complexes with various inorganic cations or anions are useful as are columns containing ions that form complexes with certain organic molecules.

## Liquid-Liquid Chromatography

Liquid-liquid chromatography (partition chromatography) employs a liquid moving phase and a second liquid phase held immobile by adsorption as a thin film on a solid support. Since the chemical influence of the solid support may be largely ignored, the adhering film behaves essentially like a liquid stationary phase. A substance added to such a column will be distributed (partitioned) between the two liquid phases. The distribution law is identical with that pertaining to the distribution of a solute between two immiscible solvents.

$$[\text{Solid} + \text{Film of Solvent}_1] + [\text{Solution of Compound in Solvent}_2]$$
$$\downarrow\uparrow K$$
$$[\text{Solid} + \text{Film of Solution of Compound in Solvent}_1] + [\text{Solvent}_2]$$

A convenient form of partition chromatography involves the use of a paper strip as the solid support. When the paper is treated with a mixture of two insoluble or slightly soluble solvents the more polar solvent is adsorbed on the paper to form the stationary liquid phase. In one technique the sample is placed as a spot near the bottom of a dry paper strip and the strip dipped into a shallow pool of the mixed solvents. As the solvents ascend by capillary action, the more polar solvent is adsorbed and the sample is partitioned between the moving and stationary liquid phases. Partition chromatography on paper strips is useful in separating amino acids by means of water-butanol solvent mixtures.

## Gas-Liquid Chromatography

In gas-liquid chromatography (vapor phase chromatography or VPC) the stationary phase is a film of liquid adsorbed on a solid support and the moving phase is a mixture of vaporized sample and a *carrier gas*, usually helium or nitrogen. The pertinent equilibrium is the distribution of sample between solution in the liquid film and vapor in the moving carrier gas. The rate at which the sample progresses through the column is determined principally by the rate of flow of carrier gas and the equilibrium vapor pressure

$$\text{[Solid + Film of Nonvolatile Solvent] + Vapor of Sample + Carrier Gas} \underset{K}{\rightleftharpoons} \text{[Solid + Film of Solution] + Carrier Gas}$$

of the sample in contact with the solution. Gas-liquid chromatography is superficially similar to fractional distillation: in both, separation depends upon differences in vapor pressure of the components of a mixture. In fractional distillation the counterflow of rising vapor and descending liquid establishes (ideally) equilibrium between the components at each point in the column. In gas-liquid chromatography there is a uni-directional flow of carrier gas and the components move independently of each other. An expression for the number of plates required to obtain a 95 percent pure sample with 80 percent recovery from a 50:50 mixture is:

$$\text{Number of required plates} = \frac{2.0}{(\log \alpha)^2}$$

Comparison of this expression with the analogous one for fractional distillation shows that, aside from a small difference in the constant, they differ by the exponent of the log $\alpha$ term. Since $\alpha$, the relative volatility, is close to unity for any mixture likely to be fractionally distilled or chromatographed, the log $\alpha$ term is near zero. Gas-liquid chromatography therefore requires many more theoretical plates to achieve a separation than does fractional distillation. The HETP of gas-liquid chromatography columns is usually much smaller than those of distillation columns so that the same length of gas-liquid chromatography column contains many more theoretical plates. Moreover, since separations by gas-liquid chromatography do not depend on gravity return of a counterstream, it is possible to use long lengths of column coiled into a small volume (50-ft columns are common). Another important factor aiding separations by gas-liquid chromatography is the availability of a wide range of liquid phases, one or more of which may give a greatly enlarged $\alpha$ value. This flexibility is inherently absent in fractional distillation. Hundreds of different liquid phases have been employed in gas chromatography and since these can be supported on over half a dozen different solid phases it is apparent that the practical art of column selection is complex. Table 11 lists three general purpose liquid phases with a few of their characteristics. A widely used solid support suitable for both polar and nonpolar samples is Chromosorb W, a white, flux-calcined diatomite. The liquid phase is dispersed on the surface of the support by adding the support to a solution of the liquid phase in a volatile solvent and allowing the solvent to evaporate slowly. The most serious deficiencies of gas-liquid chromatography are the related restrictions that the sample be readily vaporized and that it be small (typical sample sizes are 0.1 to 10 *micro*liters). Elaborate instruments are available that handle larger samples (0.1 to 10 *milli*liters) but they are expensive.

**TABLE 11**
**Liquid Phases for Gas-Liquid Chromatography**

| LIQUID PHASE | TEMPERATURE LIMIT | APPLICATION |
|---|---|---|
| Carbowax 20M | 250° | Separation of high boiling polar compounds |
| Silicone oil DC-550 | 275° | For compounds of intermediate polarity |
| Silicone gum rubber GE SE-30 | 375° | Separation of non-polar compounds |

## Laboratory Practice

**Thin-layer Chromatography (TLC).** Thin-layer plates may be either prepared in the laboratory or purchased. Unless a large number of plates is to be used or some special adsorbent or binder is required it is not much more expensive (and considerably faster) to purchase the plates. One convenient type comes as 20 × 20-cm sheets consisting of a 100-micron thick layer of adsorbent bound to a 200-micron thick sheet of plastic. With reasonable care these can be cut with ordinary scissors into 2 × 10-cm strips suitable for analytical separations.

In liquid-solid chromatography it is found that the resolution obtained depends on the ratio of solid to sample. For difficultly separated mixtures ($R_f$ values differing by 0.1 or less) the ratio should be at least 200:1. With more easily separated mixtures this ratio may be reduced proportionately. Because of the small amount of solid on a TLC plate the sample spot should be applied with a micro-capillary tube prepared by drawing out an ordinary capillary tube in a soft flame. In order to simplify the later calculation of $R_f$ values the plate should be marked lightly with two pencil lines 0.5 cm from each end. A micro drop of a solution of the sample in a volatile solvent is placed precisely on one of the two lines. If only one sample is being analyzed the drop should be centered between the edges; if more are to be analyzed on the same plate the spots should be placed symmetrically along the starting line. With 2 × 10-cm plates not more than three samples should be applied. When a low concentration of any component is being sought it is necessary to superpose additional drops on the first spot until sufficient sample has accumulated. The solvent should be evaporated between additions.

A convenient developing chamber for TLC plates can be prepared from an ordinary wide-mouthed screw cap bottle. The inside of the bottle is lined with a folded circle of filter paper, which acts as a wick to transfer the developing solvent to the upper portions of the chamber. It is essential that a gap in the paper be left near the top of the bottle so that the approach of the solvent front to the upper line on the plate can be seen without removing the cap. In practice sufficient solvent is added to the bottle to saturate the liner and leave a layer 2–4 mm deep at its shallowest point. The spotted end of the plate is centered in the bottom of the chamber with the upper edge leaned against the wall. The bottle is capped and gently set aside until the rising solvent front has just reached the upper line. The plate is then removed and the solvent allowed to evaporate from it.

If one or more of the components to be identified is colorless a convenient visualization technique is to place the plate in a second screw cap bottle containing a few crystals of iodine. Iodine vapor is absorbed on the plate wherever there is a concentration of organic material to produce a brown spot (the commercial plastic plates do not absorb a significant amount of iodine under these conditions). After the color has developed the plate is removed and a circle penciled around each spot. On exposure to air the brown iodine spots evaporate gradually.

**Column Chromatography.** A simple apparatus for liquid-solid column chromatography is a glass tube that has been constricted at one end (Figure 34). For separation of 0.1 to 0.5-g samples, a convenient tube size is 60 cm of 15-mm diameter tubing. This size will hold about 50 g of solid support and give a 100:1 ratio of packing to sample. Other samples sizes may be used with appropriately scaled apparatus.

A small wad of cotton or (preferably) glass wool is pushed into the tube with a wooden dowel until the wad rests on the constriction. With the tube clamped in an upright position a 1-cm supporting layer of clean coarse sand is then poured into the tube. Columns may be packed quickly by pouring in the solid support but a packing that gives superior separations may be prepared by using a slurry of the solid in the desired eluent. The slurry is added slowly through a funnel to a column that has been stoppered temporarily at the bottom with a medicine dropper bulb. One advantage of the slurry method is that the solid settles slowly giving a more uniform packing. Nonuniform packing usually permits channels to develop in the packing and these seriously diminish the resolution. Another advantage of the slurry method is that any heat of adsorption of the solvent on the support will be given off before the solid is added to the tube. When solid supports containing many active adsorption sites are packed dry and then wetted with solvent the heat evolved is frequently sufficient to expand the packed column and cause channels to develop.

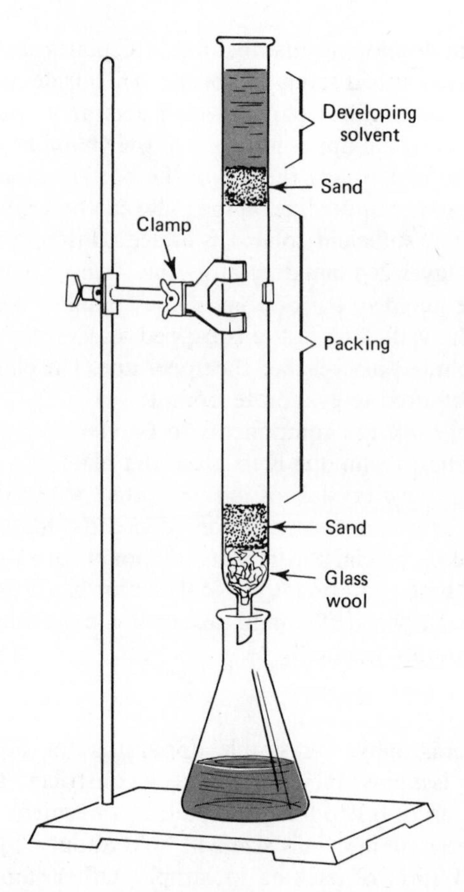


**Figure 34.** Apparatus for Column Chromatography

Alumina frequently produces tightly packed columns that have excessively slow flow-rates unless pressure is applied to force the eluent through the column. This problem can be minimized by adding about 10 percent Celite to the alumina in order to produce a coarser column. When the slurry method is applied to mixtures of adsorbents they should be added in particularly small portions in order to prevent segregation of the solids as they sink through the solvent.

After the addition of the solid support has been completed a second 1-cm layer of clean sand is added followed by a small circle of filter paper just large enough to touch the wall of the chromatography column. The sand and filter paper prevent the upper layer of support from being disturbed during subsequent operations. Small irregularities at the top of the column produce large distortions in the shape of each band of sample by the time it reaches the bottom of the column and may cause closely spaced bands to overlap.

The sample is normally added in a solution as concentrated as possible in order that narrow bands be formed. When the solution is ready, the bulb is removed from the column and the excess solvent allowed to drain. At the moment the solvent level reaches the top of the packing, the solution of the sample is added and allowed to penetrate into the column. After the sample has penetrated and before the column can become dry the residue adhering to the walls is washed down with a few drops of solvent. Enough eluent is then added to fill the tube.

As the eluent flows through the column the sample is separated gradually into bands that descend at different rates. It will be necessary from time to time to refill the tube with more eluent to keep the column from running dry. If all of the components of the sample are colored it will be obvious when to change receivers for each component. When one or more of the components is colorless it is necessary to collect fixed volumes of eluent in tared flasks, evaporate the solvent, and reweigh the flasks. A graph of the weights of sample eluted in successive fractions plotted against the accumulated volume of eluent reveals information about the number of components and the degree of separation. As shown in Figure 35, a large number of fractions yield more detailed information about the number of components.

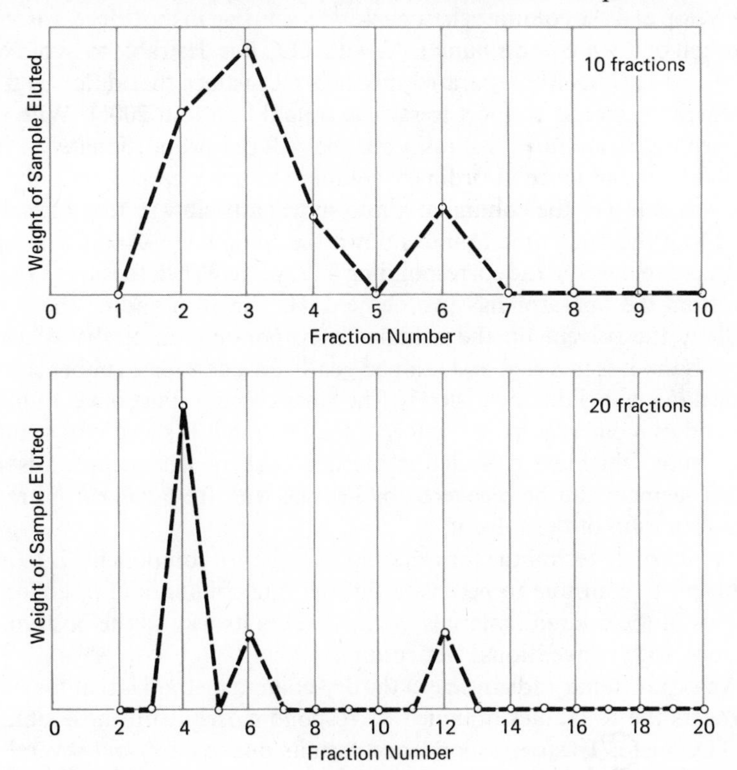


**Figure 35.** Effect of Number of Fractions on Chromatographic Resolution

**Dry-Column Chromatography.** Because of the superior resolution of mixtures using thin-layer chromatography (TLC) there have been numerous attempts to scale up the apparatus to achieve separations of preparative size samples. In general such "thick-layer" methods do not separate as well and the apparatus becomes unmanageably large for samples of more than about 0.5 g. The recently introduced technique of "dry-column" chromatography appears to combine the resolving ability of TLC with the large sample capacity of column chromatography and is rapidly becoming one of the most powerful separation techniques available to the organic chemist.

An activated chromatographic adsorbent is composed of adsorption sites with a wide range of activity, which leads to different migration rates and promotes poor resolution. When the adsorbent is partially deactivated by addition of small quantities of water, the more active sites are preferentially deactivated and better resolution is obtained from the leveled range of activity. Commercial or homemade TLC plates do not need deactivation because of their natural exposure to humid air. In the dry-column technique the adsorbent is deactivated by addition of a prescribed amount of water (and of the eluent if it is polar). The resulting adsorbent remains a free-flowing solid. In the dry-column method there is little danger that channels will develop and the column is best packed by pouring in the "dry" adsorbent (the origin of the method's name). As with TLC it is desirable to avoid overloading; with difficultly separated mixtures ($R_f$ values that differ by 0.1 or less) the ratio of solid support to sample should be about 200:1. With more easily separated mixtures this ratio may be reduced proportionately.

Unlike the practice in ordinary column chromatography the solvent is added to the top of the column at a rate sufficiently slow so that the solvent takes about the same time to move down the column (10–20 min) as would be required to develop the corresponding TLC plate. When the eluent reaches the bottom the operator has two choices. He can stop adding the solvent and allow the solvent on the column to evaporate completely. When the column is dry it is inverted and tamped gently on a cork or a rubber stopper. The packing will slide down slowly (the glass column must have a uniform bore) and as it emerges from the top of the column it is sliced into segments with a knife, which are collected in marked beakers. The sample adsorbed on each segment can be recovered by heating with fresh solvent, filtration, and evaporation of the solvent.

An alternate technique for isolating the mixture components is to allow the solvent to continue to percolate through the column and to collect the solutions of the desired materials as each makes its way to the bottom, just as is done with conventional wet columns.

An extraordinary advantage of the dry-column method is that for similar adsorbents the $R_f$ values obtained correspond closely with those obtained on a TLC plate. This correspondence permits one to carry out several trial separations using TLC and then to transfer the best conditions to a dry-column separation of a preparative size sample.

The correspondence between the $R_f$ values of the two techniques also can be used advantageously to calculate the volume of solvent required to elute a given component from the column. From the definition of $R_f$ as the ratio of migration distance of the sample to the solvent front it follows that the volume of solvent required to bring a sample to the bottom of the column is just $1/R_f$ times the volume of solvent required initially to just wet the entire column. For example, consider the case where a particular component has an $R_f$ value of 0.2 and a volume of solvent, $V$, is required to just wet the column. After one volume of solvent has passed down the column the component in question will be 0.2 of the way down; no solvent will yet have emerged. After a second volume, $V$, has been added, the component will have moved to a point 0.4 of the way down the column; one volume of solvent will have emerged. After five volumes of solvent have passed down the column the component will have reached the bottom of the column; four volumes of solvent will have emerged. An additional small volume of solvent will elute the component from the column. Because of the spread in the sample along the column and the consequent lack of a unique value for $R_f$ one should collect the emerging solvent a little before and a little after the main peak.

With easily separated mixtures the solid extrusion technique is recommended for its speed and conservation of solvent. With difficultly separated mixtures the alternative *liquid-flow* method will enhance the resolution and is preferred.

EXERCISE

# 8

# Chromatography

The experiments presented here are designed to illustrate three of the more commonly used chromatographic techniques. The materials to be separated in these initial experiments are highly colored because this permits the operator to see the band development and to determine immediately the consequences of careless or hasty technique.

**(A) Separation of a Dye-Mixture by Liquid-Solid Chromatography.** Insert a small wad of glass wool onto the constricted end of a 60-cm length of 15-mm diameter tubing and clamp the tube in an upright position (see Figure 34). Add a 1-cm layer of coarse sand to the tube. In a 100-ml beaker prepare a slurry of 5 g of alumina[1] and 5 g of Celite[2] in 50 ml of warm water and transfer the slurry in small batches to the tube (swirl between additions). The water that filters through the sand and glass wool should be collected and used to transfer any column material that remains in the beaker. After the packing has settled, add a second 1-cm layer of sand, followed by a small filter paper circle.[3]

When the last drop of water penetrates the column add 10 drops of the dye solution[4] to the top of the column. When the dye solution has penetrated add a few drops of water to wash down any dye adhering to the walls. After the wash water has penetrated fill the tube with water and allow the chromatogram to develop.

[1] Merck Reagent Grade Alumina, 71707, is suitable.

[2] The 50:50 alumina-Celite mixture is much more porous than the more usual 90:10 mixture and removes the need to use pressure or vacuum.

[3] These circles can be prepared conveniently from a larger piece of filter paper using a cork borer as a cutter and a large cork as a cutting surface.

[4] A suitable dye solution can be prepared by dissolving 0.1 g each of crystal violet, auramine hydrochloride and malachite green in 100 ml of water. Alternatively one of the dark colored commercial food colors may be analyzed.

**(B) Separation of Plant Pigments by Thin-Layer Chromatography.** In a clean porcelain mortar place 1 g of spinach, 1 g of clean sand, 5 ml of acetone and 5 ml of petroleum ether. Grind the spinach until the green chlorophyll appears to have been extracted completely. Decant the solution into a small beaker.

On a 2 × 10-cm thin-layer plate[5] draw two horizontal pencil lines across the plate 7 mm from each end. In the center of one line place a microdrop (see page 116) of the chlorophyll extract. Blow gently on the spot so that the solvent evaporates quickly. Repeat the addition of the extract several times until a distinct green spot is visible. The additions should superpose as closely as possible.

Add sufficient developing solvent[6] to an 8-oz wide-mouthed screw cap bottle containing a filter paper lining (see page 117) until a layer 3 mm deep is produced. Center the spotted plate in the bottle with the upper edge leaning against the side and screw the cap tightly into the bottle. When the solvent front reaches the upper pencil line, remove the plate and allow the solvent to evaporate. It may be necessary to repeat the development in order to resolve the chlorophylls.

**(C) Dry-Column Chromatography.** Insert a small wad of glass wool into the constricted end of a 60-cm length of 15-mm glass tubing and clamp the tube in an upright position (see Figure 34). Add a 1-cm layer of coarse sand to the tube followed by 25 g of deactivated Silica gel.[7] A second 1-cm layer of sand is added followed by a filter paper disc.[3]

Prepare 5 ml of leaf extract as in part B and place it in a 10-ml graduated cylinder. Add 5 ml of water and stir the mixture with a glass rod. Transfer the upper dark-green pentane layer to the column by means of a dropper. When the last drop has been adsorbed the chromatogram is developed by the periodic addition of a few milliliters of the developing solvent.[6] The frequency of additions is adjusted so that the solvent front moves down the column at a regular rate and reaches the bottom about 20 min after the first drop has been added. Compare the pattern of bands with the spots obtained by thin-layer chromatography.

[5] Eastman Chromagram Sheet, Type 6060 or 6061, is suitable. It can be cut conveniently with ordinary scissors. In humid climates it is desirable to activate the plates by heating them in an oven at 100° for 15 to 30 min.

[6] Suitable developing solvent is prepared by mixing 8 volumes of petroleum ether with 2 volumes of acetone. Lower percentages of acetone tend to improve the separation of chlorophylls *a* and *b* at the expense of poorer separation from the yellow xanthophylls.

[7] The Silica Gel must be deactivated at least a day in advance by adding 2.5 ml of water and 2.5 ml of acetone (used in the developing solvent) to 25 g of Silica Gel (Fisher 60–200 mesh, Grade 950 is suitable) in a 125-ml Erlenmeyer flask. The flask is stoppered tightly, shaken several times to disperse the adsorbed liquids, and put aside for 24 hr.

## Questions

**1.** Arrange the following compounds in the sequence of their elution from a silica gel column, using benzene as eluent: $CH_3(CH_2)_{10}CH_3$, $CH_3CO_2H$, $CH_3CH_2CH_2OH$, $CH_3COCH_2CH_3$.

**2.** Suggest suitable liquid phases for separation of carboxylic acids by liquid-liquid chromatography. (Hint: consider the results of experiment 6 on extraction.)

**3.** Silicone oil exhibits approximately ideal behavior as a liquid phase in gas-liquid chromatography, so that relative volatility values, $\alpha$, obtained from fractional distillation can be employed in estimating the number of VPC theoretical plates required. (a) If a mixture of two liquids requires 50 theoretical plates for adequate separation by fractional distillation, how many VPC plates would be required for separation using silicone oil as the liquid phase? (b) With silicone oil as the liquid phase, what would be the expected elution order of: acetone, *n*-butyl alcohol, benzene, and pentane?

# Identification of Organic Compounds

# Determination of Structure

Structure elucidation is an important activity of organic chemists that includes problems presenting a broad range of difficulty. At one extreme chemists are frequently faced with simple questions such as identifying a functional group, while at the other extreme they may wish to elucidate the complete structure of a new organic compound isolated from natural sources or produced in the course of laboratory studies.

A complete structure elucidation begins with the isolation of the new substance in a pure state and qualitative tests to disclose the presence of elements such as nitrogen, sulfur, halogens, etc. Quantitative analysis furnishes the weight composition of the substance from which one can calculate an empirical formula, which gives the atomic ratios of the elements present. Determination of the molecular weight permits the assignment of a definite molecular formula, which expresses the actual number of atoms of each element present in the compound.

The next stage in structure elucidation involves identification of the functional groups and other characteristic structural features present in the molecule. The methods used today are both chemical and physical.

Finally, the partial information is pieced together to give the total structure. Where possible the assigned structure is confirmed by an independent synthesis.

A detailed illustration of structure elucidation is presented at the end of this section. Since the advent of computers, with their ability to handle large amounts of information, the direct X-ray structure determination of crystalline solids has become increasingly important, particularly with large organic molecules containing many functional groups. The X-ray method requires several months of work and is not applicable to liquids or amorphous solids, so that for not too complex molecules the approach outlined in this section remains the method of choice.

## Qualitative Detection of Elements

**Carbon, Hydrogen, Oxygen.** The elements that are most commonly present in organic compounds are carbon, hydrogen, and oxygen. The presence of carbon and hydrogen may be detected qualitatively by heating the substance in a tube with dry powdered copper oxide, whereby carbon dioxide and water are formed. Carbon is detected by passing the evolved gases into an aqueous solution of calcium or barium hydroxide, in which the carbon dioxide produces a precipitate of the carbonate. Hydrogen is detected by the condensation of droplets of water in the cool upper portion of the reaction tube.

There is no satisfactory qualitative test for the presence of oxygen in organic compounds. In order to determine whether or not oxygen is present, it is necessary to have recourse to quantitative analysis. If the sum of the percentages of all the known constituent elements does not amount to 100 percent, the deficit is taken as the percentage of oxygen.

**Nitrogen, Halogens, Sulfur.** The qualitative detection of these elements in organic compounds is more difficult than in inorganic compounds because most organic compounds are not appreciably ionized in solution. Since the tests used in qualitative inorganic analysis are based upon ionic reactions, they cannot be applied directly to organic compounds. For example, sodium chloride or bromide gives an immediate precipitate of the silver halide when treated with an aqueous solution of silver nitrate; carbon tetrachloride, carbon tetrabromide, and most organic halides, do not produce a precipitate of the silver halide when treated with aqueous silver nitrate solution. These organic halides fail to respond to the common test for halides because they do not furnish an appreciable amount of halide ions in the solution.

For qualitative detection it is necessary, therefore, to convert elements such as nitrogen, sulfur, and halogens into ionized inorganic substances in order to apply suitable tests. This conversion may be accomplished by several methods, of which the most general is that of fusion with metallic sodium. In this way there are formed sodium cyanide, sodium halides, sodium sulfide, etc., as indicated below. The resulting ionized compounds may then be detected by applying the usual inorganic tests.

$$\text{Organic compound containing C, H, O, N, S, Cl} + \text{Na} \xrightarrow[\text{temperature}]{\text{high}} \text{NaCN} + \text{NaCl} + \text{Na}_2\text{S} + \text{NaOH} + \text{etc.}$$

***Nitrogen.*** The filtered alkaline solution is treated with aqueous ferrous sulfate and ferric chloride, boiled for a few moments and acidified with hydrochloric acid. If nitrogen is present, a precipitate of prussian blue results:

$$2NaCN + FeSO_4 \longrightarrow Fe(CN)_2 + Na_2SO_4$$

$$4NaCN + Fe(CN)_2 \longrightarrow Na_4Fe(CN)_6$$

$$3Na_4Fe(CN)_6 + 4FeCl_3 \longrightarrow \underset{\text{(prussian blue)}}{Fe_4[Fe(CN)_6]_3} + 12NaCl$$

***Sulfur.*** A fresh portion of the filtered alkaline solution is acidified with acetic acid and treated with an aqueous solution of lead acetate. If sulfur is present, a dark-brown precipitate of lead sulfide results.

Sulfur may also be detected by the use of a solution of sodium nitroprusside, which gives a deep reddish-violet coloration with solutions of sulfides. The following equation has been given for the reaction:

$$Na_2S + Na_2Fe(CN)_5NO \longrightarrow Na_4Fe(CN)_5S + NO$$

***Halogens.*** A fresh portion of the filtered alkaline solution is acidified with nitric acid and boiled for a short while in order to expel hydrogen cyanide and hydrogen sulfide (if nitrogen and sulfur are present). The resulting solution, which contains free nitric acid, is treated with aqueous silver nitrate, and if halogens are present a precipitate of silver halide results. The usual qualitative methods are employed in testing for the individual halogens, or in detecting two or more halogens in the presence of each other.

In place of the sodium fusion method, certain other methods may be used for the conversion of the elements into ionized inorganic substances. For the detection of halogens, heating with pure calcium oxide (quicklime) is sometimes employed. By this means the calcium halides are produced and can be detected in the usual way. Fusion with sodium peroxide in the Parr peroxide bomb is a very useful and generally applicable method for the qualitative and quantitative determination of carbon, halogens, sulfur, phosphorus, etc., in organic substances.

## Quantitative Organic Analysis

**Carbon and Hydrogen.** The quantitative determination of carbon and hydrogen in organic compounds is effected by combustion of the substance in an atmosphere of purified air or oxygen, and weighing the resulting water and carbon dioxide by absorption in suitable reagents (the method of Liebig). For this purpose the combustion is carried out in a glass tube packed with copper oxide, which is maintained at a high temperature to insure complete oxidation of the organic matter. The gases from the combustion tube are passed through an absorption train which consists of a tube containing a water absorbent, a tube containing a carbon dioxide absorbent, and a safety tube containing absorbents for an additional quantity of carbon dioxide and water. Increases in weight of the first and second absorption tubes give the weights of water and carbon dioxide respectively, and from these data and the weight of the original substance the percentages of hydrogen and carbon may be calculated.

***Oxygen.*** Although methods have been developed for the direct determination of oxygen,[1] the amount present is usually found by determining all of the other elements present and assuming the deficit to be oxygen.

***Nitrogen.*** The most general method for the quantitative determination of nitrogen is the Dumas method, which consists of combustion of the organic substance in an atmosphere of carbon dioxide and determining the volume of gaseous nitrogen produced. For this purpose the combustion is carried out in a tube packed with copper oxide and containing a gauze of metallic copper (to reduce oxides of nitrogen to free nitrogen). The gaseous nitrogen is collected over a strong solution of potassium hydroxide, which serves to remove carbon dioxide and water, and its volume is measured carefully at a definite temperature and pressure. From these data the weight of nitrogen is obtained.

Nitrogen in the form of amino groups may be determined conveniently by the Kjeldahl method. The organic substance is digested with concentrated sulfuric acid and potassium sulfate, in the presence of a small amount of cupric or mercuric sulfate as oxidation catalyst, and the amino nitrogen is converted to ammonium sulfate. The ammonia is then liberated by the addition of alkali and distilled into an excess of standard acid solution. The excess of standard acid is determined by titration with standard alkali, and the quantity of nitrogen may then be calculated from the amount of acid required to neutralize the ammonia produced.

Compounds containing nitrogen in the form of nitro groups, azo groups, and other reducible forms may be converted by preliminary reduction to amino compounds, and the nitrogen determined by the Kjeldahl method.

**Halogens, Sulfur, etc.** The only special feature of the quantitative determination of these elements in organic compounds is the necessity of a preliminary decomposition to form ionized substances. When this decomposition has been effected, the usual quantitative methods, gravimetric and volumetric, may be employed.

**Molecular Weight Determination.** Many techniques are available for determining the molecular weight of a compound. Most of these involve solutions of the unknown and depend on Raoult's law for equating the weight of dissolved unknown with the number of moles present (freezing point depression and boiling point elevation).

[1] J. B. Niederl and V. Niederl, *Micromethods of Quantitative Organic Analysis*, 2nd edition, John Wiley and Sons, Inc., New York (1948), page 207.

With molecules containing an easily characterized functional group sometimes it is possible to determine quantitatively the amount present in a given weight of the unknown and hence the equivalent weight. The molecular weight will be an exact multiple of this quantity.

Occasionally a special technique like the X-ray determination of the crystal lattice dimensions of a unit cell (repeating three dimensional structure) combined with density measurements is used to determine the combined molecular weights of the molecules in a unit cell. Division of this quantity by the number of molecules per unit cell gives the molecular weight.

Mass spectroscopy affords a rapid, precise determination of molecular weights up to about 500 mass units on samples of only 1 milligram or less. With certain instruments of extremely high mass resolution it is possible to make use of the non-integral values of atomic weights of even the pure isotopes ($C^{12}$ = 12.00000 basis, $H^1$ = 1.007825, $O^{16}$ = 15.99491, etc.) to determine empirical formulas directly.

## Preliminary Identification Tests

**Ignition Test.** From the behavior of a small sample when heated in a flame it is possible to determine whether a solid has an accessible melting point, and whether a solid or a liquid is volatile, forms volatile decomposition products, or is explosive. Combustion of the sample to give a sooty flame indicates the presence of unsaturation, aromatic groups or long aliphatic chains. A residue indicates the presence of a metal, usually as the salt of an acid.

**Physical Constants.** Melting points and boiling points are characteristic properties of pure materials. The boiling point of an unknown liquid is related approximately to its molecular weight; the melting point of a solid is partly determined by molecular weight but more importantly by the presence or absence of polar groups that interact strongly in the crystal lattice.

**Solubility Tests.** The classification of an unknown according to its solubility behavior (see Solubility Relationship, page 104 and the solubility chart on page 132) when combined with a qualitative determination of the elements present greatly limits the number of functional groups that need be considered. These may be further differentiated by either chemical tests or spectroscopic features but most effectively by a combination of both. Systematic solubility classification was introduced by Kamm[2] and his scheme forms the basis of the flow chart presented on the following pages. The

[2] Kamm, *Qualitative Organic Analysis*, John Wiley and Sons, Inc., New York (1922).

## Solubility Classification Scheme*

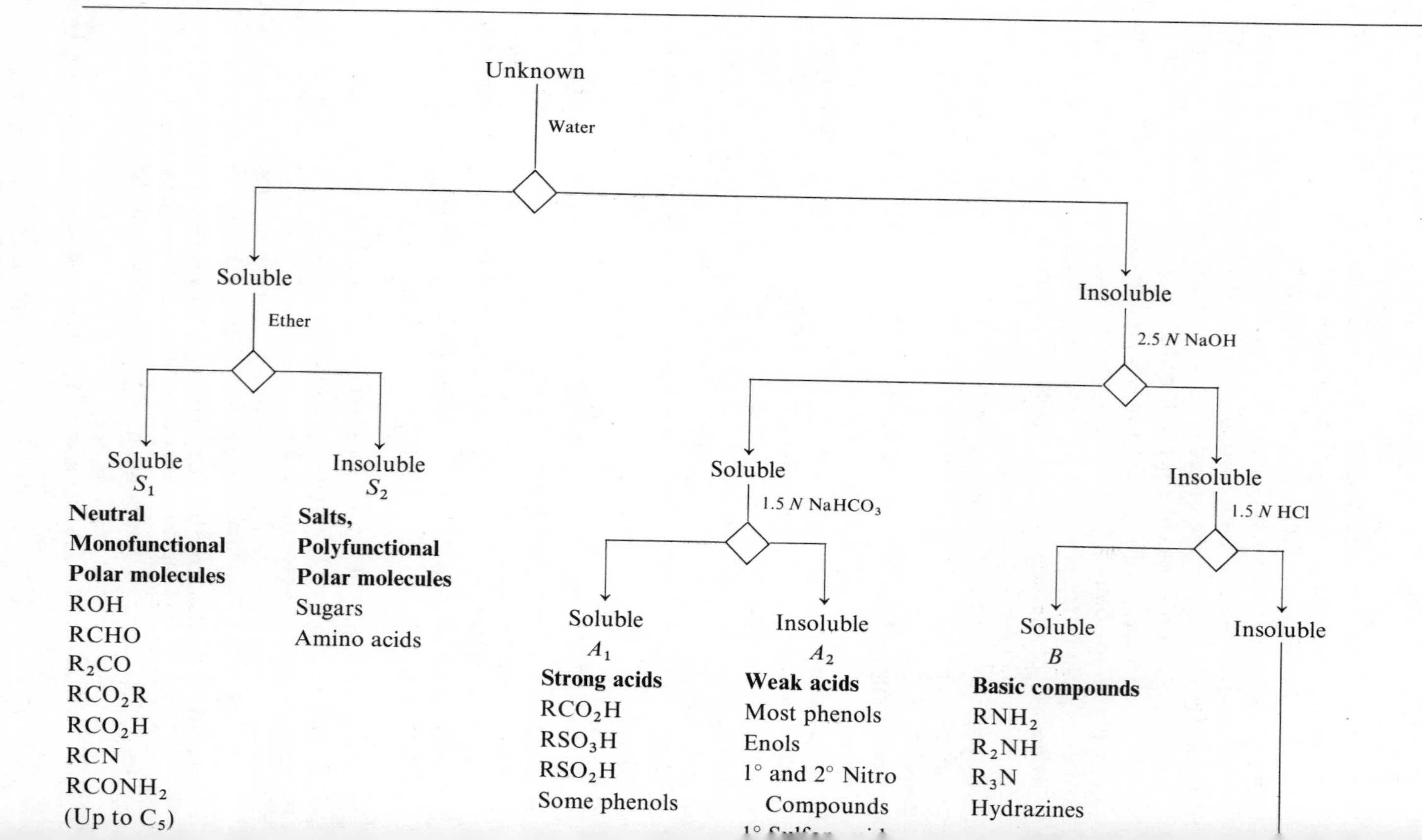

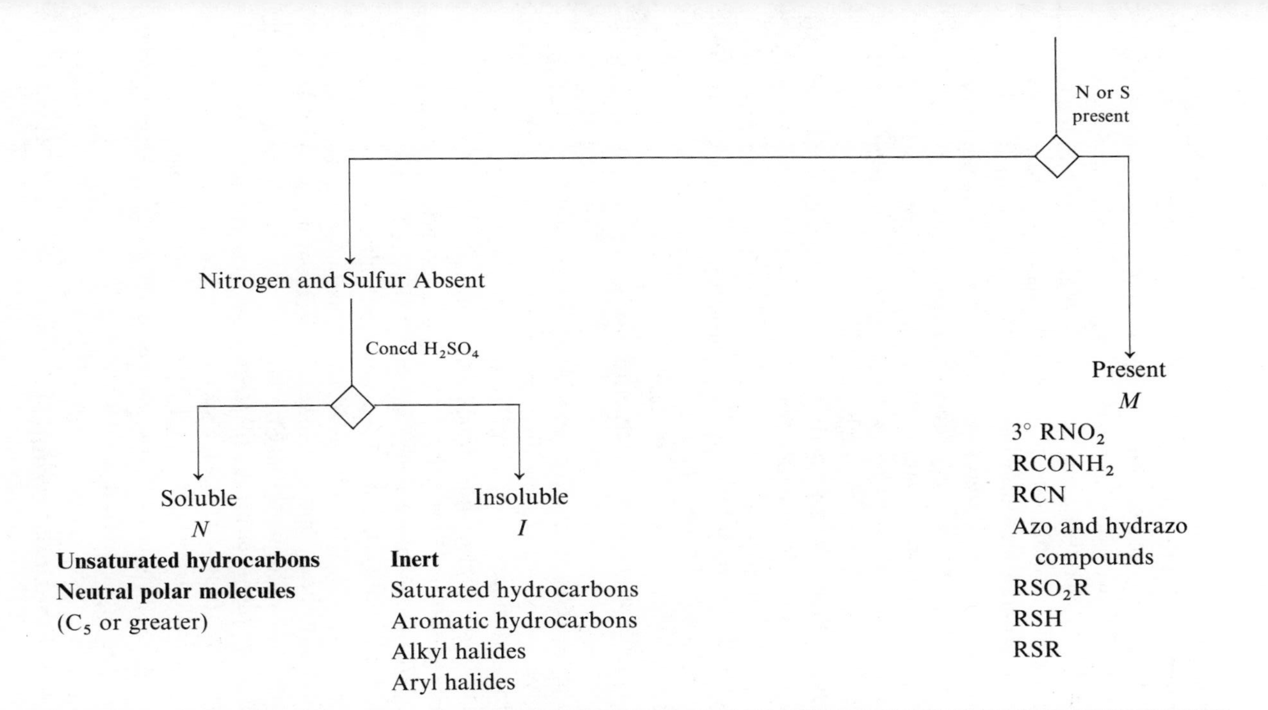


* The diamonds indicate decisions to be made in response to the chemical test shown at the upper right-hand edge. Possible answers are given at the ends of the bent arrows leading from the right and left apices.

classes of molecules included in this chart are not exhaustive, but do represent the more commonly encountered types.[3]

Molecules belonging to solubility group $S_1$ include polar molecules, usually with five or fewer carbon atoms. Particularly effective water solubilizing groups such as carboxylate anions or quaternary ammonium cations confer water solubility to molecules with as many as twenty to twenty-five carbon atoms; polyfunctional molecules may have proportionately larger numbers of carbon atoms. Salts and polyfunctional molecules are insoluble in ether and comprise group $S_2$. Any functional group may be present in a molecule belonging to groups $S_1$ and $S_2$. The aqueous solution from the water solubility test should be tested with pH paper to determine if the dissolved molecule is acidic, neutral, or basic.

Acidic species with more than five (but fewer than about twenty) carbon atoms will be soluble in 2.5 *N* sodium hydroxide. Strong acids, $A_1$ (carboxylic acids, sulfonic, and sulfinic acids, some phenols bearing electron-withdrawing groups) will also be soluble in 1.5 *N* aqueous sodium bicarbonate but weak acids, $A_2$ (most phenols, enols, primary, and secondary nitro compounds, primary sulfonamides), will not dissolve.

Molecules soluble in 1.5 *N* hydrochloric acid, group *B*, will be either amines or hydrazines.

In classifying the remaining molecules, which are insoluble in all of the aqueous test solvents, it is useful to distinguish between those containing nitrogen or sulfur and those that do not. Those with these elements present, *M*, include tertiary nitro compounds, amides, nitrites, azo and hydrazo compounds, sulfones and sulfonyl derivatives, mercaptans, sulfides, and sulfates.

When nitrogen and sulfur are absent the molecules are further classified according to their solubility in cold, concentrated sulfuric acid. Soluble molecules, group *N*, may contain any of the functional groups (except those with N and S) found in the $S_1$ and $S_2$ classes. Molecules falling into class *N* will have more than five carbon atoms. Compounds of one additional group falling into class *N* are the unsaturated hydrocarbons, which react with concentrated sulfuric acid to produce polar species that are soluble. The inert molecules, group *I*, include the saturated hydrocarbons, aromatic hydrocarbons (not as reactive towards sulfuric acid as olefins) and alkyl and aryl halides.

## Identification of Functional Groups

Until recent years the only methods available for identification of functional groups were chemical test reactions. In the last two decades there

[3] For more exhaustive treatments of solubility classification and lists of molecular classes see Cheronis and Entrikin, *Identification of Organic Compounds*, Interscience Publishers, New York (1963), or Shriner, Fuson, and Curtin, *The Systematic Identification of Organic Compounds*, 5th edition, John Wiley and Sons, Inc., New York (1964).

have become readily available various spectroscopic instruments that give rapid identification of many functional groups and other structural features. The chemical and instrumental procedures largely complement each other; instrumental methods are precise but normally do not give a complete identification of an unknown molecule so that chemical tests still must be used. Moreover even when an instrument does give an identification, confirmation by some chemical procedure is highly desirable and is considered normal practice. Organic chemists must be familiar with both approaches to identification and recognize the scope and limitation of each.

**Spectroscopic Methods.** Electromagnetic radiation passing through molecules is transmitted, scattered,[4] and absorbed. The absorption phenomenon is valuable for chemists as a result of relationships existing between the wavelength of absorbed radiation and molecular structure. A model of absorption is desirable for understanding these relationships. For the *infrared* region of absorption (wavelengths of about 1 to 50 microns), molecules can be treated as masses (atoms) connected by springs (bonds). It is a property of mechanical systems that they have characteristic vibrational frequencies (resonances) that absorb energy from applied oscillatory forces. Small masses or stiff springs give rise to high vibrational frequencies; it is a consequence of the small weights of atoms and the tightness of bonds that molecular vibrations occur in the infrared region ($6 \times 10^{12}$ to $3 \times 10^{14}$ cycles/sec). The extent of absorption per molecule depends on the change of dipole moment during the vibration; a large change in dipole moment accompanies intense absorption.

Even for the simple molecules the number of vibrational frequencies is large. Fortunately, certain functional groups and structural units have characteristic absorption frequencies that change little from molecule to molecule. For many structural units, the frequency shifts that do occur can be related to variations in the neighboring structure. Table 12 lists a few structural units and their characteristic absorption ranges in the infrared region.[5] Further information on the characterization of functional groups by infrared spectroscopy is included in the discussion of specific groups under Chemical Methods (page 140); the infrared spectrum of methyl benzoate is given as a typical example.

In the visible and ultraviolet regions (wavelengths of about 150 to 800 millimicrons) a different method for absorption is required. The frequencies are so high ($\sim 4 \times 10^{14}$ to $2 \times 10^{15}$ cycles/sec) that the massive atoms do not respond to the rapid oscillation of the electric field and, instead,

[4] Incoherent light scattering is used to elucidate polymer structure; coherent light scattering (Raman effect) provides structural information similar to that obtained by infrared absorption.

[5] Two excellent reference books on correlation of structure with infrared absorption frequencies are L. J. Bellamy, *The Infra-red Spectra of Complex Molecules*, 2nd edition, John Wiley and Sons, Inc., New York (1958) and K. Nakanishi, *Infrared Absorption Spectroscopy*, Holden-Day, Inc., San Francisco (1962).

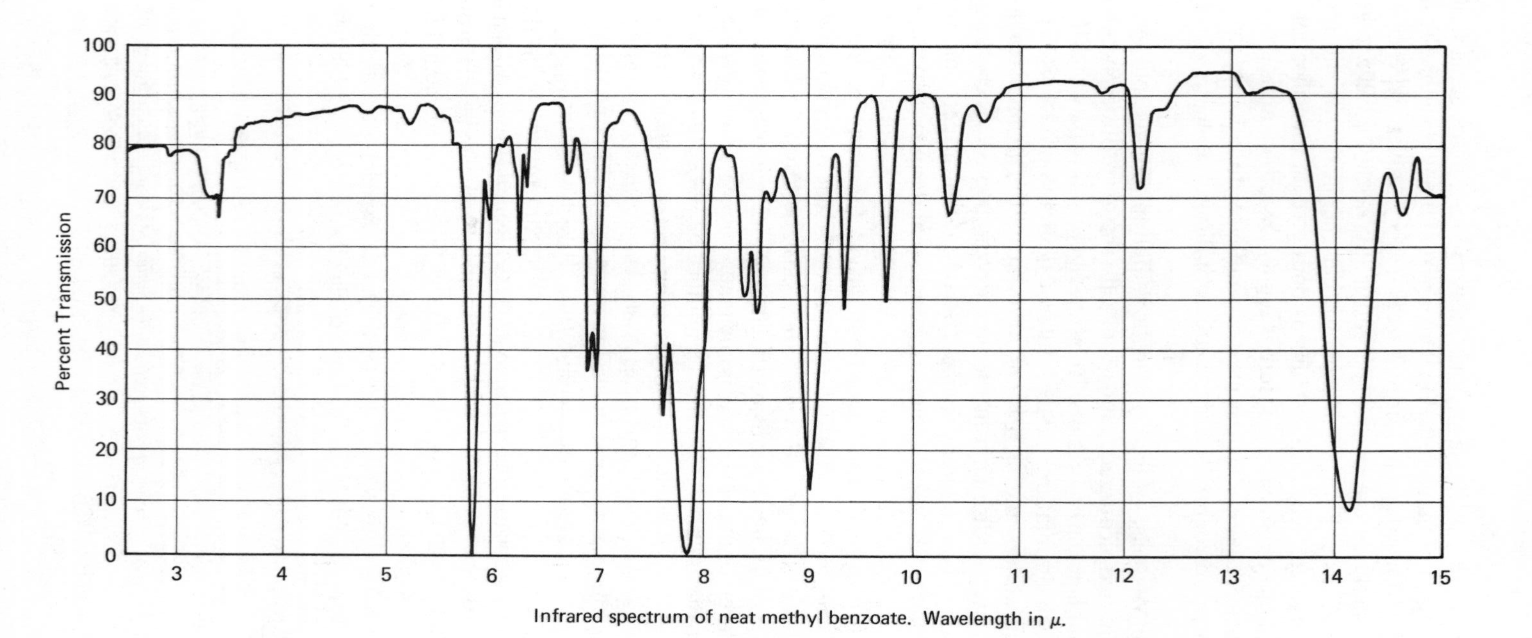


Infrared spectrum of neat methyl benzoate. Wavelength in $\mu$.

**TABLE 12**
**Infrared Fundamental Absorption Regions**

| | | FREQUENCY, $CM^{-1}$ | WAVELENGTH, $\mu$ |
|---|---|---|---|
| OH | Alcohol (free) | 3650–3580 | 2.74–2.79 |
| | (Hydrogen bonded) | 3550–3210 | 2.82–3.12 |
| | Acid | 2700–2500 | 3.70–4.00 |
| NH | Amine | near 3400 | near 2.94 |
| CH | Alkane | 2960–2850 | 3.37–3.50 |
| | Alkene | 3095–3010 | 3.23–3.32 |
| | Alkyne | 3300 | 3.03 |
| | Aromatic | near 3030 | near 3.30 |
| C≡C | Alkyne | 2260–2140 | 4.42–4.76 |
| C=C | Alkene | 1680–1620 | 5.95–6.16 |
| | Conjugated diene | near 1600 | near 6.25 |
| | Aromatic | near 1600 and near 1500 | near 6.25 and near 6.66 |
| C=O | Aldehyde | 1740–1720 | 5.75–5.81 |
| | Ketone | 1725–1675 | 5.79–5.97 |
| | Acid | 1725–1700 | 5.79–5.87 |
| | Ester | 1750–1720 | 5.71–5.86 |

it is the electrons that vibrate. Since all electrons have the same mass, only variations in electron binding determine absorption frequencies. Structural units containing loosely bound electrons, such as occur in double bonds, have absorption frequencies that can be determined easily. Table 13 lists several structural units that have characteristic absorption regions.[6] The ultraviolet spectrum of methyl benzoate is given.

The nucleus of a hydrogen atom is surrounded by a weak magnetic field and is said to have a nuclear magnetic moment. When hydrogen atoms or materials containing hydrogen atoms are placed in a strong magnetic field the nuclear magnetic moments line up either parallel or antiparallel to the direction of the magnetic lines of force. The two possible orientations differ in energy and if the nuclei are exposed to electromagnetic radiation of the correct frequency (energy) the parallel moments absorb energy, flip over and become antiparallel moments. This absorption phenomenon is called *nuclear*

[6] An excellent reference book on correlation of structure with ultraviolet absorption frequencies is A. E. Gillam and E. S. Stern, *An Introduction to Electronic Absorption Spectroscopy in Organic Chemistry*, 2nd edition, Edward Arnold Ltd, London (1957).

**TABLE 13**
**Ultraviolet Absorption Regions**

| STRUCTURAL UNIT | WAVELENGTH AT MAXIMUM ABSORPTION, $m\mu$ |
|---|---|
| Isolated double bond | 180–195 |
| Isolated carbonyl group | 270–290 |
| Conjugated dienes *cis* | near 240 |
| *trans* | near 220 |
| Alkylbenzenes | 255–270 |

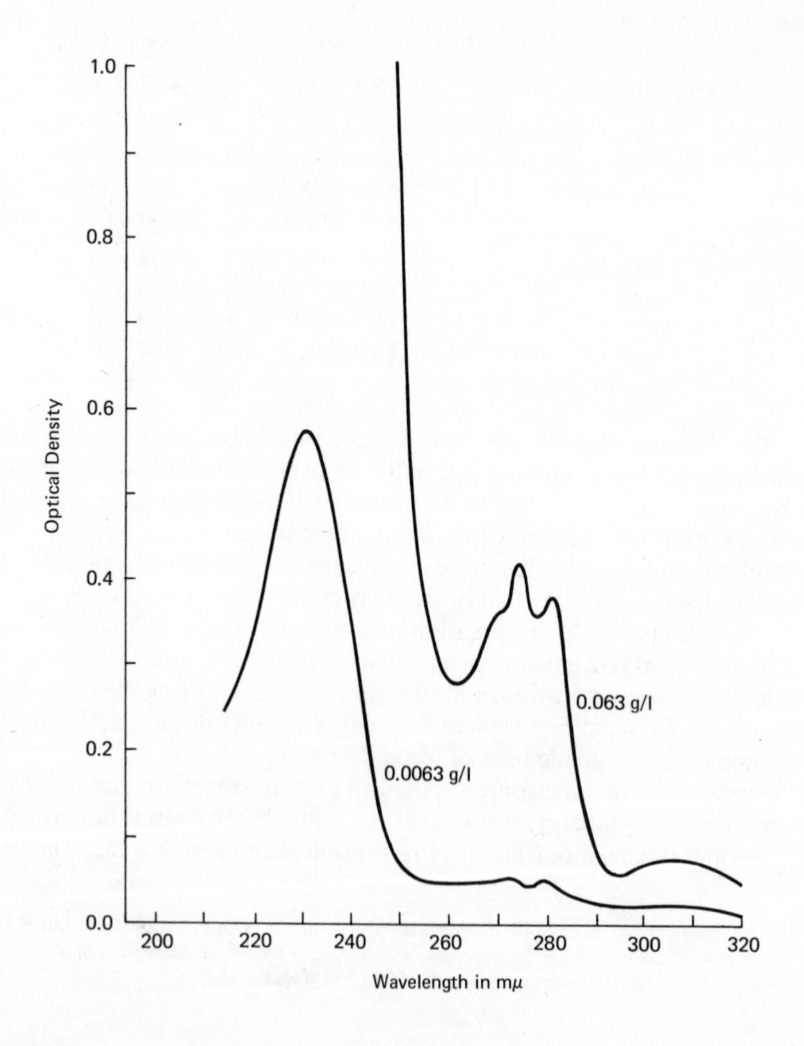

*magnetic resonance* (*nmr*).[7] The difference in energy between parallel and antiparallel orientations and hence the frequency required for absorption is largely determined by the electron density surrounding the nucleus. A high electron density leads to a small energy difference and a low absorption frequency.

Because of the correlation of electron density with structure, nmr is a powerful tool for determining the structural environment of hydrogen atoms in organic molecules. The difference in resonance for several proton environments is listed in Table 14. The frequency is relative to a tetramethylsilane standard and is expressed in parts per million (chemical shift). With suitable experimental refinements it is possible to use nmr techniques to determine the ratios of the number of hydrogen atoms with the same environment occurring in a molecule. It is sometimes possible to determine distances between hydrogen atoms. Certain other nuclei also have nuclear magnetic moments and show the nmr phenomenon. Unfortunately $C^{12}$, the common isotope of carbon, is not among them.

**Chemical Methods.** The chemical approach to functional group analysis of a pure compound begins with the classification of the unknown according to its solubility behavior (see Solubility Tests, page 131). The solubility classification combined with a qualitative determination of the elements present greatly limits the number of functional groups that need be considered.

Thoughtful application of chemical test reactions gradually narrows the range of possibilities until the functional groups present in the unknown have been identified. The significance of the systematic procedure for identification of functional groups cannot be appreciated until the student has become familiar with the typical reactions of the functional groups.

In the following presentation of test reactions for individual classes of molecules any characteristic infrared or other spectroscopic absorptions are included. Information about the preparation of derivatives is also included. The list of functional groups is not exhaustive but does include the more commonly encountered types.

**Alcohols.** Infrared spectra show characteristic absorption at 3650–3580 $cm^{-1}$ and 3500–3210 $cm^{-1}$. Primary, secondary, and tertiary alcohols may be differentiated by the Lucas test (Experiment 3). Alcohols may be converted to solid ester derivatives by reaction with an appropriate acid chloride (Experiment 35, page 355) or to α-naphthyl carbamates with α-naphthyl isocyanate (Experiment 3).

[7] Excellent reference books on this subject are J. D. Roberts, *Nuclear Magnetic Resonance*, McGraw-Hill Book Co., Inc., New York (1959) and L. M. Jackman, *Applications of Nuclear Magnetic Resonance Spectroscopy in Organic Chemistry*, Pergamon Press, New York (1959).

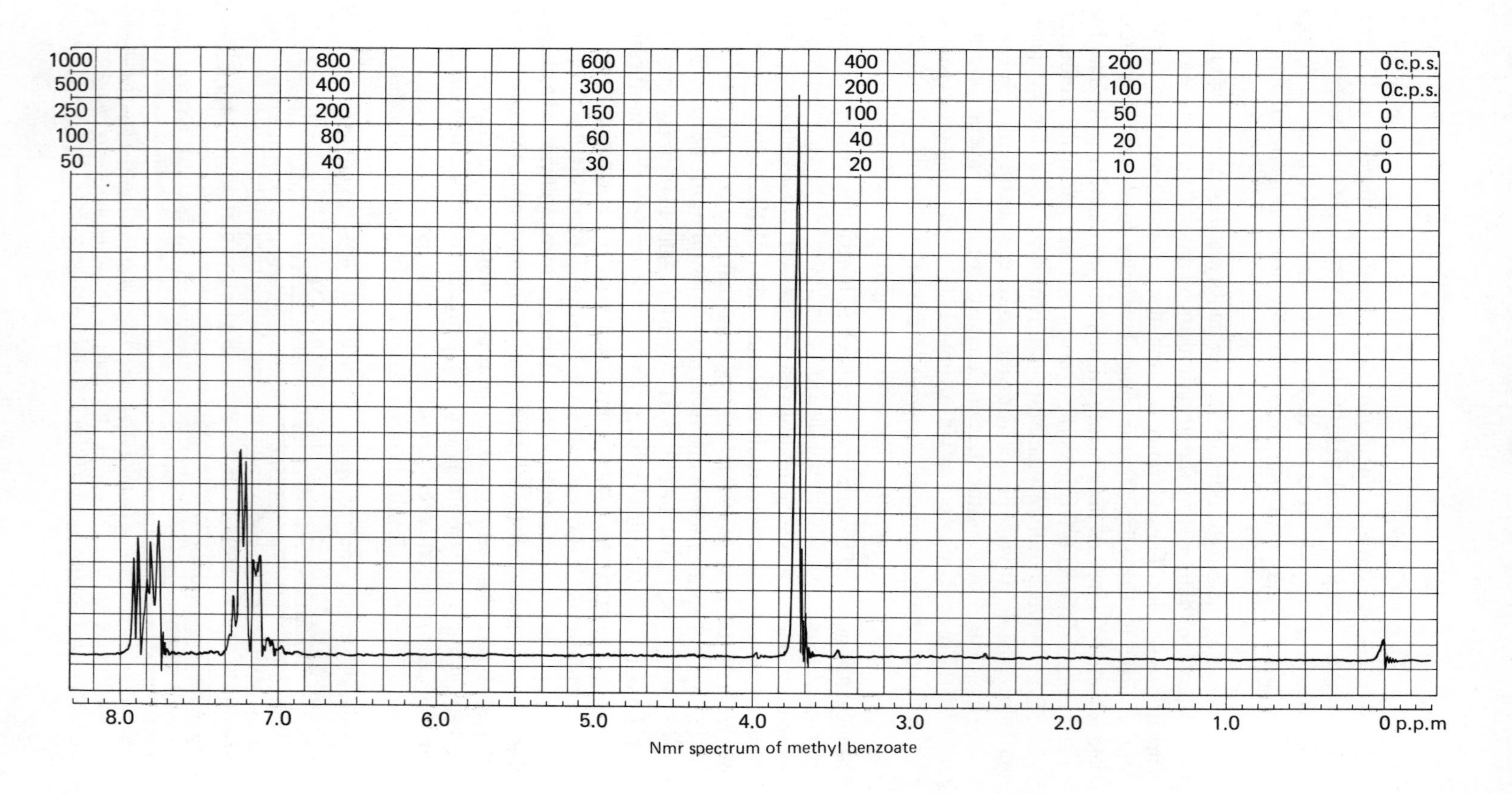


Nmr spectrum of methyl benzoate

**TABLE 14**
**Nuclear Magnetic Resonance Chemical Shifts**

| STRUCTURAL ENVIRONMENT | | CHEMICAL SHIFT, PPM FROM TETRAMETHYLSILANE |
|---|---|---|
| $-\overset{\vert}{\underset{\vert}{C}}-CH_3$ | (saturated) | 0.2 |
| $-CH_2-$ | (saturated) | 1.2–1.3 |
| $-\overset{\vert}{C}H-$ | (saturated) | 1.4–1.6 |
| $X-\overset{\vert}{\underset{\vert}{C}}-CH_3$ | (X = halogen, OH, N) | 1.0–2.0 |
| $>C=C<CH_3$ | | 1.6–1.9 |
| $ArCH_3$ | | 2.1–2.5 |
| $O=\overset{\vert}{C}-CH_3$ | | 2.1–2.6 |
| $-C{\equiv}C-H$ | | 2.4–3.1 |
| $-O-CH_3$ | | 3.5–3.8 |
| $>C=CH_2$ | (nonconjugated) | 4.6–5.0 |
| $>C=C<^{H}$ | | 5.2–5.7 |
| $Ar-H$ | | 6.6–8.0 |
| $-C(=O)H$ | | 9.8–10.8 |

**Aldehydes and Ketones.** Both classes of molecules have carbonyl absorption near 1725 $cm^{-1}$, but only aldehydes have a CH stretching absorption near 2705 $cm^{-1}$ (aliphatic) or 2730 $cm^{-1}$ (aromatic). In the nmr, aldehydes have a characteristic absorption near 10.0 ppm. Aldehydes and ketones may be detected by their reaction with hydroxylamine hydrochloride, semicarbazide hydrochloride, phenylhydrazine hydrochloride and 2,4-dinitrophenylhydrazine as described in Experiment 9. These same reagents yield solid derivatives for final identification. Aldehydes may be distinguished from ketones by the Tollens test or the Fuchsin-Aldehyde reagent (Experiment 7). Aldehydes give a methone derivative (Experiment 7).

**Amides.** In the infrared primary, secondary and tertiary amides absorb near 1650 $cm^{-1}$. Primary amides show two $-NH$ bands near 3500 $cm^{-1}$ and 3400 $cm^{-1}$ while secondary amides show one band near 3250 $cm^{-1}$. Amides are hydrolyzed by refluxing with base to release ammonia (primary amides) or amines (secondary and tertiary amides) as described in Experiment 13. The carboxylic acid and nitrogenous components can be isolated and identified separately.

**Amines.** Primary amines show two infrared absorption bands near 3500 and 3400 $cm^{-1}$; secondary amines show one band near 3400 $cm^{-1}$. Amines are most easily detected by their solubility behavior and may be further classified by their reaction with benzenesulfonyl chloride (Experiment 12). The preparation of solid derivatives of primary and secondary amines is described in Experiment 12 and melting points are listed in Table 20. Tertiary amines may be characterized as quaternary ammonium salts.

**Anhydrides.** The infrared spectra of symmetrical and unsymmetrical anhydrides show two carbonyl bands near 1870 and 1760 $cm^{-1}$. Anhydrides may be detected by the hydroxamate test (page 230). Solid derivatives such as the amide or anilide may be prepared by treatment with ammonia or amines (page 238).

**Carboxylic Acids.** In addition to a carbonyl absorption band near 1700 $cm^{-1}$, acids show a characteristically broad absorption band in the range of 3000–2500 $cm^{-1}$. The solubility behavior is diagnostic for acids. Acids may be characterized further by conversion to an amide (Experiment 13a), an anilide (Experiment 15, page 261), or by the determination of its neutralization equivalent (page 358) or Duclaux constant (Experiment 10, page 229).

**Esters.** This class of molecules shows a carbonyl absorption near 1740 $cm^{-1}$. They may be detected by the sensitive hydroxamate test (page 230) and identified further by determination of the saponification equivalent (page

228) as well as by hydrolysis, followed by isolation and identification of the acid and alcohol components.

**Hydrocarbons.** There are no satisfactory tests for alkanes; they are usually detected by the absence of other functional groups and characterized by their physical constants. Alkenes and alkynes give positive tests for unsaturation; terminal alkynes react with cuprous chloride (Experiment 1). Alkenes show medium intensity absorption in the infrared near 1670 $cm^{-1}$. Alkynes have a distinctive absorption near 2200 $cm^{-1}$. Aromatic hydrocarbons are difficultly detected by chemical tests and the infrared spectra do not show truly distinctive absorption. However the nmr spectra of aromatic hydrocarbons are characteristic with absorptions in the range of 6.6–8.0 ppm. Protons attached to alkenes and alkynes usually stand out in nmr spectra (see Table 14). Aromatic hydrocarbons may be identified by nitration (Experiment 18) to yield solid nitro derivatives.

**Nitriles.** These molecules have distinctive infrared absorption near 2250 $cm^{-1}$. Nitriles are usually identified chemically by hydrolysis to ammonia and the corresponding carboxylic acid, which is then identified (Experiment 15).

**Phenols.** In addition to absorption near 3620 and 3350 $cm^{-1}$, phenols absorb in the ultraviolet region (Table 8) and possess typical aromatic proton nmr absorptions (Table 9). Phenols give a positive enol test (page 349) and may be identified as bromo derivatives (Experiment 34) or, like the alcohols, by conversion to urethans (Experiment 3).

**Sulfonic Acids.** These acids and their salts are the most common organic compounds containing sulfur. They may be identified as S-benzylthiouronium salts (page 289) or like carboxylic acids, they may be converted to amides (page 289).

## Illustrative Structure Proof

One of the pungent principles of garden cress is spilanthol, a colorless high-boiling liquid (bp 165°/1 mm) that has a molecular formula of $C_{14}H_{25}ON$. Spilanthol is insoluble in water, dilute acid, and dilute base; it therefore belongs to Group M of the solubility scheme (page 132), which classifies it as an amide or some oxygenated derivative of a nitrile. Spilanthol must be unsaturated because it decolorizes a carbon tetrachloride solution of bromine immediately. On hydrogenation spilanthol takes up two moles of hydrogen to give tetrahydrospilanthol, a colorless crystalline solid, mp 37°, with a molecular formula of $C_{14}H_{29}ON$.

Hydrolysis of tetrahydrospilanthol gives two new materials, an acidic fragment ($C_{10}H_{20}O_2$) and a basic fragment ($C_4H_{11}N$). The balanced equation for this degradation corresponds to the hydrolysis of an amide to form a carboxylic acid and an amine.

$$C_{14}H_{29}ON + H_2O \longrightarrow C_{10}H_{20}O_2 + C_4H_{11}N$$

The acidic fragment proved to be capric acid, I,

$$\underset{\text{I}}{CH_3(CH_2)_8-\overset{\overset{\displaystyle O}{\|}}{C}-OH} \qquad \underset{\text{II}}{H_2N-CH_2-CH(CH_3)_2}$$

since the *p*-toluidide of the unknown had the same melting point as the *p*-toluidide of authentic capric acid and showed no depression of melting point when mixed with the known sample. The basic fragment was shown to be isobutylamine, II, by preparation of a *p*-toluenesulfonamide that was identical in all respects to an authentic sample. With the identification of the hydrolysis fragments tetrahydrospilanthol was assigned structure III.[8]

$$\underset{\text{III}}{CH_3-(CH_2)_8-\overset{\overset{\displaystyle O}{\|}}{C}-\underset{H}{N}-CH_2-CH(CH_3)_2}$$

The structure of tetrahydrospilanthol was confirmed by synthesis from the acid chloride of capric acid and isobutylamine. The synthetic material was identical to the material obtained from the natural product.

Evidence that the two double bonds in spilanthol are conjugated comes from the observation that spilanthol forms a Diels-Alder adduct with maleic anhydride. The location of the double bonds was determined by oxidative ozonolysis of spilanthol, which gave as isolable materials butyric acid and succinic acid. These products are consistent only with the conjugated diene structure shown as IV.

$$CH_3-CH_2-CH_2-CH=CH-CH=CH-CH_2-CH_2-\overset{\overset{\displaystyle O}{\|}}{C}-\underset{H}{N}-CH_2-CH(CH_3)_2$$

IV, Spilanthol

Although the structure of spilanthol was determined entirely by chemical methods it is worthwhile to consider how the problem might have been approached if spectroscopic instruments had been readily available. An infrared spectrum of spilanthol would have revealed the probable presence of a primary amide functional group (absorption near 6.06 $\mu$ and near 2.85 $\mu$) and a conjugated diene (absorption near 6.25 $\mu$). An ultraviolet spectrum would have confirmed the presence of conjugated double bonds (absorption

[8] Asahina and Asano, *J. Pharm. Soc. Japan*, 1922, No. 480, 1; Asano and Kanematsu, *Ber.*, **65**, 1602 (1932).

near 227 m$\mu$). An nmr spectrum of spilanthol would be complicated because of the many hydrogen atoms present but the four hydrogen atoms on the two double bonds would stand out clearly as would the amide hydrogen and the tertiary hydrogen of the isobutyl group. It can be seen that spectroscopy affords many clues to the structure of spilanthol but that further elucidation by chemical methods is necessary. The role of spectroscopy here would have been to suggest the proper chemical tests and to act as confirmatory evidence for the assigned structure.

## Questions

**1.** Why do most organic halogen compounds, before heating with lime or metallic sodium, give no precipitate with silver nitrate solution?

**2.** What is the meaning of the word halogen and why is it applied to chlorine, bromine, and iodine?

**3.** By what fault in the procedure might a compound which contains nitrogen but not halogen, be found to give a positive test for halogen?

**4.** What important classes of organic compounds contain nitrogen? sulfur?

**5.** Define and illustrate the terms: empirical formula, molecular formula, isomers, homologous series.

**6.** Predict the solubility class of the following compounds:

(a) Aniline
(b) Benzamide
(c) Butyl amine
(d) Citric acid
(e) Cyclohexanone
(f) Diethyl ether
(g) Methylene chloride
(h) Nitrobenzoic acid
(i) Pentane
(j) Propanol

**7.** By use of the solubility classification solvents, indicate how the following binary mixtures could be separated and each component recovered:

(a) Dextrose and butyl alcohol
(b) Benzoic acid and benzamide
(c) *n*-Butyl bromide and di-*n*-butyl ether
(d) Nitrobenzene and aniline

# EXERCISE 9

# Determination of Structure

## 9a Qualitative Detection of Elements

**Carbon and Hydrogen.** In a porcelain crucible, place 1–2 g of finely powdered cupric oxide and dry by heating for a few minutes with a Bunsen flame. While the copper oxide is still warm, carefully mix it with about one-tenth of its bulk of powdered sugar and place the mixture in a Pyrex test tube fitted with a cork and right-angled delivery tube. Arrange the delivery tube so that the gases evolved are passed into about 5 ml of limewater (calcium hydroxide solution) in a test tube. Heat the mixture of copper oxide and sugar, and observe the result. Note the water which condenses in the cool portion of the tube. Remove the delivery tube from the limewater *before* allowing the copper oxide to cool so that the limewater will not be drawn back into the hot reaction tube.

**Halogens. (A) Beilstein's Test.** Heat a copper wire spiral to redness until the flame is no longer colored. Cool, moisten the spiral with about two drops of carbon tetrachloride ($CCl_4$), heat again and observe. The color is given to the flame by the vapor of the copper halide formed.

**(B) Calcium Oxide Test.** In a test tube heat a small lump of pure quicklime (halogen free) to a high temperature, and while still hot add about two drops of carbon tetrachloride ($CCl_4$). When cold, boil with 5–10 ml of distilled water, transfer to a 100-ml beaker and dissolve in dilute nitric acid (1 volume of concentrated acid to 1 volume of pure water). If a clear solution is not obtained, filter the solution before proceeding. Add 2–3 ml of a dilute solution of silver nitrate (5–10 percent) and observe.

**Sodium Fusion Method.** Support a small test tube (about 50 × 8 mm) by inserting it through a small hole in a piece of asbestos board so that the tube is supported by its rim. Drop in a piece of bright sodium (metal) no larger than a small pea, and heat until the sodium melts and its vapors fill the lower

➤ CAUTION: Metallic sodium must be manipulated with great care. Never use large quantities of metallic sodium and do not touch the sodium with your fingers—use tongs or pincers. Do not throw small residual pieces of sodium into water, or into a sink or waste crock; place sodium residues in the bottle provided for this purpose.

part of the tube. Remove the flame momentarily, and add suddenly a halogen, sulfur, or nitrogen compound (if a mixture of sulfanilic acid and *p*-dichlorobenzene is used, it is possible to make all three tests from one fusion). If a solid, use only a small pinch of the material; if a liquid, use only a few drops. A spontaneous exothermic reaction takes place. Return the flame and heat the tube to redness. Without removing the tube from the asbestos board, and while it is still hot, touch the bottom of the tube to the surface of about 15 ml of distilled water contained in a beaker. The tube will usually crack and the water will then react more or less violently with the excess of sodium. When the reaction subsides, crush the end of the tube in the beaker, and then heat to boiling on a wire gauze. Filter through a small filter paper and apply one or all of the following tests to the solution.

➤ CAUTION: Carry out the sodium fusion with care, and be especially cautious in decomposing the fused mass with water. If too large an amount of sodium is used, the treatment with water may result in a small explosion. Safety glasses *must* be worn for this experiment—and should be worn at all times in the laboratory.

*1. Sulfur.* Acidify a 3-ml portion of the solution with acetic acid, boil, and test the evolved gases for hydrogen sulfide with a strip of filter paper moistened with lead acetate solution (10 percent). To another portion of the solution add one to two drops of a solution of sodium nitroprusside.[9] A deep coloration indicates sulfur.

*2. Nitrogen.* To a 3-ml portion of the solution add five drops of a freshly prepared ferrous sulfate solution, one drop of ferric chloride solution, and five drops of potassium fluoride solution (10 percent). Add enough sodium hydroxide solution (usually 1–2 ml of a 10 per cent solution) to produce a distinct alkalinity and heat to boiling (*caution—bumping!*). *If sulfur is not present* (as shown by test 1), cool and acidify with dilute sulfuric acid (20–25 percent). A precipitate of prussian blue, indicative of the presence of

[9] Sodium plumbite solution, prepared by adding an excess of dil sodium hydroxide to 1–2 ml of lead acetate solution, may be used instead of the nitroprusside. Sodium plumbite gives with soluble sulfides a precipitate of lead sulfide.

nitrogen, may be detected after standing for a short time, by filtering through a small filter.

Since the presence of sulfur often obscures the nitrogen test, it is advisable, when sulfur is present, to modify the procedure in the following way: To a portion of the solution add five drops of freshly prepared ferrous sulfate solution and enough sodium hydroxide to produce a distinct alkalinity (usually 1–2 ml of a 10 percent solution). Heat to boiling (*caution—bumping!*) and filter from the precipitate of iron sulfide. Acidify with dilute sulfuric acid (20–25 percent) and add five drops of potassium fluoride solution (10 percent) and a drop of ferric chloride solution to obtain the precipitate of prussian blue.

***3. Halogens.*** Acidify a 3-ml portion of the solution with dilute nitric acid (1 volume of concentrated acid to 1 volume of water) and *if nitrogen or sulfur is present*, as shown by tests 1 and 2, boil gently for 5 or 10 min to remove any hydrogen cyanide or hydrogen sulfide which may have been formed. Add about 5 ml of a dilute solution of silver nitrate (5–10 percent) and continue the gentle boiling for a few minutes. A heavy precipitate indicates the presence of halogen; if there is only a faint turbidity, it is probably due to the presence of impurities in the reagents.

## 9b Solubility Tests

In a small test tube place 0.2 ml of a liquid (0.1 g of a solid) compound and add in small portions a total of 3 ml of water. Between each addition stir the sample vigorously with a rounded stirring rod; with solids it is desirable to crush the crystals in order to increase their surface area. If the sample has dissolved test the aqueous solution with a wide-range indicating paper to determine if the solution is acidic or basic. Record your observations in your notebook.

Follow the solubility scheme illustrated on pages 132 and 133 using fresh 0.2 ml samples and 3-ml portions of the appropriate solvents. Keep a careful record of each test.

Carry out the above classification tests with samples known to fall into each solubility group.[10] Obtain two unknowns from your instructor and classify them according to their solubility.

[10] Possible samples are $S_1$, acetone, methyl alcohol; $S_2$, sucrose, sodium acetate; $A_1$, benzoic acid, benzenesulfonic acid; $A_2$, benzene sulfonamide, *o*-cresol; *B*, dimethylaniline, *n*-octylamine; *N*, cyclohexanol, benzaldehyde; *I*, benzene, *n*-butyl bromide; *M*, acetanilide, nitrobenzene.

# Preparation and Reactions of Typical Organic Compounds

# General Remarks

In performing experiments the student should bear in mind that the preparations and reactions studied for a particular compound are usually examples of methods and reactions that can be applied to an entire class of compounds (often to entire homologous series). After the completion of an experiment it is well for the student to consider the general utility and the limitations of the method of preparation used, and of any reactions or tests that were performed.

In many experiments a definite product is prepared and purified. The preparation thus obtained should be placed in an appropriate bottle (wide-mouth bottles for solids and narrow-mouth bottles for liquids) of suitable size, labeled with the information given below, and presented with the notebook and report card when the experiment is submitted for final approval.

## Labels for Preparation Bottles

| | |
|---|---|
| Experiment number and name of product.... | Expt 2b 2-Pentene |
| Mp or bp as actually observed............. | bp 35–39° |
| Student's name.......................... | Alan Balch |
| Actual yield (in g) and tare of bottle......... | Yield, 12 g Tare, 37.5 g |

The purity of the product and the general appearance of the preparation bottle serve as an index of the neatness and skill of the student in laboratory manipulations. Preparations should not be submitted in flasks, test tubes, etc., but in clean dry bottles or specimen tubes (for very small quantities) fitted with glass stoppers or *clean corks of the proper size*.[1] The use of dirty

[1] For the preparation bottle, the cork should be selected of such size that it will extend from one-third to one-half of its length (but not farther) into the mouth of the bottle.

corks or bottles, and of glass stoppers or corks that do not fit the container properly, are indications of carelessness and poor laboratory technique. Rubber stoppers should never be used for preparation bottles, as they may lead to contamination of the product.

It is advantageous to collect the purified preparation directly in a dry bottle, previously weighed together with the stopper or cork. The weight of the empty bottle with its stopper or cork is called the tare. This weight should be recorded in the laboratory notebook and on the label of the bottle.

For liquid preparations the observed boiling point should be recorded directly in the laboratory notebook, and copied on the label and the report card. For solids (except certain very high-melting substances), the melting point should be determined and recorded directly in the notebook, and copied on the label and the report card.

## Calculation of Yields

The yield (sometimes called the actual yield) is the amount of the pure product actually obtained in the experiment. The theoretical yield (sometimes called the calculated yield) is the amount that could be obtained under theoretically perfect or ideal conditions; that is, the main reaction is assumed to proceed to completion without side reactions or mechanical losses, so that the starting materials are entirely converted into the desired product and no material is lost in isolation and purification.

The percentage yield is obtained by comparing the actual yield with the theoretical yield, in the following manner:

$$\text{Percentage yield} = \frac{\text{actual yield}}{\text{theoretical yield}} \times 100 \text{ percent}$$

The percentage yield is the measure of the overall efficiency of the preparation since many factors, such as incomplete reactions, side reactions, and mechanical losses, affect the actual yield. In elementary laboratory work it is usually considered just as important to obtain a pure product as to obtain a satisfactory yield. The student may be required to repeat a preparation unless a certain minimum yield is attained, in order to emphasize the development of skill and care in manipulation and in following laboratory directions. Nevertheless, purity of product must not be sacrificed for the obtainment of a larger quantity of the product in an impure state.

In actual laboratory work, the best results are not always secured by using the reagents in the exact proportions indicated in the equation for the reaction. In some cases it may be essential to use the reacting materials in the exact proportions demanded by the equation, but in many cases it is advantageous to use an excess of one of the reacting materials. In a given preparation the choice depends upon a number of factors, such as the relative cost and availability of the reactants, saving of time, and ease of purification. The student is expected to consider why the particular proportion of reagents is used.

If a preparation involves two reacting substances and the amounts actually used are not in the exact proportions demanded by the equation, it is necessary to determine by calculation which of the reactants is the limiting factor for the calculation of the theoretical yield, as shown by the example given below. In this connection the terms *mole* and *moles used* are commonly employed. A *mole* is a gram-molecule, and is equal to the molecular weight expressed in grams.[2] The term *moles used* is employed to express the number of moles or the fraction of a mole of a particular compound actually used in an experiment. The number of moles is equal to the weight of the substance divided by the molecular weight.

**Theoretical and Percentage Yields.** Suppose that methyl ethyl ether, $CH_3{-}O{-}C_2H_5$, has been prepared by the action of methyl iodide upon sodium ethoxide (the Williamson synthesis of ethers). This preparation is carried out by reacting metallic sodium with ethanol and treating the resulting sodium ethoxide with methyl iodide:

$$C_2H_5{-}OH + Na \longrightarrow C_2H_5{-}ONa + \tfrac{1}{2}H_2 \qquad (1)$$

$$C_2H_5{-}ONa + CH_3{-}I \longrightarrow C_2H_5{-}O{-}CH_3 + NaI \qquad (2)$$

The equations 1 and 2 may be summarized[3] as follows:

$$\underset{1\text{ mole}}{C_2H_5{-}OH} + \underset{:\ 1\text{ gram-atom}\ :}{Na} + \underset{1\text{ mole}\ \longrightarrow}{CH_3{-}I} \rightarrow \underset{1\text{ mole}}{C_2H_5{-}O{-}CH_3} + \underset{:\ 1\text{ mole}\ :}{NaI} + \underset{\frac{1}{2}\text{ mole}}{\tfrac{1}{2}H_2} \qquad (3)$$

From this it is evident that the proportions demanded by the equation are 1 mole of $C_2H_5{-}OH$ : 1 gram-atom of Na : 1 mole of $CH_3{-}I$, and the resulting products would be 1 mole of $C_2H_5{-}O{-}CH_3$ : 1 mole of NaI : $\frac{1}{2}$ mole of $H_2$.

Suppose that the following quantities of the reagents are actually used in a laboratory preparation:

$$92.0\text{ g of absolute ethanol} = \frac{92}{46}\text{ moles} = 2.0\text{ moles } C_2H_5{-}OH$$

$$5.5\text{ g of metallic sodium} = \frac{5.5}{23}\text{ mole} = 0.24\text{ gram-atom Na}$$

[2] The term mole is used most frequently in indicating the proportions of various reagents, and any convenient weight unit might be chosen. For laboratory purposes the gram is most convenient and is chosen for that reason. For some purposes the unit weight is taken as a kilogram or a milligram, and the terms kilogram-mole or millimole are used. Obviously, a kilogram-mole is equal to 1000 gram-moles, and a millimole is equal to 0.001 gram-mole.

[3] For some substances used in an elemental state (sodium, magnesium, zinc, phosphorus, sulfur, etc.) it is convenient and preferable to express the quantities as *gram-atoms* rather than moles. But hydrogen and the halogens ($H_2$, $Cl_2$, $Br_2$, etc.) are expressed usually as gram-molecules (moles).

$$28.4 \text{ g of methyl iodide} = \frac{28.4}{142} \text{ mole} = 0.20 \text{ mole } CH_3{-}I$$

The amounts of the reagents are converted from grams into moles (by dividing by the molecular weights) in order to compare them with the molar proportions expressed in the equation. The *relative* proportions actually used are thus 10 of $C_2H_5{-}OH$:1.2 of Na:1.0 of $CH_3{-}I$. Obviously, ethanol and sodium are used in excess and methyl iodide is the limiting reactant that will determine the theoretical yield.

From equation 2 or 3, it can be seen that 1 mole of methyl iodide reacting with a sufficient quantity of sodium ethoxide will produce, under perfect conditions, exactly 1 mole of methyl ethyl ether. Consequently, the maximum amount of methyl ethyl ether that could be produced in the above preparation (the theoretical yield) is 0.2 mole. By multiplying this fraction of a mole by the weight of 1 mole of methyl ethyl ether (60 g), the theoretical yield is converted into grams. The weight of 1 mole of methyl ethyl ether is the gram-molecular weight; this is determined by multiplying the number of atoms of each element in the compound by its atomic weight, and taking the sum of these weights.

$$\text{Theoretical yield} = 0.2 \times 60 \text{ g} = 12.0 \text{ g}$$

If the actual yield of methyl ethyl ether in the above preparation was 8.2 g, the percentage yield would be:

$$\text{Percentage yield} = \frac{8.2}{12.0} \times 100 = 68.3 \text{ percent}$$

***Problem:*** In another preparation of methyl ethyl ether suppose that the amounts of the reagents actually used were: 6.9 g of metallic sodium, 46 g of absolute ethanol and 49.7 g of methyl iodide. Calculate the theoretical yield, in grams. (Answer = 18.0 g.)

Assuming that the actual yield in the above preparation was 15.0 g, calculate the percentage yield. (Answer = 83.3 percent.)

## Notebooks[4] (See also page 4)

For experiments dealing with syntheses and reactions of typical organic compounds, the preparation of the notebook involves several features not required for the exercises on laboratory operations. The most important of these is a table of physical constants of the substances being manipulated

[4] The following specific directions for the preparation of notebooks and the general laboratory procedure have been used in the elementary courses in organic chemistry at Cornell University. For the particular conditions which obtain in other laboratories the instructor may wish to alter these directions or substitute others.

in the experiment (see paragraph 4 of this section). By having this information available the student will be able to understand more readily the reasons for the particular procedure followed in the experiment, and will often be able to overcome independently any small difficulties that may arise in the course of the experiment. The student is expected to proceed in the following manner:

1. Read the laboratory directions *for the entire experiment* and note particularly any cautions for handling materials.

   To aid in understanding the reasons for the procedure followed in the experiment it is helpful to consult the textbook or lecture notes for a discussion of the particular class of compounds that is to be studied. Consideration should be given to important general principles, such as the law of mass action, influence of solvents and catalysts on rate of reaction, etc.
2. In the notebook give a concise statement of the type of reaction which is to be carried out, such as "Conversion of an alcohol to an alkene," or "Oxidation of a secondary alcohol to a ketone." Write *balanced* equations, using condensed structural formulas, for the *main reaction* or sequence of reactions involved in converting the starting materials to the final products. Along the reaction arrow indicate the conditions used—temperature, solvent, catalyst (if any), etc.
3. Write balanced equations for *significant side reactions* that may divert an appreciable amount of the starting materials and lead to formation of by-products or accessory products that must be removed in the purification of the main product.

   Write balanced equations for any *test reactions* that are used to test for completion of the reaction, to detect the presence of an impurity, to confirm the identity of the product by conversion to a derivative, etc.

   It is helpful also to indicate schematically the successive steps involved in the *purification procedure*, starting with the substances present in significant amounts in the reaction mixture at the completion of reaction and showing what substances are removed at each step. The general solubility table on page 132 and the table of physical constants are useful in this connection.
4. Prepare a table of the physical constants of all organic and inorganic substances which enter into the main and side reactions, and are produced in these reactions. The form shown in the "Sample Notebook Pages" may be used. The physical constants of common organic and inorganic compounds may be found in the chemical handbooks.[5]

[5] Such as the Chemical Tables from the *Handbook of Chemistry and Physics*, Chemical Rubber Publishing Co., Cleveland, Ohio; Lange's *Handbook of Chemistry*, Handbook Publishers, Inc., Sandusky, Ohio; etc. The most complete and authoritative information on organic

Include in the table the weight (in g) of each reactant and the number of moles, or fraction of a mole, actually used. From these data and the balanced equation for the main reaction, determine which starting material is the limiting factor and calculate the theoretical yield (in g) based on this reactant.

5. After preparing the notebook in accordance with the foregoing instructions, submit it to the instructor for preliminary approval *before starting to perform the experiment.*[6] In certain experiments, the student will be expected to arrange the apparatus for the experiment and secure the approval of the instructor for the setup.[7]
6. Perform the experiment *according to the laboratory directions*, and promptly record your observations (including physical constants as actually observed, actual yields, etc.) directly into the notebook. Always observe and record the boiling point of a liquid preparation, and the melting point of a solid organic preparation, provided the melting point is not above 250°. Special types of melting point apparatus are necessary for samples that melt at high temperatures.

   Record the actual yield, and calculate the percentage yield. Record any general observations and conclusions drawn from your performance of the experiment.
7. Write balanced equations illustrating other important methods of preparing the product, and write answers to the questions given at the end of each experiment. In your answers use condensed structural formulas and make complete statements.
8. *Promptly* on completion of the experiment, submit the completed notes and the report card, together with any substance prepared, to the instructor for final approval. The instructor will examine the notes and the product prepared, and may ask questions designed to test the student's knowledge of the fundamental principles involved in the experiment, and to test his ability to make generalizations and to apply generalizations to specific cases.

➤ It is essential to make efficient use of the time assigned for laboratory work: the student is expected to plan his laboratory schedule and to make preliminary preparations before coming to the laboratory.

compounds is found in Beilstein's monumental work, *Handbuch der Organischen Chemie*, of which the most recent edition comprises about thirty large volumes, and three series of supplementary volumes.

[6] If an experiment is performed without adequate preliminary preparation of the notebook and preliminary approval by the instructor, the student may receive only partial credit for the experiment or may be required to repeat the experiment. The laboratory instructors should reserve the first half-hour of each laboratory period for the preliminary approval of notebooks.

[7] At the time of the preliminary approval, the instructor will countersign a stockroom order for the required reagents that are not on the laboratory side shelf, and will issue a report card for the experiment.

Since many experiments require that the reactants be refluxed for several hours, the student should plan to perform other laboratory work while the operation of refluxing is being carried out.

## Laboratory Directions

The laboratory directions given in this manual are deliberately complete. In advanced work and research one frequently must follow directions that assume a general knowledge of manipulative technique and of the chemistry involved. As an example consider the preparation of *n*-butyl bromide discussed above and described in detail on page 198. Were this preparation to appear in a technical journal the description might be:

> A mixture of 60 ml of water, 130 g (1.3 mole) of concentrated sulfuric acid (cool), 37 g (0.5 mole) of *n*-butyl alcohol, and 87 g (0.65 mole) of sodium bromide was refluxed for two hours, and then distilled until no more product was collected. The crude distillate was washed with water, cold 80 percent sulfuric acid, saturated sodium bicarbonate, and finally with water. After drying over calcium chloride, the product was distilled, yield 45 g, bp 99–103°.

The journal directions assume that the worker realizes that the sodium bromide should be pulverized before addition and that good mixing is essential. Typically, experimental sections of professional journals do not specify the amounts of washing reagents or drying agents. In fact, many journal authors might have condensed the experimental description to read:

> *n*-Butyl alcohol (0.5 mole, 37 g) was converted to the bromide by heating with sodium bromide (0.65 mole) and 70 percent sulfuric acid (190 g). The crude bromide was washed, dried, and redistilled; yield 45 g, bp 99–103°.

The reader would have to understand the chemistry well enough to isolate the bromide by distillation from the reaction mixture as well as to use the washes specified in the complete directions in the given order. The ability to fill in experimental detail is an important part of being a good laboratory worker and the student is advised, as he carries out the experiments in this manual, to ask himself at each stage why a certain operation or reagent is used.

## Sample Notebook Page

---

*EXPERIMENT* 5a *n-BUTYL BROMIDE*

Conversion of a Primary Alcohol to an Alkyl Bromide

*Main Reactions:*

$$CH_3CH_2CH_2CH_2{-}OH + HBr \xrightarrow[\text{reflux}]{H^+} CH_3CH_2CH_2CH_2{-}Br + H_2O$$

$$NaBr + H_2SO_4 \longrightarrow HBr + NaHSO_4$$

*Side Reactions:*

No important organic side reactions

$$2NaBr + 3H_2SO_4(\text{concd}) \longrightarrow Br_2 + SO_2 + 2H_2O + 2NaHSO_4$$

*Purification:*

$$\left.\begin{matrix} C_4H_9{-}Br,\ C_4H_9{-}OH \\ NaHSO_4,\ H_2SO_4 \\ (Br_2),\ H_2O \end{matrix}\right\} \xrightarrow{\text{distillation}} \left.\begin{matrix} C_4H_9{-}Br \\ C_4H_9{-}OH \\ (Br_2),\ H_2O \end{matrix}\right\} \xrightarrow{NaHSO_3} \left.\begin{matrix} C_4H_9{-}Br \\ C_4H_9{-}OH \\ (H_2O) \end{matrix}\right\} \xrightarrow{H_2SO_4}$$

$$\begin{matrix} C_4H_9{-}Br \\ \text{purified by} \\ \text{distillation} \end{matrix} \xleftarrow{CaCl_2} \left\{\begin{matrix} C_4H_9{-}Br \\ (H_2O) \end{matrix}\right. \xleftarrow[(2)\ H_2O]{(1)\ NaHCO_3} \left\{\begin{matrix} C_4H_9{-}Br \\ (H_2O) \\ (H_2SO_4) \end{matrix}\right.$$

*Test Reactions:* None

---

## Answers to Questions:

**1.** Strong sulfuric acid removes butyl alcohol more effectively than water does because it reacts chemically to convert the alcohol essentially completely to compounds that are much more soluble in sulfuric acid than in butyl bromide. Water removes the alcohol merely by simple solubility and is not very effective since butyl alcohol is much less soluble in water than in butyl bromide (unfavorable distribution coefficient toward water).

**2.** Ionic addition of HBr to 1-butene gives 2-bromobutane (Markovnikov rule) but free radical addition in the presence of peroxides (Kharasch) gives 1-bromobutane.

**3.** $n\text{-}C_4H_9{-}OH + SOCl_2 \longrightarrow n\text{-}C_4H_9{-}Cl + SO_2 + HCl$

$6\ n\text{-}C_4H_9{-}OH + 2P + 3I_2 \longrightarrow 6\ n\text{-}C_4H_9{-}I + 2H_3PO_3$

**4.** The $S_N2$ reaction with cyanide anion will be favored by using a high concentration of sodium cyanide and a solvent medium that is not highly polar. If the solvent were a good ionizing medium it would favor solvolysis ($S_N1$ mechanism).

## Sample Notebook Page

PHYSICAL CONSTANTS

| SUBSTANCE | MOL WT | GRAMS USED | MOLES USED | SP G 20° | MP | BP | SOLUBILITY (IN G PER 100 ML) | | |
|---|---|---|---|---|---|---|---|---|---|
| | | | | | | | *Water* | *Ethanol* | *Ether* |
| $n\text{-}C_4H_9\text{–}OH$ | 74 | 37 | 0.50 | 0.810 | −80° | 117° | 9 | ∞ | ∞ |
| $H_2SO_4$ | 98 | | 1.25 | 1.83 | 11° | d340° | ∞ | reacts | sol. |
| $NaBr \cdot 2H_2O$ | 139 | 87 | 0.62 | — | 51° | — | 80 cold | slight | insol. |
| $n\text{-}C_4H_9\text{–}Br$ | 137 | (68.5) | (0.50) | 1.277 | −112° | 102° | insol. | ∞ | ∞ |

*Quantities used:* $NaBr \cdot 2H_2O$ $\quad \dfrac{87\text{ g}}{139} = 0.62\text{ mole}$

$H_2SO_4$ (96%) $\quad \dfrac{129\text{ g} \times 0.96}{98} = 1.25\text{ mole}$

$n\text{-}C_4H_9\text{–}OH$ $\quad \dfrac{37\text{ g}}{74} = 0.50\text{ mole} = \text{limiting factor}$

*Theoretical yield:* based on *n*-butyl alcohol (0.5 mole)

$= 0.5 \times 137\text{ g} = 68.5\text{ g}$ *n*-butyl bromide

**19 March 1969:** Reaction mixture refluxed two hours; organic layer had slight brown color. After distillation the organic layer in distillate was colorless. Product was washed and allowed to stand over $CaCl_2$ in corked flask.

**21 March 1969:** Product filtered into dry 100-ml distillation apparatus and distilled: atmospheric pressure = 745 mm

Tare of receivers (without corks): A = 39.5 g, B = 42.0 g.
Fractions collected: A, up to 99°; B, from 99 to 102°.
Receiver A, 40.3 g; net weight (40.3–39.5) = 0.8 g.
Receiver B, 90.5 g; net weight (90.5–42.0) = 48.5 g.

*Percentage yield:* $\dfrac{48.5}{68.5} \times 100\% = 70.8\%$ of theoretical

*Other Methods of Preparation:*

$$3\ n\text{-}C_4H_9 - OH + PBr_3 \longrightarrow 3\ n\text{-}C_4H_9 - Br + H_3PO_3$$

*Write legibly in ink*

**ORGANIC CHEMISTRY**

COURSE NO. ................SECTION ..............................................EXPERIMENT NO. ...............

NAME ............................................................................ DATE .............................

*Last Name* *First Name*

PROCESS OR REACTION STUDIED ..............................................................................

.....................................................................................................................

PRODUCT PREPARED.........................................................................................

| PHYSICAL CONSTANTS | | | YIELD |
|---|---|---|---|
| | Observed in Laboratory | Found in Literature | |
| | | | Based upon .......................................... |
| BP ............ | ...................... | ...................... | Theoretical Yield (in g) .............................. |
| MP............ | ...................... | ...................... | Actual Yield (in g) .................................. |
| Color ......... | ...................... | ...................... | Percentage Yield ............................ percent |

| | |
|---|---|
| Preliminary Approval ................................ | GRADE: Manipulation ................................ |
| REMARKS: | Oral Quiz ................................ |
| | FINAL GRADE ........................................ |
| | INSTRUCTOR ........................................ |

Sample Report Card
(Actual size 4 × 6 inches)

## Synthetic Sequences

Several sequences of preparations may be carried out using materials prepared in various experiments. These illustrate stepwise syntheses so frequently used in laboratory work. Sequences may also be devised with other starting materials, using as a model a general procedure given for a similar compound. Frequently the small scale preparation of a characteristic derivative is given.

*n*-propyl bromide—*n*-propylmagnesium bromide—2,2-dimethylpentanol (this could be altered by treating the Grignard with ethyl formate to give 4-heptanol)

*n*-butyl bromide—*n*-valeronitrile—*n*-valeric acid—(*n*-valeranilide)

*n*-butyl bromide—ethyl *n*-butylacetoacetate—methyl *n*-pentyl ketone

*n*-butyl bromide—diethyl *n*-butylmalonate—caproic acid—(caproanilide)

nitrobenzene—aniline—acetanilide—*p*-nitroacetanilide—*p*-nitroaniline (this could be extended to *p*-nitrostilbene, Experiment 32)

*m*-dinitrobenzene—*m*-nitroaniline—*m*-nitrophenol
nitrobenzene—hydrazobenzene—benzidine and azobenzene
benzoin—benzil—benzilic acid—diphenylacetic acid
benzil—benzilmonohydrazone (can be extended to desoxybenzoin, or diphenylketene, page 371)

acetanilide—*p*-bromoacetanilide—*p*-bromoaniline—*p*-bromophenylurea (this could be altered by conversion to *p*-bromobenzaldehyde: *Organic Syntheses*, **46**, 13)

acetanilide—*p*-acetamidobenzenesulfonyl chloride—*p*-acetamidobenzenesulfonamide—sulfanilamide

*p*-acetotoluidide—bromo-*p*-acetotoluidide—2-bromo-4-methylaniline—*m*-bromotoluene (this could be altered by conversion of 2-bromo-4-methylaniline to 3-bromo-4-methylbenzaldehyde: *Organic Syntheses*, **46**, 13)

acetophenone—benzalacetophenone—benzalacetophenone epoxide—benzoylphenylacetaldehyde

# Accessory Laboratory Operations

## Drying Agents

The removal of admixed or dissolved water from starting materials and finished preparations is an important feature of organic laboratory work. In general water must be regarded as an objectionable impurity since it may bring about an undesired hydrolytic reaction or exert an unfavorable catalytic effect. The presence of water sometimes retards a desired reaction and may inhibit it completely, as in the formation of Grignard reagents. On the other hand, it would be superfluous to remove the last traces of water from a substance that is to be brought into contact with aqueous reagents.

An organic solid may be dried by spreading it in thin layers exposed to the air at room temperature, but this method usually allows at least a small amount of moisture to remain. More effective drying is obtained by heating the substance in thin layers in a drying oven at a temperature below the melting or decomposition point, or by placing in a desiccator over drying agents such as anhydrous calcium chloride, solid sodium hydroxide, phosphorus pentoxide, etc. The use of concentrated sulfuric acid in desiccators is dangerous.

An organic liquid, or an organic solid dissolved in an organic solvent, is usually dried by placing the liquid in direct contact with a solid inorganic drying agent. After mixing thoroughly and allowing the mixture to stand, the dried liquid is filtered to remove the spent drying agent and may then be distilled, etc. *If, in any drying operation, sufficient water is present to cause the separation of an aqueous layer, the organic liquid should decanted (or separated) and treated with a fresh portion of the drying agent.*

**Selection of a Drying Agent.** The selection of an appropriate drying agent involves consideration of the properties of the substance to be dried, and the characteristics of the various drying agents. The latter must remove water efficiently and must not dissolve in the liquid or react with it in any way.

The efficiency of a drying agent refers to the completeness with which it can remove water, and depends essentially on the amount of water existing in equilibrium with the hydration product of the drying agent under the conditions used. Drying capacity relates to the amount of water which can be withdrawn (by hydration or chemical reaction) by a given weight of the drying agent under the conditions used. For example, magnesium sulfate is an efficient drying agent because the vapor pressure of water in equilibrium with its product of hydration ($MgSO_4 \cdot 7H_2O$) is quite low at ordinary temperatures; it has an exceedingly high drying capacity since the anhydrous salt can take up 105 percent of water by weight to form the heptahydrate. Often it is advantageous to eliminate a large part of the admixed water by means of a drying agent of high capacity (calcium chloride, magnesium sulfate) and then complete the drying with a more efficient agent.

The most efficient drying agents remove water by chemical reaction (sodium metal, phosphorus pentoxide, quicklime, etc.) and it is not essential nor always desirable to remove the spent drying agent before distilling the dried liquid. Many drying agents (calcium chloride, magnesium sulfate, sodium hydroxide, etc.) remove water by the formation of hydrates. Since the hydrates generally liberate water when heated, even at moderate temperatures, it is always desirable and usually necessary to remove the spent drying agent before distilling the dried liquid. The most common drying agents are listed below, together with information concerning their properties. Table 15 indicates drying agents that are suitable for various classes of organic compounds.

*Anhydrous calcium chloride* is widely used because of its high drying capacity and cheapness. It is moderately efficient but not very rapid in action, consequently it should remain in contact with the liquid for an ample time, with occasional shaking. Fused calcium chloride, crushed into small pieces, is more satisfactory than the porous form used for desiccators since the latter tends to retain mechanically an appreciable quantity of liquid. Calcium chloride forms a series of hydrates: a dihydrate, a tetrahydrate, and a hexahydrate (mp 30°).

Alcohols and amines cannot be dried with calcium chloride since they react with it to form molecular compounds analogous to hydrates. Certain aldehydes, ketones, acids, amides, and esters also form loose complexes with calcium chloride, but this tendency is greatly reduced if an organic solvent is present. It is inadvisable to use calcium chloride for drying acidic substances (acids, phenols) since the technical product often contains small amounts of basic salts.

*Anhydrous magnesium sulfate*[1] has a large drying capacity and is rapid and

[1] The anhydrous salt can be prepared readily by heating Epsom salts ($MgSO_4 \cdot 7H_2O$) to dryness in a porcelain dish or casserole over a wire gauze, and stirring with a spatula or glass rod until a dry powder is obtained.

**TABLE 15**
**Common Drying Agents for Organic Compounds**

| | |
|---|---|
| Hydrocarbons (and some Ethers) | Anhydrous calcium chloride or sulfate, sodium (metal), phosphorus pentoxide |
| Alcohols | Anhydrous potassium carbonate, anhydrous sulfates (*not* calcium chloride), quicklime |
| Organic Bases (Amines) | Solid sodium hydroxide or potassium hydroxide, quicklime |
| Organic Acids | Anhydrous magnesium sulfate, sodium sulfate, calcium sulfate |
| Alkyl Halides and Aryl Halides | Anhydrous calcium chloride, anhydrous sulfates, phosphorus pentoxide |
| Aldehydes | Anhydrous magnesium sulfate, calcium sulfate, sodium sulfate |
| Ketones | Anhydrous potassium carbonate, anhydrous sulfates |

efficient. It is insoluble and chemically inert, and can therefore be used for almost all types of organic compounds.

*Anhydrous sodium sulfate* is an inert and inexpensive drying agent but it acts slowly and not very effectively. It forms a heptahydrate and a decahydrate (which decomposes at 32°).

*Anhydrous calcium sulfate* in a form suitable for drying purposes has been introduced under the trade name Drierite.[2] This reagent is extremely rapid and efficient, is chemically inert, and insoluble in organic liquids. When a large amount of water must be removed it is good practice to use a preliminary drier of larger capacity (magnesium sulfate, calcium chloride, etc.) since Drierite takes up only 6.6 percent by weight of water, forming the hydrate $CaSO_4 \cdot \frac{1}{2} H_2O$.

*Anhydrous potassium carbonate* may be used for drying neutral and basic compounds (alcohols, ketones, esters, nitriles, amines, etc.) but not for acids or phenols. It has good drying capacity but is rather slow and only moderately efficient. It is used frequently in salting out water-soluble alcohols, amines, and ketones from aqueous solutions, and as a preliminary drier. Potassium carbonate forms a "trihydrate" ($2K_2CO_3 \cdot 3H_2O$) and a dihydrate ($K_2CO_3 \cdot 2H_2O$).

*Solid sodium hydroxide* and *potassium hydroxide* are rapid and effective drying

Occasionally samples of "anhydrous magnesium sulfate" may contain as much as 20 percent of water. It is advisable to test the drying agent and dry it further before use, if necessary.

[2] Drierite must not be confused with Dehydrite (magnesium perchlorate). Although Dehydrite and Desicchlora (barium perchlorate) are excellent drying agents for certain purposes, they should not be used for organic materials owing to the danger of violent explosions.

agents for organic bases (amines). Their use is quite restricted since they react with many organic compounds in the presence of water and are soluble in certain organic liquids. Technical flake sodium hydroxide (ordinary solid household lye, in small cans) is convenient and inexpensive.

*Calcium oxide* (*quicklime*) is a slow but efficient drying agent, which acts by combination. The oxide and resulting hydroxide are stable to heat and practically nonvolatile, consequently the spent drying agent need not be removed. It is used frequently for drying low molecular weight alcohols and amines, but cannot be used for acidic or alkali-sensitive substances.

*Sodium* (*metal*) is most effective in the form of fine wire, which is forced directly into the liquid by means of a sodium press. Sodium removes water very efficiently but is so active chemically that its use is restricted to the most inert types of compounds (saturated and aromatic hydrocarbons, and ethers). A preliminary drying should always be carried out before using sodium.

Sodium must be manipulated with great care, and used only in small amounts. The metal is stored under a hydrocarbon oil (kerosene or xylene) to prevent reaction with air and moisture. Sodium should not be handled with the fingers but with tongs or pincers. Waste or residual pieces of sodium must never be thrown into a waste jar or sink, and should not be destroyed by throwing them into water. Sodium residues should be placed under kerosene in a bottle provided for this purpose.

*Phosphorus pentoxide* is exceedingly efficient and rapid; it combines chemically with water to give $HPO_3$, $H_4P_2O_7$, and finally $H_3PO_4$. It can be used for drying hydrocarbons, low-boiling ethers, alkyl and aryl halides, simple esters and nitriles, but reacts with most other compounds. Since phosphorus pentoxide is relatively expensive and is difficult to handle (causing severe burns on the skin), it is used only when extreme dryness is required and a preliminary drying agent is used to remove most of the water.

*Molecular sieves* are extremely effective drying agents (see Inclusion Compounds, page 94) suitable for drying all of the organic molecules listed in Table 15. Fresh molecular sieves are conditioned by heating at 320° for 3 hr; used sieves can be reconditioned. Sieves of Type 4A, one of the most commonly used grades, are available as powder, as $\frac{1}{16}$-in and $\frac{1}{8}$-in pellets and as beads. Solvents may be dried by passing them through a 2-ft column of $\frac{1}{16}$-in pellets followed by distillation from fresh, powdered sieves.

In many cases it is possible to apply the principle of steam distillation for drying organic liquids and solids. For example, if wet benzene or a wet benzene solution of an organic substance is distilled, the first portion of the distillate consists of a mixture of benzene and water. This mixture distills at a temperature below the boiling point of benzene (80°); when the water has been removed benzene distills, and finally the pure dry organic liquid (or solid) remains. This method is simple and very useful for drying benzene itself and benzene solutions of many organic substances. Carbon tetrachloride, toluene, xylene, and other water-insoluble liquids may be dried in a similar way and are used occasionally in place of benzene for drying other organic compounds.

In certain instances, especially with the simpler aliphatic alcohols, drying by distillation with benzene is complicated by the formation of a ternary azeotropic mixture of benzene-water-alcohol, and a binary azeotropic mixture of benzene-alcohol. In such cases the amount of benzene used must be carefully controlled to avoid a large excess.

The commercial use of distillation with benzene to prepare absolute ethanol from the constant-boiling (azeotropic) mixture of ethanol and water is an example of the latter type. The ternary azeotropic mixture, benzene-water-ethanol, boils at 64.85° and contains 74.1 percent by weight of benzene, 7.4 percent of water, and 18.5 percent of ethanol. When the water has been removed, a binary azeotropic mixture, containing 67.63 percent by weight of benzene and 32.37 percent of ethanol, distills at 68.24°. When the benzene has been removed, the residue consists of practically pure absolute ethanol which distills at 78.37°.

## Heating and Cooling Baths

Steam baths are suitable for heating or distilling when the required temperature is less than 100°. A general precaution to be observed in using a steam bath is to avoid excess steam. Not only are clouds of steam annoying, but condensation of escaping steam on the apparatus being heated may permit moisture to enter into the reaction vessel, which could be deleterious for some reactions.

For temperatures above 100°, an oil bath heated by a flame or an electrical hot plate[3] is suitable. When a flame is used care must be taken not to ignite the hot oil vapors. An alternate method of electrical heating is to place a coil of resistance wire directly in the bath oil. This method does not have the flammability hazard but it suffers from the disadvantage that the hot heating coils decompose the oil and make frequent replacement necessary.

For very high temperatures a sand bath heated by a flame has many desirable features. In addition to being inexpensive a sand bath has a high heat capacity, which gives a steady temperature; it is not subject to fire hazards and it does not decompose on prolonged heating.

Another heating device used frequently in advanced work is a Glas-Col heating mantle, which is constructed by weaving nichrome wire heating elements into fiber glass cloth to form a hemispherical jacket that fits snugly against the lower part of a flask. Other shapes are available for special types of apparatus. These mantles, when equipped with thermostatic control, are particularly useful for heating reaction mixtures over a long period. They are

[3] A relatively inexpensive heating unit with thermostatic control is an electrically heated deep-fry pan such as Sears No. 34H6932 or Wards No. L86B46200.

relatively expensive (about $16 for the mantle for a 500-ml flask) and a different mantle is required for each size of flask.

**TABLE 16**
**Cooling Baths**

| INGREDIENTS | LOWEST TEMPERATURE OBTAINED |
|---|---|
| 1 part sodium chloride<br>3 parts crushed ice | −20° |
| 1 part ammonium chloride<br>1 part sodium nitrate<br>1 part cold water | −20° |
| 3 parts powdered calcium chloride<br>2 parts crushed ice | −50° |

The common cooling bath is a slush of crushed ice and water. Because of the inversion of density of water at 4° an ice bath should be stirred well if it is desired to maintain the whole bath at 0°. Temperatures below 0° can be obtained by mixing inorganic salts with ice or cold water. Table 16 lists the proportions of ingredients to be mixed in order to obtain the stated temperature. It is important in using ice-salt baths that the mixture be stirred well. Temperatures down to about −80° can be maintained by addition of pieces of dry ice to acetone or other low melting heat transfer liquids.

## Refluxing

In preparative organic work frequently it is necessary to maintain a reaction at an approximately constant temperature for a long period of time with a minimum of attention. The simplest procedure for reactions carried out in solution is to boil the solution and condense the vapors in such a manner as to return them to the reaction flask (refluxing). The temperature in the flask remains nearly constant at approximately the boiling point of the solvent (the actual temperature is elevated above the solvent boiling temperature to an extent governed by the concentration of non-volatile solutes). This operation is so common and important that the apparatus has been pictured in two ways in Figure 36. The principle and technique of refluxing should be understood clearly.

Another common requirement in preparative chemistry is to add a liquid to a reaction mixture while the mixture is refluxing. In Figure 37 are illustrated two assemblies suitable for this purpose. The first, appropriate to

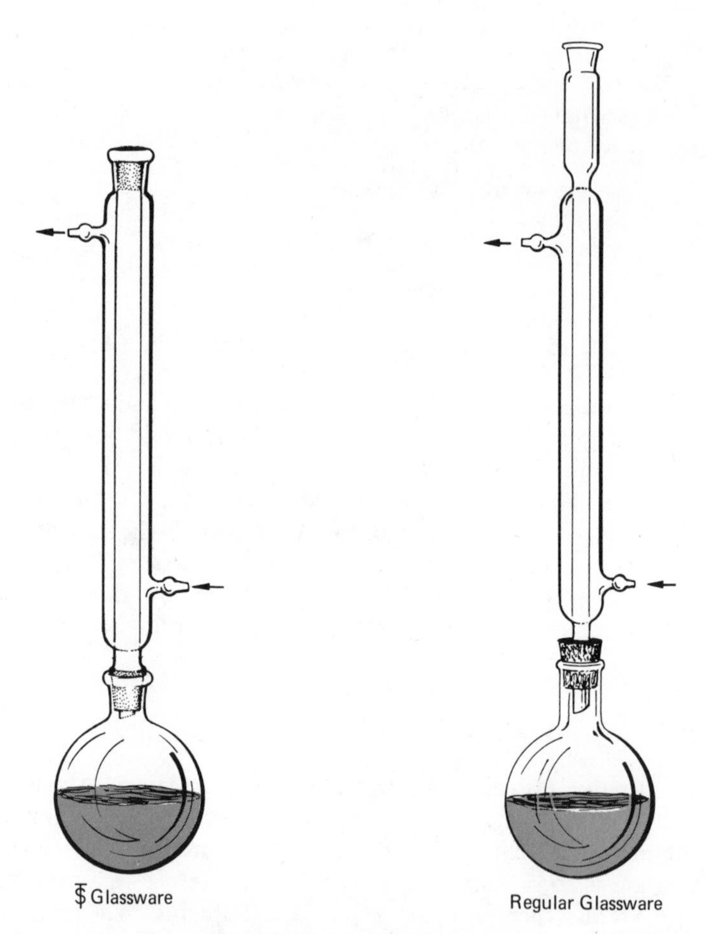


**Figure 36.** Reflux Assembly

ground glass equipment, uses a Claisen adapter to create a second opening to the reaction vessel. The second, for use with regular glassware and corks, allows the added liquid to run through the condenser. It is essential that the cork be slotted in order to prevent pressure from building up inside the equipment. For this reason the second arrangement can *not* be used with ground glass equipment, which makes tight seals.

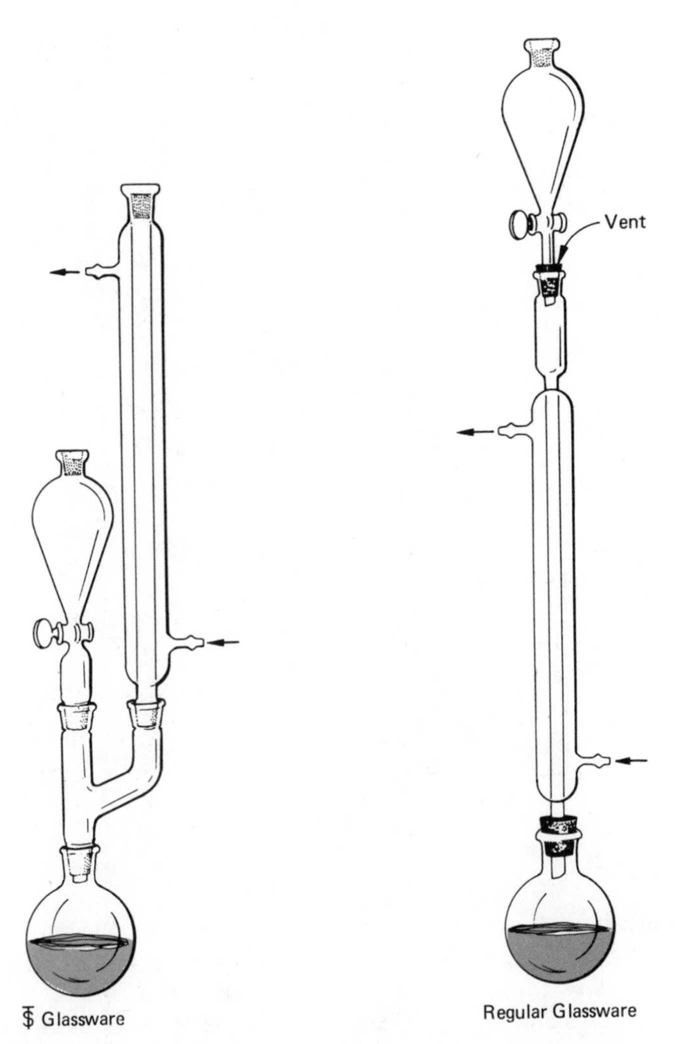


**Figure 37.** Assembly for Simultaneous Reflux and Addition

## Gas Absorption Traps

In some organic preparations noxious gases are liberated that must not be allowed to escape into the laboratory. Two common methods for trapping gases soluble in water are pictured in Figure 38. The principal precaution to be observed in trapping water-soluble gases is to construct the apparatus in

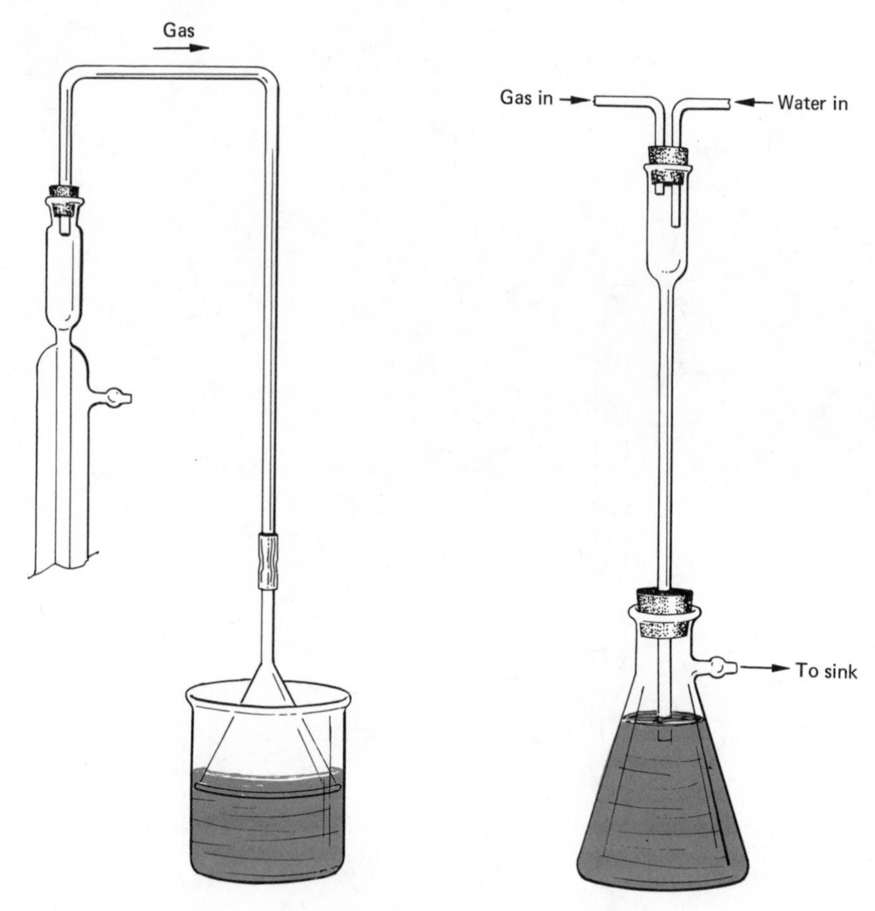


**Figure 38.** Gas Absorption Traps

such a manner that a drop of pressure in the reaction flask will not suck water back into the reaction flask. In the inverted funnel method the lower edge of the funnel should not dip into the water more than 1 or 2 mm. With the filter flask assembly the inlet from the reaction flask should protrude no more than 2 or 3 mm through the stopper; the lower end of the air condenser should not be more than 4–5 mm below the side arm of the filter flask.

## Mechanical Stirring

In many organic preparations it is desirable to stir the reaction mixture. This is particularly true of heterogeneous reactions or large scale reactions where heat transfer from the central portions of the liquid can be slow enough to cause an excessive temperature rise. Inexpensive motors of the brush type give off sparks and are exceedingly hazardous in an organic laboratory con-

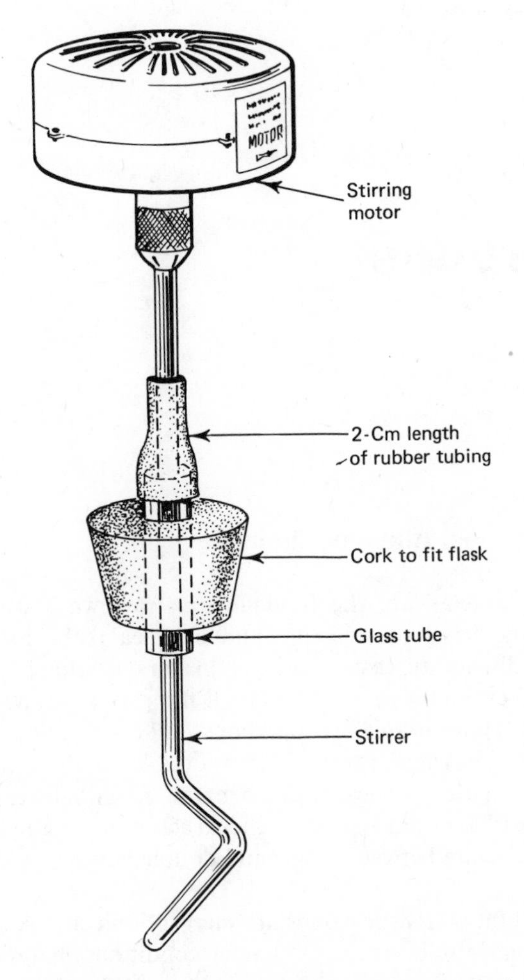


**Figure 39.** Stirring Assembly with Vapor Seal

taining vapors of volatile organic chemicals. Nonsparking motors of the induction type normally have high rotation speeds that make them unsuitable; they have low torque at speeds lower than their design speed so that attempts to slow down induction motors by adding a friction device usually are not successful. Geared induction motors are satisfactory but expensive.

An inexpensive air-driven stirrer is available.[4] When the reaction must be stirred while it is refluxing a vapor seal such as is shown in Figure 39 is necessary. In use a lubricant (glycerin or stopcock grease) is required to prevent the rubber tubing from sticking to the stirrer shaft.

[4] Air Driven Stirrer, distributed by Will Scientific, Inc. (Rochester, N.Y., 14603), $12.00.

# EXPERIMENT 1

# Hydrocarbons

## Alkanes, Alkenes, Alkynes, Benzene

Hydrocarbon systems are the fundamental framework of organic compounds. On the basis of structure and chemical behavior they may be classified broadly into three categories: (1) **saturated** aliphatic hydrocarbons—alkanes and cycloalkanes—in which all the carbon atoms are joined by single bonds; (2) **unsaturated** hydrocarbons—alkenes, cycloalkenes, alkynes, alkadienes, etc.—in which carbon-carbon double bonds or triple bonds are present; (3) **aromatic** hydrocarbons—benzene, naphthalene, etc. (arenes)—in which there is a special type of cyclic structure having hybridized bonds that are intermediate between single and double bonds.

**Alkanes.** Saturated hydrocarbons are inert to ionic attack by typical electrophilic or nucleophilic reagents. Under conditions that give rise to free radicals they undergo a variety of *free radical substitution* reactions. Ultraviolet illumination, peroxidic catalysts, and high temperatures initiate free

$$Br_2 \xrightarrow[\text{heat}]{\text{light}} 2Br\cdot$$

$$Br\cdot + CH_4 \longrightarrow HBr + CH_3\cdot$$

$$CH_3\cdot + Br_2 \longrightarrow CH_3{-}Br + Br\cdot\text{, etc.}$$

$$\text{overall reaction} = CH_4 + Br_2 \longrightarrow CH_3{-}Br + HBr$$

radical chain reactions such as halogenation, nitration, and oxidation. At high temperatures carbon-carbon bonds are ruptured ("cracking") and smaller hydrocarbon fragments are formed. For example, the nitration of propane, with nitric acid in the vapor phase at 450°, gives a mixture of nitromethane, nitroethane, and the two nitropropanes.

Saturated hydrocarbons having a tertiary hydrogen, $R_3C-H$, are more reactive than other types. They undergo oxidation and substitution reactions more readily and will add to an alkene in the presence of acid catalysts. Isobutane reacts with ethylene in the presence of sulfuric acid to give neohexane (2,2-dimethylbutane).

$$(CH_3)_3C-H + CH_2=CH_2 \xrightarrow{H_2SO_4} (CH_3)_3C-CH_2CH_3$$

The typical reactions of **alkenes** (and of alkynes) are ionic addition reactions with electrophilic reagents, such as halogens, acids, and oxidizing agents. Halogen acids and sulfuric acid add to alkenes to form alkyl halides and alkyl hydrogen sulfates, respectively. Hypochlorous acid leads to chloroalkanols (chlorohydrins) and cold aqueous permanganate effects hydroxylation of the double bond to form vicinal diols (1,2-glycols). Other reactions of alkenes are discussed in Experiment 2.

The triple bond of **alkynes**, *unlike the double bond of the alkenes*, will undergo nucleophilic addition reactions, which generally do not go beyond the addition of one molecule of the addendum. In the presence of alkaline catalysts acetylene adds one molecule of ethanol to form vinyl ethyl ether ($CH_2=CH-OEt$) and with hydrogen cyanide gives acrylonitrile ($CH_2=CH-CN$).

The triple bond is capable of adding *two* molecules of the typical electrophilic reagents, stepwise, to give at the first stage an alkene derivative. Under controlled conditions the reaction often can be stopped at the first step and

$$H-C\equiv C-H \xrightarrow{HCl} \underset{\text{Vinyl chloride}}{CH_2=CHCl} \xrightarrow{HCl} CH_3-CHCl_2$$

$$CH_3-C\equiv C-H \xrightarrow{HCl} CH_3-CCl=CH_2 \xrightarrow{HCl} CH_3-CCl_2-CH_3$$

this affords a route to vinyl compounds ($CH_2=CH-X$) used for the manufacture of synthetic polymers (see Experiment 56).

Salts of mercury, silver, and cuprous copper are effective catalysts for numerous addition reactions. In the presence of acidic mercuric sulfate solution acetylene adds water to produce acetaldehyde, which can be converted to acetic acid and other industrial chemicals. Addition of acetic acid to acetylene furnishes vinyl acetate, $CH_2=CH-O-CO-CH_3$, and ethylidene diacetate, $CH_3-CH(O-COCH_3)_2$. The latter can be pyrolyzed to yield acetaldehyde and acetic anhydride.

Alkynes having hydrogen bonded directly to the carbon-carbon triple bond, as in $HC\equiv CH$ and $R-C\equiv CH$, are very weakly acidic and form metallic derivatives such as sodium acetylide. A useful test reaction for the presence of reactive acetylenic hydrogen is the formation of water-insoluble silver or cuprous derivatives, by reaction with ammoniacal solutions of silver nitrate or cuprous chloride. In a dry state silver and copper acetylides are dangerously explosive.

The presence of the group H−C≡C− is detected readily by its infrared absorption at ~2100 $cm^{-1}$ (see Determination of Structure).

**Benzene.** The most characteristic reactions of aromatic hydrocarbons (**arenes**) are ionic substitution reactions with electrophilic reagents: nitration, sulfonation, halogenation, and Friedel-Crafts alkylations and acylations.[1] Concentrated sulfuric acid or Lewis acids (anhydrous aluminum chloride, ferric halides, boron trifluoride) are used frequently to facilitate such substitutions.

Under conditions giving rise to *free radical* attack chlorine or bromine may undergo addition to the aromatic ring or effect substitution in the side chain. With ultraviolet illumination or free radical initiators, chlorine adds to benzene to furnish a mixture of stereoisomeric hexachlorides, $C_6H_6Cl_6$. Under similar conditions toluene and chlorine react to give side chain substitution products, $C_6H_5{-}CH_2{-}Cl$, $C_6H_5{-}CHCl_2$, and finally $C_6H_5{-}CCl_3$.

**Test Reactions.** The tests may be performed with examples of the four classes of hydrocarbons: (1) the saturated hydrocarbon heptane (or pentane or cyclohexane),[2] (2) the unsaturated hydrocarbon 2-pentene (or methylpentenes or cyclohexene),[3] (3) an acetylene derivative such as 1-heptyne or 1-hexyne,[4] (4) the aromatic hydrocarbon benzene.

Tests(A) and (B) are commonly used as *tests for unsaturation*, and test (C) is also used to a limited extent for the same purpose. Test (D) is diagnostic for the group H−C≡C−.

Record the results in a neat tabular form like that below.

### Hydrocarbon Tests

| REAGENT | HEPTANE OR PENTANE | 2-PENTENE OR CYCLOHEXENE | 1-HEXYNE | BENZENE |
|---|---|---|---|---|
| Bromine in $CCl_4$ | | | | |
| Bromine in $CCl_4$ + light | | | | |
| Permanganate solution | | | | |
| Concd sulfuric acid | | | | |
| Cuprous ammonium chloride | | | | |

[1] These reactions are discussed in Experiments 17–25.

[2] Heptane or pentane purified by permanganate is available from chemical supply firms. Purified petroleum ether or ligroin also is suitable.

[3] Preparation of these alkenes is described in Experiment 2.

[4] As each test requires only 5–6 drops of the alkyne, only about 1 ml of 1-hexyne is needed

***(A) Bromine Test.*** Add $\frac{1}{2}$ ml of hydrocarbon to 3–4 ml of 2 percent bromine in carbon tetrachloride. Shake the tube and observe after 2 or 3 min. If the solution has not been decolorized prepare a second tube and place one tube in a dark place and the other in bright sunlight or ultraviolet illumination. Compare the tubes after they have stood for about 10 min. If an ordinary light bulb is used the tube must be placed very close to the light.

***(B) Permanganate Test (Baeyer's Test).*** Add $\frac{1}{2}$ ml of hydrocarbon to 3 ml of 0.5 percent potassium permanganate solution to which 3 ml of 10 percent sodium carbonate solution has been added and shake the tube well.

For a solid hydrocarbon dissolve about 0.1 g in 1 ml of pure, *alcohol-free* acetone and add this solution dropwise to the permanganate. The solvent must be tested beforehand for purity.

***(C) Sulfuric Acid Test.*** Add $\frac{1}{2}$ ml of hydrocarbon, *cautiously* and *with gentle shaking*, to about 3 ml of concentrated sulfuric acid. Does the hydrocarbon dissolve? Is heat evolved?

***(D) Cuprous Ammonium Chloride Test.*** Add 2–3 drops of hydrocarbon to 3 ml of cuprous ammonium chloride solution, prepared as described below.[5] If a copper derivative is formed collect it on a small Büchner or Hirsch funnel with suction and wash it with water. Transfer the moist filter paper and precipitate to a beaker and destroy the copper derivative by boiling it with about 20 ml of dilute hydrochloric acid (1 volume of concentrated acid to 3 volumes of water). This procedure regenerates the parent alkyne.

➤**CAUTION:** Copper derivatives of alkynes may become explosive in a dry state. Do not place such compounds in the waste jars or in your laboratory desk. Destroy such material by heating with acid as described above.

## Questions

**1.** Write structural formulas and systematic (IUPAC or Geneva) names for: (a) the five isomeric hexanes; (b) the five isomeric heptanes having a maximum straight chain of five carbons.

**2.** Compare the three isomeric pentanes with respect to their reactivity toward chlorination. Explain.

per student. Phenylacetylene (ethynylbenzene) may be used also to illustrate the formation of a copper derivative with ammoniacal cuprous chloride. 1-Hexyne is available from Farchan Research Laboratories, 4702 East 355th Street, Willoughby, Ohio, 44094.

[5] **Preparation of Cuprous Ammonium Chloride Solution:** Dissolve 3 g of copper sulfate crystals (pentahydrate) and 0.8 g of sodium chloride in 10 ml of hot water and add with shaking a solution of 0.7 g of sodium bisulfite in 8 ml of 5 percent sodium hydroxide solution. Cool the mixture to room temperature and wash the precipitated cuprous chloride by decantation. Dissolve the washed material in 15 ml of a mixture of equal volumes of concentrated aqueous ammonia and water.

**3.** What types of organic substances other than unsaturated compounds will decolorize aqueous permanganate solution? will dissolve in cold concentrated sulfuric acid?

**4.** Write structural formulas and systematic names for the three isomeric $C_5$ alkynes. Which one of the isomers will not form a cuprous derivative?

**5.** Write equations for the action of the following reagents on 1-butyne (ethylacetylene): (a) hydrogen bromide in excess, stepwise; (b) an acidic aqueous solution of mercuric sulfate; (c) sodium amide ($NaNH_2$), followed by ethyl bromide.

**6.** Indicate a method for converting propene to propyne.

**7.** Suggest a procedure for separating the components of a mixture containing propane, propylene, and propyne.

# EXPERIMENT 2

# Alkenes

## Conversion of Alcohols to Alkenes

Alkenes are obtained from alcohols or alkyl halides by elimination reactions. Alcohols undergo elimination of water by heating with sulfuric or phosphoric acid, or by passing the alcohol vapor over alumina or silica catalysts at high temperatures. Alkyl halides undergo loss of halogen acid (dehydrohalogenation) by heating with a solution of potassium hydroxide in ethanol.

Alkenes are produced in enormous quantities in the petroleum industry by the pyrolysis ("cracking") of alkanes at 400–600°, in the presence of metallic oxide catalysts. Large alkane molecules undergo rupture of carbon-

$$C_{12}H_{26} \xrightarrow{500^\circ} C_6H_{12} + C_6H_{14},\ C_5H_{10} + C_7H_{16}, \text{ etc.}$$

carbon bonds to form a mixture of smaller alkanes and alkenes. Catalytic dehydrogenation of alkanes is also an important industrial method for producing alkenes.

The ease of dehydration of alcohols varies with the structural class, according to the sequence: tertiary > secondary > primary. *t*-Butyl alcohol is converted rapidly to isobutylene by 40–50 percent sulfuric acid at 85°; *s*-butyl alcohol requires 60–65 percent acid at 100° for alkene formation, and *n*-butyl alcohol 75–80 percent acid at 135–140°. Phosphoric acid has the advantage of causing less oxidative degradation than strong sulfuric acid but the rate of reaction is slower and higher temperatures are required.

The specific examples given in this experiment illustrate the dehydration of secondary alcohols by two slightly different procedures: 65 percent sulfuric acid in excess at 80–120° (2-pentene and methylpentenes) and concentrated sulfuric acid in catalytic amount at higher temperature (cyclohexene).

**Carbonium Ion Rearrangements.** The ease of dehydration of an alcohol is related to the stability of the carbonium ion resulting from rupture of the carbon-oxygen bond. It is unlikely that free carbonium ions are formed. The actual intermediate is probably a highly solvated cation, but one can use the free carbonium ion as a model for its chemical behavior.

The stability and ease of formation of carbonium ions follow the sequence—tertiary > secondary > primary. Elimination of a proton from the atom adjacent to the electron-deficient carbon atom leads to an alkene. If two different alkenes can be formed, usually there is selectivity in the mode

$$CH_3CH_2CH_2-\overset{+}{C}H-CH_3 \longrightarrow \underset{\text{mainly}}{CH_3CH_2CH{=}CH-CH_3}$$

of elimination and one of the isomers predominates. 2-Pentanol, for example, yields almost entirely 2-pentene and very little 1-pentene.

In many instances the double bond is formed at a position removed from the carbon atom that was bonded to the hydroxyl group and occasionally the carbon skeleton itself is altered during the reaction. The formation of 2-pentene from 1-pentanol is the consequence of migration of hydrogen

$$CH_3CH_2CH_2CH_2\overset{+}{C}H_2 \longrightarrow CH_3CH_2CH_2-\overset{+}{C}H-CH_3 \longrightarrow CH_3CH_2CH{=}CH-CH_3$$

with its bonding electrons (hydride shift), thereby converting the primary 1-pentyl carbonium ion into the more stable secondary pentyl carbonium ion *before* alkene formation takes place.

The occurrence of a similar migration of a methyl group with its bonding electrons (methide shift) is observed in the dehydration of neopentyl alcohol,

$$(CH_3)_3C-\overset{+}{C}H_2 \longrightarrow (CH_3)_2\overset{+}{C}-CH_2CH_3 \longrightarrow (CH_3)_2C{=}CH-CH_3$$

$(CH_3)_3C-CH_2-OH$. The shift of a methyl group converts the primary neopentyl carbonium ion into the more stable tertiary pentyl structure and leads to the formation of 2-methyl-2-butene.

Addition of a carbonium ion to an alkene furnishes a new carbonium ion and leads stepwise to alkene polymers of increasing molecular weight (dimers, trimers, and higher polymers). Usually small amounts of such dimers and trimers are formed in the preparation of alkenes from alcohols.

**Geometrical Isomerism.** Alkenes in which two different groups are attached to *each* of the carbon atoms of the double bond, as in 2-butene or 2-pentene, are capable of existing in two stereoisomeric forms arising from differences in the spatial distribution of the groups (*cis* and *trans* configurations). In these molecules the bonding of the atoms immediately surrounding the double bond is planar. The groups are held in relatively fixed positions: an

$$\underset{cis}{\begin{matrix} CH_3 & & C_2H_5 \\ & C=C & \\ H & & H \end{matrix}} \qquad \underset{trans}{\begin{matrix} CH_3 & & H \\ & C=C & \\ H & & C_2H_5 \end{matrix}}$$

energy barrier of 35–40 kilocalories restricts rotation of the groups about the double bond. The *cis* and *trans* stereoisomers exhibit discernible differences in physical properties and also in their rates of reaction with a given reagent.

**Reactions.** Alkenes are used in the laboratory and in industry as starting materials or intermediates for the synthesis of many important compounds. These transformations generally involve the characteristic addition reactions of the carbon-carbon double bond (ionic mechanism), with electrophilic reagents such as sulfuric acid or halogen acids, hypochlorous acid, chlorine or bromine, and oxidizing agents.

With unsymmetrical alkenes such as propylene and isobutylene, the mode of addition is highly selective: the electron-deficient fragment of the reagent adds to the carbon atom bearing the larger number of hydrogen atoms (the Markovnikov rule). This orientation of addition is attributed to

$$CH_3{-}CH{=}CH_2 + HBr \longrightarrow CH_3{-}CHBr{-}CH_3$$

electron-release by the alkyl group(s), which increases electron density at the less highly substituted carbon atom. An exception to the rule is the addition of hydrogen bromide to terminal alkenes in the presence of peroxides. Under these conditions (free radical mechanism) propylene gives 1-bromopropane since the reaction involves attack of the $CH_2$ group by an electrophilic bromine *atom*.

Substitution reactions of alkenes occur under special conditions and at high temperatures. Propylene and isobutylene react with chlorine at 500° to give allyl chloride, $CH_2{=}CH{-}CH_2Cl$, and methallyl chloride, respectively. These reactive alkenyl chlorides are useful synthetic intermediates.

Alkenes are oxidized readily by palladous or platinum salts to carbonyl compounds. Conversion of ethylene to acetaldehyde by this means recently has become an important industrial process; the palladium is recovered and

$$CH_2{=}CH_2 + PdCl_2 + H_2O \longrightarrow H^+[C_2H_4{\cdot}PdCl_2OH]^- \longrightarrow CH_3CH{=}O + 2HCl + Pd^0$$

is converted easily to palladous chloride for reuse. Propylene gives acetone, and 1- and 2-butenes give methyl ethyl ketone.

## 2a Methylpentenes[1]

Arrange a distillation assembly similar to that shown in Figure 9 (page 25)[2] using a 100-ml boiling flask. Fit the lower end of the condenser with an adapter that protrudes into a 50-ml Erlenmeyer flask cooled in an ice-water bath. Arrange an oil bath for heating.

Place 20 ml of water in a 125-ml Erlenmeyer flask and add *carefully*, with swirling, 0.4 mole (39.5 g, 21.5 ml) of concentrated sulfuric acid. Cool the diluted acid (~65 percent $H_2SO_4$) to 20–25° and add slowly 0.15 mole (15.3 g, 19 ml) of 4-methyl-2-pentanol,[3] with good mixing. Transfer the solution to the reaction flask through a funnel, add a few small boiling chips, and put the oil bath in place.

Heat the bath gradually until the temperature of the oil reaches 95–100° and continue to heat slowly until distillation of the volatile product ceases. About 1 hr is required. Transfer the distillate to a separatory funnel, shake well with 5–6 ml of 10 percent sodium hydroxide solution and allow the layers to settle. Draw off the lower aqueous layer, wash the hydrocarbon layer with 5–6 ml of water, and finally pour the hydrocarbon through the mouth of the funnel (why?) into a dry 25-ml Erlenmeyer flask. Add 1–2 g of anhydrous calcium chloride and allow the flask to stand, with occasional swirling, for about 20 min; longer standing does no harm.

➔ **CAUTION:** The methylpentenes are volatile liquids and are flammable. Take care to avoid fire hazards and to minimize loss by evaporation.

Carefully decant the dried product into a distilling flask of the proper size, add a few small boiling chips, and distill from a water bath. Collect in a dry weighed bottle the portion boiling at 54–68°. The yield is 9–10 g. The product consists of three structural isomers:[1] 4-methyl-2-pentene, 4-methyl-1-pentene (bp 53.9°), and 2-methyl-2-pentene (bp 67.3°). The first of these exists in two stereoisomeric forms (*cis*-, bp 56.4°; *trans*-, bp 58.6°). 2-Methyl-2-pentene results from molecular rearrangement (hydride shift) within the intermediate secondary carbonium ion to form a more stable tertiary car-

[1] Analysis by vapor phase chromatography in the Cornell laboratory has shown that the mixture of methylpentenes obtained in this preparation contains approximately 78–79 percent *trans*-4-methyl-2-pentene and 15–16 percent 2-methyl-2-pentene, with small amounts of 4-methyl-1-pentene and *cis*-4-methyl-2-pentene. See also, *J. Chem. Educ.*, **46**, 765 (1969).

[2] If ordinary glassware and corks are used the assembly will be that of Figure 10, page 26, with a 125-ml distilling flask and an ice-cooled Erlenmeyer flask as receiver.

[3] The inexpensive practical grade of 4-methyl-2-pentanol is quite satisfactory. It is readily available: Eastman Organic Chemicals, #P3607; Matheson, Coleman and Bell, #MX1295.

bonium ion before the transformation into an alkene occurs. Elaborate fractional distillation techniques are required to separate the mixture into the individual components.

**Tests for Unsaturation.** Carry out tests using 1–2 drops of the methylpentene mixture and 1–2 ml of the following reagents, as described in Experiment 1: (A) 2 percent bromine solution in carbon tetrachloride; (B) 0.5 percent aqueous permanganate (Baeyer's test); (C) cold concentrated sulfuric acid.

*Questions follow Experiment 2c.*

## 2b 2-Pentene

### (β-Amylene)[4]

Arrange a distillation assembly as described in Experiment 2a. Prepare the diluted sulfuric acid ($\sim$65 percent) as indicated there and add to it slowly, 0.20 mole (17.5 g, 21.5 ml) of 3-pentanol,[5] instead of the 4-methyl-2-pentanol.

Heat the reaction mixture and collect the product according to the procedure in Experiment 2a. Since 2-pentene is extremely volatile take care to minimize loss by evaporation. Wash the crude product with 10 percent sodium hydroxide solution and with cold water, and dry it over 1–2 g of anhydrous calcium chloride.

➛**CAUTION:** 2-Pentene is highly flammable. Avoid fire hazards.

Carefully decant the dried product into a distilling flask of the proper size, add a few small boiling chips, and distill from a water bath. Collect in a dry weighed bottle the portion boiling at 35–41°. The yield is 8–9 g. The 2-pentene obtained in this experiment is a mixture of the two geometrical isomers (*cis*-, bp 36.7°; *trans*-, bp 36.1°) in which the stereoisomer of *trans*-configuration predominates.[6] Some of the structural isomer, 1-pentene, is present as a result of molecular rearrangement (hydride shift) within the intermediate carbonium ion before its transformation into an alkene.

[4] Amylene is an old name used for any of the $C_5$ alkenes or mixtures of them, that often are prepared by dehydration of the amyl alcohols (pentanols). Olefin is a general name used for any alkene or cycloalkene.

[5] The practical grade of 3-pentanol is quite satisfactory. 2-Pentanol, practical grade, may be used in place of 3-pentanol since it also furnishes 2-pentene.

[6] Analysis by vapor phase chromatography in the Cornell laboratory indicated that the product contains about 95 percent *cis*- and *trans*-2-pentenes and not more than 5 percent 1-pentene.

**Tests for Unsaturation.** Carry out tests using 1–2 drops of the alkene and 1–2 ml of the following reagents, as described in Experiment 1: (A) 2 percent bromine solution in carbon tetrachloride; (B) 0.5 percent aqueous permanganate (Baeyer's test); (C) cold concentrated sulfuric acid.

*Questions follow Experiment 2c.*

## 2c Cyclohexene

Place 0.2 mole (20 g, 21 ml) of cyclohexanol (a technical grade is satisfactory) and 2 ml of concentrated sulfuric acid[7] in a 100-ml round-bottomed flask, and mix thoroughly. Provide the flask with a short fractionating column (see Figure 14, page 42) and fit the lower end of the condenser with an adapter which passes well into a 50-ml Erlenmeyer flask cooled in an ice-water bath.

➤**CAUTION:** Cyclohexene is a flammable liquid. Avoid fire hazards.

Add two boiling chips and heat the flask gently over a wire gauze with a small flame, so that the cyclohexene and water distill through the column. The temperature at the top of the column should not be allowed to rise above 100°. Continue the distillation until only a small residue remains.

To the distillate add 1–2 ml of sodium carbonate solution to neutralize traces of acid (test with litmus paper). Transfer to a separatory funnel, allow the layers to separate, and draw off the lower aqueous layer. Pour the hydrocarbon layer through the mouth of the funnel (why?) into a small dry flask or bottle, add 3–4 g of anhydrous calcium chloride and allow to stand for 20 min, with occasional shaking. Decant the dried product into a small distilling flask, add a few small boiling chips, and distill from a water bath. Collect the portion boiling at 80–85° in a weighed bottle. If an appreciable low-boiling fraction is obtained, dry this again and redistill. The yield is 8–10 g.

**Tests for Unsaturation.** Carry out tests using 1–2 drops of cyclohexene and 1–2 ml of the following reagents, as described in Experiment 1: (A) 2 percent bromine solution in carbon tetrachloride; (B) 0.5 percent aqueous permanganate (Baeyer's test); (C) cold concentrated sulfuric acid.

[7] In place of 2 ml of concentrated sulfuric acid, 10 ml of syrupy (85 percent) phosphoric acid may be used. Cyclohexene can be prepared equally well using 65 percent sulfuric acid as described in Experiment 2a. For 20 g of cyclohexanol use a solution of 21 ml (39 g) of concentrated sulfuric acid in 20 ml of water.

## Questions

**1.** Write structural formulas and systematic names for: (a) all of the isomeric pentenes ($C_5H_{10}$); (b) all of the isomeric methylcyclohexenes.

**2.** What alkene will be the main product when each of the following alcohols is dehydrated (*cf.* carbonium ion rearrangements): (a) 3-methyl-1-butanol? (b) 3-methyl-2-butanol? (c) 3,3-dimethyl-2-butanol?

**3.** What product is formed from cyclohexene by addition of hypochlorous acid? by addition of ozone and subsequent hydrolysis of the ozonide in the presence of a mild reducing agent?

**4.** What isomeric hexenes (propylene dimers) can be formed by addition of the 2-propyl carbonium ion to propylene and subsequent loss of a proton? How could the structure of these hexenes be established?

**5.** Would nitric acid be an acceptable substitute for sulfuric acid for conversion of an alcohol to an alkene? Would concentrated hydrochloric acid do?

# EXPERIMENT 3

# Alcohols

## Fermentation of Sugars

The fermentation processes involved in bread making, wine making, and brewing are among the oldest chemical arts. For many years it was believed that the transformation of sugar by yeasts into ethanol and carbon dioxide was inseparably connected with the life process of the yeast cell. This view was abandoned when Eduard Buchner (Nobel Laureate, 1907) demonstrated that yeast juice will bring about alcoholic fermentation in the absence of any yeast cells. The fermenting activity of yeast is due to a remarkably active catalyst of biochemical origin, the enzyme zymase. It is now recognized that most of the chemical transformations that go on in living cells of plants and animals are brought about by enzymes.

Enzymes show an extraordinary specificity—a given enzyme can act only on a specific compound, or closely related group of compounds. Thus, zymase acts only on a few specific sugars and not on all carbohydrates; the digestive enzymes of the alimentary tract are equally specific in their activity.

In the alcoholic fermentation of disaccharides such as maltose or sucrose ($C_{12}H_{22}O_{11}$) the first step is an hydrolysis to hexoses ($C_6H_{12}O_6$) by the enzyme maltase or invertase, which is present in yeast in addition to zymase. The hexoses are then converted by zymase into alcohol and carbon dioxide. Pasteur observed that growth and fermentation were promoted by the addition of small amounts of mineral salts to the nutrient medium. Later it was found that, prior to fermentation, the hexose sugars combine with phosphoric acid and the resulting hexose-diphosphate is then split up into carbon dioxide and ethanol.

Fractional distillation of the fermentation liquor gives an ethanol-water azeotrope, bp 78.15°, containing 95.6 percent ethanol. Commercially the water is removed from the azeotrope by distillation with benzene and essentially anhydrous ethanol, bp 78.37°, is obtained.

Yeasts, molds, and bacteria are used commercially for the large scale production of various organic compounds. An important example, in addition to ethanol, is the anerobic fermentation of starch by certain bacteria to yield *n*-butyl alcohol, acetone, ethanol, carbon dioxide and hydrogen.

In the United States industrial ethanol is manufactured mainly from ethylene, a product of the "cracking" of petroleum hydrocarbons. By reaction with concentrated sulfuric acid ethylene is converted to ethyl hydrogen sulfate, which is hydrolyzed to ethanol by dilution with water. Isopropyl, *s*-butyl, *t*-butyl and higher secondary and tertiary alcohols also are produced on a large scale from alkenes derived from the cracking process.

Mixtures of $C_5$ alcohols are manufactured by hydrolysis of $C_5$ alkyl chlorides obtained by chlorination of *n*-pentane and isopentane (from petroleum). Higher alcohols, $C_{12}$–$C_{18}$, are manufactured by catalytic reduction of fatty esters.

Laboratory syntheses of alcohols often make use of the reaction of an organomagnesium halide (Grignard reagent) with an aldehyde, ketone, or ester; examples are given in Experiments 4 and 41.

**The Naming of Alcohols.** Several types of names are used: (a) common or trivial names of the *alcohol* or the *carbinol* system (carbinol = methanol); (b) Geneva or IUPAC names based on the longest chain to which the hydroxyl function is attached; (c) hybridized names using parts of both systems.

(a) Alcohol is used as a separate word following the name of the appropriate hydrocarbon radical denoting *all* of the carbon atoms and their structural arrangement. Carbinol (or methanol) is joined directly to the names of the attached hydrocarbon radicals; the sum of the carbon atoms named in the substituent radical(s) will be *one less* than the total in the molecule because one carbon is included in carbinol.

The difference between carbinol and alcohol names arises because alcohol is a generic name for a whole class of compounds rather than a parent structure. On the other hand, carbinol is the name of a specific parent compound, in which the hydrocarbon radical(s) may replace one or more hydrogen atoms. The same principles are used in naming many other compounds: separate words—*s*-butyl chloride, dimethyl ether, diethyl ketone; combined form—isobutylacetylene, dimethylamine, *t*-butylacetic acid, *n*-propylbenzene.

Prefixes are needed in the alcohol and carbinol names to specify the structure of the hydrocarbon radicals: *n*- for normal (straight chain, OH at terminal), *s*- or *sec*- for secondary ($R_2CH-$), and *t*- or *tert*- for tertiary ($R_3C-$). The prefixes iso, for a chain with the terminal group $(CH_3)_2CH-$, and neo, for a system with a quaternary carbon group such as $R_3C-CH_2-$, are joined directly to the radical name as in isopropyl and neopentyl. This system would require 17 prefixes for the $C_6$ alcohols.

(b) In the Geneva (IUPAC) system alcohols are named as alkanols. The longest hydrocarbon chain *including the carbon bearing the hydroxyl function*, together with any substituent side chains, is given its appropriate systematic name and the -ane ending is changed to -anol. The position of the hydroxyl is indicated by the smallest appropriate number along the chain. The number is often placed before the name with the simpler alcohols, as in 2-butanol, but in more complex structures it is better to place the number close to the name it pertains to, as in 3-chloro-2-methylpentanol-2.

(c) Hybridized names generally combine a prefix used in the alcohol system with a systematic alkanol name. Examples are *s*-butanol instead of *s*-butyl alcohol or 2-butanol, and *t*-butanol instead of *t*-butyl alcohol or 2,2-dimethylbutanol-1. In some cases these hybrids have the advantage of brevity without loss of clarity.

In this manual examples of the various systems of nomenclature will be used from time to time as appropriate.

## 3a Ethanol by Fermentation

Place 80 g of sucrose (common granulated sugar) in a 1-liter Florence flask, add about 700 ml of water, 70 ml of a solution of Pasteur's salts,[1] and one-half cake of compressed yeast rubbed to a thin paste with 30–40 ml of water. Shake vigorously and close the flask with a rubber stopper holding a delivery tube arranged so that any gas evolved must bubble through about 10 ml of a clear solution of barium hydroxide in a test tube. Protect the barium hydroxide from carbon dioxide of the air by a 4–5 mm layer of paraffin oil or kerosene. Allow the mixture to stand at a temperature of about 25° until fermentation is complete, as indicated by the cessation of gas evolution (usually about a week is required).

Without stirring up the yeast any more than is necessary, decant the liquid through a plug of cotton or glass wool into a 1-liter round-bottomed flask. Connect the flask by means of a bent glass tube (about 8 mm in diam) with a condenser set downward for distillation, add two tiny boiling chips and distill off 200–250 ml of liquid. Discard the residue remaining in the flask. Transfer the distillate to a 500-ml round-bottomed flask, arrange a fractionating column and distill slowly (see Exercise 2). Collect the following fractions: A, 78–82°; B, 82–88°; C, 88–95°. Discard the residue containing the fusel oil.

Redistill the fractions A, B, and C to separate as large a quantity of 95 percent ethanol as possible. Salt out the ethanol from fraction B by adding solid anhydrous potassium carbonate until the liquid is saturated.[2] Shake well, decant the liquid from any solid material into a small dry separatory

[1] Pasteur's salts solution consists of: potassium phosphate 2.0 g, calcium phosphate 0.20 g, magnesium sulfate 0.20 g, ammonium tartrate 10.0 g, in 860 g of water.

[2] Fraction C of the second distillation will be quite small *if the fractionation has been carried out properly* and may be neglected. If fraction C at this point is relatively large it should be subjected to further *careful* fractionation.

funnel, and allow the layers to separate. Draw off the lower aqueous layer and pour the upper layer into fraction A. Redistill the fraction A from a small distilling flask and collect the fraction boiling at 78–82° in a clean weighed receiver.

Carry out the *Iodoform Test* and *Esterification Test* (see Experiment 3b) with small portions of the product.

### Questions

**1.** What is fusel oil? Write the structural formulas and Geneva names of the most important constituents of fusel oil.

**2.** What is proof spirit? What is meant by 180-proof?

**3.** Using as many systems as seem appropriate, name all of the alcohols having the carbon skeleton of: (a) 3-methylpentane; (b) neohexane.

**4.** When benzene is used to dry 95.6 percent ethanol two azeotropes are involved: a ternary azeotrope, bp 65°, containing benzene, 74 percent by weight, ethanol 18.5, and water 7.5; a binary azeotrope, bp 68°, containing benzene, 68 percent, and ethanol 32 percent.

Assuming a safety factor of 10 percent excess, calculate the amount of benzene needed to dry 1 kg of 95.6 percent ethanol. How much absolute alcohol would be obtained?

## 3b Reactions of Alcohols

Although a few specific alcohols are suggested for each of the test reactions it is not intended that the tests be limited to these examples. In addition to the alcohols mentioned it is desirable to have others available for use as unknowns.

For the tests it is recommended that the specified quantities be measured in a graduated cylinder or a small pipette and not by rough visual estimation, to insure satisfactory results.

***(A) Iodoform Test.*** Try this test with methanol, ethanol, 1-butanol, and 2-butanol. Iodoform is not formed directly from the alcohol but from the carbonyl compound produced by oxidizing action of the hypoiodite of the reagent (see Experiment 9, part F).

In a test tube mix 0.5 ml of the liquid to be tested with 5 ml of water and 5 ml of 10 percent sodium hydroxide solution. Add dropwise with shaking a 10 percent solution of iodine in potassium iodide,[3] until a definite brown color persists (indicating an excess of iodine).

[3] The iodine-potassium iodide solution is prepared by dissolving 10 g of iodine crystals in a solution of 20 g of potassium iodide in 80 ml of water and stirring until the iodine has dissolved.

With some compounds a precipitate of iodoform appears almost immediately in the cold. If iodoform does not appear within 5 min, warm the solution to 60° in a beaker of water. If the brown color is discharged add more of the iodine solution until the brown color persists for 2 min. Add a few drops of sodium hydroxide solution to remove the excess iodine, dilute the mixture with 5–10 ml of water and allow it to stand for 10 min.

Iodoform crystallizes as lemon-yellow hexagons having a characteristic odor. The identity of the crystals may be confirmed by collecting them with suction, drying, and taking the melting point (119°).

***(B) Esterification Test.*** Carry out this test with ethanol, *s*-butanol, and *t*-butanol. Primary alcohols, and most secondary alcohols, of low molecular weight are converted into volatile esters (acetates), that have a characteristic odor different from that of the alcohol. Tertiary alcohols are not esterified under these conditions but may be converted to alkenes.

Mix 0.5 ml of the liquid to be tested with 0.5 ml of glacial acetic acid and add cautiously 3–4 drops of concentrated sulfuric acid. Mix the liquid thoroughly and warm the tube gently for a few minutes. Add 2–3 ml of water and *cautiously* note the odor of the vapor.

***(C) Oxidation Test.*** To 5 ml of a 1 percent solution of sodium dichromate, add one drop of concentrated sulfuric acid and mix thoroughly by shaking. Add 2 drops of the liquid to be tested, and warm gently. Note the odor, and observe any change in color of the solution.

Carry out this test with ethanol, *s*-butanol and *t*-butanol.

***(D) Lucas' Test.*** The reagent used is concentrated hydrochloric acid containing one mole of fused zinc chloride to one mole of the acid.[4] The differentiation is based upon the rate of formation of the alkyl chlorides from the alcohols with the test reagent, and with concentrated hydrochloric acid.

To 0.5 ml of the alcohol add quickly 3 ml of the hydrochloric acid-zinc chloride reagent at 26–27°. Close the tube with a cork and shake; then allow the mixture to stand. Observe carefully for the first five minutes, and again after one hour. Note the time required for reaction to take place, as indicated by the cloudy appearance of the solution, and whether or not two immiscible layers finally separate. If the result is positive, carry out a second test using concentrated hydrochloric acid alone, instead of the test reagent, and observe.

Perform the Lucas test with *n*-butanol, *s*-butanol, *t*-butanol, and isopentyl alcohol. Other alcohols that are available may also be tested (*e.g.*, an unknown alcohol furnished by the instructor), bearing in mind the limitations mentioned in the following discussion.

[4] Lucas' reagent is prepared by dissolving 34 g of anhydrous (fused) zinc chloride in 27 g (23 ml) of concentrated hydrochloric acid, with stirring and external cooling to avoid loss of hydrogen chloride. The resulting solution has a volume of about 35 ml and is sufficient for ten individual tests. The reagent should be reasonably fresh in order to obtain reliable results.

The lower alcohols through $C_5H_{11}-OH$ are soluble in Lucas' reagent. Tertiary alcohols react at once to give the alkyl chloride, as indicated by the development of a cloudy appearance; the chloride is sparingly soluble in the reagent and soon forms a distinct layer. Secondary alcohols react within five minutes to give a cloudy appearance, and a distinct layer appears later. Saturated primary alcohols do not react at ordinary temperatures and the solution remains clear. If small amounts of secondary or tertiary alcohols are present as impurities in a primary alcohol, some cloudiness may be observed, but a distinct layer of alkyl halide will not separate on standing. Separation into two distinct layers on standing will occur only if the test liquid contains an appreciable quantity of secondary or tertiary alcohol.

The differentiation of secondary and tertiary alcohols is confirmed by a supplementary test with concentrated hydrochloric acid alone. A tertiary alcohol reacts rapidly to form the chloride, which forms a separate layer in a few minutes. The solution of a secondary (or primary) alcohol remains clear, since neither will form an alkyl chloride under the conditions of the supplementary test.

Isopropyl alcohol does not always show the characteristic behavior of a secondary alcohol, due presumably to the circumstance that isopropyl chloride is very volatile (bp 36°) and may escape from the solution. Certain very reactive primary alcohols, such as allyl alcohol, and higher molecular weight compounds behave in a nontypical manner.[5]

***(E) Reaction with 1-Naphthylisocyanate*** *(Optional)*. This reagent, $1-C_{10}H_7-N{=}C{=}O$, reacts with alcohols to form 1-naphthylcarbamates, $C_{10}H_7-NH-CO-OR$, which are useful for identification. Since the reagent forms a sparingly soluble diarylurea, Ar−NH−CO−NH−Ar, in the presence of water, the reaction should be carried out with an essentially anhydrous sample of the alcohol in a *dry* test tube.

Mix 1 ml (0.8 g) of the alcohol with 0.5 ml (0.6 g) of 1-naphthylisocyanate (*caution—irritating vapor!*) and warm the mixture in a beaker of boiling water for 5–10 min. Cool the tube and scratch the inside wall with a glass rod to induce crystallization. Dissolve the product in 5 ml of hot ligroin, bp 100–120° (*flammable solvent*), which leaves any diarylurea undissolved. Filter the hot solution, allow the filtrate to cool, and collect the crystals. Recrystallize the product from hot ligroin and take the melting point (see Table 17 (page 190)).

Tertiary alcohols react sluggishly. For these use a 2 ml sample of the alcohol, 1 ml of the isocyanate, and add 0.2 g of anhydrous sodium acetate as catalyst. Heat the mixture, with protection from moisture, for 3–4 hr in a boiling water bath. Usually a large amount of dinaphthylurea is formed.

***(F) Examination of an Unknown Alcohol*** *(Optional)*. Obtain from the instructor a 5 ml sample of an unknown alcohol and carry out tests to determine as many structural features as possible.

[5] Lucas, *J. Amer. Chem. Soc.*, **52**, 803 (1930).

**TABLE 17**
**Melting Points of 1-Naphthylcarbamates of Alcohols**

| ALCOHOL | °C | ALCOHOL | °C | ALCOHOL | °C |
|---|---|---|---|---|---|
| Methyl | 124 | *s*-Butyl | 97 | 3-Pentyl | 95 |
| Ethyl | 79 | Isobutyl | 104 | Isopentyl | 68 |
| *n*-Propyl | 80 | *t*-Butyl | 101 | *t*-Pentyl | 71 |
| Isopropyl | 106 | *n*-Pentyl | 68 | Cyclopentyl | 118 |
| *n*-Butyl | 71 | 2-Pentyl | 75 | Cyclohexyl | 128 |

## Questions

**1.** Why does methanol fail to give the iodoform test?

**2.** Write the structural formula for each of the following alcohols and indicate if it would give a positive iodoform test: (a) isobutyl alcohol; (b) isopropyl alcohol (2-propanol); (c) neopentyl alcohol (2,2-dimethyl-1-propanol); (d) methyl-*n*-butylcarbinol (2-hexanol).

**3.** How could one ascertain whether the alcohol or the carboxylic acid furnishes the hydroxyl group that appears in the molecule of water produced in formation of an ester?

EXPERIMENT

# 4

# The Grignard Reaction

## Synthesis of an Alcohol

Organomagnesium halides, Grignard reagents, are formed by direct reaction of alkyl or aryl halides with magnesium metal in the presence of a solvent such as anhydrous ether or tetrahydrofuran. Unless the reactants, solvent and apparatus are carefully dried and the magnesium is pure and relatively free of oxide coating, the reaction does not start easily. Addition of a small crystal of iodine aids in inducing reaction, probably by activating the surface of the metal. Alkyl and aryl bromides usually are the preferred halides. With

$$C_2H_5{-}Br + Mg + (C_2H_5)_2O \longrightarrow C_2H_5{-}MgBr\cdot 2(C_2H_5)_2O$$

chlorides the reaction is more difficult to initiate; with iodides there is greater tendency to favor a side reaction—coupling at the metal surface to form the hydrocarbon R−R (Wurtz reaction).

Although Grignard reagents exist as coordination compounds in a complex equilibrium mixture, the simplified expression R−MgX is used for convenience. In the course of reaction the highly polarized carbon-magnesium linkage is cleaved so that the alkyl (or aryl) group is transferred *with its bonding electrons* to an electrophilic center. Owing to their high reactivity Grignard reagents are among the most versatile compounds available for laboratory syntheses. Organolithium compounds also are used.

Compounds containing active hydrogen (water, alcohols, ammonia, acetylene, acids) convert a Grignard reagent to the parent hydrocarbon, R−H. It is for this reason that materials used must be scrupulously pure and dry. Halogens and oxygen also react with R−MgX. Halides of mercury, cadmium, tin, boron, and many other elements, react with R−MgX to furnish organic derivatives of the less reactive elements.

The most important synthetic applications of the Grignard reagent involve two different types of reaction, both of which effect transfer of the alkyl (or aryl) group of R−MgX to a carbon atom of the reactant: (1) addition to the carbonyl function of aldehydes, ketones, esters, amides, acid halides and carbon dioxide, and to the cyano group of nitriles; (2) replacement of the alkoxyl group of esters, and acetals, and of the halogen atom of reactive

$$CH_3{-}CH{=}O \xrightarrow{C_2H_5{-}MgX} \underset{\displaystyle OMgX}{CH_3{-}CH{-}C_2H_5} \xrightarrow[H^+]{H_2O} \underset{\displaystyle OH}{CH_3{-}CH{-}C_2H_5}$$

$$CH_3{-}CO{-}OC_2H_5 \xrightarrow{2\ C_2H_5{-}MgX} \underset{\displaystyle OMgX}{CH_3{-}C(C_2H_5)_2} \longrightarrow \underset{\displaystyle OH}{CH_3{-}C(C_2H_5)_2}$$

organic halides. The halomagnesium complex produced in the reaction is usually hydrolyzed by cold dilute mineral acid to disengage the organic product. For acid-sensitive compounds a strong aqueous solution of ammonium chloride can be used for the hydrolysis.

## 2-Methyl-2-Pentanol

**Preparation of *n*-Propylmagnesium Bromide Solution.**[1] Arrange an assembly of apparatus like that in Figure 37 (page 169) with upright condenser, separatory funnel for addition, and a 500-ml reaction flask. Prepare a bath of ice and water to permit rapid cooling if the reaction becomes vigorous. Make certain that no flame is nearby to ignite the ether vapor.

➤**CAUTION:** Ether is extremely volatile and highly flammable. Take great care to avoid fire hazards.

In the reaction flask place 0.33 gram-atom (8 g) of clean magnesium turnings and add a *small* crystal of iodine. Weigh out 0.33 mole (40.5 g, 30 ml) of pure *n*-propyl bromide[2] in a clean dry flask and mix it with 100 ml of *absolute* ether.[3] Pour a portion of the bromide solution into the separatory

[1] In Grignard reactions it is essential that the reagents be free of ethanol and water, and the apparatus perfectly clean and dry. Rubber stoppers must not be used because they contain extractable deleterious sulfur compounds.

[2] *n*-Propyl bromide may be prepared conveniently from *n*-propyl alcohol by the method described for *n*-butyl bromide (Experiment 5a). The bromide should be carefully purified and dried, and collected over a small boiling point range, bp 69–70° at 760 mm.

[3] Ether used as the solvent must be carefully purified to remove impurities (ethanol and water) present in some technical grades of ether. The procedure used in the preparation of absolute, anhydrous ether consists in: (1) washing with strong calcium chloride solution;

funnel and allow 10–15 ml to flow onto the magnesium in the flask. Under favorable conditions the reaction will start within a few minutes, accompanied by vigorous boiling of the ether. As soon as this occurs, introduce 25 ml of ether directly through the top of the condenser to moderate the vigor of the reaction.

If the reaction does not start promptly, warm the flask gently in a bath of tepid water, and be prepared to moderate the reaction by cooling and addition of absolute ether if the reaction starts suddenly. If necessary add another small crystal of iodine, or a small quantity of a previously prepared Grignard reagent, to initiate reaction.[4] It is absolutely essential for the success of the experiment that the reaction begin before the remainder of the propyl bromide solution is added to the magnesium.

When the initial vigorous reaction has slowed down, allow the remainder of the propyl bromide solution to flow dropwise into the reaction mixture, at such a rate that the ether refluxes gently without external heating. Swirl the contents of the flask at frequent intervals. After all the bromide has been added, reflux the mixture gently for 0.5 hr on a water bath. Do not heat so vigorously that ether vapors traverse the condenser. At this point the magnesium will have dissolved essentially completely and the solution should have a volume of not less than 100 ml. If necessary add more anhydrous ether to bring the volume to about 100–125 ml.

**Reaction of the Grignard Reagent with Acetone.** Cool the reaction flask containing the Grignard reagent in an ice-salt mixture to as low a temperature as possible. The yield in the next step is increased by carrying out the reaction at low temperature, with good mixing and slow addition of the acetone. In a small dry flask mix 0.35 mole (20 g, 26 ml) of dry acetone (dried at least overnight over anhydrous potassium carbonate or magnesium sulfate) with 40 ml of absolute ether. Transfer the solution to the separatory funnel and allow it to drop *very slowly* into the cooled Grignard solution, while swirling and shaking the flask to insure good mixing and effective cooling. Each drop of the solution reacts vigorously, producing a hissing sound and forming a white precipitate that usually redissolves on shaking. After all of the acetone has been added, remove the cooling bath and allow the mixture to stand at room temperature with occasional shaking for 20 min or longer.

(2) preliminary drying with calcium chloride; (3) drying with phosphorus pentoxide; (4) distillation, with precautions to protect the product from atmospheric moisture. In some laboratories absolute ether is dried over metallic sodium instead of phosphorus pentoxide.

Technical ether of high quality, such as that from Carbide and Carbon Chemicals Co., contains no ethanol or water, and in fact is satisfactory for direct use in this experiment provided it has not been exposed to air, from which it absorbs moisture. It is advisable to store this ether over phosphorus pentoxide or metallic sodium.

[4] Particularly sluggish Grignard reactions frequently can be initiated by addition of a few drops of 1,2-dibromoethane.

Pour the reaction mixture slowly and carefully, with stirring, onto a mixture of chipped ice and dilute sulfuric acid (prepared by adding 0.4 equivalent (20 g, 11 ml) of concentrated sulfuric acid to 100 ml of water, and adding about 100 g of chipped ice). Rinse the reaction flask with a little of the dilute sulfuric acid and a little ordinary ether, and add these washings to the main product. Transfer the mixture to a separatory funnel and separate the two layers; *save both layers*. Extract the aqueous layer with two 50–ml portions of ordinary ether and combine the ether extracts with the ether layer from the first separation. The aqueous layer may now be discarded.

Wash the ether layer with 25 ml of water to which 2–3 ml of strong sodium bisulfite solution has been added,[5] then with 25 ml of cold water to which 5 ml of saturated sodium bicarbonate solution has been added, and separate the layers carefully. Dry the ethereal solution overnight or longer, over anhydrous magnesium sulfate or potassium carbonate, filter from the drying agent and remove the solvent by distilling the dried solution from a flask fitted with a fractionating column, using a water bath or a steam bath.

Transfer the residue to a small distilling apparatus, heat on a wire gauze over a small flame and collect the material boiling above 75° and below 120°, in fractions collected over 15° temperature ranges. Collect the product over the range 120–125°. If an appreciable quantity of low-boiling material is obtained in the fraction above 105°, dry this again and redistill it, collecting additional product in the 120–125° range. Additional product may also be obtained by redistilling the higher-boiling fraction collected over the range 125–130°. The yield is 12–20 g. The reported boiling point of 2-methyl-2-pentanol is 123–124° at 760 mm.

Test the fractions boiling below 105° for unsaturation with 0.5 percent aqueous permanganate solution (Baeyer's test) or with 2 percent bromine in carbon tetrachloride (see Experiment 1). The distillation residue consists mainly of polymers formed from alkenes produced by thermal dehydration of the tertiary alcohol. This alcohol can be characterized as the 1-naphthylcarbamate by reaction with 1-naphthyl isocyanate (see Experiment 3b, part E). The yield of the derivative is poor since tertiary alcohols react slowly and tend to form alkenes.

## Questions

**1.** Write the formulas of two alkenes that may be formed as accessory products in this experiment.

**2.** Write equations for two methods of preparing 2-methyl-2-pentanol using methylmagnesium bromide.

**3.** Explain the role of ether in the formation of Grignard reagents.

[5] Sodium bisulfite removes traces of free iodine that would catalyze alkene formation when the product is distilled. Any acid remaining in the product has a similar effect.

**4.** Write equations for the action of the following substances on *n*-propylmagnesium bromide: carbon dioxide; ethyl formate; *n*-butyraldehyde; iodine; acetylene; ethanol.

**5.** Outline a satisfactory method of purifying the following materials for use in a Grignard reaction: *n*-butyl bromide; acetone; ethyl acetate.

# EXPERIMENT 5

# Alkyl Halides

## Conversion of Alcohols to Alkyl Halides

In the laboratory alkyl halides usually are prepared from the corresponding alcohols, many of which are readily available from commercial sources. Alkyl halides can be obtained also by addition of halogen acids to alkenes and by direct chlorination or bromination of alkanes, but these methods are less convenient and generally produce mixtures of isomers that are difficult to separate with laboratory fractionating columns.

Alcohols are converted to alkyl halides by means of the halogen acids, or halides of phosphorus ($PCl_3$, $PBr_3$). Thionyl chloride ($SOCl_2$) is useful for the preparation of alkyl chlorides. The ease of reaction—tertiary > secondary > primary—follows the same sequence as the formation of carbonium ions, since the mechanism involves a carbonium intermediate. For this reason rearrangements are observed analogous to those occurring in the formation of alkenes from alcohols. Reactivity of the halogen acids toward alcohols,

$$\text{R–OH} \xrightarrow{\text{H}^+} \text{R–OH}_2^+ \longrightarrow \text{H}_2\text{O} + \text{R}^+ \xrightarrow{\text{X}^-} \text{R–X}$$

like the reactivity toward alkenes, diminishes in the order—HI > HBr > HCl. Consequently, conditions suitable for conversion of alcohols to alkyl halides are determined by the structure of the alcohol *and* the specific halogen acid to be used.

Tertiary alcohols react readily with any of the three halogen acids, even at 25° and in the absence of catalysts. Secondary alcohols react more slowly; moderate heating and acidic catalysts (50 percent sulfuric acid, zinc chloride) are used to promote the conversion to bromides and chlorides. Primary alcohols require more vigorous conditions and more active catalysts (65–70 percent sulfuric acid for conversion to bromides). The Lucas test for differentiating the classes of alcohols (Experiment 3b) is based upon relative rates of conversion to alkyl chlorides.

Primary and secondary alkyl bromides ($C_2-C_5$) are prepared by heating the alcohol with concentrated hydrobromic acid, or a hydrobromic-sulfuric acid mixture. A convenient procedure is to use sodium bromide and an excess of strong sulfuric acid, but this method is not suitable for higher bromides since the high concentration of salts present greatly reduces the solubility of the alcohol in the reaction medium. For secondary alcohols a lower concentration of sulfuric acid (50 percent) is used; stronger acid is unnecessary and promotes a side reaction (alkene formation). For higher molecular weight alcohols the action of anhydrous hydrogen bromide at 100–120° in the absence of a solvent, or use of phosphorus tribromide, is a satisfactory procedure. Water-soluble tertiary alcohols are converted rapidly to bromides merely by shaking with concentrated aqueous hydrobromic acid.

Hydrochloric acid containing dissolved zinc chloride (a Lewis acid) is used for the conversion of primary and secondary alcohols to alkyl chlorides. Primary alcohols require heating with a saturated solution of zinc chloride in concentrated hydrochloric acid. Thionyl chloride is a useful reagent for the preparation of primary alkyl chlorides.

Alkyl iodides are obtained readily from alcohols by means of strong aqueous hydriodic acid but a more economical method is to treat the alcohol

$$6\,R{-}CH_2OH + 3I_2 + 2P \longrightarrow 6\,R{-}CH_2I + 2H_3PO_3$$

with iodine and phosphorus. Alkyl iodides may be prepared also by reaction of alkyl chlorides or bromides with sodium iodide in acetone solution (Finckelstein reaction).

Alkyl iodides are used much less frequently in synthetic work than are the alkyl bromides and chlorides, because the latter are usually more readily available and less expensive. However, methyl iodide and ethyl iodide are often used in preference to the corresponding bromides or chlorides in ordinary laboratory work because the latter are so extremely volatile. The boiling points of the methyl halides are: methyl chloride, −24°; bromide, 5°; iodide, 42°.

**Reactions.** Alkyl halides, especially alkyl bromides, are of great importance in laboratory syntheses. They undergo displacement (substitution) reactions with nucleophilic reagents (alkoxides, phenoxides, cyanide anion and ammonia) to produce ethers, nitriles, amines, etc. They effect carbon-alkylation of the sodium derivatives of malonic and acetoacetic esters, and react with magnesium to furnish Grignard reagents (see Experiment 4).

Substitution reactions follow two diverse mechanisms, influenced by the structure of the alkyl group, the nature of the nucleophilic reagent and the experimental conditions. Reactions of the primary halides follow second order kinetics ($S_N2$ mechanism)—the rate depends upon the concentrations of both reactants. For $S_N2$ reactions the relative reactivities of the halides are—methyl > primary > secondary ≫ tertiary—and steric factors (size and shape of the reactants) play a significant role.

Tertiary halide reactions follow first order kinetics ($S_N1$ mechanism)—the rate depends only on the concentration of the halide and is independent of the concentration of the nucleophile. The rate-determining step is an ionic

$$(CH_3)_3C{-}Cl \xrightarrow[\text{slow}]{} (CH_3)_3C^+ + Cl^- \xrightarrow[\text{fast}]{OH^-} (CH_3)_3C{-}OH$$

cleavage of the carbon-halogen bond to form an alkyl carbonium ion, which reacts rapidly in the second step. For $S_N1$ reactions the relative reactivities of the halides are: tertiary ≫ secondary > primary > methyl.

Secondary halides react by either mechanism, or a combination of both, and the mechanism is influenced by experimental conditions. The $S_N2$ process is favored by powerful nucleophilic reagents (strong bases), high concentrations of the reagent, and a weakly polar solvent. The $S_N1$ mechanism is favored by weakly nucleophilic reagents and low concentrations, and especially by reaction media of high solvating power. Molecular rearrangements occur frequently in $S_N1$ reactions, through the intervention of an alkyl carbonium ion, but not in $S_N2$ reactions.

Identification of alkyl halides is often based upon physical properties: boiling point, density, refractive index, and spectroscopic data. Confirmation of identity may prove difficult, especially for chlorides and for tertiary halides. Alkyl chlorides are usually converted first to the iodides by reaction with potassium iodide in acetone.

A useful method is to form a Grignard reagent and convert it to an anilide or naphthylamide by reaction with an aryl isocyanate.

$$R{-}Br + Mg \xrightarrow{\text{ether}} R{-}MgBr \xrightarrow{C_6H_5NCO} R{-}CO{-}NHC_6H_5$$

## 5a *n*-Butyl Bromide

### (1-Bromobutane)

In a 500-ml round-bottomed flask place 60 ml of water and add in small portions, with cooling, 1.3 mole (130 g, 70 ml) of concentrated sulfuric acid. To the cold diluted acid add 0.5 mole (37 g, 46 ml) of *n*-butyl alcohol,[1] with good mixing and cooling. While shaking the flask to avoid formation of

[1] *n*-Propyl bromide can be made by the same procedure, using an equivalent amount of *n*-propyl alcohol (30 g, 37 ml) instead of butyl alcohol. For the preparation of *n*-pentyl bromide, using 44 g (54 ml) of *n*-pentyl alcohol, the amount of concentrated sulfuric acid should be increased to 140 g (76 ml). This method is unsuitable for the preparation of secondary bromides.

lumps, add 0.6 mole (84 g) of finely pulverized sodium bromide crystals ($NaBr \cdot 2H_2O$)[2] in small portions. Good mixing is essential.

Introduce a few boiling chips into the flask, attach an upright condenser, and connect the top of the condenser to a gas absorption trap to absorb hydrogen bromide that evolves during the reaction (see traps shown in Figure 38, page 170). Heat the flask *gently* over a wire gauze with a small flame, while swirling the contents constantly, until most of the sodium bromide has dissolved. Adjust the heating so that the mixture boils gently, and swirl the contents occasionally, for 0.5–1 hr.

Allow the flask to cool slightly and disconnect the condenser. Arrange the apparatus for distillation, add 25–30 ml of water and a few boiling chips, and distill the mixture until a test portion of the distillate contains little or no water-insoluble material (20–30 min).

Transfer the distillate to a separatory funnel, add 40–50 ml of water, and shake well. Draw off *carefully* the butyl bromide layer[3] into another separatory funnel and discard the aqueous layer. To remove unchanged butyl alcohol, wash the crude alkyl bromide by shaking it thoroughly with about 25 ml of *ice-cold* 80 percent sulfuric acid (prepared by adding 20 ml of concentrated acid to 6 ml of water). Separate the sulfuric acid carefully and completely (which layer ?).[3] Wash the butyl bromide layer with 25 ml of saturated sodium bicarbonate solution (*caution—foaming!*); shake the mixture gently at first before inserting the stopper. Insert the stopper, shake more vigorously, invert the separatory funnel, and release the internal pressure by opening the stopcock. Finally, wash the product with water and draw it off carefully.[4] Collect the butyl bromide in a small dry Erlenmeyer flask, add 6–8 g of anhydrous calcium chloride, cork the flask firmly and shake vigorously. Through a funnel decant (or filter) the dried product into a distilling flask of proper size, add a few small boiling chips and distill over a wire gauze. Collect in a weighed bottle the portion boiling at 99–103°. If an appreciable low-boiling fraction is obtained, dry this portion with 3–4 g of anhydrous calcium chloride and redistill. The yield is 44–48 g.

*Questions follow Experiment 5c.*

[2] Instead of sodium bromide and sulfuric acid, a hydrobromic-sulfuric acid solution may be used. This is prepared by adding 60 g (33 ml) of concentrated sulfuric acid, with cooling, to 0.65 mole (110 g, 74 ml) of 48 percent hydrobromic acid in a 500-ml flask. *n*-Butyl alcohol, 0.5 mole (37 g, 40 ml), is added and the reaction carried out as described previously.

[3] A general procedure to determine which layer is the organic material is to separate the layers carefully and add a few drops of each layer separately to 5 ml of water in a test tube. The organic layer will be insoluble in the water, but the usual washing liquids (sulfuric acid, sodium bicarbonate solution, etc.) will be soluble in water.

[4] Careful separations are necessary. The aqueous layer removes sodium sulfate or sodium carbonate, which would produce troublesome precipitates when calcium chloride is added as drying agent.

# 5b *s*-Butyl Chloride

## (2-Chlorobutane)

In a 250-ml round-bottomed flask, place 0.67 mole (90 g) of anhydrous zinc chloride and 0.67 mole (63 g, 55 ml) of concentrated hydrochloric acid that has been cooled to 5°. Shake the mixture and keep it cool until the zinc chloride has dissolved, then add 0.33 mole (25 g, 31 ml) of *s*-butyl alcohol. Attach a reflux condenser, add two small boiling chips, and boil the mixture briskly on a wire gauze for 2 hr. Cool the reaction mixture and transfer to a separatory funnel. Draw off and discard the lower, aqueous layer.

Transfer the crude butyl chloride to a 250-ml round-bottomed flask and add an equal volume of concentrated sulfuric acid. Attach a reflux condenser, heat until the butyl chloride boils *gently*, and continue the boiling for 15 min. This treatment eliminates impurities that would not be removed by ordinary distillation. Cool the flask slightly, remove the condenser, set it downward for distillation, and connect it with the flask by a wide, bent glass tube. Distill off the *s*-butyl chloride (avoiding excessive heating of the residual sulfuric acid) and transfer the distillate to a separatory funnel. Wash the product with water, decant into a small dry flask, and shake it with 5–6 g of anhydrous calcium chloride. Decant the dried liquid through a funnel into a small dry distilling flask, add a few small boiling chips and redistill. Collect the fraction boiling at 67–69° in a weighed receiver. The yield is 18–24 g. If an appreciable low boiling fraction is obtained, dry this again and redistill it.

*Questions follow Experiment 5c.*

# 5c *t*-Butyl Chloride

## (2-Methyl-2-Chloropropane)

In a 125-ml separatory funnel, place 0.8 mole (82 g, 70 ml) of concentrated hydrochloric acid that has been cooled to 5°. Add 0.25 mole (18.5 g, 24 ml) of *t*-butyl alcohol and shake the mixture occasionally during 20 min. From time to time relieve internal pressure by inverting the funnel and slowly opening the stopcock.

Allow the mixture to stand undisturbed until the layers have separated sharply, then separate and discard the spent hydrochloric acid (which layer?). Wash the product with 10–15 ml of water, then with 5 percent sodium bicarbonate solution, and again with water. Transfer the chloride to a small dry flask and dry it with 5–6 g of anhydrous calcium chloride. Decant the dried liquid through a funnel into a dry distilling flask, add a few small boiling chips and distill. Collect the fraction boiling at 49–52° in a weighed bottle. The yield is 15–18 g. If an appreciable low-boiling fraction is obtained, dry this again with calcium chloride, and redistill it.

## Questions

**1.** Why is strong sulfuric acid more effective than water in removing a small amount of an alcohol from an alkyl halide?

**2.** What alkyl halide is formed by ionic addition of hydrobromic acid to 1-butene? Would the same bromide be formed in the presence of peroxides (free radical catalysts)?

**3.** How may *n*-butyl alcohol be converted to: *n*-butyl chloride; *n*-butyl iodide; *n*-butyl bromide, without using hydrobromic acid?

**4.** Show how *n*-butyl bromide could be converted to: *n*-octane; *n*-butylmagnesium bromide; *n*-butyl methyl ether; *n*-butane.

**5.** What experimental conditions would be suitable for conversion of *s*-butyl bromide to α-methylbutyronitrile by reaction with sodium cyanide ($S_N2$ mechanism)? Explain.

**6.** Could *n*- and *s*-butyl alcohols be converted to the chlorides merely by shaking with concentrated hydrochloric acid, the way *t*-butyl is? (see Lucas' test, Experiment 3b)

**7.** What is the relative reactivity of the four isomeric butyl chlorides in $S_N2$ reactions? in $S_N1$ reactions?

**8.** In what type of reactions of isobutyl chloride might you expect to observe molecular rearrangements? Explain.

## 5d Ethyl Iodide

### (Iodoethane)

In a 250-ml round-bottomed flask, place 0.08 gram-atom (2.5 g) of red phosphorus and 0.5 mole (25 g, 31 ml) of ordinary ethanol (95 percent by volume, 92.5 percent by weight), and add gradually, during the course of 15 min, 0.2 gram-atom (25.5 g) of pulverized iodine. During the addition of the iodine, swirl the flask frequently and keep the mixture cool by dipping in cold water.

➤**CAUTION:** Handle phosphorus and iodine carefully. They will stain the skin and may cause severe burns.

Phosphorus is easily flammable in the dry state and is usually stored in a moist condition.

Cork the flask loosely and allow the reaction mixture to stand at least 4 hr. There is no objection to the reaction mixture standing longer. To complete the reaction attach an upright condenser and heat the mixture *very gently* for 2 hr on a steam bath. During this heating it is necessary to keep the condenser well cooled and to avoid violent boiling of the mixture in order to minimize loss of the volatile ethyl iodide.

Cool the contents of the flask thoroughly, remove the condenser, set it downward for distillation, and connect the end of the condenser to an adapter that dips just below the surface of about 200 ml of ice and water. Add 10–15 ml of water to the reaction flask and distill the ethyl iodide, by heating the flask in a steam bath or a rapidly boiling water bath.

➤**CAUTION:** Do not overheat the residue since this will lead to the formation of iodine vapors that will discolor the product. Overheating of the residual phosphorous acid (to about 200°) will cause formation of phosphine, which inflames spontaneously in air.

Transfer the distillate to a separatory funnel and decant the ethyl iodide, which is usually colored brown by the presence of free iodine. Wash the ethyl iodide with about 100 ml of water to which 5–10 ml of 10 percent sodium hydroxide solution has been added (to remove free iodine), and finally with a little water. Separate the heavy, almost colorless oil, and allow it to dry for about $\frac{1}{4}$ hr with 2–3 g of anhydrous calcium chloride, in a corked flask or bottle.

Through a funnel decant the dried product into a small dry distilling flask. Distill slowly and carefully from a water bath, and collect in a dry weighed bottle the portion which boils at 70–75°. The yield is 15–20 g. If the product is pure and dry, it will be colorless and perfectly clear. If the product is turbid, this indicates the presence of moisture: dry it again with 1–2 g of anhydrous calcium chloride and redistill. Ethyl iodide usually becomes violet or brown in color on standing.

## Questions

**1.** Can ethyl iodide be prepared by the action of iodine on ethane?

**2.** How could ethyl iodide be prepared from ethyl chloride?

**3.** Write equations indicating how each of the following compounds may be obtained from ethyl iodide: *n*-butane; ethylene; a Grignard reagent.

**4.** Why is ethyl iodide more convenient for laboratory syntheses than ethyl chloride of bromide?

# EXPERIMENT 6

# The Williamson Synthesis of Ethers

Symmetrical aliphatic ethers may be prepared from the simpler primary and secondary alcohols by heating with sulfuric acid, but dehydration to the alkene is an important competing reaction. The sulfuric acid process is unsuited to the preparation of ethers from tertiary alcohols and of unsymmetrical ethers.

The Williamson synthesis, using an alkyl halide and a metallic alkoxide, is of broader scope and can be used to obtain symmetrical or unsymmetrical (mixed) ethers. For the latter type, either of two combinations of reactants

$$(CH_3)_2CH{-}Br + Na^+[OCH_3]^- \searrow$$

$$(CH_3)_2CH{-}O{-}CH_3 + NaBr$$

$$[(CH_3)_2CH{-}O]^-Na^+ + CH_3{-}Br \nearrow$$

is conceivable. The proper choice depends mainly upon the structure of the alkyl halides involved. In this synthesis competition arises between the displacement reaction ($S_N2$ type) to form an ether ($1° > 2° \gg 3°$ halides) and dehydrohalogenation to form an alkene ($3° \gg 2° > 1°$ halides). Thus, tertiary alkyl halides are quite unsuitable for the reaction, but ethers having a tertiary alkyl group can be prepared from a tertiary alkoxide and a primary halide.

The Williamson synthesis is an excellent method for the preparation of alkylaryl ethers. Primary and secondary alkyl halides react readily with sodium or potassium phenoxides to form the ethers.

**Reactions.** Ethers are inert toward many reagents. This property, and their good solvent properties, make them useful for extraction purposes and as indifferent media for reactions.

Diethyl ether is slightly soluble in water but dissolves in cold concentrated hydrochloric acid or sulfuric acid through formation of an oxonium

$$(C_2H_5)_2O + HCl \rightleftharpoons (C_2H_5)_2\overset{+}{O}{-}H + Cl^-$$

$$(C_2H_5)_2O + BF_3 \longrightarrow (C_2H_5)_2\overset{+}{O}{-}\overset{-}{B}F_3$$

salt. It also forms coordination compounds with Lewis acids ($BF_3$, $ZnCl_2$, $AlCl_3$, etc.). The ether linkage is resistant to attack by ammonia, strong alkalies, and other nucleophilic reagents.

Dimethyl and diethyl ether can be converted by reaction with boron trifluoride to the corresponding trialkyloxonium salts,[1] which are crystalline solids. These compounds are powerful alkylating agents.

Heating with concentrated hydriodic acid cleaves aliphatic ethers to form alkyl iodides, which may be used to identify the alkyl group(s) present. Unsymmetrical ethers yield a mixture of two alkyl iodides that usually can be separated by fractional distillation. Hydrobromic acid cleaves ethers more slowly and hydrochloric acid is relatively ineffective.

Aliphatic ethers, especially those having secondary alkyl groups, are attacked by atmospheric oxygen. Long exposure to air leads to accumulation of peroxides that are dangerously explosive. The peroxidic compounds can be removed by reducing agents (acidified ferrous sulfate solution) or by adsorption on alumina.

## Methyl *n*-Butyl Ether

### (1-Methoxybutane)

Mix thoroughly in a 100-ml round-bottomed flask, 0.3 mole (16 g) of anhydrous sodium methoxide and 0.2 mole (27.5 g, 21.5 ml) of *n*-butyl bromide. Add 15 ml of methanol and connect the flask *at once* to a reflux condenser. Shake to mix the reactants, and when the initial reaction has subsided, reflux the mixture for 1.5 hr on the steam bath. Cool the mixture to about 50°, add 15 ml of water through the top of the condenser, and shake to dissolve the sodium bromide. Remove the condenser and arrange the apparatus for distillation. Distill over a wire gauze until only 10–15 ml of aqueous residue remains in the flask.

Transfer the distillate to a separatory funnel and wash it with three 15-ml portions of 25 percent aqueous calcium chloride solution, to remove methanol. Dry the product over 3–4 g of anhydrous calcium chloride for 10 hr or longer. Filter from the spent drying agent directly into a small distilling

[1] Meerwein, *Organic Syntheses*, **46**, 113 (1966).

flask. Add two small boiling chips and distill over a wire gauze, using a small flame. Collect the product[2] boiling at 68–75°. The yield is 12–14 g.

➤**CAUTION:** Handle sodium methoxide carefully. Any material spilled on the hands should be washed off promptly with a large quantity of water.

## Questions

**1.** How may sodium methoxide be prepared? What reaction occurs if it is exposed to atmospheric moisture?

**2.** How could diethyl ether be prepared by the Williamson synthesis?

**3.** Why is the Williamson synthesis preferable to the sulfuric acid method for the preparation of an unsymmetrical (mixed) ether?

**4.** Why is it not feasible to prepare methyl *t*-butyl ether from *t*-butyl bromide and sodium methoxide? Suggest an alternative procedure.

[2] Methyl *n*-butyl ether and methanol form an azeotrope, bp 56.3°, containing 35 percent methanol. An appreciable low-boiling fraction will be obtained if methanol has not been removed. This may be washed with calcium chloride solution, dried and redistilled.

# EXPERIMENT 7

# Aldehydes

Formaldehyde is essentially unique among aliphatic aldehydes. Lacking an $\alpha$-carbon atom, it cannot undergo an aldol-type condensation with itself; but it does enter into mixed aldol condensations with other aldehydes and ketones through its very reactive carbonyl group. In carbonyl reactions it is more reactive than other aldehydes, or ketones, and undergoes some reactions that occur difficultly (or not at all) with other carbonyl compounds.

Formaldehyde (bp $-21°$) is quite soluble in water and is commonly available as a 37 percent aqueous solution containing about 10 percent methanol (formalin). In aqueous solution it exists almost entirely as the hydrate, $HO{-}CH_2{-}OH$, and low polymers of the type $HO{-}CH_2{-}O{-}CH_2{-}OH$. Evaporation of the aqueous solution gives a water-insoluble *linear* polymer ("paraformaldehyde") of the structure $HO{-}(CH_2{-}O{-})_n{-}H$, having an average molecular weight of about 1000. Large quantities of formaldehyde are used in the manufacture of synthetic resins and plastics (Bakelite-type phenolic resins, Delryn, etc.).

Acetaldehyde is a typical representative of the homologous series of aliphatic aldehydes. Unlike formaldehyde, acetaldehyde and its homologs undergo self-addition (aldol condensation) under the influence of *dilute* aqueous alkalies. Acetaldehyde furnishes acetaldol, $CH_3{-}CHOH{-}CH_2{-}CH{=}O$, which loses water readily to form crotonaldehyde, $CH_3{-}CH{=}CH{-}CH{=}O$. Catalytic hydrogenation of crotonaldehyde furnishes a synthetic route to *n*-butyl alcohol, starting from acetaldehyde.

Concentrated aqueous alkalies convert acetaldehyde and similar aliphatic aldehydes into colored, oily condensation products known as aldehyde resins. Formaldehyde, and benzaldehyde, on warming with strong

$$2H{-}CH{=}O + NaOH \longrightarrow CH_3{-}OH + H{-}CO_2Na$$

 alkalies undergo an intermolecular oxidation-reduction process (Cannizzaro reaction). Formaldehyde can be used in a *crossed* Cannizzaro reaction with another aldehyde; formaldehyde acts as the hydrogen donor and reduces the other reactant.

In the presence of basic catalysts, alkynes having reactive acetylenic hydrogen undergo addition to the carbonyl group of aldehydes and ketones. For example, acetylene reacts with two molecules of formaldehyde to give $HO{-}CH_2{-}C{\equiv}C{-}CH_2{-}OH$, from which 1,4-butanediol and other industrial chemicals are manufactured.

Formaldehyde, acetaldehyde, and the higher homologs undergo polymerization in the presence of strong acids to form cyclic trimers, which are derivatives of 1,3,5-trioxane.

With ammonia, acetaldehyde gives an unstable addition product that undergoes transformation to a 1,3,5-triazane derivative and other complex condensation products. Formaldehyde and ammonia react to form hexamethylenetetramine ($C_6H_{12}N_4$), a crystalline polycyclic compound. Aromatic aldehydes behave differently toward ammonia; benzaldehyde gives a crystalline linear condensation product (hydrobenzamide).

$$C_6H_5{-}CH{=}N{-}CH(C_6H_5){-}N{=}CH{-}C_6H_5$$

The most useful general test reaction for distinguishing between aldehydes and ketones is based upon the ability of aldehydes to be oxidized by mild reagents to the corresponding acids (silver mirror test; Fehling's test). Another useful differentiation is based upon differences in reactivity. Aldehydes, *but not ketones*, undergo condensation with the reactive methylene group of 5,5-dimethyl-1,3-cyclohexanedione (called methone or dimedone). Two molecules of methone react with one molecule of an aldehyde to furnish the crystalline methone derivative.

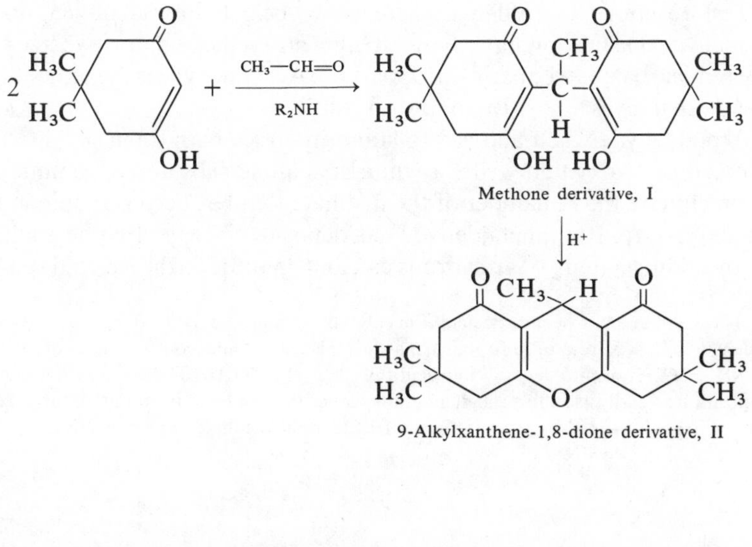

$$2\ \text{(methone)} + \xrightarrow[R_2NH]{CH_3{-}CH{=}O}$$

Methone derivative, I

$$\downarrow H^+$$

9-Alkylxanthene-1,8-dione derivative, II

The methone derivatives (but not those of formaldehyde and *o*-hydroxybenzaldehydes) are cyclized readily to xanthene derivatives (II) by heating with dilute acid in ethanol.

## Acetaldehyde

Acetaldehyde is extremely volatile (bp 20°) and cannot be stored satisfactorily unless it is refrigerated or sealed in glass ampoules. It is convenient to prepare acetaldehyde as needed, by depolymerization of the cyclic trimer, paraldehyde (bp 125°), which can be handled and stored easily. Anhydrous (gaseous) formaldehyde is obtained in the laboratory by heating the solid linear polymer, paraformaldehyde, previously dried thoroughly. In some reactions the solid polymer can be used directly in the reaction mixture and formaldehyde is generated by warming with the reactant.

Arrange the apparatus as shown for fractional distillation in Figure 14 (page 42) using a 100-ml boiling flask and as receiver a 125-ml Erlenmeyer flask. Cool the receiver in an ice-water bath but not in an ice-salt mixture. To diminish evaporation losses a *loose* plug of cotton may be placed between the adapter and the receiver. The cotton must not be allowed to come in contact with the distillate.

In the 100-ml boiling flask place 0.15 mole (20 g, 20 ml) of paraldehyde and add 4–5 drops of concentrated sulfuric acid, or other depolymerizing agent.[1] Support the flask on a wire gauze and heat gently with a small flame so that the temperature of the distilling vapor does not rise above 35°. Avoid overheating as this can cause carbonization and formation of sulfur dioxide, which may cause the acetaldehyde in the distillate to repolymerize.

Collect about 10–12 ml of distillate; this will be practically pure acetaldehyde. Stopper the receiver loosely with a cork, keep it in the ice bath and take out small portions for test reactions and derivatives.

Test reactions for aldehydes are given below. For identification of acetaldehyde one may prepare the crystalline *bis*-methone derivative (part D) or the semicarbazone (Experiment 9, part C). Acetaldehyde is the only simple aldehyde that gives a positive iodoform test.

Acetaldehyde when allowed to stand in an ice bath often polymerizes spontaneously to metaldehyde, a white crystalline substance. Examine for such precipitate the remainder of the distillate that has been standing in the freezing mixture. If no metaldehyde has deposited it may often be made to form by adding a lump of anhydrous calcium chloride to the ice-cold sample

[1] The small amount of sulfuric acid acts catalytically and causes the depolymerization of paraldehyde. The 4–5 drops of sulfuric acid may be replaced advantageously by 1–2 g of sulfamic acid ($NH_2{-}SO_3H$) or *p*-toluenesulfonic acid ($CH_3{-}C_6H_4{-}SO_3H$). These acids do not cause carbonization so easily as sulfuric acid, and thus reduce the danger of liberating sulfur dioxide. If sulfur dioxide is formed it causes the acetaldehyde in the distillate to repolymerize.

and allowing it to stand for some time at a low temperature, especially in an ice and salt mixture. Do not conserve a sample of acetaldehyde from one laboratory period to another.

## Reactions of Aldehydes

The reactions may be carried out with any of the following: aqueous formaldehyde solution; acetaldehyde (pure liquid or concentrated aqueous solution); benzaldehyde or furfural. For comparison a ketone, such as acetone or cyclohexanone, may be used.

For satisfactory results it is essential to carry out the tests in thoroughly *clean* test tubes.

***(A) Tollens' Reagent.*** This test involves reduction of silver oxide to metallic silver and oxidation of the aldehyde to the carboxylic acid. The reagent is an alkaline solution of silver ammonium hydroxide, $[Ag(NH_3)_2]^+OH^-$.

In a *clean* test tube place about 5 ml of silver ammonium hydroxide solution,[2] add a few drops of the aldehyde and warm the tube gently. If the tube is clean a silver mirror will be formed; if not, a black precipitate of finely divided silver will appear.

A suspension of silver oxide[3] in dilute alkali is a good reagent for oxidation of aldehydes, particularly unsaturated or phenolic aldehydes.

Fehling's or Benedict's solution (alkaline cupric tartrate or citrate) also may be used as a test for aldehydes or for reducing sugars (Experiment 16, part A). In these tests a reddish-yellow precipitate of cuprous oxide is formed.

***(B) Fuchsin-Aldehyde Reagent.***[4] Add 1 ml of Schiff's fuchsin-aldehyde reagent to a few drops of the aldehyde in 4–5 ml of water and observe the development of color. Ketones do not give this test when perfectly pure

[2] **Preparation of Tollens' Reagent.** In a thoroughly clean test tube place 2 ml of a 5 percent solution of silver nitrate, and add a drop of dilute aqueous sodium hydroxide (10 percent). Add a very dilute solution of ammonia (about 2 percent) drop by drop, with constant shaking, until the precipitate of silver oxide just dissolves. In order to obtain a sensitive reagent it is necessary to avoid a large excess of ammonia.

This reagent should be freshly prepared just before use and should not be stored, since the solution decomposes on standing and deposits a highly explosive precipitate, silver fulminate.

[3] Thomason and Kubler, *J. Chem. Educ.*, **45**, 546 (1968); Campaigne and LeSuer, *Organic Syntheses*, Collective Volume **IV**, 919 (1963).

[4] **Preparation of Schiff's Fuchsin-Aldehyde Reagent.** Dissolve 0.5 g of pure fuchsin (*p*-rosaniline hydrochloride) in 500 ml of distilled water, and filter the solution. Saturate 500 ml of distilled water with sulfur dioxide, mix thoroughly with the filtered fuchsin solution, and allow to stand overnight. This produces a practically colorless and very sensitive reagent. If the solution is not colorless, it may be shaken with a small amount of decolorizing carbon and filtered.

but the color reaction is very sensitive and responds to mere traces of an aldehyde.

***(C) Methone Derivatives** (Optional).* In a large test tube place 0.50 g (~3.6 mmoles) of methone,[5] 5 ml of 50 percent aqueous ethanol, and *not more than* 0.10 ml of acetaldehyde (~1.8 mmoles). Add *one drop* of a secondary amine as catalyst (diethylamine or piperidine), and introduce two small boiling chips. Attach a small reflux condenser and heat the tube in a beaker about one-fourth filled with water. Apply heat gently until the reaction mixture reaches its boiling point. After the mixture has boiled gently for 5–10 min, remove the tube and add water dropwise until a turbidity develops. Allow the solution to cool, with occasional shaking. Methone derivatives often crystallize slowly; if necessary, cork the tube and allow it to stand overnight or longer. Collect the crystals with suction and wash them with cold 50 percent aqueous ethanol. The yield is about 0.30 g. After drying the crystals, take the melting point.

Instead of acetaldehyde, an equivalent amount (~1.8 mmoles) of another aldehyde may be used. The reported melting points of several methone derivatives (I) and the corresponding cyclized xanthene-1,8-dione derivatives (II) are given in Table 18.

**TABLE 18**
**Methone Derivatives (I) and Xanthenedione Derivatives (II) of Aldehydes**

| ALDEHYDE | I (MP, °C) | II (MP, °C) | ALDEHYDE | I (MP, °C) | II (MP, °C) |
|---|---|---|---|---|---|
| Formaldehyde | 190–191 | (171)[6] | Isovaleraldehyde | 154–155 | 170–172 |
| Acetaldehyde | 141–142 | 176–177 | *n*-Hexaldehyde | 107–108 | |
| Propionaldehyde | 157–158 | 141–143 | Benzaldehyde | 194–195 | 204–205 |
| *n*-Butyraldehyde | 134–135 | 135–136 | Anisaldehyde | 142–143 | 241–243 |
| Isobutyraldehyde | 153–154 | 154–155 | Piperonal | 177–178 | 218–220 |
| *n*-Valeraldehyde | 107–109 | 112–113 | Furfural | 159–160 | |

The methone derivative (except that of formaldehyde)[6] undergoes cyclization very readily. For this purpose dissolve about 0.2 g in 6–10 ml of 80 percent ethanol, add one drop of concentrated hydrochloric acid, and boil the solution gently for 5 min. Add water dropwise to the hot solution until a

[5] 5,5-Dimethyl-1,3-cyclohexanedione (methone) can be purchased from Distillation Products Industries (Eastman Organic Chemicals, #1259). For convenience and to avoid waste this reagent may be issued in 0.5 g samples in small tubes, or 5 ml quantities of a 10 percent solution in 50 percent ethanol.

[6] For cyclization the formaldehyde derivative requires 6–8 hours' heating on a steam bath with ten times its weight of concentrated sulfuric acid. After pouring into water and neutralizing with sodium carbonate, the product is collected with suction and recrystallized from ethanol.

faint turbidity develops and allow the liquid to cool. Collect the crystals of the cyclized product (II) and take its melting point.

## Questions

**1.** Write equations for the action of the following reagents on formaldehyde: (a) hydrogen cyanide; (b) ethylmagnesium bromide, followed by dilute acid; (c) acetylene, in the presence of a basic catalyst; (d) methanole, with acid catalyst.

**2.** Why does formaldehyde differ from typical aliphatic aldehydes in its behavior toward strong aqueous alkalies?

**3.** Show the steps in a series of reactions for the conversion of isobutyraldehyde to 2,2-dimethyl-1,3-propanediol, by treatment with excess formaldehyde and a basic catalyst (mixed aldol condensation + Cannizzaro reaction).

**4.** What product is obtained by self-condensation of propionaldehyde, followed by elimination of water from the aldol?

EXPERIMENT

# 8

# Ketones

## Chromic Acid Oxidation

Chromic acid mixture (dichromate and 40–50 percent sulfuric acid) oxidizes a primary alcohol stepwise to the aldehyde and the corresponding carboxylic acid. The simpler aldehydes (acetaldehyde, propionaldehyde) can be prepared by introducing the dichromate solution dropwise *into* a hot acidified solution of the alcohol, so that the oxidizing agent is not present in excess, and distilling the volatile aldehyde away from the reaction mixture as rapidly as it is produced. Even so some of the aldehyde is oxidized to the acid; also, a part of the aldehyde reacts with the alcohol to form an acetal, $R{-}CH(OCH_2{-}R)_2$.

$$Na_2Cr_2O_7 \quad + \quad 2H_2SO_4 \longrightarrow 2NaHSO_4 \quad + \quad H_2Cr_2O_7 \underset{}{\overset{H_2O}{\rightleftarrows}} 2H_2CrO_4$$

$$3R{-}CH_2{-}OH + 2H_2CrO_4 + 3H_2SO_4 \longrightarrow RCH{=}O + Cr_2(SO_4)_3 + 8H_2O$$

Oxidation of secondary alcohols to ketones by means of chromic acid mixture is generally a satisfactory method of preparing ketones, since the latter do not undergo further oxidation easily. The oxidation is exothermic and the temperature must be controlled to avoid a violent reaction. For water-insoluble compounds, chromic acid ($CrO_3$ + $H_2O$) in an acetone medium (Jones' reagent) or in glacial acetic acid may be used. Other variations are the pyridine complex ($CrO_3{:}2C_5H_5N$) and di-*t*-butyl chromate. Tertiary alcohols are relatively stable to cold chromic acid but under more vigorous conditions tertiary alcohols (and ketones) are degraded by cleavage of carbon-carbon bonds.



Chromic acid oxidation of secondary alcohols occurs through the chromic ester of the alcohol.[1] Removal of a proton from the $R_2CH{-}O{-}$ group of the chromic ester gives the ketone and an unstable tetravalent chromium compound. Through a series of interactions involving a highly reactive pentavalent chromium compound (designated for convenience by the symbol $[HCrO_3]$) more of the chromic ester is oxidized to ketone and the chromium is reduced to the trivalent state. The equations shown for the reaction are a simplified version of the overall process.

$$R_2CH - OH + H_2CrO_4 \overset{H^+}{\rightleftharpoons} H_2O + \underset{\text{(chromate ester)}}{R_2CH{-}O{-}CrO_3H} \xrightarrow{\text{slow}} R_2C{=}O + H_2CrO_3 + H_2O$$

$$H_2CrO_4(Cr^{+6}) + H_2CrO_3(Cr^{+4}) \xrightarrow{\text{fast}} H_2O + 2[HCrO_3](Cr^{+5})$$

$$[HCrO_3] + \begin{matrix}\text{chromate}\\ \text{ester}\end{matrix} + H_2SO_4 \xrightarrow{\text{fast}} R_2C{=}O + Cr_2(SO_4)_3 + H_2O$$

On a larger scale ketones (and aldehydes) are prepared in excellent yields by dehydrogenation of the corresponding alcohol at elevated temperatures (250–350°) over a metal catalyst, such as platinum, silver, copper, and copper-zinc alloy. Ketones may be prepared also by several other procedures: pyrolysis of the calcium or barium salt of an organic acid, or passing the vaporized acid over an oxide catalyst ($ThO_2$, MnO); addition of a Grignard reagent to a nitrile; and the acetoacetic ester synthesis.

Diaryl ketones may be prepared by chromic acid oxidation of diarylmethanes ($Ar{-}CH_2{-}Ar$) or diarylcarbinols ($Ar{-}CHOH{-}Ar$), preferably in glacial acetic acid solution. Chromic oxidation of methyl side chains can be arrested at the aldehyde stage by operating in the presence of acetic anhydride, which serves to protect the aldehyde from further oxidation by conversion to the diacetate, $Ar{-}CH(OCOCH_3)_2$. The latter can be hydrolyzed readily to regenerate the aldehyde.[2]

Apart from their relative stability toward oxidizing agents, ketones differ from aldehydes in their inability to react directly with alcohols, in the presence of acid catalysts, to form the analogs of acetals (ketals, of the type $R_2C(OCH_3)_2$) and in their failure to react with certain active methylene compounds (methone test, Experiment 7, part C).

The reactivity of ketones falls off as the number of substituents in the $\alpha$-position increases. Acetone and other methyl ketones undergo nearly all of the addition and condensation reactions that occur with typical aldehydes. Several reactions of the carbonyl group that furnish crystalline derivatives suitable for identification purposes are described in Experiment 9.

Diethyl ketone, di-isopropyl ketone, and diaryl ketones are much less reactive than the methyl ketones and undergo carbonyl reactions with greater difficulty (or not at all, with certain reagents).

[1] Westheimer, "The Mechanisms of Chromic Acid Oxidation," *Chem. Revs.*, **45**, 419 (1949).

[2] *Organic Syntheses*, Collective Volume **II**, 441 (1943).

# 8a Cyclohexanone

In a 400-ml beaker, dissolve 0.07 mole (21 g) of sodium dichromate dihydrate in 120 ml of water. Add carefully 0.3 mole (30 g, 17 ml) of concentrated sulfuric acid, with stirring, and cool the deep orange-red solution to 30°. Place 0.2 mole (20 g, 21 ml) of cyclohexanol and 60 ml of water in a 500-ml Erlenmeyer flask and *to it* add the dichromate solution in one portion. Swirl the mixture to insure thorough mixing and observe its temperature. The mixture rapidly becomes warm; when the temperature reaches 55°, cool the flask in a basin of cold water, or under the tap, and regulate the amount of cooling so that the temperature does not rise above 60° nor fall below 55°. Continue external cooling only so long as necessary to maintain this temperature; when the temperature of the mixture no longer rises above 60° on removal of external cooling, allow the flask to stand with occasional shaking for 1 hr.

Pour the mixture into a 500-ml round-bottomed flask, add an additional 120 ml of water and, by means of a short unpacked fractionating column, attach a condenser set downward for distillation. Add two boiling chips and distill the mixture until about 100 ml of distillate, consisting of water and an upper layer of cyclohexanone, has been collected.

Saturate the aqueous layer with salt (20–25 g will be required), separate the cyclohexanone layer, and extract the aqueous layer with 15 ml of methylene chloride, or pentane (*flammable!*). Combine the solvent extract with the cyclohexanone layer and dry it with ~6 g of anhydrous magnesium sulfate. Filter the dried solution into a distilling flask of suitable size, attach a condenser, and distill off the solvent from a water bath. Distill the residual cyclohexanone, using a wire gauze, and collect the fraction boiling at 151–155° (mainly 152–154°). The yield is 12–15 g.

Conversion of cyclohexanone to the oxime and phenylhydrazone is described in Experiment 9, parts B and D.

*Questions follow Experiment 8b.*

# 8b Diethyl Ketone

## (3-Pentanone)

In a 500-ml round-bottomed flask place 50 ml of water and add carefully, with cooling, 0.2 mole (20 g, 11 ml) of concentrated sulfuric acid. To the cold diluted acid add 0.3 mole (26.5 g, 32.5 ml) of 3-pentanol,[3] and swirl the flask to obtain good mixing. To prepare the chromic acid oxidizing solution, dissolve 0.1 mole (30 g) of sodium dichromate dihydrate in 50 ml of water, add carefully 0.2 mole (20 g, 11 ml) of concentrated sulfuric acid, and cool the solution to room temperature.

Introduce the oxidizing solution *into* the pentanol solution in small portions, with swirling, and observe the temperature. By intermittent cooling as needed, in a pan of water, maintain the internal temperature at 25–30°. If the temperature is kept too low the oxidizing agent may accumulate in the solution and react suddenly with great vigor. When all of the oxidizing agent has been added, and the temperature no longer rises spontaneously, stopper the flask and allow it to stand at room temperature for 1 hr or longer, with occasional swirling.

Fit the flask with a short fractionating column (it need not be packed) and attach a water-cooled condenser arranged for distillation. Add 150 ml of water and a boiling chip, and distill the mixture until a test portion of the distillate is essentially free of oily droplets. Do not collect an excessive amount of aqueous distillate; the ketone is appreciably soluble in water and a larger portion will be lost in the solution. The azeotrope of 3-pentanone and water, containing 14 percent of water, distills at 82.9°; stop the distillation when the temperature of the distilling vapor has risen a few degrees above this point.

To the distillate add 0.5 g of solid sodium carbonate to neutralize any acidity, and salt out the dissolved ketone by adding 2 g of clean sodium chloride for each 10 ml of water present. Draw off the aqueous layer and transfer the ketone to a small dry Erlenmeyer flask. Add 4–5 g of anhydrous potassium carbonate and shake well. If an aqueous layer is formed, draw this off and add a fresh portion of the drying agent. Filter (or decant) the dried liquid into a small distilling flask, add two small boiling chips and distill over a wire gauze. Collect the product boiling at 99–104°. The yield is 12–15 g. The product should be colorless.[4]

If a moist low-boiling fraction of product is obtained it may be dried and redistilled, or used for preparation of the semicarbazone or 2,4-dinitrophenylhydrazone (see Experiment 9, parts C and E).

[3] Methyl *n*-propyl ketone (2-pentanone) may be prepared by the same method; in the preliminary distillation of the product the azeotrope, containing 20 percent water, distills at 83.3°. The procedure given here is an adaptation of that described by Yohe, Louder, and Smith, *J. Chem. Educ.*, **10**, 374 (1933).

[4] Yellow discoloration of the product may arise from the presence of a little of the intensely yellow diketone, 2,3-pentanedione. This impurity is alkali-sensitive and can be removed by adding about 0.5 g of crushed sodium hydroxide and redistilling.

## Questions

**1.** Compare the behavior of the following alcohols toward mild oxidation with chromic acid solution: (a) 2-methyl-2-butanol; (b) 3-methyl-2-butanol; (c) 3-methyl-1-butanol; (d) 2,2-dimethylpropanol. What are the common names of these pentyl alcohols?

**2.** What products are formed by oxidation of cyclohexanone with aqueous permanganate?

**3.** What products are formed by vigorous chromic acid oxidation of 3-pentanone?

**4.** If impure 3-pentanol is used in Experiment 8b some 2-pentanone may be present in the final product. Suggest a simple test for detecting this impurity. Would the presence of 2,3-pentanedione[4] interfere with your test reaction?

# EXPERIMENT 9

# Carbonyl Reactions of Aldehydes and Ketones

Test reactions that may be used to distinguish between aldehydes and ketones are given in Experiment 7: (A) Tollens' reagent; (B) Fuchsin-Aldehyde reagent; (C) Methone derivatives. The latter are satisfactory crystalline derivatives for identification (see Table 18, page 210).

For the following experiments it is recommended that the quantities specified be measured in a small graduated cylinder or a pipette and not by rough visual estimation. In performing reactions, the proper results may not be obtained unless particular proportions of the reactants are used. For test purposes a smaller quantity of material than that specified may be used *provided that all* of the reagents are reduced proportionately.

Several of the reactions described below, particularly the preparations of the semicarbazone and 2,4-dinitrophenylhydrazone, are used for the purpose of identifying an unknown aldehyde or ketone. The reported melting points of the commonly used carbonyl derivatives of several aldehydes and ketones are given at the end of the experimental section (Table 19).

**(A) Sodium Bisulfite Reaction.** The formation of sodium bisulfite addition compounds, $R-CHOH-SO_3Na$, is a general reaction of aldehydes. Typical aliphatic methyl ketones and cycloalkanones (cyclopentanone, cyclohexanone) also form similar adducts. Aryl methyl ketones and sterically hindered alkyl methyl ketones react poorly or not at all. Bisulfite adducts are used occasionally for the purification of reactive carbonyl compounds and for removing them from mixtures containing other types of compounds. They are not very useful derivatives for identification purposes.

***Butyraldehyde (or Benzaldehyde).*** In a 125-ml Erlenmeyer flask place 10 ml of a *saturated* solution of sodium bisulfite and add slowly 4 g (5 ml) of *n*-butyraldehyde (freshly distilled). Shake the mixture thoroughly to complete the reaction. After several minutes add about 50 ml of ethanol, shake well, and chill the flask thoroughly in an ice-salt mixture. The sodium bisulfite addition compound crystallizes in lustrous white plates. Collect the crystals with suction, wash with ethanol, finally with ether, and allow them to dry. Do not attempt to determine the melting point of this salt.

Treat a small amount of the sodium bisulfite adduct with 5–10 ml of aqueous sodium carbonate (10 percent) and warm. Note the odor. Repeat with 5–10 ml of dilute hydrochloric acid in place of sodium carbonate solution and cautiously test the odor.

***Acetone and Sodium Bisulfite.*** Add 4 g (5 ml) of acetone drop by drop to 10 ml of a *saturated* aqueous sodium bisulfite contained in an Erlenmeyer flask immersed in an ice bath. Shake the solution thoroughly and allow the product to crystallize. After a few minutes add about 25 ml of ethanol to promote the precipitation. Collect the crystals with suction, wash with ethanol, finally with ether, and allow them to dry. Do not attempt to determine the melting point.

Treat a small amount of the sodium bisulfite addition compound with 5–10 ml of aqueous sodium carbonate (10 percent) and warm. Note the odor. Repeat with 5–10 ml of dilute hydrochloric acid in place of sodium carbonate solution and cautiously test the odor.

**(B) Formation of Oximes.** ***Acetone Oxime.*** In a small Erlenmeyer flask mix together 5 g of hydroxylamine hydrochloride dissolved in 10 ml of water, 3 g of sodium hydroxide dissolved in 10 ml of water, and add 6.5 g (8 ml) of acetone. Cork the flask firmly, shake thoroughly, and leave overnight, during which time the crystalline oxime separates. If no crystals appear, scratch the inside of the flask with a glass rod and at the same time cool in an ice bath; or add a seed crystal of the oxime and allow to stand again. Collect the crystals with suction, dry quickly, and take the melting point. Do not leave acetoxime exposed to the air overnight as it sublimes readily. Boil a little acetoxime with 5–10 ml of dilute hydrochloric acid and note the odor.

Furfural may be converted to the oxime by using instead of acetone, a suspension of 5.5 g (5 ml) of furfural in 10 ml of water. Furfuraldoxime exists in *cis* and *trans* forms, geometrical isomers, having different physical properties ($\alpha$-, mp 74°; $\beta$-, mp 89°). The $\beta$-form is usually obtained by the oximation procedure given above.

***Cyclohexanone Oxime.*** Dissolve 5 g of hydroxylamine hydrochloride and 7.5 g of sodium acetate crystals ($NaC_2H_3O_2 \cdot 3H_2O$) in 20 ml of water in a 50-ml Erlenmeyer flask. Warm the solution to about 40° and add 5 g (5.3 ml) of cyclohexanone. Cork the flask securely and shake it vigorously for a few minutes. The oxime separates as a crystalline solid. Cool the flask in an ice bath, filter the crystals with suction, and wash them with a little ice-cold water. Press as dry as possible on the funnel, spread the product on a piece of filter paper to dry in the air, and take the melting point. The yield is 4.5–5 g.

***Oximation of an Unknown Carbonyl Compound.*** For identification purposes oximation may be carried out on a smaller scale in the following way. Dissolve 1 g of hydroxylamine hydrochloride in 6 ml of water, add 4 ml of 10 percent aqueous sodium hydroxide, and introduce 0.5 ml (0.4 g) of the aldehyde or ketone. If the compound does not dissolve completely add *just enough* ethanol, dropwise with shaking, to obtain a clear or only faintly turbid solution. Heat the solution in a boiling water bath for 10–20 min, cool it in an ice bath, and induce crystallization by scratching the walls of the container with a glass rod. If necessary allow the solution to stand overnight or longer. Collect the crystals on a small suction filter and wash them with 1 ml of ice-cold water. Recrystallize the oxime from a little water or aqueous ethanol, reserving a few tiny crystals for seeding.

Many aldehydes and ketones give oximes that are liquids at ordinary temperatures (e.g., *n*-butyraldehyde, methyl ethyl ketone), or are very difficult to obtain in crystalline form. For such compounds the semicarbazones or 2,4-dinitrophenylhydrazones are likely to be suitable crystalline derivatives for characterization.

**(C) Semicarbazones.** ***n-Butyraldehyde Semicarbazone.*** Prepare a solution of 1 g of semicarbazide hydrochloride ($NH_2{-}CO{-}NH{-}NH_2 \cdot HCl$) and 1.5 g of sodium acetate crystals in 10 ml of water, in a test tube. Add 1 g (1 ml) of *n*-butyraldehyde, close the tube with a cork, and shake vigorously. Allow the mixture to stand, with occasional vigorous shaking, until the product has crystallized completely (cool in an ice bath, if necessary, to promote crystallization). Filter the crystals, wash with a little cold water, and dry in the air. Take the melting point.

Acetone, furfural, and cyclohexanone, may be converted to crystalline semicarbazones in the same manner. For less reactive carbonyl compounds it is advantageous to heat the reaction mixture for a few minutes in a beaker of boiling water. For water-insoluble carbonyl compounds, dissolve 1 g of the substance in 10 ml of ethanol, add water dropwise until the solution becomes turbid, then add 1 g of semicarbazide hydrochloride and 1.5 g of sodium acetate crystals. Shake vigorously, and warm to 100° if necessary.

**(D) Phenylhydrazones.** ***Furfural Phenylhydrazone.*** Prepare a solution of phenylhydrazine reagent[1] by dissolving 2 g of phenylhydrazine hydrochloride ($C_6H_5{-}NH{-}NH_2 \cdot HCl$) in 18 ml of water, and adding 3 g of sodium acetate crystals and one drop of glacial acetic acid. If the resulting solution

[1] A satisfactory reagent may also be made by dissolving 2 g (2 ml) of phenylhydrazine (the free base is a liquid) in a solution of 2 g (2 ml) of glacial acetic acid in 20 ml of water, and clarifying with a little decolorizing charcoal. This procedure is less desirable owing to the poisonous character of phenylhydrazine (vapor or liquid). If the liquid comes in contact with the skin it must be washed off at once, using 5 percent acetic acid, followed by soap and water.

is turbid, add a *small pinch* of decolorizing carbon, shake vigorously, and filter. This reagent deteriorates rapidly and should not be stored.

Divide the solution into two equal portions, and to one-half add a solution of 0.5 ml of furfural ($C_4H_3O{-}CH{=}O$) in 6–8 ml of water. Cork the tube and *shake vigorously* until the product has crystallized. Collect the crystals with suction (do not allow them to stand overnight, as discoloration occurs readily), and wash them thoroughly with water. Recrystallize from dilute ethanol (see solvent-pairs, page 85), dry the crystals as quickly as possible, and take the melting point.

***Cyclohexanone Phenylhydrazone.*** To the other half of the phenylhydrazine reagent, add a solution of 0.5 ml of cyclohexanone in 6–8 ml of water. Isolate the phenylhydrazone in the same way, recrystallize from dilute ethanol, and take the melting point. The product decomposes on standing.

**(E) 2,4-Dinitrophenylhydrazones.** ***Cyclohexanone Dinitrophenylhydrazone.*** In a 200-ml Erlenmeyer flask place 1 g of cyclohexanone, 1 g of 2,4-dinitrophenylhydrazine, and 75 ml of ethanol (95 percent). Attach a reflux condenser and bring the mixture to the boiling point. Extinguish the flame, allow the mixture to cool slightly, remove the condenser momentarily, and add 1.5 ml of concentrated hydrochloric acid (directly into the flask). The color changes to yellow-orange and the solution usually becomes clear. Adjust the condenser and boil the mixture gently for about 5 min. If the solution is not clear, filter it through a fluted filter and collect the filtrate in an Erlenmeyer flask. Close the flask *loosely* with a cork and allow the solution to cool slowly. Rapid chilling may cause the product to separate as an oil. Collect the crystals by suction filtration, wash with a little cold ethanol, and recrystallize from ethanol.

**(F) Iodoform Test.** The formation of iodoform is an example of the haloform reaction, which is characteristic of methyl ketones, $CH_3{-}CO{-}R$, and of secondary alcohols of the type, $CH_3{-}CHOH{-}R$. Only one aldehyde ($CH_3{-}CH{=}O$) and only one primary alcohol ($CH_3{-}CH_2OH$) give a positive haloform test. The reaction involves formation of a trihalomethyl carbonyl compound, $X_3C{-}CO{-}R$, which is cleaved by sodium hydroxide to form $X_3C{-}H$ and $R{-}CO_2Na$. A few methyl ketones fail to give a positive reaction.[2]

In a small Erlenmeyer flask (or large test tube) dissolve 0.5 ml (0.4 g) of the compound in 5 ml of water and add 5 ml of 10 percent aqueous sodium hydroxide. Add dropwise, with shaking, a 10 percent solution of iodine in potassium iodide,[3] until a definite brown color persists (indicating an excess

[2] For a review of this test see Fuson and Bull, *Chem. Revs.*, **15**, 275 (1934).

[3] The potassium iodide-iodine solution is prepared by adding 10 g of iodine crystals to a solution of 20 g of potassium iodide in 80 ml of water and stirring the mixture until the iodine has dissolved.

of iodine). If a precipitate of iodoform does not appear after 5 min, warm the solution to 60° in a beaker of water. If the brown color is discharged add more of the iodine solution until the brown color persists for 2 min. Add a few drops of aqueous sodium hydroxide to remove the excess iodine, dilute the mixture with water, and allow it to stand for 10 min. Iodoform precipitates in yellow hexagonal crystals having a characteristic odor. It is prudent to confirm the identity of the crystals by collecting them with suction, drying and taking the melting point (119°).

Perform the test with acetone and with *n*-butyraldehyde. For compounds that are not appreciably soluble in water, the sample may be dissolved in dioxane or *pure* methanol instead of water, before adding the sodium hydroxide and iodine solutions. Before using a solvent it should be tested to see if iodoform-producing impurities are present.

**(G) Identification of an Unknown Carbonyl Compound.** Obtain from the instructor a 5-ml sample of a carbonyl compound selected from among those listed in Table 19, and proceed to identify it. If the sample were not known definitely to be an aldehyde or ketone, it would be necessary to ascertain this by means of a preliminary test with a reagent such as semicarbazide or 2,4-dinitrophenylhydrazine.

Perform a series of experiments in systematic fashion, along these lines: (a) find out whether the substance is an aldehyde or a ketone (see Experiment 7); (b) see whether or not it contains the grouping $CH_3{-}CO{-}R$; (c) prepare a crystalline derivative, such as the semicarbazone or dinitrophenylhydrazone; (d) prepare a second derivative, selected from Table 19, that will establish the identity with certainty and take its melting point. It may be helpful in planning the choice of derivatives, to determine the boiling point of the unknown, using the small scale distillation procedure described in the chapter on distillation (see Figure 11, page 27).

In doubtful cases it is customary to secure an authentic sample of the compound corresponding to the proposed structure and compare its derivatives with those of the unknown.

## Questions

**1.** What tests may be used to distinguish aldehydes from ketones?

**2.** List some properties that are desirable in a derivative to be used for identification of an unknown compound.

**3.** How may a carbonyl compound be regenerated from its semicarbazone?

**4.** Suggest a procedure for isolating *n*-butyraldehyde from an ethereal solution containing *n*-butyl alcohol and *n*-butyric acid as impurities, without resorting to fractional distillation.

## TABLE 19
## Melting Points of Derivatives of Aldehydes and Ketones*

| ALDEHYDES† | BP | OXIME‡ | SEMI-CARBAZONE | PHENYL-HYDRAZONE | 2,4-DNP§ |
|---|---|---|---|---|---|
| Acetaldehyde | 21° | 47° | 162° | 57°; 99° | 168°; 157° |
| Propionaldehyde | 50° | 40° | 154°; 89° | liquid | 154° |
| Isobutyraldehyde | 64° | liquid | 125° | liquid | 187° |
| *n*-Butyraldehyde | 74° | liquid | 104° | liquid | 123° |
| Isovaleraldehyde | 92° | 48° | 107° | liquid | 123° |
| *n*-Valeraldehyde | 103° | 52° | 108° | | 107° |
| Furfural | 161° | 89°; 74° | 202° | 97° | 230°; 214° |
| Benzaldehyde | 179° | 35°; 130° | 222° | 158° | 237° |
| KETONES | | | | | |
| Acetone | 56° | 59° | 187° | 42° | 126° |
| Methyl ethyl | 80° | liquid | 136° | liquid | 117° |
| Methyl isopropyl | 94° | liquid | 113° | liquid | 117° |
| Methyl *n*-propyl | 102° | | 110° | liquid | 144° |
| Methyl isobutyl | 119° | 58° | 134° | liquid | 95° |
| Methyl *n*-butyl | 129° | | 122° | liquid | 106° |
| Methyl *n*-pentyl | 151° | liquid | 127° | | 89° |
| Diethyl | 102° | 69° | 139° | liquid | 156° |
| Di-*n*-propyl | 145° | liquid | 133° | liquid | 75° |
| Cyclopentanone | 131° | 56° | 205° | 50° | 142° |
| Cyclohexanone | 155° | 90° | 166° | 77° | 162° |
| Acetophenone | 200° | 59° | 198° | 105° | 240° |

* Two values are given for certain derivatives that may be encountered in polymorphic modifications or as *cis* and *trans* geometrical isomers.

† Methone derivatives may be useful for some aldehydes (see Experiment 7, part C).

‡ Oximes of many aliphatic carbonyl compounds often separate as oily liquids that resist persistent efforts to induce crystallization. Moreover, the simpler oximes are appreciably soluble in water and in organic solvents.

§ A few 2,4-dinitrophenylhydrazones exist in a red and a yellow form, of different melting points. Mixtures of the two have lower or intermediate melting points.

# EXPERIMENT 10

# Esters

## Esterification and Saponification

Esters may be prepared by direct esterification of an acid with an alcohol in the presence of an acid catalyst (sulfuric acid, hydrogen chloride) and by alcoholysis of acid chlorides, acid anhydrides, and nitriles. Occasionally they are prepared by heating the metallic salt of a carboxylic acid with an alkyl halide or alkyl sulfate.

Typical examples of direct esterification with acid catalysts involve the following mechanism: (1) protonation of the carboxyl group, (2) addition of the alcohol, (3) elimination of water and deprotonation. Since the reaction is reversible, the equilibrium must be shifted forward to obtain good con-

$$CH_3{-}CO_2H \overset{H^+}{\rightleftharpoons} \left[ CH_3{-}C\begin{matrix} {/\!\!/}OH \\ \diagdown OH \end{matrix} \right]^+ \underset{C_2H_5OH}{\rightleftharpoons}$$

$$\left[ \begin{matrix} & OH & \\ & | & \\ CH_3{-} & C & {-}OH \\ & | & \\ H{-} & O & {-}C_2H_5 \end{matrix} \right]^+ \overset{-H^+}{\rightleftharpoons} CH_3{-}CO_2C_2H_5 + H_2O$$

version to the ester. The use of an excess of one of the initial reactants, or removal of the products, will serve this purpose.

The composition of the equilibrium mixture is given approximately by the mass law, shown in equation 1, where $K_E$ is the equilibrium constant for esterification and the symbols [ester], [water], etc., refer to *concentrations* expressed in moles per liter or as mole-fractions. Starting with one mole of acetic acid and one mole of ethanol (a total of two moles) the equilibrium

$$K_E = \frac{[\text{ester}][\text{water}]}{[\text{acid}][\text{alcohol}]} \tag{1}$$

$$K_E = \frac{[0.33][0.33]}{[0.17][0.17]} = 3.77 \tag{2}$$

mixture is found experimentally to contain 0.66 mole of ethyl acetate (and an equimolar amount of water). Thus, the mole-fractions of ester and water are 0.66/2 and the mole-fractions of unesterified acid and alcohol are 0.34/2. Putting these equilibrium concentrations into the mass law expression (equation 2) gives a $K_E$ value of 3.77 for this particular esterification.

Inspection of the equilibrium expression shows that the use of an excess of the alcohol (or an excess of the organic acid) will increase the amount of ester formed. Calculations based upon the $K_E$ value of 3.77 indicate that the use of two moles of ethanol to one of acetic acid will bring about an 80 percent conversion of the acid to ethyl acetate, and three moles of ethanol will effect almost 90 percent conversion to ester. The choice of reactant to be used in excess will depend upon factors such as availability, cost, and ease of removal of excess reactant from the product.

Under ideal conditions the composition of an equilibrium mixture is not affected by the presence or absence of a catalyst, but experiments have shown that the observed $K_E$ values may increase as much as two-fold if a relatively large amount of the acid catalyst is used. In these situations the "catalyst" changes the environment within the system and removes product through its hydration by the water formed in the reaction.

Driving an esterification to completion by removal of the water formed in the reaction is a common practice, especially in larger scale preparations. One method consists in distilling off a water-alcohol azeotrope, treating the azeotropic mixture with a drying agent, and returning the alcohol to the reaction mixture. Another procedure is to add benzene or a similar hydrocarbon and distill out a ternary azeotropic mixture, benzene-alcohol-water.

The *rate of reaction* is influenced significantly by the structure of the acid, and steric factors play an important part. Increasing the number of bulky substituents in the α- or β-position of the acid brings about a marked reduction in the rate constant for esterification. The reaction rates for two series of acids follow the sequences shown below. Specific rates of esterification with methanol at 40°, relative to acetic acid, are shown in the second

$$H{-}CO_2H > CH_3{-}CO_2H > (CH_3)_2CH{-}CO_2H \gg (CH_3)_3C{-}CO_2H$$

| | |
|---|---|
| $C_2H_5{-}CH_2{-}CO_2H$ (0.51) | $(CH_3)_3C{-}CO_2H$ (0.037) |
| $(CH_3)_3C{-}CH_2{-}CO_2H$ (0.023) | $(C_2H_5)_3C{-}CO_2H$ (0.00016) |

series (in parentheses). Esters of sterically hindered acids are prepared by methods other than direct esterification: conversion of the acid to the acid chloride, followed by treatment with an alcohol; or reaction of a salt of the acid with an alkyl halide, in the presence of a secondary amine as catalyst.

Acid chlorides and anhydrides react rapidly with primary and secondary alcohols to give the corresponding esters. In the absence of a base, acid chlorides convert tertiary alcohols into alkyl chlorides; but in the presence of a tertiary amine (pyridine, triethyl amine), tertiary alcohols furnish esters. Acid anhydrides are less reactive than acid chlorides but react with most alcohols upon heating. Acetylations with acetic anhydride are promoted by acid catalysts (sulfuric acid, zinc chloride) and by base catalysts (sodium acetate, tertiary amines).

Hydrolysis of an ester is the reverse of esterification. With an acid catalyst, even in the presence of a large amount of water, an appreciable amount of the ester may be present in the equilibrium mixture. Hydrolysis by strong alkalies, *saponification*, is more rapid, and is more effective because hydroxyl ion reacts with the organic acid and drives the reaction to completion. Saponification affords a means of establishing the structure of an

$$CH_3{-}CO_2C_2H_5 + OH^- \longrightarrow [CH_3{-}CO_2]^- + C_2H_5{-}OH$$

unknown ester, through identification of the resulting alcohol and organic acid.

Since many esters are insoluble in water, solutions of potassium hydroxide or sodium hydroxide in 85–90 percent aqueous methanol or ethanol are used frequently for saponifications. A high-boiling solvent such as diethylene glycol ($HO{-}CH_2CH_2{-}O{-}CH_2CH_2{-}OH$, bp 245°) is advantageous for the saponification of esters of high molecular weight, and esters of sterically hindered acids or tertiary alcohols. Compounds of the latter group are very resistant to saponification.

**Saponification Equivalent.** An informative preliminary step in the identification of an unknown ester is the determination of its saponification equivalent, which is the weight of the ester (in grams) that reacts with one gram-molecule of alkali. This is determined by heating a weighed sample of

$$\begin{array}{c}\text{Saponification}\\ \text{Equivalent}\end{array} = \frac{\text{weight of ester (in grams)}}{\text{ml of 1 } N \text{ alkali}} \times 1000$$

the ester with an excess of standardized potassium hydroxide solution (usually in aqueous methanol or ethanol) and titrating the excess alkali with standardized hydrochloric acid, using phenolphthalein as indicator. The saponification equivalent[1] expresses the molecular weight of the ester divided

[1] The saponification equivalent of an ester is different from the *saponification number*, commonly used in industry for fats and fatty oils, which is defined as the number of milligrams

by the number of ester groups in the molecule. For esters of dibasic acids or of bifunctional alcohols it is one-half the molecular weight, for trifunctional esters one-third the molecular weight, etc.

**Glyceryl Esters. Fats and Fatty Oils.** Fats and fatty oils represent one of the three main groups of foodstuffs; the others are carbohydrates and proteins. Apart from their use as foods, fats, and fatty oils are used in enormous quantities in the manufacture of household and industrial products.

Fats and fatty oils are mixtures of esters of the higher fatty (aliphatic) acids, $C_6$ to $C_{24}$, with the trifunctional alcohol glycerol,

$$HO{-}CH_2{-}CHOH{-}CH_2{-}OH$$

The nature of the acyl groups present affects the physical and chemical properties of the glycerides. Fats, which are solid or semi-solid at room temperature, are made up largely of the glycerides of long chain *saturated* acids, chiefly the straight chain $C_{16}$ and $C_{18}$ acids (palmitic and stearic acids). Typical fatty oils consist mainly of glycerides of *unsaturated* fatty acids, which are chiefly $C_{18}$ acids containing one, two, or three double bonds per molecule (oleic, linoleic and linolenic acids). A few fatty oils owe their liquid character to the presence of glycerides of the lower saturated fatty acids, $C_6{-}C_{14}$; for example, coconut oil contains large amounts of the glycerides of lauric and myristic acids ($C_{12}$ and $C_{14}$), as well as glycerides of lower acids ($C_{10}$, $C_8$, and $C_6$).

The term saponification came from the ancient art of making soap, by heating fats and fatty oils with potassium hydroxide solution obtained by leaching wood ashes (pot ash) with slaked lime. The term is now used to designate the general process of hydrolysis of an ester with caustic alkalies. Soaps are the sodium (or potassium) salts of the higher aliphatic acids.

Vegetable oils that contain a substantial proportion of glycerides of fatty acids having two or more double bonds have the property of drying to a hard durable film on exposure to air. Linseed oil from flaxseed is one of the principal drying oils used in the manufacture of paints and varnishes, and materials such as oilcloth and linoleum. Many synthetic substitutes for natural drying oils have been developed, such as modified alkyd, epoxy, and acrylate polymers that become cross-linked on exposure to air to form high molecular weight polymers.

In soap manufacture *glycerol* is recovered by evaporation and vacuum distillation of the aqueous solution remaining after the soap has been salted

of potassium hydroxide required to saponify *one gram* of the fat or fatty oil (mixtures of glyceryl

$$\begin{matrix}\text{Saponification}\\ \text{Number}\end{matrix} = \frac{\text{ml of 1 } N \text{ alkali} \times 56.1}{\text{weight of ester (in grams)}}$$

esters of higher aliphatic acids). In the above formula the factor 56.1 is the molecular weight of potassium hydroxide.

out by means of sodium chloride. It is used in large amounts for making alkyd resins, nitroglycerine (glyceryl trinitrate), and cosmetic preparations. Because it is hygroscopic and has a high boiling point (280° at 760 mm) glycerol finds application as a moistening agent for tobacco, printing inks, and textile processing. Since 1950 glycerol has been manufactured by a synthetic route, starting from propylene.

**Detergents and Wetting Agents.** Water and oil do not mix with each other because water does not wet an oily surface. The forces of attraction between water and oil molecules are much smaller than the attraction of water for other water molecules, and not large enough to overcome the surface tension of the water. In the presence of small amounts of a third substance that contains a hydrophilic (water-attracting) *and* an oleophilic (oil-attracting) group in the same molecule, the surface tension is reduced so that droplets of water and oil can become commingled to form an emulsion. Detergents are cleansing agents that remove films of oil and dirt adhering to a surface by emulsification with water. Sodium and potassium soaps, the oldest detergents, are effective because of the presence of a long hydrocarbon group (oleophilic) and a carboxylate anion (hydrophilic). Synthetic detergents of similar type are produced commercially on a large scale. Examples are the sodium salts of long chain alkyl sulfates ($C_{12}H_{25}{-}O{-}SO_3Na$) and branched chain alkylbenzenesulfonates ($C_{12}H_{25}{-}C_6H_4{-}SO_3Na$). An important advantage of the synthetic detergents is that they do not give precipitates with hard water, because their calcium, magnesium, and heavy metal salts are appreciably soluble in water.

## 10a *n*-Butyl Acetate

**(A) Esterification of Acetic Acid.**[2] In a 100-ml round-bottomed flask provided with a water-cooled reflux condenser, mix *thoroughly* 0.15 mole (11 g, 15 ml) of *n*-butyl alcohol (1-butanol), 0.30 mole (18 g, 17 ml) of glacial acetic acid, and 3–4 g (2 ml) of concentrated sulfuric acid. Reflux the mixture on a wire gauze for 2 hr. Cool the contents of the flask slightly, remove the condenser and set it downward for distillation.

➤**CAUTION:** *n*-Butyl acetate is a flammable liquid. Avoid fire hazards.

Add a boiling chip and distill the reaction mixture until the residue in the distilling flask amounts to only a few milliliters (avoid overheating the resi-

[2] Other esters such as *n*-propyl acetate or ethyl propionate may be prepared from the appropriate alcohol and acid combination by using proportionate molar quantities of the reactants.

due). The distillate consists of butyl acetate, together with *n*-butyl alcohol, acetic acid, sulfurous acid, and water. Place the distillate in a 500-ml Erlenmeyer flask or beaker, cool well by immersion in cold water or in an ice bath, and add carefully in small portions (*caution—foaming!*) saturated aqueous sodium carbonate, until testing with blue litmus paper shows that the acid present is neutralized completely.

Transfer the mixture to a separatory funnel, remove the lower layer as completely as possible, and wash the *n*-butyl acetate layer with about 10 ml of water. Separate the water carefully and dry the *n*-butyl acetate with 5–6 g of anhydrous magnesium sulfate (or 8–10 g of Drierite).[3] Decant through a funnel into a dry distilling flask. Add a small boiling chip, distill on a wire gauze, and collect in a weighed bottle the portion boiling at 119–125°. The yield is 12–14 g.

**(B) Saponification of *n*-Butyl Acetate** *(Optional)*. In a 250-ml round-bottomed flask provided with a reflux condenser, place 5 g of *n*-butyl acetate. To this add 25 ml of 10 percent aqueous sodium hydroxide and about 45 ml of water. Add two small boiling chips and boil until the odor of *n*-butyl acetate can no longer be detected (about 2 hr). Set the condenser downward for distillation, distill off about 30 ml of liquid, and examine the distillate. What does it contain? Add enough clean salt (sodium chloride) to saturate the liquid (about 10 g will be required), shake thoroughly, and allow it to stand undisturbed for a short while.

Cool the residual solution in the flask and acidify with dilute sulfuric acid (about 10 percent). Distill off about 25 ml of liquid and examine the distillate. Test the odor and the reaction to litmus paper. What does the solution contain?

**(C) Saponification Equivalent of an Ester** *(Optional)*. The procedure given here is intended primarily to illustrate the general method used and will not furnish accurate results. For greater precision a smaller sample of the ester is weighed carefully ($\pm 0.001$ g) and more dilute standard solutions of base and acid are used (0.1–0.2 *N*).

Prepare an ethanolic solution of potassium hydroxide by dissolving 3 g of potassium hydroxide pellets in 10 ml of water and adding 50 ml of ethanol; if sediment is present allow the solution to stand until it has settled. Withdraw *carefully* by means of a pipette, two 25-ml portions of the solution and place them in separate 125-ml Erlenmeyer flasks. From your preparation of *n*-butyl acetate withdraw a 1-ml sample (0.88 g) by means of a small pipette and add it to one of the flasks. Attach a reflux condenser, add a boiling chip, and boil the solution gently for 30 min. Meanwhile titrate the 25-ml portion

[3] Anhydrous calcium chloride is not suitable because it forms addition complexes with esters.

of potassium hydroxide solution in the other flask, using standardized hydrochloric acid about 0.5 *N*, with phenolphthalein as indicator.

When the saponification has been completed, cool the solution and rinse the condenser tube with 5–10 ml of water (collect the rinsing water directly in the flask). Titrate the alkali remaining in the solution against standardized hydrochloric acid, as in the previous titration. The difference in the volumes of acid required in the two titrations represents the amount of alkali consumed in the saponification. Using the normality factor of the standard acid, determine the volume of 1 *N* alkali consumed and calculate the saponification equivalent by the formula given in the discussion section. The result should correspond to the molecular weight (within about 5 percent) if the sample of butyl acetate was fairly pure.

**(D) Duclaux Constants** *(Optional)*. These constants are used as an aid in the identification of the lower aliphatic acids ($C_1$–$C_6$). The values depend on differences in volatility when a dilute aqueous solution of the acid is distilled under specified conditions. Since the attractive forces between water and a carboxylic acid molecule decrease with increasing molecular weight, the volatility observed *under these conditions* increases in going from formic acid to the higher acids.

**Duclaux Constants of Volatile Aliphatic Acids**

| | PERCENTAGE OF TOTAL ACID IN DISTILLATE* | | |
|---|---|---|---|
| | *First 10 ml* | *Second 10 ml* | *Third 10 ml* |
| Formic acid | 3.9 | 4.4 | 4.5 |
| Acetic acid | 6.8 | 7.1 | 7.4 |
| Propionic acid | 11.9 | 11.7 | 11.3 |
| *n*-Butyric acid | 17.9 | 15.9 | 14.6 |
| Isobutyric acid | 25.0 | 20.9 | 16.0 |
| *n*-Valeric acid | 24.5 | 20.6 | 17.0 |
| *n*-Caproic acid | 33.0 | 24.0 | 19.0 |

* Based on distillation of 100 ml of 1–2 percent solution.

***Procedure.*** For the necessary titrations prepare a dilute solution of base (about 0.1 *N*) by dissolving 1 g of sodium hydroxide pellets in 250 ml of water. The titer need not be known accurately since the values involve ratios of acidity. Dissolve 2–3 g of one of the acids listed above[4] in 150 ml of water, and mix the solution thoroughly. Withdraw 10 ml of the acid solution with a pipette, titrate this against the sodium hydroxide solution (phenolphthalein indicator), and record the volume of base required.

[4] This may be a known acid or an unknown furnished by the instructor.

Place exactly 100 ml of the acid solution in a 250-ml flask and arrange the apparatus for distillation with a short water-cooled condenser set at an angle of about 45°. Close the mouth of the flask with a clean rubber stopper or cork. Use a 10-ml graduated cylinder to collect measured portions of the distillate.

Add a few boiling chips and heat the flask over a wire gauze. It is essential that the distillation proceed at a *constant rate*: not so rapidly that drops of the distillate cannot be counted nor so slowly that there is an appreciable interval between drops. Collect the first three 10-ml portions of distillate separately and transfer them to labeled Erlenmeyer flasks. Titrate each portion against the sodium hydroxide solution prepared for the original titration, and record the volumes of base required. Compare your results with the numbers shown in the table.

$$\text{Duclaux Numbers} = \frac{\text{ml alkali for 10 ml of distillate} \times 100\%}{\text{ml alkali for 10 ml original solution} \times 10}$$

*Questions follow Experiment 10b.*

**(E) Hydroxamate Test for Esters.** (*Optional*). In a 75-mm test tube place a few drops of *n*-butyl acetate or other liquid ester (or about 50 mg of a solid), add 1 ml of a 5 percent solution of hydroxylamine hydrochloride in ethanol and 0.5 ml of 10 percent aqueous sodium hydroxide, and heat the solution to boiling. Allow the solution to cool slightly and acidify by adding 1 ml of 10 percent hydrochloric acid. Add a small amount of ethanol if the solution is turbid and then 2 drops of 3 percent ferric chloride solution. If necessary add a few more drops of ferric chloride solution to develop the color fully. A red to violet color appears through formation of a ferric hydroxamate: $Fe(\text{-O-NHCOR})_3$. Note that phenols and aliphatic enols give a somewhat similar color reaction with ferric chloride.

This test may be used also for an acid anhydride or acid chloride by converting it first to an ester.

# 10b Methyl Benzoate

In a 100-ml round-bottomed flask place 0.1 mole (12.2 g) of benzoic acid and 1 mole (32 g, 40 ml) of methanol. Carefully pour 3 ml of concentrated sulfuric acid down the wall of the flask and swirl the flask to obtain good mixing. Add two small boiling chips, attach an upright condenser, and reflux the mixture for 1 hr in a water bath. Cool the solution to room temperature and decant it from the boiling chips into a separatory funnel containing about 60 ml of methylene chloride (dichloromethane).[5] Rinse the reaction flask with 10–15 ml of methylene chloride and pour this into the separatory funnel. Shake the mixture vigorously and separate the organic liquid from the aqueous layer, which contains sulfuric acid and methanol. Wash the organic liquid with 25 ml of water and then with 25 ml of 5 percent aqueous sodium carbonate (*caution—foaming!*) to remove unesterified benzoic acid. Shake the mixture gently at first, then insert the stopper and shake more vigorously; invert the separatory funnel and release internal pressure by opening the stopcock. Separate the layers carefully and reserve the aqueous portion for recovery of benzoic acid.[6]

Shake the organic layer with a second 25-ml portion of sodium carbonate solution and separate the layers carefully (the aqueous layer may be discarded). Finally, wash the organic layer with water, separate the layers, and place the organic layer in a dry Erlenmeyer flask. Add 8–10 g of anhydrous magnesium sulfate,[3] cork the flask firmly, shake the mixture thoroughly and allow it to stand for at least 20 min.

Filter the liquid into a dry flask, attach a condenser, add a boiling chip, and distill off the methylene chloride from a water bath or steam bath. Use a long condenser and collect the recovered solvent in a receiver cooled in an ice bath. When no more solvent distills over, decant the residual ester into a small dry distilling flask. Fit the side-arm with a short air-cooled condenser tube (or an ordinary condenser without water in the jacket),[7] add a boiling chip and distill the product by heating the flask directly with a small luminous flame, kept in motion. Collect in a dry weighed bottle the fraction boiling above 190° (mainly in the range 192–196°, uncor). The yield is 9–10 g.

The saponification equivalent of the product may be determined by the procedure given for butyl acetate, using a 1-ml sample (1.09 g). Nitration of methyl benzoate furnishes a crystalline derivative, methyl *m*-nitrobenzoate, mp 78° (see Experiment 18d).

## Questions

**1.** What is the purpose of adding sulfuric acid in the preparation of *n*-butyl acetate? Would any of the ester be formed in the absence of sulfuric acid?

[5] Methylene chloride is less flammable than ether or benzene, but either of these solvents may be used if methylene chloride is not available. Benzene often gives troublesome emulsions during the washing operations.

[6] The unconverted benzoic acid may be recovered by careful acidification of the sodium carbonate extract with hydrochloric acid (*caution—foaming!*). The precipitated benzoic acid is collected with suction, washed with a little water and dried. If an appreciable amount of the acid is recovered, this may be taken into account in calculating the yield of ester.

[7] For substances that boil above 160–170° an air-cooled condenser is used since the hot vapors may crack a water-cooled condenser.

**2.** What procedures may be used to drive an esterification toward completion?

**3.** Assuming a $K_E$ value of 4, calculate the percentage conversion of butyl alcohol to ester (at equilibrium) with the molar ratio of reactants used in this experiment. (Answer = ~85%.)

**4.** Assuming a $K_E$ value of 3, calculate the percentage conversion of benzoic acid to methyl benzoate (at equilibrium) with the molar ratio of reactants used in this experiment. (Answer = ~96%.)

**5.** What is formed by the action of ammonia on methyl benzoate?

**6.** Suggest a method for preparing *t*-butyl benzoate from benzoic acid and *t*-butyl alcohol.

**7.** Why is the hydrolysis of an ester by an alkali called saponification? Why is this preferable to the use of an acid catalyst such as sulfuric acid?

**8.** In what way would the saponification equivalent of a sample of *n*-butyl acetate be affected by the presence of the following impurities: (a) water? (b) butyl alcohol? (c) acetic acid?

**9.** What is the value of the saponification equivalent of the following compounds:

(a) $CH_3CO-O-CH_2-CO-OCH_3$
(b) $CH_3-O-CH_2CH_2-CO-OCH_3$
(c) tripalmitin (glyceryl tripalmitate)

EXPERIMENT

# 11

# Acylating Agents

## Acid Chlorides and Anhydrides

Aliphatic acids may be converted to acid chlorides conveniently by means of phosphorus trichloride or thionyl chloride (sulfurous oxychloride, $SOCl_2$). Phosphorus pentachloride is a more active reagent but more disagreeable to handle; it is employed mainly with hindered acids and aromatic acids.

$$3R{-}CO{-}OH + PCl_3 \longrightarrow 3R{-}CO{-}Cl + H_3PO_3$$

$$R{-}CO{-}OH + SOCl_2 \longrightarrow R{-}CO{-}Cl + SO_2 + HCl$$

$$R{-}CO{-}OH + PCl_5 \longrightarrow R{-}CO{-}Cl + POCl_3 + HCl$$

The choice of reagent is determined mainly by the reactivity of the acid and the physical properties of the reactants and products, which affect the ease of separation and purification of the acid chloride. Other considerations include the cost and efficiency of the reagent (in terms of chlorine atoms convertible to acid chloride), and convenience of the manipulations involved. All of the reagents and most acid chlorides must be handled carefully as they burn the skin, and their vapors are irritant and harmful to breathe.

Phosphorus trichloride (bp 76°) is an inexpensive and fairly satisfactory reagent. The acid chloride may be distilled or decanted from the phosphorous acid before final purification. The reaction gives yields of about 70 percent; hydrogen chloride is evolved in appreciable amounts as a result of side reactions.

$$R{-}CO{-}OH + PCl_3 \longrightarrow R{-}CO{-}O{-}PCl_2 + HCl, \text{etc.}$$

$$H_3PO_3 + PCl_3 \longrightarrow (HO)_2P{-}O{-}PCl_2 + HCl, \text{etc.}$$

Although thionyl chloride (bp 79°) is somewhat more costly than phosphorus trichloride, it has several advantages and is employed frequently.

Aliphatic and aromatic acids can be converted to the acid chlorides in high yields. The other products of the reaction are gaseous and escape from the reaction mixture, and excess thionyl chloride is removed easily by distillation or evaporation under reduced pressure. Often the residual acid chloride may be used directly without further purification. Under unfavorable conditions,

$$R{-}CO{-}Cl + HO{-}CO{-}R \longrightarrow R{-}CO{-}O{-}CO{-}R + HCl$$

or with an insufficient amount of thionyl chloride, a portion of the acid chloride may be converted to the anhydride. By employing only 0.5 mole of thionyl chloride per mole of the carboxylic acid, this reaction becomes a good method for conversion of acids to acid anhydrides.

Phosphorus pentachloride is used principally for the preparation of aroyl chlorides. Since the aroyl chlorides are not hydrolyzed readily, the reaction mixture may be poured into cold water to dispose of the phosphorus oxychloride and any excess of the pentachloride. Phosphorus oxychloride (or

$$2R{-}CO_2Na + POCl_3 \longrightarrow 2R{-}CO{-}Cl + NaCl + NaPO_3$$

pentachloride) or thionyl chloride will effect the conversion of salts of carboxylic acids to the acid chlorides.

**Acylating Agents.** The high reactivity of acid halides and anhydrides makes them effective reagents for introducing an acyl group (R−CO−) into compounds having an active hydrogen atom. Esters are formed by reaction with alcohols and phenols; amides are produced by reaction with ammonia and with primary and secondary amines.[1] Aromatic compounds may be acylated to form ketones (the Friedel-Crafts reaction). The mechanism of the acylation of hydroxyl and amino groups involves addition of the nucleophilic

$$R{-}CO{-}Cl + H{-}OCH_3 \longrightarrow R{-}CO{-}OCH_3 + HCl$$

$$R{-}CO{-}Cl + 2H{-}N(CH_3)_2 \longrightarrow R{-}CO{-}N(CH_3)_2 + (CH_3)_2\overset{+}{N}H_2\,Cl^-$$

$$(R{-}CO{-})_2O + H{-}C_6H_5(+AlCl_3) \longrightarrow R{-}CO{-}C_6H_5 + R{-}CO{-}OH$$

oxygen or nitrogen atom to the electrophilic carbon atom of the carbonyl group, followed by elimination of hydrogen chloride or $R{-}CO_2H$.

The lower aliphatic acid chlorides are hydrolyzed rapidly by water and must be used under anhydrous conditions to obtain satisfactory results. Higher aliphatic acid chlorides and aroyl chlorides, and many acid anhydrides, may be employed in the presence of water for some acylations.

$$C_6H_5{-}CO{-}Cl + H_2N{-}C_6H_5 + NaOH \xrightarrow{H_2O} C_6H_5{-}CO{-}NH{-}C_6H_5 + NaCl + H_2O$$

[1] Tertiary amines, such as trimethyl- and triethylamine, react with aliphatic acid chlorides in an unusual way. Acid chlorides of the type $R{-}CH_2{-}COCl$ and $R_2CH{-}COCl$ undergo dehydrohalogenation to form highly reactive *ketenes*, $R{-}CH{=}C{=}O$ and $R_2C{=}C{=}O$. The ketenes usually are difficult to isolate because they readily form cyclic dimers by self-addition under the conditions of the reaction.

Acylations of hydroxyl and amino compounds with aroyl chlorides can be done effectively in the presence of an aqueous base (the Schotten-Baumann reaction).

Acetyl chloride is a highly active acetylating agent. Its reactions with water, alcohols, and amines are vigorous and strongly exothermic. Moreover, in the acylation of an amine, half of the amine is diverted from amide formation through conversion to the hydrochloride; in some instances the hydrogen chloride formed may have a deleterious effect. Acetic anhydride reacts less vigorously and the acetic acid produced in the reaction usually has no harmful effects. Acetic anhydride is usually the preferred acetylating agent.

Ketene, $CH_2{=}C{=}O$, which may be regarded as a special type of acid anhydride (or anhydro acid) is an elegant acetylating agent that can be used

$$CH_2{=}C{=}O + H_2N{-}R \longrightarrow CH_3{-}CO{-}NH{-}R$$

for acid-sensitive materials. At ordinary temperatures ketene is gaseous (bp $-56°$) and because it reacts by an addition process, there are no accessory products. Ketene is made commercially by pyrolysis of acetic acid or acetone at 650–700°. Higher ketenes are not readily available.

## 11a Acetyl Chloride

It is important to have all apparatus used in this preparation perfectly dry, since both acetyl chloride and phosphorus trichloride are decomposed by water.

➢**CAUTION:** Do not allow phosphorus trichloride or acetyl chloride to come in contact with the skin. These substances produce dangerous burns which heal very slowly. Their vapors are irritant and harmful to breathe.

Arrange an assembly like that shown in Figure 40, with a 100-ml flask for the reactants. Provide the exit tube from the receiver with a calcium chloride tube half-filled with anhydrous calcium chloride, to protect the distillate from moisture. By means of a cork holding a glass tube, connect the open end of the calcium chloride tube to a funnel inverted *over* 200–300 ml of water in a beaker, to absorb the hydrogen chloride formed in secondary reactions. The funnel must not dip below the surface of the water, or the latter may be drawn up into the dry apparatus (see Figure 38).

Place 0.4 mole (24 g, 23 ml) of glacial acetic acid in the distilling flask, reattach the adapter and separatory funnel, and add slowly through the funnel 0.15 mole (21 g, 13 ml) of phosphorus trichloride. Cool the flask by immersion in a water bath. Mix the reactants thoroughly and, after allowing

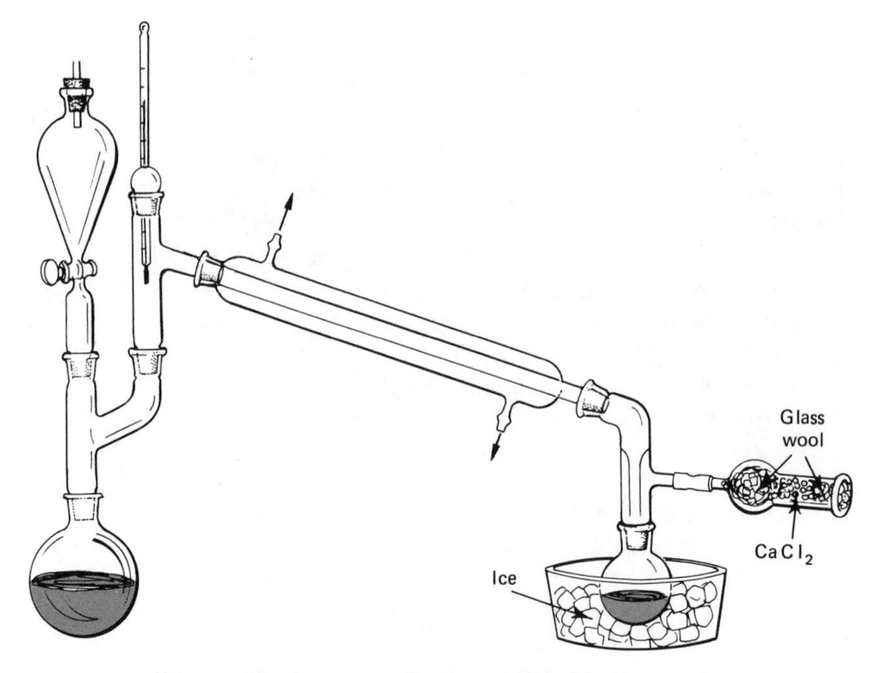


**Figure 40.** Apparatus for Acetyl Chloride Preparation

the mixture to stand for 10 min, heat the water bath to 40–50° and maintain it at this temperature for 30 min, with occasional swirling of the flask. During the heating the liquid usually separates into two layers. Acetyl chloride forms the upper layer.

Distill the acetyl chloride by heating the water bath to boiling and maintaining it at that temperature as long as any liquid passes over. Cool the receiver in an ice bath during the distillation. The syrupy residue in the distilling flask is phosphorous acid, which is discarded.

➤**CAUTION:** The reaction mixture must not be overheated since this will lead to formation of phosphine, which is spontaneously flammable in contact with air.

To the distillate add *two drops* of glacial acetic acid, to destroy mixed phosphorous-acetic anhydrides that would cause turbidity to develop on standing. Redistill the acetyl chloride[2] from a distillation apparatus arranged as above, except that the separatory funnel is replaced by a thermometer. Collect separately in a dry receiver the portion boiling at 50–56° and transfer

[2] If the acetyl chloride is to be converted to acetic anhydride (Experiment 11b) it need not be redistilled; do *not* add the two drops of acetic acid as directed for redistillation. Reserve 3–4 ml of the unpurified acetyl chloride for the test reactions.

it to a dry weighed glass-stoppered bottle. Acetyl chloride attacks corks and rubber stoppers. The yield is 20–25 g.

**Reactions of Acetyl Chloride.** *(A) Hydrolysis.* Add cautiously 2–3 drops of acetyl chloride to 5 ml of ice-cold water. Shake the tube carefully and observe the result.

*(B) Acetylation of an Alcohol.* Place 2 ml of ethanol in a test tube and cool the tube in cold water. Add cautiously, dropwise and with shaking, 1 ml of acetyl chloride. When the reaction is complete add about 2 ml of saturated salt solution (sodium chloride), shake the mixture, and allow the layers to separate. Note the odor of the product.

*(C) Acetylation of an Amine.* In a large dry test tube place a solution of 1 g (1 ml) of aniline in 5–6 ml of ether (or benzene) and *very cautiously* add 0.5 ml of acetyl chloride, drop by drop, with shaking. After the vigorous exothermic reaction is over, mix the materials with a glass rod, and add 15–20 ml of water. Break up any lumps of the product and collect it on a small suction filter. Wash the crystals with a little very dilute hydrochloric acid, to remove any unreacted aniline, and then with a little water. Recrystallize the acetanilide from hot water, with addition of a little decolorizing carbon.

## Questions

**1.** What reagent and procedure would you choose to prepare *pure* propionyl chloride (bp 80°) from propionic acid (or sodium propionate)?

**2.** Compare the boiling point of propionic acid with that of its methyl ester, acid chloride, anhydride, amide, and nitrile. Which of the derivatives has the highest melting point?

**3.** What method would you choose to convert (a) a *limited* amount of an amine to an amide? (b) a *limited* amount of an organic acid to an anilide?

**4.** Acid chlorides react with tertiary alcohols in an atypical way, to form the tertiary alkyl chloride. Suggest an explanation for this behavior. What reaction conditions are needed to convert a tertiary alcohol to an ester (see Experiment 10)?

# 11b Acetic Anhydride

The apparatus for this experiment is similar to that used for the preparation of acetyl chloride (Figure 40). The same precautions for exclusion of moisture must be observed but it is unnecessary to provide the inverted funnel arrangement, since hydrogen chloride is not evolved.

In the dry 100-ml reaction flask place 0.3 mole (25 g) of finely pulverized *anhydrous* sodium acetate.[3] Arrange in place the condenser, dry receiving flask and drying tube; the receiver need not be cooled. Check to insure that all connections are tight.

Chill the reaction flask in a bath of cold water and add dropwise, through the separatory funnel, 0.25 mole (19.5 g, 18 ml) of acetyl chloride. After the addition has been completed, remove the water bath and shake the flask to obtain good mixing of the reactants. Recheck the connections for tightness.

Dry the outside of the flask with a towel and heat it directly with a luminous flame (air inlet of the burner closed). Keep the flame constantly in motion and take care to *avoid local overheating*. Continue the heating until no more distillate comes over but do not overheat the solid residue.

To the distillate add 2–3 g of finely powdered anhydrous sodium acetate, to react with a small amount of acetyl chloride that may be present. Add a boiling chip and redistill the crude acetic anhydride, using a wire gauze. Collect the product distilling at 132–138° in a dry glass-stoppered bottle. The yield is 15–20 g.

**Reactions of Acetic Anhydride.** ***(A) Hydrolysis.*** Add 2–3 drops of acetic anhydride to 5 ml of water and shake the tube carefully. Warm the tube gently and observe the result.

***(B) Alcoholysis.*** Place 2 ml of ethanol in a large test tube and add 1 ml of acetic anhydride. Does a reaction occur? By means of a thin capillary tube introduce very cautiously into the liquid a *minute* amount of concentrated sulfuric acid and observe the result. After the reaction has occurred add a small pinch of solid sodium carbonate and about 2 ml of saturated salt solution (sodium chloride). Shake the mixture and allow the layers to separate. Note the odor.

***(C) Aminolysis.*** In a large dry test tube place 0.5 ml of aniline and add 0.5 ml of acetic anhydride, dropwise. Does a reaction occur? Heat the solution gently and then add 15–20 ml of water and shake the mixture vigorously. Break up any lumps of the product and collect it on a small suction filter. Wash the crystals with a little very dilute hydrochloric acid, to remove any unreacted aniline, and then with a little water. Recrystallize the crude acetanilide from hot water, with the addition of a little decolorizing carbon.

[3] Commercial anhydrous sodium acetate usually contains some moisture. To remove moisture fuse 35–40 g of commercial anhydrous sodium acetate (mp 325°) in a porcelain dish and stir the material until no more water vapor is evolved. Avoid overheating the molten salt since it decomposes at higher temperatures. Allow the fused mass to cool, pulverize it quickly in a mortar, and place it in a tightly corked bottle until needed. Reserve 2–3 g of the dry material for use in purifying the crude anhydride.

## Questions

1. Outline a series of reactions used for the manufacture of acetic anhydride, starting from acetylene.

2. What is the relative reactivity of the following carboxylic acid derivatives toward water (and nucleophilic reagents in general): acid anhydrides, esters, acid chlorides, acid fluorides, nitriles?

3. What product is formed by reaction of ketene ($CH_2{=}C{=}O$) with hydrogen bromide? with anhydrous propionic acid? What compounds would be obtained by slow fractional distillation of the product from propionic acid, in the presence of a little sodium acetate as catalyst?

EXPERIMENT

# 12

# Amines

Amines are classified as primary, secondary, and tertiary (1°, 2°, 3°) according to the number of alkyl or aryl groups attached to the nitrogen atom. All classes of amines have an unshared electron-pair on the nitrogen atom and are bases (nucleophiles). The unshared electron-pair is responsible for the ability of an amine to form salts with acids and coördination complexes with metal cations, and to undergo alkylation with alkyl halides (nucleophilic displacements).

The relative availability of the unshared pair for combination with an electrophile (base strength or nucleophilicity) is strongly influenced by the nature of the groups attached to the nitrogen atom. In general, alkyl groups, through electron-release, enhance the basic strength, while electron-withdrawing groups such as aryl and acyl markedly diminish it. A few basic dissociation constants, $K_b$, are shown below; the numbers in parentheses are p$K_b$ values ($-\log K_b$).

| | |
|---|---|
| $NH_3$ ................ | $0.18 \times 10^{-4}(4.74)$ |
| $CH_3NH_2$ ............ | $4.38 \times 10^{-4}(3.36)$ |
| $(CH_3)_2NH$ ........... | $5.12 \times 10^{-4}(3.29)$ |
| $(CH_3)_3N$ ............ | $0.53 \times 10^{-4}(4.28)$ |
| $C_6H_5NH_2$ ........... | $3.8 \times 10^{-10}(9.42)$ |
| $C_6H_5NHCH_3$ ........ | $7.1 \times 10^{-10}(9.15)$ |
| $(C_6H_5)_2NH$ .......... | $7.6 \times 10^{-14}(13.1)$ |
| $C_6H_5NHCOCH_3$ ..... | $4.1 \times 10^{-14}(13.4)$ |

Replacement of active hydrogen occurs with the $NH_2$ and NH groups of primary and secondary amines. Such reactions may serve as test reactions to distinguish between the three classes of amines and also for the identification of unknown amines. Thus, 1° and 2° amines are distinguished from 3° amines by their ability to form acetyl, benzoyl, benzenesulfonyl, and other acyl derivatives. Primary amines can be differentiated from secondary amines by virtue of the solubility of the benzenesulfonyl derivatives of the former in aqueous alkali (Hinsberg test).

With some reagents there are differences not only between 1°, 2°, and 3° amines but also differences between aliphatic and aromatic amines. With nitrous acid a 1° aliphatic amine reacts to evolve nitrogen quantitatively and give a mixture of alcohols and alkenes; a 1° arylamine in acidic medium gives an aryldiazonium salt. Aliphatic *and* aromatic 2° amines give N-nitroso derivatives on treatment with nitrous acid. With 3° aliphatic amines nitrous acid merely forms nitrite salts but with 3° arylamines, such as dimethylaniline, C-nitrosation occurs at the *para* position of the ring.

## Reactions of Amines

The reactions described next may be performed directly on an unknown amine furnished by the instructor but it is instructive to carry out first a few tests with known compounds. Aniline, methylaniline, and dimethylaniline may be used to illustrate the behavior of a 1°, 2°, and 3° amine in parts A, B, and C. Cyclohexylamine and diethylamine (or morpholine) may be used if desired to exemplify a 1° and 2° amine of aliphatic character.

Some of the following tests are carried out with small quantities of the reactants. These should be measured with reasonable care to obtain satisfactory results. A small pipette or a glass tube drawn to a small bore at one end is useful. If bottles fitted with medicine droppers are available they are very satisfactory for this purpose. Freshly distilled samples of amines should be used to obtain the best results.

**(A) Solubility.** Test the solubility of the amine in water, 5 percent hydrochloric acid, and 5 percent aqueous sodium hydroxide by adding about two drops of the amine to 5 ml of the test liquid in a test tube.

**(B) Acetylation with Acetic Anhydride.**[1] To 1 ml of the amine in a large *dry* test tube add cautiously 3 ml of acetic anhydride and observe any temperature change. Boil the solution *gently* over a very small flame for 3–4 min, cool the solution slightly, and cautiously add about 10 ml of cold water. Heat the solution to the boiling point, and allow the tube to cool with shaking. If a precipitate is not formed, neutralize the solution with ammonia and observe the result. If an oily liquid appears attempt to induce crystallization by scratching with a glass rod.[2]

[1] Acetylation can be carried out effectively in an aqueous medium (Lumière-Barbier method) by the general procedure described for aniline in Experiment 22b, part A.

[2] The acetyl derivatives of some arylamines crystallize with difficulty; those of low molecular weight aliphatic amines are liquids at room temperature.

If a crystalline acetyl derivative is formed, collect it on a suction filter and wash with a little cold water. After drying in the air, take the melting point.

**(C) Acylation with Benzenesulfonyl Chloride. The Hinsberg Test.**[3] To 8–10 drops of the amine in a large test tube add 10 ml of 10 percent aqueous sodium hydroxide and 14–15 drops of benzenesulfonyl chloride. Shake the tube thoroughly and note any reaction. Warm the mixture *very gently* with shaking (*do not boil*) until the disagreeable odor of benzenesulfonyl chloride can no longer be detected. The reaction mixture should be alkaline at this point. Cool the tube to room temperature, shake well and note any precipitate (liquid or solid).

Acidify the reaction mixture with concentrated hydrochloric acid, cool to room temperature, shake well and observe the result. If a solid precipitate is formed collect it on a suction filter and wash it with water. Crystallize the product from 3–5 ml of about 70 percent ethanol and take its melting point. Outline the behavior of a typical 1°, 2°, and 3° amine in the above procedure.

The arylsulfonyl derivatives of aliphatic and aromatic amines are usually good crystalline derivatives for identification purposes. If the benzenesulfonamide is a liquid, another reagent such as *p*-toluenesulfonyl chloride, *p*-nitrobenzenesulfonyl chloride, or *p*-bromobenzenesulfonyl chloride may be used.

**(D) Benzoylation by the Schotten-Baumann Method.** To 15 ml of 5 percent aqueous sodium hydroxide in a large test tube or small Erlenmeyer flask, add 0.5 ml (0.5 g) of the amine and 1 ml (1.2 g) of benzoyl chloride (*caution—irritating vapor!*). Stopper the tube firmly and shake it vigorously; release internal pressure by removing the stopper cautiously. Continue to shake the tube until the odor of benzoyl chloride can no longer be detected. Collect the precipitate on a small suction filter and wash it with water, followed by a little dilute hydrochloric acid. Recrystallize the benzoyl derivative from methanol or aqueous ethanol, and take its melting point.

The benzoyl derivatives of some secondary amines crystallize with difficulty. The oily liquid should be scratched in the presence of a little solvent with a glass rod. A few crystals should be reserved for seeding when the material is recrystallized.

**(E) Tertiary Amines.** Since tertiary amines lack replaceable hydrogen on the nitrogen atom they cannot be acylated but may be converted to salts with acids (hydrochlorides, arylsulfonates) and to quaternary ammonium salts.

[3] In parts C and D, *n*-propylamine, *s*-butylamine, *t*-butylamine, and isobutylamine, and aqueous solutions (20–30 percent) of methylamine, dimethylamine, and ethylamine, are suitable examples to illustrate the preparation of crystalline derivatives of aliphatic amines.

The latter are formed by treatment with reactive halides (methyl or ethyl iodide, benzyl chloride) or a sulfonic ester (methyl *p*-toluenesulfonate).[4]

Tertiary arylamines such as dialkylanilines may be converted to crystalline derivatives through reactions involving the aromatic ring. These include C-nitrosation with nitrous acid, coupling with a diazonium salt to form a dialkylamino derivative of azobenzene, ring halogenation and nitration.

**Quaternary Ammonium Salts.** In a dry test tube mix 0.5 ml (0.5 g) of the 3° amine and 0.5 ml of methyl iodide (bp 43°). Warm the tube *very gently*, to avoid loss of methyl iodide, for about 5 min. Cork the tube loosely, allow it to cool and stand for 10 min, and add 2–3 ml of dry ether or benzene. Collect the crystals of the methiodide with suction and wash them with a little solvent. Quickly place the product in a stoppered vial, since many quaternary salts rapidly absorb moisture from the air. Absolute ethanol or methanol may be used for recrystallization. Many quaternary salts decompose at or near the melting point.

Quaternary ethiodides may be prepared in a similar way by using ethyl iodide (bp 72°) instead of methyl iodide.

Melting points of the methiodides of a few 3° amines are: dimethylaniline, 220° dec (ethiodide, mp 136°); pyridine, 117° (ethiodide, mp 90°); quinoline, 133°, hydrate 72° (ethiodide, mp 159°); trimethylamine, 230° dec; tri-*n*-propylamine, 207° (ethiodide, mp 238° dec).

**(F) Examination of an Unknown Amine.** Obtain from the instructor a sample of an unknown amine and apply the Hinsberg test to determine the class to which the amine belongs (1°, 2°, or 3°). Record the result in your report of this experiment.

If the amine is a 1° or 2° amine it may be possible to identify it by the preparation of two or more crystalline derivatives (see Table 20). Tertiary amines are more difficult to identify.

## Questions

**1.** Write equations showing the action of sodium nitrite and hydrochloric acid upon: aniline; methylaniline; dimethylaniline; methylamine; dimethylamine.

**2.** How would ammonia behave toward benzenesulfonyl chloride in the Hinsberg test? *p*-methylaminobenzoic acid ($CH_3NH-C_6H_4-CO_2H$)?

[4] More complete lists of derivatives and various procedures for preparing derivatives of tertiary amines will be found in Shriner, Fuson, and Curtin, *The Systematic Identification of Organic Compounds*, John Wiley, New York, 5th edition (1964); Cheronis and Entrikin, *Semimicro Qualitative Organic Analysis*, Interscience Publishers, Inc., New York, 2nd edition (1957).

**3.** Explain the reactions involved in the carbylamine (isocyanide) test for primary amines. (In this test the amine is treated with chloroform and alkali.)

**4.** What product is formed by the action of phenyl isothiocyanate ($C_6H_5{-}N{=}C{=}S$) on aniline? on methylaniline?

**TABLE 20**
**Derivatives of Primary and Secondary Amines**

| AMINE (BP) | ACETYL | BENZOYL | BENZENE-SULFONYL | *p*-TOLUENE SULFONYL |
|---|---|---|---|---|
| Methylamine (−6°) | oil | 80° | 30° | 75° |
| Ethylamine (17°) | oil | 71° | 58° | 63° |
| *n*-Propylamine (49°) | 47° | 84° | 36° | 52° |
| *t*-Butylamine (45°) | | 134° | | |
| *s*-Butylamine (69°) | | 76° | 70° | 55° |
| Isobutylamine (69°) | 107° | 57° | 53° | 78° |
| *n*-Butylamine (77°) | oil | 42° | | 44° |
| Ethylenediamine (116°) | 172° di | 249° di | 168° di | 160° di |
| Cyclohexylamine (134°) | 104° | 149° | 89° | |
| Benzylamine (185°) | 60° | 105° | 88° | 116° |
| Aniline (185°) | 114° | 160° | 112° | 103° |
| *o*-Toluidine (199°) | 112° | 142° | 124° | 108° |
| *m*-Toluidine (203°) | 66° | 125° | 95° | 114° |
| *p*-Toluidine (200°; mp 45°) | 148° | 158° | 120° | 117° |
| 2,4-Dimethylaniline (212°) | 130° | 192° | 129° | |
| 2,5-Dimethylaniline (215°) | 139° | 140° | 139° | 119° |
| *o*-Anisidine (225°) | 85° | 65° | 89° | 127° |
| *p*-Anisidine (243°; mp 57°) | 130° | 157° | 95° | 114° |
| *o*-Phenetidine (229°) | 79° | 104° | 102° | 164° |
| *p*-Phenetidine (254°) | 135° | 173° | 143° | 107° |
| 1-Naphthylamine (mp 50°) | 159° | 160° | 167° | 157° |
| 2-Naphthylamine (mp 112°) | 132° | 162° | 102° | 133° |
| Dimethylamine (7°) | oil | 41° | 47° | 79° |
| Diethylamine (55°) | oil | 42° | 42° | 60° |
| Piperidine (105°) | oil | 48° | 93° | 96° |
| Morpholine (130°) | oil | 75° | 118° | 147° |
| N-Methylaniline (192°) | 102° | 63° | 79° | 94° |
| N-Ethylaniline (205°) | 54° | 60° | oil | 87° |
| N-Methyl-*p*-toluidine (208°) | 83° | 53° | 64° | 60° |

EXPERIMENT

# 13

# Acid Amides

Carboxylic acids and their functional derivatives may be converted to amides by various methods of ammonation or ammonolysis. These reactions of ammonia are analogous to the processes of hydration, hydrolysis, and alcoholysis. Acetamide is prepared in this experiment by heating ammonium acetate in the presence of acetic acid. The reaction is reversible and amide formation is promoted by adding acetic acid and also by removal of water from the system (as in esterification). Acids may be ammonated by heating with ammonium carbonate or urea as the source of ammonia.

$$[CH_3CO-\overset{-}{O}]\overset{+}{N}H_4 \underset{}{\overset{\Delta}{\rightleftarrows}} CH_3CO-OH + NH_3 \overset{\Delta}{\rightleftarrows} CH_3CO-NH_2 + H_2O$$

The amide group is resonance stabilized: for acetamide the resonance stabilization energy amounts to 11 kcal/mole. Amides can exist in a tauto-

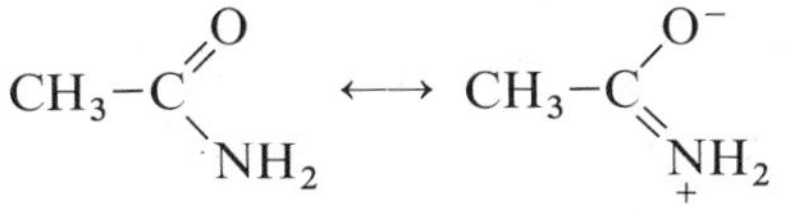

$$CH_3-C(=O)-NH_2 \longleftrightarrow CH_3-C(-O^-)=\overset{+}{N}H_2$$

meric imidol form (I), which is known definitely in the form of O-substituted derivatives, the imido esters (II). Acetamide ($K_b$ $10^{-15}$) is a weaker base than

$$\underset{\text{I}}{CH_3-C(-OH)=NH} \qquad \underset{\text{II}}{CH_3-C(-OEt)=NH}$$

water but forms salts with strong acids under anhydrous conditions.

The simple amides, $R-CO-NH_2$, are crystalline solids, with the exception of formamide, mp $+2°$. N-Alkyl and dialkylamides are liquids;

dimethylformamide (DMF) and dimethylacetamide are extremely useful solvents.

Amides having an NH group are associated through hydrogen-bonding. The extent of association decreases with introduction of an alkyl group on the nitrogen atom and with increasing molecular weight of the acyl group. The boiling points of the amides are much higher than those of the corresponding acids although the amines have lower boiling points than the corresponding alcohols: acetic acid, bp 118°, acetamide, bp 223°; methanol, bp 65°, methylamine, bp −6°. The boiling point of an amide is *lowered* by alkylation of the nitrogen atom.

$$\underset{\text{bp } 223^\circ}{CH_3CO{-}NH_2} > \underset{\text{bp } 206^\circ}{CH_3CO{-}NHCH_3} > \underset{\text{bp } 166^\circ}{CH_3CO{-}N(CH_3)_2}$$

Acid chlorides and acid anhydrides react rapidly with ammonia and amines to form amides; this method is rapid and convenient. Methyl and ethyl esters of carboxylic acids undergo ammonolysis to amides but much more slowly. Nitriles are converted to amides by treatment with hydrogen

$$R{-}C{\equiv}N + 2H_2O_2(+NaOH) \longrightarrow R{-}CO{-}NH_2 + O_2 + H_2O$$

peroxide and dilute alkali, and can be hydrolyzed with acid catalysts under controlled conditions to stop at the amide stage.

All plant and animal cells contain characteristic proteins, which are high molecular weight polymers of amino acids of the type $H_2N{-}CHR{-}CO_2H$. The amino acid units are combined through amide linkages, designated in this particular case as peptide linkages. Synthetic fiber-forming polymers of Nylon type are polyamides of high molecular weight prepared from aliphatic diamines and dicarboxylic acids (see Experiment 56).

Amides can be hydrolyzed by heating with aqueous acids or alkalies. Primary amides, $R{-}CO{-}NH_2$, on treatment with nitrous acid in the presence of mineral acid, evolve nitrogen quantitatively and form the corresponding acid. If one substituent is present on the nitrogen atom, a nitroso derivative is formed.

Primary amides react with hypobromite to form N-bromo amides, which are converted by alkali to isocyanates, $R{-}N{=}C{=}O$. With excess of aqueous alkali the isocyanate is converted to a primary amine, $R{-}NH_2$, and sodium carbonate. The overall reaction is known as the Hofmann degradation.

Primary amides on treatment with phosphorus pentoxide or thionyl chloride undergo elimination of water to form the nitrile.

# 13a Capryl Chloride and Capramide

A general method for the conversion of acids to amides consists in forming the acid chloride, by reaction with thionyl chloride, and treating it with ammonia (or an amine). The procedure given below, employing aqueous ammonia, is satisfactory for the preparation of amides from the higher aliphatic acids ($C_8$ and above), such as lauric, myristic, and palmitic. This procedure is not suitable for the lower members because the low molecular weight acid chlorides are hydrolyzed rapidly by water and the amides are quite soluble in water.

For the lower aliphatic acids, and higher aliphatic and aromatic acids as well, crystalline amide derivatives may be obtained by reaction of the acid chloride with an arylamine (aniline, *p*-toluidine) under anhydrous conditions. The preparation of *n*-valeranilide by a typical procedure is given in Experiment 15.

Aliphatic and aromatic acid chlorides are prepared readily in excellent yields by means of thionyl chloride (bp 79°). The reaction proceeds through a mixed anhydride; hydrogen chloride and sulfur dioxide are evolved as gases,

$$R{-}CO{-}OH + SOCl_2 \longrightarrow HCl + R{-}CO{-}O{-}SOCl \longrightarrow R{-}CO{-}Cl + SO_2$$

and the excess thionyl chloride can be removed by distillation if necessary. The residual acid chloride usually may be used directly without purification. An excess of thionyl chloride must be used or some of the acid will be converted to the anhydride.

***Capryl Chloride.*** In a large dry test tube place 2 g (0.012 mole) of capric acid (decanoic acid, $CH_3(CH_2)_8CO_2H$)[1] and 3.2 g (2 ml, 0.025 mole) of thionyl chloride. Attach a dry reflux condenser, add a small boiling chip, and reflux the mixture gently for 30 min in a hood. Small amounts of sulfur dioxide and hydrogen chloride are evolved during the heating. The excess thionyl chloride could be removed by warming under reduced pressure but that is unnecessary in this particular case.

➤**CAUTION:** Handle thionyl chloride carefully. The liquid burns the skin and the vapor is irritant and harmful to breathe.

***Capramide.*** Cool the solution and add 10–15 ml of methylene chloride or ethanol-free ether.[2] Pour the solution of capryl chloride (and excess thionyl chloride) *cautiously*, and in small proportions, into 0.3 mole (20 ml)

[1] A practical grade of capric acid is quite satisfactory.

[2] For this experiment ordinary ether may be freed of ethanol by washing with one-fifth its volume of cold 50 percent sulfuric acid, followed by an equal volume of water.

of cold, concentrated, aqueous ammonia contained in a 125-ml Erlenmeyer flask. The reaction mixture should be mixed thoroughly by swirling during the addition. The acid amide separates as a crystalline precipitate during the reaction.

After allowing the reaction mixture to stand for 20 min, with occasional shaking, collect the crystals by suction filtration. Wash the crystals with water and press them as dry as possible on the filter. Recrystallize the amide from 75 percent aqueous ethanol and determine its melting point. Capramide exists in two crystalline forms, one melting near 108° and the other near 98°. The lower-melting metastable form is encountered occasionally. The yield is 1–1.5 g.

### Questions

**1.** Why is the presence of ethanol objectionable in a solvent used to dissolve the crude capryl chloride?

**2.** What products would be formed if capryl chloride were treated: (a) with methylamine instead of ammonia? (b) hydroxylamine? (c) sodium azide NaN=N=N + heat (*cf.* Schmidt reaction)?

**3.** What products would be formed by the action of the following reagents on capramide: (a) hot sodium hydroxide solution? (b) sodium hypobromite solution? (c) phosphorus pentoxide? (d) sodium nitrite and hydrochloric acid?

## 13b Acetamide

Place 0.65 mole (50 g) of ammonium acetate and 0.7 mole (42 g, 40 ml) of glacial acetic acid in a 250-ml round-bottomed flask with a short fractionating column leading to a condenser set downward for distillation. Add a few boiling chips and heat the flask on a wire gauze, so that the mixture boils very gently and the vapors do not rise into the fractionating column. After the mixture has boiled gently for 1 hr increase the heating slightly so that the water formed in the reaction and a part of the acetic acid distill off *very slowly at a uniform rate.*[3] It is convenient to observe and control the rate of distillation by using a small graduated cylinder as the receiver.

During the first hour of distillation the temperature of the distilling vapor should not rise above 104–105°, and at the end of the hour the volume of distillate should not exceed 12–14 ml. After the first hour the vapor

[3] The success of this preparation depends upon careful fractionation to remove the water and acetic acid, without loss of ammonium acetate or acetamide. The desired result cannot be achieved by rapid, careless fractionation.

temperature rises gradually to about 118°, as the distillation progresses. Continue the fractionation very slowly and at a uniform rate, until the total distillate amounts to 50 ml.

Allow the residual acetamide in the flask to cool somewhat, and transfer it to a small dry distilling flask attached to an air-cooled condenser, set downward for distillation. Add a few boiling chips and continue the distillation. Collect the following fractions: A up to 195°, B 195–215°, C 215–225°. Fractions B and C will usually solidify completely or in part upon cooling. If either fraction does not solidify completely upon cooling, filter with suction the mixture of liquid and solid, and press the crystals as dry as possible on the filter paper. Combine the crystalline products from B and C, pulverize the crystals, and place them quickly in a dry weighed bottle, since acetamide is somewhat hygroscopic. The liquid portion may be combined with fraction A and refractionated to obtain an additional quantity of acetamide. The yield is 20–25 g.

Recrystallize 2–3 g of crude acetamide from a mixture of 3 ml of benzene and 1 ml of ethyl acetate. Warm the tube in a water bath until all has dissolved, and cool rapidly in an ice bath. Filter rapidly with suction, press the crystals between filter papers, and dry in a desiccator. The recorded melting point is 80–81°.

Carry out the test reactions with the crude acetamide.

## Reactions of Acid Amides

*(A)* Boil about 0.25 g of acetamide with 3–4 ml of 10 percent aqueous sodium hydroxide and cautiously note the odor of the vapor.

*(B)* Boil about 0.25 g of acetamide with 3–4 ml of 10 percent sulfuric acid and cautiously note the odor of the vapor.

*(C)* Dissolve a gram of acetamide in 2–3 ml of water with a spatula, add a very small amount (not over 0.5 g) of yellow mercuric oxide, and warm the tube gently. A soluble, covalent mercury derivative, $(CH_3-CO-NH)_2Hg$, is formed.

### Questions

**1.** What products would be formed upon heating methylammonium acetate in place of ammonium acetate in the above experiment?

**2.** Write equations for two other methods of preparing acetamide.

**3.** Write equations for the action of the following reagents upon acetamide: sodium hypobromite solution; phosphorus pentoxide; sodium nitrite and hydrochloric acid.

**4.** Compare the relative volatility of $R-CO_2H$, $R-CO-OCH_3$,

R−CO−Cl, R−CO−$NH_2$, and R−C≡N, where R− is a simple alkyl group. Why does the sequence not correspond to that of molecular weights?

**Acetonitrile.** ***Dehydration of an Amide*** *(Optional).* In a dry 125-ml distilling flask place 0.2 mole (30 g) of phosphorus pentoxide,[4] and add 0.3 mole (18 g) of *dry*, pulverized acetamide. Shake thoroughly to mix the two solids and connect the flask with a condenser set downward for distillation. At the lower end of the condenser, arrange a large test tube or a small flask to serve as receiver. Carefully heat the flask directly with a *small* barely luminous flame; hold the burner in the hand and keep the flame in constant motion. The reaction mixture foams during the heating.

➤**CAUTION:** Phosphorus pentoxide is extremely hygroscopic and must be kept out of contact with the air. It must be handled carefully since it will produce painful burns if allowed to come in contact with the skin.

After the mixture has been heated for 5 or 6 min, distill the acetonitrile with a barely luminous flame as before and collect the distillate in the receiver. Treat the distillate with half its volume of distilled water, and add solid anhydrous potassium carbonate in small portions until the aqueous layer is saturated (about 8–9 g of potassium carbonate will be required for every 10 ml of water). Cool the tube in a bath of ice water during the addition of potassium carbonate to prevent loss of acetonitrile.

Allow the excess of solid potassium carbonate to settle, and decant the liquid into a small separatory funnel. Draw off and discard the lower aqueous layer. Transfer the upper layer of acetonitrile (methyl cyanide) through the mouth of the funnel to a very small distilling flask and add 0.5–1 g of phosphorus pentoxide as a drying agent. Add a tiny boiling chip, distill on a wire gauze, and collect in a weighed bottle the portion boiling at 79–82°. The yield is 8–10 g.

## Questions

**1.** Write equations for another method of preparing nitriles.

**2.** Show by means of equations the reaction of acetonitrile with the following reagents: (a) lithium aluminum hydride; (b) phenylmagnesium bromide, followed by dilute acid; (c) ethanol + dry hydrogen chloride; (d) isobutylene + strong sulfuric acid, followed by hydrolysis (the Ritter reaction).

[4] Under the conditions of this reaction, phosphorus pentoxide is converted to a mixture of acids consisting largely of $H_4P_2O_7$.

**3.** When acetonitrile is treated in dry ether with sodium metal catalyst it forms a crystalline dimer, $C_4H_6N_2$, mp 52–53°. The dimer dissolves in cold concentrated hydrochloric acid to give ammonium chloride and $\beta$-keto-butyronitrile ($CH_3-CO-CH_2-CN$). Write a structure for the dimer.

# EXPERIMENT 14

# Urea

Carbonic acid forms a monoamide, carbamic acid ($H_2N{-}CO_2H$), and a diamide, carbamide or urea ($H_2N{-}CO{-}NH_2$). Urea is formed in the human body as an end product of the metabolism of amino acids derived from proteins and is excreted in the urine. The average quantity of urea excreted by an adult man is about 25–30 g per day, depending upon the protein intake. A small amount of nitrogen is excreted in the form of uric acid (a purine). In reptiles and birds uric acid is the principal end product of protein metabolism.

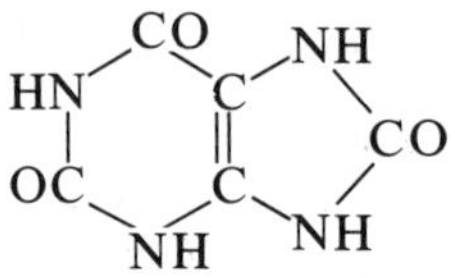

```
      CO    NH                R OC—NH
 HN      C      \              \ /    \
  |      ||      CO              C      CO
 OC      C      /              / \    /
      NH    NH                R'  OC—NH
```

Uric Acid

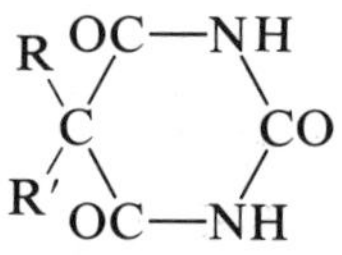

Barbituric Acids

Urea was isolated from urine by Rouelle in 1773 and synthesized by Wöhler in 1828 from ammonium cyanate. A few years earlier he had synthesized oxalic acid from cyanogen (1824). These were the first laboratory syntheses of substances that occur in living organisms. The mechanism of the conversion of ammonium cyanate to urea proceeds through dissociation

$$NH_4NCO \rightleftharpoons NH_3 + H{-}N{=}C{=}O \rightleftharpoons H_2N{-}CO{-}NH_2$$

into ammonia and isocyanic acid, followed by recombination to form the more stable product, urea.

Aliphatic and aromatic primary amines react with isocyanic acid (easily generated in the reaction mixture from potassium cyanate and acetic acid), to form N-alkyl- and N-arylureas. The product from *p*-phenetidine is an

$$p\text{-}C_2H_5O-C_6H_4-NH_2 + H-N{=}C{=}O(KNCO + CH_3CO_2H) \longrightarrow \underset{\text{Dulcin}}{C_2H_5O-C_6H_4-NH-CONH_2}$$

artificial sweetening agent (Dulcin), about 8 times as sweet as sodium N-cyclohexylsulfamate (Sucaryl, Sodium Cyclamate) and 250 times as sweet as sucrose.

Urea is manufactured industrially from carbon dioxide and excess ammonia under pressure. It is used in large amounts as fertilizer and in the manufacture of urea-formaldehyde resins.

By the action of cold concentrated sulfuric acid, urea nitrate is converted to nitrourea, $H_2N-CO-NH-NO_2$, which upon reduction furnishes semicarbazide, $H_2N-CO-NH-NH_2$. The latter is a reagent used for characterization of carbonyl compounds (Experiment 9). The reaction of urea with esters of disubstituted malonic acids leads to the corresponding barbituric acids. The sodium salts of these acids are widely used as synthetic drugs (Phenobarbital, Nembutal, etc.).

## Inclusion Compounds (Clathrate Complexes)

Many materials have the property of forming crystals that have open spaces between the molecules in the crystal lattice. Crystallization of such a material in the presence of another substance of the proper size and shape may incorporate the second component into the holes in the crystal lattice to form an *inclusion compound*.[1] In spite of the name, the enclosed molecules in inclusion compounds usually are held by weak intermolecular forces rather than stronger interactions like hydrogen bonds or chemical bonds. Since it is largely the dimensions of the included substance relative to the geometry of the crystal lattice of the host compound that determines the stability of the complex, some rather unusual combinations are observed. For example, hydroquinone forms stable inclusion compounds with argon and many other small chemically inert species. Methane is quite insoluble in water but forms an inclusion compound with ice.

Inclusion compounds formed by urea are useful for certain separations and purifications. Urea forms complexes with straight chain alkanes having more than six carbon atoms, and with many straight chain alcohols, ketones, alkyl halides, acids, and esters. Since branched chain and cyclic compounds usually do not form inclusion compounds with urea (although they do so with thiourea), this difference in behavior may be used to separate straight chain compounds from branched chain compounds. Where differences like this exist, formation of an inclusion compound becomes an effective method of purification.

[1] For a survey of inclusion compounds, with excellent diagrams, see Brown, *Scientific American*, July 1962, pages 82–92; also, Fieser, *J. Chem. Educ.*, **40**, 457 (1963); **42**, 408 (1965).

Urea complexing can be used to increase the octane number of gasoline by removal of straight chain hydrocarbons, which have low octane numbers. The method has been used also to separate oleic acid, which has one double bond, from linoleic acid, which has two double bonds. The example given below illustrates the ease with which the complexes are formed and subsequently decomposed to regenerate the included compound.

## 14a Wöhler's Synthesis of Urea

***Ammonium Cyanate and Urea.*** Dissolve 0.25 mole (20 g) of potassium cyanate (KNCO) in 100 ml of distilled water and add with swirling a solution of 0.2 mole (26 g) of ammonium sulfate in 45 ml of distilled water. After swirling the mixture vigorously, filter off the precipitated potassium sulfate with suction, transfer the filtrate to an evaporating dish or large beaker, and evaporate it to dryness on a steam bath. Stir the material frequently during the evaporation.

Powder the dry solid residue and transfer it to a 200-ml flask. Add 30 ml of methanol, heat to boiling on a steam bath, and filter the mixture rapidly on a Büchner funnel. Return the solid material to the flask and extract again with 30 ml of boiling methanol. Finally, wash the residue on the filter with 10 ml of boiling methanol. Combine the filtrates and chill *thoroughly* in an ice-salt mixture. Filter the crystals of urea rapidly with suction and allow them to dry in the air. The yield is 6–6.5 g. A small additional quantity of urea can be obtained by re-extracting the residue with the mother liquor from which the urea was isolated and concentrating the solution to half volume on a steam bath. Urea melts at 132–133°.

***Reactions of urea.*** *(A)* Dissolve about 0.25 g of urea in 5 ml of water, add one drop of concentrated hydrochloric acid and a small crystal of sodium nitrite. Write stepwise equations for the reactions that occur.

*(B)* Dissolve about 0.25 g of urea in 5 ml of water and add about 5 ml of a solution of sodium hypochlorite (Clorox) or sodium hypobromite.[2]

*(C)* Dissolve about 0.25 g of urea in a few drops of water, add 1 ml of concentrated nitric acid, and cool. What is the precipitate?

*(D)* In a test tube heat gently about 0.25 g of urea until melted and warm the liquid gently for 2 min so that bubbles are produced. Cool the tube and add 10 ml of water, followed by one drop of very dilute copper sulfate solution and one drop of 10 percent aqueous sodium hydroxide. This is known as the biuret test.

[2] Prepared by treating 5 ml of bromine water *dropwise* with 10 percent aqueous sodium hydroxide until the bromine color is just discharged.

***p-Ethoxyphenylurea: Dulcin*** *(Optional)*. The following procedure is a general method for the preparation of N-arylureas,[3] but is not suitable for the simple N-alkyl- and N,N-dialkylureas because they are very soluble in water.

In a 50-ml Erlenmeyer flask dissolve 0.01 mole (1.4 g, 1.4 ml) of *p*-phenetidine in 10 ml of water and 2 ml of glacial acetic acid. While swirling the solution vigorously add dropwise a solution of 0.02 mole (1.6 g) of potassium cyanate (or 1.3 g of sodium cyanate) dissolved in 5 ml of water. As soon as a precipitate of the arylurea begins to appear add the remainder of the cyanate solution at once and mix the contents thoroughly. Allow the mixture to stand with occasional shaking for an hour or longer, add 5 ml of water and cool the mixture in an ice bath. Collect the product on a suction filter, wash it with a little cold water, and recrystallize it from hot water (~30 ml per gram) with addition of decolorizing carbon. Avoid prolonged boiling since this leads to the formation of a little N,N′-di(ethoxyphenyl) urea. The recorded melting point of Dulcin is 173–174°.

Prolonged use of Dulcin as a sweetening agent may lead to toxic effects resulting from *p*-aminophenol that is formed by chemical transformations in the body.

***Eicosane-Urea Complex*** *(Optional)*. In a 250-ml Erlenmeyer flask dissolve 20 g of urea in 80 ml of methanol by warming the mixture gently on a steam bath and swirling the contents of the flask. Into the warm solution introduce 10 ml of a 50 percent solution of technical grade eicosane ($n$-$C_{20}H_{42}$) in xylene and continue to swirl the mixture until the complex begins to separate. Allow the solution to cool to room temperature and to stand at least an hour.

Collect the crystals and wash them on the filter with 25 ml of cold methanol, which is used also to rinse the flask in which the reaction was performed. Press the crystals as dry as possible and allow to dry for a short time in the air. The urea-eicosane complex contains approximately 15 moles of urea to 1 mole of the hydrocarbon (3.2 g per gram of eicosane). The yield is 16–20 g of the complex. Note the shape of the urea complex crystals.

Regenerate the hydrocarbon by adding the urea complex slowly to 50 ml of warm water contained in a 250-ml beaker. Cool the beaker to 15° and after the eicosane has solidified collect it on a suction filter. Wash the product with methanol and allow it to dry in the air. Take the melting point of the crystals.[4] The yield of purified eicosane is about 4 g. Note the crystal form and compare it to that of the urea complex. The melting point of eicosane is 36.5°.

[3] Kurzer, *Organic Syntheses*, Collective Volume **IV**, 49 (1963).

[4] This is difficult because of the low melting point. One technique that works well here is to prepare a second capillary tube that will fit inside the melting point tube. The fine capillary is pushed into the sample to form a plug inside its end. The fine capillary containing the plug is then dropped into the melting point tube and the melting point determined in the usual manner.

### Questions

**1.** Indicate a synthesis of urea starting from coke (carbon) not involving carbon dioxide.

**2.** Write a sequence of reactions illustrating formation of urea- formaldehyde polymers.

**3.** Indicate the structure of the products formed by reaction of isocyanic acid (HNCO) with the following: (a) hydrogen chloride; (b) *t*-butyl alcohol; (c) N,N-dimethylurea; (d) 2-methyl-2-propyl-1,3-propanediol (the product is the drug Meprobamate, or Equanil).

**4.** Write structures for the isomeric O-methyl and N-methyl derivatives of urea, and explain why the former is a very much stronger base than the latter.

**5.** Fulminic acid, to which the highly explosive covalent fulminate salts are related, is a structural isomer of isocyanic acid. Fulminic acid adds a molecule of hydrogen chloride to form a compound that yields formic acid and hydroxylamine hydrochloride upon hydrolysis. What is the structure of fulminic acid and its hydrogen chloride adduct?

## 14b Urea from Urine

***Urea Nitrate.*** Under a hood, evaporate in a porcelain dish 500 ml of fresh urine,[5] first by heating on a wire gauze over a burner (until the volume is reduced to about 150 ml) and *finally by heating on a steam bath* until a thick syrup remains. Treat the syrup and solid residue with 100 ml of ethanol and stir the mixture thoroughly. Warm on a steam bath and stir well to obtain a good extraction. Decant the hot ethanol through a fluted filter paper into a porcelain dish, and repeat the extraction as before with two additional 50-ml portions of ethanol. To the ethanolic extracts add 50 ml of water and concentrate on a steam bath to a volume of 25–30 ml. All of the ethanol *must* be removed before the next step. Cool in an ice-salt mixture and while stirring thoroughly add drop by drop, from a separatory funnel, pure concentrated nitric acid until the precipitate of urea nitrate ceases to increase in amount. Usually about 25 ml of nitric acid is required. During the addition of the nitric acid it is very important to keep the mixture well stirred and thoroughly cooled.

Collect the precipitated urea nitrate with suction (preferably with a hardened filter paper), wash with a small amount of ice-cold concentrated

[5] The urine should be preserved from decomposition by adding about 10 ml of toluene if the isolation of urea nitrate is not to be started within a few hours from the time the urine is collected. If toluene is used, it must be separated before evaporating over a burner. If the entire evaporation is effected on a steam bath, the toluene need not be removed.

nitric acid, and press the crystals as dry as possible with a clean cork or flat glass stopper. Dissolve the crude urea nitrate in about 85 ml of boiling water and add *5–10 drops* of dilute potassium permanganate solution (0.5 percent). Cool slightly, add 4–5 g of decolorizing carbon, and boil the solution gently for about 20 min. Keep the volume constant by adding a small amount of water from time to time as evaporation occurs. Filter the hot solution through a fluted filter and evaporate the filtrate on a steam bath, to a volume of about 25 ml. Cool in an ice bath, stir thoroughly, and add drop by drop about 15–20 ml of concentrated nitric acid. Collect the purified urea nitrate with suction (preferably with a hardened filter paper) and press it as dry as possible with a clean cork or flat glass stopper. Wash the crystals with a small amount of ice-cold concentrated nitric acid. The resulting crystals of urea nitrate should be practically colorless; if a colored product is obtained, the purification should be repeated. The yield is 6–10 g of the dried crystals.

***Urea.*** Weigh the crystals of purified urea nitrate and in a small beaker treat with 15 ml of distilled water. Then add powdered barium carbonate in small portions, with stirring; for each gram of urea nitrate use 1.2 g of barium carbonate, which is an excess of 50 percent over the calculated amount necessary to liberate the urea from the nitrate. It is essential to use an excess of barium carbonate or the urea will be decomposed on warming.

After the evolution of carbon dioxide has ceased, add about 1 g of decolorizing carbon and heat on the steam bath for about 15 min. Filter the hot solution through a small fluted filter and wash the carbon with a few milliliters of hot water. Transfer the filtrate (which should be clear and practically colorless) to a small porcelain dish and evaporate nearly to dryness on a steam bath. During the evaporation, crystals of barium nitrate usually separate. Extract the residue without removing the crystals with two 10-ml portions of hot ethanol. Evaporate the ethanolic solution of urea on a steam bath until crystals begin to separate; allow the solution to cool and to stand for some time in a cool place to obtain a good separation of the urea. Collect the crystals of urea from the cold solution on a small suction filter, and allow them to dry in the air. The product should be colorless; if necessary recrystallize from a small quantity of hot ethanol with the addition of about 0.2 g of decolorizing carbon. The yield is 4–6 g. Urea melts at 132–133°.

Carry out the *Reactions of Urea* given under Wöhler's synthesis. The *Questions* are found at the end of that section.

EXPERIMENT

# 15

# The Nitrile Synthesis of a Carboxylic Acid

Two important preparative methods for aliphatic carboxylic acids are the Grignard synthesis and the nitrile synthesis. Both methods make use of alkyl halides as starting materials and add one additional carbon atom to the molecule.

The Grignard synthesis involves conversion of the alkyl halide to an organomagnesium halide, followed by reaction with carbon dioxide and subsequent acidification of the halomagnesium salt. In the nitrile synthesis

$$C_4H_9{-}Br \xrightarrow[\text{Ether}]{\text{Mg}} C_4H_9{-}MgBr \xrightarrow{CO_2} C_4H_9{-}CO_2MgBr \xrightarrow{H^+} C_4H_9{-}CO_2H$$

$$C_4H_9{-}Br \xrightarrow{\text{NaCN}} C_4H_9{-}C{\equiv}N \xrightarrow[H^+]{H_2O} C_4H_9{-}CONH_2 \xrightarrow[H^+]{H_2O} C_4H_9{-}CO_2H$$

the alkyl halide is converted to the nitrile by reaction with sodium or potassium cyanide and the nitrile is hydrolyzed to the acid by heating with mineral acids or with aqueous alkalies.

The Grignard synthesis is of broader scope—it can be applied to primary, secondary, and tertiary alkyl halides, and also to aryl and allylic halides. The nitrile method gives satisfactory results with primary halides but much less favorable results with secondary halides. With secondary and tertiary alkyl halides, which have slower rates of reaction with cyanide ion, alkene formation becomes an increasingly important side reaction owing to the high alkalinity of solutions of sodium and potassium cyanide. Aryl halides are too inert to react with cyanide ion under ordinary laboratory conditions.

In the nitrile synthesis a trace of the isomeric carbylamine or isocyanide $(R{-}N{\equiv}C)$ is produced and its presence gives the crude nitrile a characteristic disagreeable odor. This impurity can be removed by shaking the product with 50 percent sulfuric acid, which hydrolyzes the isocyanide without affecting the nitrile.

## 15a *n*-Valeronitrile

In a 500-ml round-bottomed flask place 0.5 mole (27 g) of powdered sodium cyanide (95 percent), 1.5 g of potassium iodide, and 40 ml of water. Attach a reflux condenser and warm the mixture until the salts have dissolved.

➤**CAUTION:** Sodium cyanide is extremely poisonous and must be manipulated with the greatest possible care. Always wash your hands thoroughly after manipulating it. Do not handle sodium cyanide if you have an open cut on your hand. Take great care not to acidify a solution containing sodium cyanide. When the reaction has been completed pour the residual cyanide solution directly into the drain-pipe of the sink and turn on the water so that it is washed down completely into the drain-pipe.

To the warm cyanide solution add a solution of 0.45 mole (62 g, 47.5 ml) of *n*-butyl bromide in 70 ml of 2-ethoxyethanol (Cellosolve),[1] introduce a few boiling chips, and boil the mixture vigorously under reflux for 2 hr. During this time the reaction mixture becomes dark-brown in color and crystals of sodium bromide separate. Remove the reflux condenser, add 100 ml of water, and connect the reaction flask by means of a fractionating tube (without packing) to a water-cooled condenser arranged for distillation. Distill the reaction mixture at a fairly rapid rate and collect about 120 ml of distillate (combined volume of both layers). The temperature of the distilling vapor is initially about 87–88° and rises gradually to about 99°. At this point (when 120 ml of distillate has been collected) a test portion of the distilling liquid will be a water-ethoxyethanol azeotrope and should be completely soluble in water.

The residual mixture in the reaction flask contains excess sodium cyanide, sodium bromide, ethoxyethanol and colored by-products. After the mixture has cooled *pour it directly into the drain pipe of the sink* (and run water from the tap simultaneously). Rinse the flask thoroughly with water and finally with a little ethanol or acetone.

Transfer the distillate to a separatory funnel and draw off the lower aqueous layer, which may be discarded. Wash the upper layer of crude valeronitrile with 25 ml of water, and then with *cold* 50 percent sulfuric acid

[1] The use of ethoxyethanol (bp 135°) instead of methanol or ethanol, and the addition of potassium iodide, permit the reaction to be completed in a relatively short time. The procedure given here is essentially that described by Ferguson, *Chem. Abstracts*, **49**, 8094e (1955).

(prepared by adding 20 ml of concentrated sulfuric acid to 35 ml of water). Shake the mixture thoroughly to insure effective removal of the carbylamine and of ethoxyethanol. Finally, wash the organic layer with 20 ml of 5 percent aqueous sodium bicarbonate and with a little water. Transfer the product to a small dry flask and dry it with 5–6 g of anhydrous calcium chloride. Decant the dried nitrile into a small distilling flask, add a few boiling chips and distill over a wire gauze. After a small forerun collect in a dry, weighed bottle the portion boiling at 138–142°. The yield is 21–23 g. If an appreciable forerun is obtained (more than 4–6 ml collected over the range 120–137°),[2] this fraction should be washed with water, dried, and redistilled.

***Reactions.*** On heating with aqueous acids or alkalies, nitriles are converted first to amides and finally to carboxylic acids and ammonia. The reaction can be stopped cleanly at the amide stage by use of hydrogen peroxide and dilute alkali.

$$R{-}C{\equiv}N + 2H_2O_2(+NaOH) \longrightarrow R{-}CO{-}NH_2 + O_2 + H_2O$$

Direct conversion to esters (alcoholysis) can be effected by heating with

$$R{-}C{\equiv}N + H_2O + CH_3OH + HCl \longrightarrow R{-}CO_2CH_3 + [NH_4]^+Cl^-$$

methanol or ethanol and one molar equivalent of water, in the presence of mineral acid.

Nitriles react with Grignard reagents to give halomagnesium derivatives of ketimines ($R_2C{=}N{-}MgX$) that furnish ketones upon hydrolysis with aqueous acids. Reduction of a nitrile by lithium aluminum hydride, or by catalytic hydrogenation, yields the primary amine, $R{-}CH_2NH_2$.

In the presence of concentrated sulfuric acid, nitriles (including HCN) react with alkenes such as isobutylene, to form adducts that yield primary amines on hydrolysis (the Ritter reaction). This affords an excellent method

$$R{-}CN + CH_2{=}C(CH_3)_2 \xrightarrow[H_2O]{H_2SO_4} R{-}CO{-}NH{-}C(CH_3)_3 \xrightarrow{H_2O} R{-}CO_2H + H_2N{-}C(CH_3)_3$$

of obtaining primary amines that contain a tertiary alkyl group.

## 15b *n*-Valeric Acid

In a 500-ml round-bottomed flask place 0.25 mole (21 g) of *n*-valeronitrile[3] and add a solution of 0.5 mole (20 g) of sodium hydroxide pellets in 65 ml of water. Attach a good reflux condenser, add a few boiling chips, and boil the mixture vigorously for 4 hr (preferably in a hood since ammonia is

[2] Incomplete removal of ethoxyethanol or inadequate drying leads to a low-boiling fraction that contains a substantial amount of the nitrile.

[3] The quantity of valeronitrile obtained in the first step, 18–23 g, may be used for the hydrolysis without altering the quantities of reagents specified.

evolved). Dilute the solution of sodium valerate with 35 ml of water, cool it to 40–50°, and add *cautiously* through the condenser in small portions a warm solution, 0.3 mole (32 g, 17 ml), of concentrated sulfuric acid in 50 ml of water. Swirl the mixture vigorously and cool it to room temperature in an ice bath. Remove the condenser and test the solution to be certain that it is strongly acidic.

Transfer the mixture to a large separatory funnel, rinse the flask with two 20-ml portions of benzene, and pour the rinsings into the separatory funnel. Shake the mixture thoroughly and draw off the aqueous layer. Extract the aqueous layer with two 30-ml portions of benzene and combine all of the benzene solutions. Wash the benzene extracts with 20 ml of water, separate the benzene layer carefully and dry it with 5–6 g of anhydrous magnesium sulfate.[4] Filter the benzene solution into a 250-ml flask and distill off the benzene (and last traces of water) until the temperature of the distilling vapor rises to 95–100°.

Transfer the residual liquid to a small distilling flask and distill over a wire gauze, using an air-cooled condenser. Collect the *n*-valeric acid boiling at 180–186°. The yield is 14–16 g. Low-boiling material may be mixed with twice its volume of benzene, dried with a little anhydrous magnesium sulfate and redistilled.

***Reactions.*** Carboxylic acids may be converted to a variety of functional derivatives, such as esters, acid halides, acid anhydrides, and amides. For identification[5] a convenient method is to form the acid chloride, by means of thionyl chloride, and convert this to the crystalline anilide, or *p*-toluidide. Crystalline esters are also useful, such as the *p*-nitrobenzyl esters ($R-CO_2-CH_2C_6H_4NO_2$).

***n-Valeranilide (Optional).*** In a large dry test tube mix 1 g of *n*-valeric acid with 2 g (1.5 ml) of pure thionyl chloride ($SOCl_2$),[6] attach a small reflux condenser, and boil the solution gently for 30 min. After cooling, dissolve the liquid in 10 ml of *dry* benzene and pour it slowly into a solution of 2 g (2 ml) of aniline in 20 ml of benzene. Warm the mixture gently on a steam bath for a few minutes to complete the reaction.

Decant the benzene solution of the anilide into a separatory funnel and wash it thoroughly with two 5-ml portions of 5 percent hydrochloric acid,

[4] Anhydrous calcium sulfate (Drierite) may be used but not calcium chloride, since the latter contains small amounts of basic salts that will cause precipitation of calcium valerate.

[5] For preliminary identification of volatile acids Duclaux constants may be determined (see Experiment 10, part D).

[6] Another method of preparing the derivative is to place 1 g of the acid and 2 g of aniline (or *p*-toluidine) in a large dry test tube, affix a *short* air-condenser tube, and heat the mixture in an oil bath at 150–160° for 2 hr. The heating should be adjusted so that water vapor escapes during the reaction. The cooled reaction mixture is treated with 20 ml of 10 percent hydrochloric acid and any lumps of material are pulverized so that the excess of amine is extracted. The acidic extract is decanted off and the residual anilide (or *p*-toluidide) is recrystallized from dilute ethanol.

followed by 10 ml of 5 percent aqueous sodium hydroxide, and finally with pure water. Distill off the solvent from a small flask and recrystallize the crude anilide from dilute ethanol, with addition of decolorizing carbon. Dry the product thoroughly and take its melting point. The reported melting point of *n*-valeranilide is 63° (*p*-toluidide, 73°).

## Questions

**1.** Which of the following halides can be converted satisfactorily to the nitrile by reaction with sodium cyanide: (a) $(CH_3)_2CH_2CH_2{-}Br$; (b) $(CH_3)_3C{-}Br$; (c) $C_6H_5{-}Br$; (d) $C_6H_5{-}CH_2Br$?

**2.** Suggest two reactions that can be used to differentiate between *n*-valeronitrile (*n*-butyl cyanide) and the isomeric *n*-butyl isocyanide (*n*-butylcarbylamine).

**3.** Indicate how the following substances may be obtained from *n*-valeronitrile: (a) methyl *n*-valerate; (b) 2-hexanone; (c) *n*-pentylamine; (d) *n*-valeramide.

**4.** Show how *n*-valeric acid may be converted to the following: (a) 1-pentanol; (b) *n*-valeraldehyde; (c) di-*n*-butyl ketone (5-nonanone).

# EXPERIMENT 16

# Sugars

The simple sugars or monosaccharides, such as D-glucose, D-fructose, and D-galactose, are the units from which the more complex carbohydrates are built up by elimination of water (condensation polymerization). Low molecular weight polymers containing two to ten units are called oligosaccharides (oligo means few). Examples of disaccharides are sucrose, maltose, cellobiose, and lactose. Raffinose is a trisaccharide with three units: galactose-glucose-fructose. Starch and cellulose are polysaccharides of high molecular weight, built up of thousands of glucose units.

The monosaccharides are polyhydroxyaldehydes (aldoses) or polyhydroxyketones (ketoses) which exist in solution in an equilibrium involving the open chain structure and a cyclic hemi-acetal form. The latter may be a five-membered ring (furanose) or a six-membered ring (pyranose). Either of the cyclic forms can exist in two stereoisomeric configurations (anomers), designated as $\alpha$- and $\beta$-forms, which arise because a new asymmetric carbon atom is present in the ring structure. The anomers can be isolated in the form of derivatives, such as $\alpha$- and $\beta$-D-glucose penta-acetates or $\alpha$- and $\beta$-methyl-D-glucosides, when a sugar is subjected to acylation or alkylation.

The monosaccharides are mild reducing agents and so are disaccharides that have one free hemi-acetal group. If the two monosaccharide units of a disaccharide are linked together by elimination of water from the hemi-acetal groups of both units, the sugar will not have reducing properties.

## Sugars

### Monosaccharides and Disaccharides

For the following tests prepare a 2 percent solution of the monosaccharides, D-glucose (hydrate) ($C_6H_{12}O_6 \cdot H_2O$) and D-fructose ($C_6H_{12}O_6$)

and of the disaccharides, sucrose ($C_{12}H_{22}O_{11}$), maltose (hydrate) ($C_{12}H_{22}O_{11} \cdot H_2O$), and lactose (hydrate) ($C_{12}H_{22}O_{11} \cdot H_2O$) by dissolving 1 g of each pure sugar in 50 ml of distilled water. Use these solutions for the tests (A), (B), and (C).[1]

Briefly describe and explain each result, using equations where possible. For comparison of the different sugars it is advantageous to record the results in a tabular form.

**(A) Test for Reducing Sugars.** In a test tube place 5 ml of freshly mixed Fehling's solution (equal volumes of No. 1 and No. 2)[2] or 5 ml of Benedict's solution.[3]

Heat the solution to gentle boiling and add 2–3 drops of the glucose solution. Continue to boil gently and observe the result after a minute or two. Continue to add the glucose solution, 2–3 drops at a time, and heat for a short while after each addition, until the deep blue color just disappears. Fehling's solution contains a known amount of copper per milliliter of solution; suggest a method for determining the amount of glucose in a solution of unknown strength.

Repeat the test with the other sugar solutions. Discontinue the test if no reduction is observed after 5–6 drops of the sugar solution has been added.

***Hydrolysis of Sucrose.*** To 10 ml of the 2 percent solution of sucrose, add 1–2 ml of dilute hydrochloric acid and heat on a steam bath for 0.5 hr. Carefully *neutralize* with 10 percent aqueous sodium hydroxide and apply a test for reducing sugars.

## Questions

**1.** What is the precipitate which forms? Compare these results with those obtained in previous tests of simple aldehydes and ketones in Experiment 9.

[1] If other monosaccharides are available, they may be tested similarly: *d*-galactose, $C_6H_{12}O_6$ (an aldohexose), *d*-mannose, $C_6H_{12}O_6$ (an aldohexose), *d*-xylose, $C_5H_{10}O_5$ (an aldopentose), *l*-arabinose, $C_5H_{10}O_5$ (an aldopentose).

[2] **Preparation of Fehling's Solution.** Fehling's solution is made by mixing together, at the moment it is to be used, equal volumes of a solution of copper sulfate (called Fehling's solution No. 1) and an alkaline solution of sodium potassium tartrate (called solution No. 2). For qualitative work these solutions may be prepared as follows: (No. 1) Dissolve 3.5 g of pure copper sulfate crystals ($CuSO_4 \cdot 5H_2O$) in 100 ml of water. (No. 2) Dissolve 17 g of sodium potassium tartrate crystals (Rochelle salt) in 15–20 ml of warm water, add a solution of 5 g of solid sodium hydroxide in 15–20 ml of water, cool and dilute to a volume of 100 ml. The solutions are kept separately and not mixed until time for use, since the mixed solution deteriorates on standing.

[3] **Benedict's Solution.** Benedict has modified Fehling's solution and devised a single test solution that does not deteriorate on standing. For qualitative work this solution may be prepared as follows: Dissolve 20 g of sodium citrate and 11.5 g of anhydrous sodium carbonate in 100 ml of hot water in a 400-ml beaker. Add slowly to the citrate-carbonate solution, with continual stirring, a solution of 2 g of copper sulfate crystals in 20 ml of water. The mixed solution should be perfectly clear (if not, pour it through a fluted filter).

**2.** What is meant by the term *aldohexose*? *ketopentose*?
**3.** Of what value is Fehling's test in classifying an unknown sugar?

**(B) Osazone Test.** In applying the osazone test for comparison of various sugars, it is advisable to perform all tests simultaneously. Place 5 ml of each sugar solution separately in large test tubes, add to each 3 ml of *freshly prepared* phenylhydrazine reagent,[4] and 2–3 drops of saturated aqueous sodium bisulfite (to avoid oxidation). Mix thoroughly, and heat in a beaker of boiling water. Note the time of immersion and record the time in which the osazones[5] precipitate. Shake the tubes from time to time to avoid forming supersaturated solutions of the osazones. Continue the heating in the bath of boiling water for 15 or 20 min, and allow to cool slowly.

From each tube in which crystals have formed, transfer a small quantity of material to a microscope slide, and examine under the microscope (compare with Figure 41). Filter the remainder of the crystals at once with suction, wash with a little water, and let them dry. The osazones discolor on standing. The melting points of the common osazones lie close together ($\sim$200–206°) and are not useful for identification. The osazone formation and precipitation are quite characteristic.

## Questions

**1.** Compare the melting points and rates of precipitation of the osazones of the more common sugars and explain how these values might be used in the identification of an unknown sugar.[5]
**2.** Explain the fact that glucose and fructose give the *same* osazone.

**(C) Acetylation of Glucose.** In a porcelain mortar, pulverize finely 0.015 mole (2.7 g) of anhydrous glucose and mix thoroughly with 1.5 g of fused sodium acetate. Transfer the mixture to a 100-ml round-bottomed flask and add 0.15 mole (15 g, 14 ml) of acetic anhydride. Heat the mixture under a

[4] **Preparation of Phenylhydrazine Reagent.** Dissolve 4 g of phenylhydrazine hydrochloride ($C_6H_5-NH-NH_2 \cdot HCl$) in 36 ml of water and add 6 g of sodium acetate crystals and a drop of glacial acetic acid. If the resulting solution is turbid, add a small pinch of decolorizing charcoal, shake vigorously, and filter the solution. The reagent deteriorates rapidly.

A satisfactory reagent may also be made by dissolving 4 g (4 ml) of phenylhydrazine (the free base is a liquid) in a solution of 4 g (4 ml) of glacial acetic acid in 36 ml of water, and clarifying as above with a pinch of decolorizing charcoal. This procedure is less desirable owing to the poisonous character of phenylhydrazine (vapor or liquid). If the liquid comes in contact with the skin, it must be washed off at once, using 5 percent acetic acid, followed by soap and water.

[5] Consult Shriner, Fuson, and Curtin, *The Systematic Identification of Organic Compounds*, John Wiley and Sons, Inc., New York, 5th edition (1964).

The precipitate formed with D-mannose (0.5–1 min) is the sparingly soluble phenylhydrazone. Sucrose must undergo hydrolysis before precipitation of glucosazone occurs.

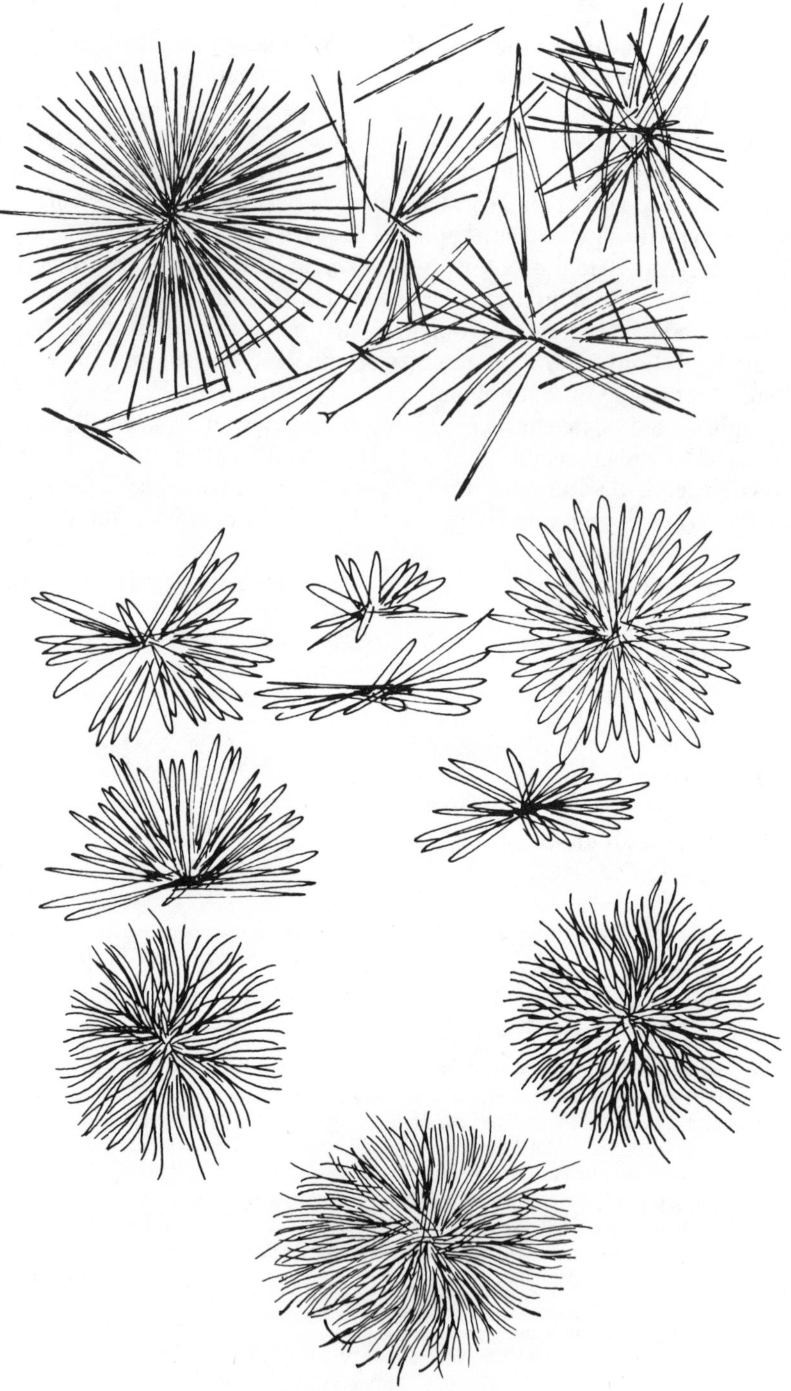

**Figure 41.** Crystals of Osazones

reflux condenser on a steam bath for 2.5 hr with occasional vigorous agitation. Pour the warm solution in a thin stream, with vigorous stirring, into 125 ml of ice-cold water in a beaker. Disintegrate any lumps of the crystalline precipitate and allow the finely divided material to stand in contact with the water, with occasional stirring, until the excess acetic anhydride has been hydrolyzed. This will require about 2 hr. It is desirable to permit the reaction mixture to stand overnight or longer. Collect the crystals with suction and press as dry as possible with a clean cork or flat glass stopper. Transfer the crystals to a beaker, mix thoroughly with about 125 ml of water, and allow to stand with occasional stirring for about 2 hr longer. Collect the crystals with suction and press as dry as possible on the filter. Recrystallize the crude $\beta$-glucose penta-acetate from about 15 ml of methanol. The yield of purified product is 3–4 g. The recorded melting point of the purified compound is 135°.

$\alpha$-Glucose penta-acetate can be prepared by acetylation with acidic catalysts. Acids effect the interconversion of $\alpha$- and $\beta$-D-glucose *and* of the penta-acetates, and at *equilibrium* in acetic anhydride, the $\alpha$-penta-acetate constitutes 90 percent of the product. With sodium acetate, a basic catalyst, the $\beta$-penta-acetate predominates because sodium acetate catalyzes interconversion of $\alpha$- and $\beta$-D-glucose but not the penta-acetates and the rate of acetylation of $\beta$-D-glucose is much faster than that of the $\alpha$-isomer.

**(D) Benzoylation of Glucose.** In a 125-ml bottle or Erlenmeyer flask place 25 ml of the 2 percent glucose solution, 14 ml of 10 percent aqueous sodium hydroxide, and 2 ml of benzoyl chloride (*caution—irritating vapor!*). Shake the bottle vigorously at frequent intervals until the odor of benzoyl chloride no longer can be detected (test cautiously); at least 15 min will be required. If necessary add a little more alkali to keep the solution alkaline. When the product has crystallized collect the crystals by suction filtration and wash them thoroughly with water. If the product tends to remain gummy, it may be allowed to stand in contact with the alkaline solution until the following laboratory period.

Recrystallize the crude product from ethanol and determine the melting point. The recorded melting point of D-glucose penta-benzoate is 179°.

## Questions

**1.** What is the significance of the fact that five acetyl or benzoyl groups can be introduced into the molecule of glucose or fructose?

**2.** What is the significance of the fact that only eight acetyl or benzoyl groups (and not ten) can be introduced into the molecule of the disaccharide, sucrose?

**3.** Give an accurate definition of the term carbohydrate.

**4.** What sugars are present in the solution after hydrolysis of sucrose? Will the sugars form an osazone with phenylhydrazine?

**5.** Why is the hydrolysis of sucrose sometimes referred to as the inversion of sucrose? What is invert sugar?

**6.** What agents other than acids promote the hydrolysis of complex carbohydrates?

# EXPERIMENT 17

# The Friedel-Crafts Reaction

## Alkylation of Benzene

The Friedel-Crafts reaction, discovered in 1877, has become the most important method for introducing alkyl and acyl groups into benzene and other aromatic compounds. It depends upon the remarkable catalytic activity of anhydrous aluminum chloride. Other catalysts having similar activity (boron trifluoride and hydrogen fluoride) and useful modifications of the reaction have been developed in more recent studies.

Typical alkylation reactions make use of alkyl halides (including aralkyl types such as $Ar{-}CH_2Cl$, $Ar{-}CHCl_2$) and alkenes. Halogen atoms directly attached to an aromatic ring are inert and do not take part in the reaction.

$$C_6H_6 + CH_3CH_2Cl\ (AlCl_3) \longrightarrow C_6H_5{-}CH_2CH_3 + HCl$$

$$C_6H_6 + CH_2{=}CH_2\ (AlCl_3 + HCl) \longrightarrow C_6H_5{-}CH_2CH_3$$

Aliphatic halides with two or more reactive halogen atoms furnish diarylated alkanes and higher arylated compounds.

The role of the catalyst is to generate a highly reactive intermediate (carbonium ion) that attacks the mobile electrons of the aromatic system. Subsequent transformations lead to formation of the alkylbenzene and hydrogen chloride, and regeneration of the catalyst. Further alkylation tends

$$C_2H_5{-}Cl + AlCl_3 \rightleftarrows C_2H_5^+[AlCl_4]^-$$

$$C_2H_5^+[AlCl_4]^- + C_6H_6 \rightleftarrows [C_2H_5{-}C_6H_6]^+[AlCl_4]^-$$

$$[C_2H_5{-}C_6H_6]^+[AlCl_4]^- \rightleftarrows C_2H_5{-}C_6H_5 + AlCl_3 + HCl$$

to occur because the ease of reaction (nucleophilic activity) increases with successive attachment of alkyl groups into the aromatic ring. Monoalkylation is favored by using a large excess of benzene.

As might be expected from the carbonium ion mechanism, primary alkyl halides (and some secondary halides) give rise to products of rearrangement. *n*-Propyl halides give almost entirely isopropylbenzene (cumene), *n*-butyl halides *s*-butylbenzene, and isobutyl halides *t*-butylbenzene. Similar products of rearrangement are formed from analogous pentyl and higher halides.

In industry alkenes are used for alkylation of benzene, under conditions different from the typical Friedel-Crafts' synthesis. The catalysts used may be a liquid complex of aluminum chloride-hydrogen chloride-hydrocarbon, phosphoric acid supported on a solid substrate, anhydrous hydrogen fluoride, or concentrated sulfuric acid. Propylene and benzene, over a phosphoric acid catalyst at 250° and under high pressure, give isopropylbenzene (cumene). Cumene is an intermediate for an efficient process for the manufacture of phenol (via cumene hydroperoxide). Cyclohexene and benzene, with sulfuric acid as catalyst, give cyclohexylbenzene.

In laboratory preparations using reactive tertiary halides, it is convenient to form anhydrous aluminum chloride directly in the reaction mixture, from amalgamated aluminum metal (as in Experiment 17b).

## 17a Ethylbenzene

In a dry 500-ml round-bottomed flask fitted with an upright condenser, place 0.4 mole (44 g, 31 ml) of ethyl bromide and 2 moles (156 g, 175 ml) of benzene. To provide for the acid vapors evolved during the reaction, fit the top of the condenser with a drying tube filled with calcium chloride and connect the open end of the drying tube to a gas absorption trap (Figure 38, page 170).

Add 0.1 mole (13 g) of pulverized *anhydrous* aluminum chloride and allow the reaction to proceed in the cold, shaking the flask occasionally. When the evolution of halogen acid has slowed down, warm the flask gently on a steam bath, and finally reflux the mixture on a steam bath for about an hour. Cool the reaction mixture and pour it with stirring into a mixture of 150 g of ice, 100 g of water, and 24 g (20 ml) of concentrated hydrochloric acid. Stir thoroughly to dissolve completely the aluminum compounds and transfer the mixture to a large separatory funnel. Separate the benzene layer, dry it with 10–15 g of anhydrous calcium chloride, and decant the dried liquid into a 500-ml round-bottomed flask. Fit the flask with a good fractionating column and fractionally distill the mixture (see Fractional Distillation, Figure 14). It is advantageous to distill very slowly at a regular rate and to wrap the column with asbestos paper.

Change the fractionating column to a 100-ml round-bottomed flask and carefully refractionate all of the fractions boiling above 85°. Collect the ethylbenzene fraction from 132–138°. In order to obtain most of the ethylbenzene, it is advisable to repeat the fractionation of the foreruns. The yield is 25–30 g.

*Questions follow Experiment 17b.*

## 17b *t*-Butylbenzene

In a dry 500-ml round-bottomed flask provided with a Claisen adapter holding a small separatory funnel in the central opening and a reflux condenser, place 25 ml of dry benzene. To provide for hydrogen chloride evolved during the reaction, fit the top of the condenser with a drying tube filled with calcium chloride and connect the open end of the drying tube to a gas absorption trap (Figure 38, page 170). In a small dry flask dissolve 0.25 mole (23 g, 27 ml), of *t*-butyl chloride in 30 ml (total, 0.6 mole; 47 g, 55 ml) of benzene.

Shake 0.04 gram-atom (1 g) of thin, narrow strips of bright aluminum metal for 3–4 minutes with 10 ml of 5 percent aqueous mercuric chloride (*caution—dangerous poison!*), pour off the aqueous solution, and rinse the metal with several portions of water. Wash the metal with small portions of ethanol and finally with dry benzene. Transfer the amalgamated aluminum to a dry test tube and add 2–3 ml of the benzene solution of *t*-butyl chloride. When reaction has started and hydrogen chloride is being evolved copiously, transfer the contents of the tube quickly to the reaction flask. Pour the remainder of the *t*-butyl chloride solution into the separatory funnel and allow about 10 ml to flow into the reaction mixture. If necessary, warm the flask *gently* to start the reaction but do not add more of the alkyl chloride until the reaction has started. Heating is undesirable (because of the volatility of the alkyl halide) and will not be necessary if the reactants are dry and the aluminum has been activated properly. When the evolution of hydrogen chloride slackens, introduce the remainder of the *t*-butyl chloride solution slowly, over a period of about 20 min, to maintain a steady rate of reaction. The reaction mixture should not warm up appreciably. After all the chloride has been added, allow the mixture to stand for 20 min or longer, with occasional shaking, and then warm it *gently* on a steam bath for 20 min.

Wash the cooled reaction mixture twice with 50-ml portions of dilute hydrochloric acid (7 ml of concentrated acid in 100 ml of water), separate the benzene layer carefully, and dry it with 10–12 g of anhydrous calcium chloride. Decant the dried liquid into a dry 250-ml round-bottomed flask, attach a short fractionating column and a condenser, and fractionally distill the mixture (see Fractional Distillation, Figure 14). Wrap the column with asbestos paper to reduce heat losses and distill slowly at a regular rate. Collect the *t*-butylbenzene fraction from 164–170°. The yield is 14–22 g. Low-boiling fractions collected between 85° and 164° (and also the intermediate fraction up to 195°) may be refractionated to obtain a small additional amount of product. *p*-Di-*t*-butylbenzene can be isolated from the still residue by the procedure described in the next paragraph.

**Di-*t*-butylbenzene.** Transfer the cooled residue from the fractionation to a 25-ml distilling flask and attach a short air-cooled condenser tube. Distill the material slowly by heating the flask directly with a *small*, barely luminous flame, kept constantly in motion, and collect an intermediate fraction up to 195°. Then distill more rapidly and collect the crude disubstituted product from 195–240°. Mix this fraction with twice its volume of methanol and cool in an ice bath. Collect the crystals of *p*-di-*t*-butylbenzene on a small suction filter and wash them sparingly with cold methanol. The yield is 1–3 g. The product melts at 76–78°.

*p*-Di-*t*-butylbenzene can be purified by means of the thiourea inclusion complex, using a solution of about 2 g of thiourea in 15–18 ml of methanol for each gram of the hydrocarbon. The channel of the thiourea matrix is nearly twice the diameter of that of urea, which forms inclusion compounds with straight chain aliphatic compounds (see page 253).

## Questions

**1.** Why is such a large excess of benzene used in this experiment?

**2.** What products would be formed by the reaction of the following alkenes with benzene, in the presence of aluminum chloride: ethylene? isobutylene? cyclohexene?

**3.** Explain why *n*-propyl bromide reacts with benzene in the presence of aluminum chloride, to give mainly isopropylbenzene. Suggest a method for obtaining *n*-propylbenzene.

**4.** Write equations showing how the following compounds can be prepared by the Friedel-Crafts reaction: diphenylmethane; triphenylmethane; *p*-chloroethylbenzene.

## Friedel-Crafts Acylation

Acylation of aromatic compounds by means of the Friedel-Crafts reaction and its modifications is the chief synthetic method for the preparation of aromatic ketones. Aliphatic and aromatic acid chlorides, in the presence of anhydrous aluminum chloride, react with aromatic compounds to furnish alkyl aryl ketones and diaryl ketones.

In acylations with acid chlorides a slight excess over one mole of aluminum chloride is used, since a 1:1 addition compound is formed by reaction of aluminum chloride with the ketone produced in the reaction. Acid anhydrides react in a similar way to produce ketones, usually in better yields

$$C_6H_6 + CH_3{-}CO{-}Cl + AlCl_3 \longrightarrow C_6H_5{-}\underset{\displaystyle CH_3}{\underset{|}{C}}{=}\overset{+}{O}{-}\overset{-}{A}lCl_3 + HCl$$

$$C_6H_6 + (CH_3{-}CO)_2O + 2AlCl_3 \longrightarrow C_6H_5{-}\underset{\displaystyle CH_3}{\underset{|}{C}}{=}\overset{+}{O}{-}\overset{-}{A}lCl_3 + CH_3CO_2{-}AlCl_2 + HCl$$

than are obtained from acid chlorides, but it is necessary to use two moles of aluminum chloride because the organic acid formed in the reaction reacts with aluminum chloride.

The function of the catalyst is to generate a reactive acyl cation (acylonium ion), which attacks the aromatic system in the same manner as other active electrophiles. Acylation does not occur in systems that have a

$$CH_3{-}CO{-}Cl + AlCl_3 \rightleftharpoons CH_3{-}\overset{+}{C}{=}O[AlCl_4]^-$$

deactivating substituent, such as $NO_2$, $CO_2CH_3$, $CO{-}CH_3$, or $C{\equiv}N$. For this reason Friedel-Crafts acylations do not go beyond the introduction of more than one acyl group, since the carbonyl group of the ketone deactivates the molecule for further substitution.

Friedel-Crafts acylations with anhydrides of dibasic acids, such as phthalic, maleic, and succinic anhydrides,[1] furnish ketonic acids that are important synthetic intermediates. Benzene and phthalic anhydride give *o*-benzoylbenzoic acid, which can be decarboxylated to obtain benzophenone or cyclized to produce anthraquinone (see Experiment 44). Derivatives of anthraquinone are used in the manufacture of important vat dyes.

The Fries reaction is a modification of the Friedel-Crafts reaction applicable to the preparation of *o*- and *p*-hydroxyaryl ketones. Esters of phenols with aliphatic or aromatic carboxylic acids, when heated with anhydrous aluminum chloride, undergo a rearrangement in which the O-acyl group enters the *ortho* or *para* position of the ring.[2] The distribution of isomers is influenced by the conditions of reaction; at lower temperatures the *para*

$$C_6H_5{-}O{-}CO{-}R \xrightarrow{AlCl_3} o\text{-}(R{-}CO)C_6H_4{-}OH \text{ and } R{-}CO{-}C_6H_4{-}OH\ (p)$$

[1] Berliner, *Organic Reactions*, **5**, 229 (1949).
[2] Blatt, "The Fries Reaction," *Organic Reactions*, **1**, 342 (1942).

isomer is favored and at higher temperatures the *ortho* isomer predominates.

Aromatic systems containing one or more alkoxyl groups (ethers of mono- and dihydric phenols, and naphthols) are acylated readily by heating with aliphatic acid anhydrides in the presence of iodine (1 mole percent) as catalyst.[3] This method is simple and gives satisfactory yields. Alkoxy derivatives of benzophenone may be obtained by using aroyl chlorides but the yields are lower.

For the acylation of aromatic compounds that cannot conveniently be used in excess as the reaction medium, or for solids such as naphthalene and biphenyl, it is desirable to employ an inert solvent. Tetrachloroethane and nitrobenzene have been used but their boiling points are relatively high, making removal from the product troublesome. 1,2-Dichloroethane (bp 84°) is well suited as an inert solvent but the reaction temperature must be kept below 50° to avoid reaction of the dichloroethane itself. The procedure given below, using acetylation of biphenyl as the example, is a general one that can be used for naphthalene, alkoxybenzenes, and others.[4] With naphthalene this solvent of low polarity strongly favors acylation at the 1-position whereas the highly polar nitrobenzene favors the 2-position.

## 17c Acetophenone

Provide a 500-ml round-bottomed flask with a Claisen adapter holding a small separatory funnel in the central opening and a reflux condenser in the other. Fit the top of the condenser with a drying tube filled with calcium chloride and connect the open end of the drying tube to a gas absorption trap (Figure 38, page 170).

In the flask place 70 g (80 ml) of pure anhydrous benzene and add 0.5 mole (66 g) of pulverized *anhydrous* aluminum chloride (using precautions to avoid exposure of the latter to atmospheric moisture). Through the separatory funnel add slowly a solution of 0.2 mole (21 g, 20 ml) of pure acetic anhydride in 25 ml (total, 1.2 moles; 93 g, 105 ml) of benzene. The addition should be carried out over a period of 0.5 hr while the contents of the flask are agitated thoroughly. The reaction mixture becomes hot and a vigorous evolution of hydrogen chloride takes place.

After all of the acetic anhydride has been added and the reaction has commenced to slacken, heat the mixture for 0.5 hr on a steam bath. Cool and pour the contents of the flask into a well-stirred mixture of 120 ml of concentrated hydrochloric acid and 120 g of ice. If necessary add a little

[3] Chodroff and Klein, *J. Amer. Chem. Soc.*, **70**, 1647 (1948); Dominguez and collaborators, *ibid.*, **76**, 5150 (1954).

[4] *Organikon: Organisch-chemisches Grundpraktikum*, VEB Deutscher Verlag der Wissenschaften, Berlin (East Germany), 7th edition (1967), pp. 306–307.

more water and acid to dissolve any basic aluminum salts that are precipitated. Add 25 ml of ether, shake the mixture thoroughly in a large separatory funnel, and separate the benzene layer. Extract the aqueous portion with 25 ml of ether and add this to the benzene solution. Wash the benzene solution (including the ether extract) with 40 ml of 10 percent aqueous sodium hydroxide, separate the benzene layer carefully and dry it with 10–12 g of anhydrous magnesium sulfate. Recover the solvent by distillation from a steam bath and pour the residual oil into a small distilling flask. Add a boiling chip and distill on a wire gauze, using an air-cooled condenser. Since there is likely to be some benzene remaining in the residual oil, take care during the initial heating to avoid igniting the benzene vapor. Collect the acetophenone boiling at 195–202°. The yield is 14–18 g.

***Reactions.*** Acetophenone possesses a reactive carbonyl group and an active methyl group. It forms the usual carbonyl derivatives (oxime, semicarbazone, etc.) and also enters into aldol-type condensations with aromatic aldehydes to form $\alpha,\beta$-unsaturated ketones, such as benzalacetophenone (Experiment 36). On heating with zinc chloride it undergoes self-condensation to form $\beta$-methylbenzalacetophenone (dypnone).

When acetophenone is condensed with an *o*-hydroxybenzaldehyde, in the presence of hydrogen chloride, the resulting *o*-hydroxybenzalacetophenone undergoes cyclization to form 2-phenylbenzopyrylium (flavylium) chloride. Polyhydroxy derivatives of this parent structure are present in the red and blue pigments of flowers in the form of glycosides (anthocyanins).

$$o\text{-}HO-C_6H_4-CH=CH-CO-C_6H_5 \xrightarrow{HCl} \text{2-phenylbenzopyrylium } (O^+)\text{-}C_6H_5 \; Cl^-$$

Acetophenone reacts with chlorine or bromine to form phenacyl halides, $C_6H_5-CO-CH_2-X$. Sodium hypochlorite converts acetophenone to sodium benzoate and chloroform (the haloform reaction). This reaction may be used in a two-step synthesis of substituted benzoic acids: Friedel-Crafts acetylation to form a substituted acetophenone, followed by hypochlorite (or hypobromite) oxidation to the carboxylic acid.

An unusual reaction takes place upon heating acetophenone with ammonium polysulfide. It is converted by dehydrogenation and rearrangement to the amide of phenylacetic acid (the Willgerodt reaction).[5] Propiophenone and *n*-butyrophenone undergo a similar transformation leading to homologs

$$C_6H_5-CO-CH_3 + 2(NH_4)_2S + S \xrightarrow{180^\circ} C_6H_5-CH_2CO-NH_2 + 3NH_4SH$$

$$C_6H_5-CO-CH_2CH_2CH_3 \longrightarrow C_6H_5-CH_2CH_2CH_2-CO-NH_2$$

[5] Carmack, *Organic Reactions*, **3**, 83 (1946); Wegler, Kühle and Schäfer, *Newer Methods of Preparative Organic Chemistry*, **3**, 1 (1964).

of phenylacetamide in which the amide group appears at the end of the side chain.

**Conversion of Acetophenone to Benzoic Acid** *(Optional)*. In a 125-ml Erlenmeyer flask place 40 ml of commercial 5 percent sodium hypochlorite solution[6] and 1 g (1 ml) of acetophenone, and swirl the mixture vigorously. The solution becomes warm and the odor of chloroform can be detected (sniff cautiously). Continue to shake the mixture and warm the flask in a water bath at 60–65° for 5–10 min. This should volatilize most of the chloroform. To destroy an excess of hypochlorite add a few drops of acetone to the warm solution and swirl the flask. Add a small amount of decolorizing carbon, heat the solution to boiling, and filter through a fluted filter paper. To the hot filtrate add 5 ml of concentrated hydrochloric acid, stir the mixture vigorously, and cool it in an ice bath. Collect the benzoic acid and wash it with a little cold water. The yield is 0.5–0.6 g.

*Questions follow Experiment 17d.*

# 17d 4-Acetylbiphenyl

The preparation of methyl *p*-xenyl ketone is a specific example of a general procedure for acylation of aromatic compounds in the presence of an inert solvent.

Assemble an apparatus with a 250-ml round-bottomed flask fitted with a Claisen adapter; place a separatory funnel in the central opening of the adapter and an upright condenser in the side-arm. Attach a drying tube filled with calcium chloride at the top of the condenser and connect the open end of the drying tube to a gas absorption trap (Figure 38, page 170) to dispose of the hydrogen chloride evolved during the reaction.

In the flask place 40 ml of 1,2-dichloroethane and 0.12 mole (16 g) of pulverized *anhydrous* aluminum chloride. Replace the condenser and add dropwise by means of the separatory funnel 0.11 mole (8.7 g, 8 ml) of acetyl chloride, while cooling the flask in an ice bath. Over a period of about 15 min add a solution of 0.10 mole (15.5 g) of biphenyl in 50 ml of 1,2-dichloroethane. Allow the reaction mixture to stand for an hour, with occasional shaking, and then warm the flask gently in a water bath at 45–50° for about 20 min.

To decompose the ketone-aluminum chloride complex, pour the cooled reaction mixture cautiously into a well-stirred mixture of 50 g of chipped ice and 50 g (40 ml) of concentrated hydrochloric acid. Transfer the mixture to

[6] A stabilized solution of sodium hypochlorite (5.2 percent) sold as a laundry bleach, such as *Clorox*, is satisfactory for this preparation.

a separatory funnel, shake well to extract the aluminum chloride from the organic layer, and separate the layers. Rinse the reaction flask with some of the aqueous layer and then extract it with 25 ml of dichloroethane. Wash the combined organic extracts with water, then with 5 percent aqueous sodium hydroxide, and finally with water. Dry the dichloroethane solution with 8–10 g of anhydrous magnesium sulfate and distill off the solvent (bp 84°). Recrystallize the crude ketone from ethanol, with addition of decolorizing carbon. The yield is 10–11 g. The reported melting point of 4-acetylbiphenyl is 120–121°.

## Questions

**1.** What products will be formed from the following reactants, in the presence of aluminum chloride: (a) benzene (in excess) + $ClCH_2CH_2Cl$; (b) benzene (in excess) + $CCl_4$; (c) chlorobenzene and propionic anhydride; (d) toluene and phthalic anhydride; (e) *p*-tolyl benzoate (Fries rearrangement)?

**2.** Indicate how the following substances may be prepared by means of the Friedel-Crafts reaction: (a) diphenylmethane; (b) desoxybenzoin (benzyl phenyl ketone); (c) *m*-bromobenzophenone; (d) *p*-nitrobenzophenone.

**3.** Write equations for reactions suitable for conversion of acetophenone to the following compounds: (a) phenylacetic acid; (b) 1,1-diphenylethanol; (c) α-hydroxy-α-phenylpropionic acid; (d) *m*-aminoacetophenone.

**4.** How could 4-acetylbiphenyl be converted to: (a) 4-ethylbiphenyl? (b) biphenyl-4-carboxylic acid? (c) 4′-amino-4-acetylbiphenyl?

EXPERIMENT

# 18

# Nitration of Aromatic Compounds

Nitration is the most important example of electrophilic aromatic substitution. Aromatic nitro compounds have limited usefulness as such, mainly as high explosives or booster charges (TNT, Tetryl), but they are exceedingly useful as intermediates for the preparation of the corresponding amines and of related compounds (benzidine, *p*-aminophenol, etc.). Nitrobenzene (bp 209°) is a good solvent for organic substances and also dissolves many inorganic compounds ($AlCl_3$, $ZnCl_2$). It is used occasionally as a reaction medium and also as a solvent for recrystallization. Most nitro compounds are dangerously poisonous and must be handled carefully.

Nitrations may be effected by means of pure nitric acid,[1] mixtures of concentrated nitric and sulfuric acids, and solutions of nitric acid in glacial acetic acid or acetic anhydride. Selection of the appropriate nitrating agent and the conditions of reaction is based upon factors such as the reactivity of the compound to be nitrated, its solubility in the nitrating medium,[2] and the ease of isolation and purification of the product.

The mechanism of nitration has been studied extensively and it is known that the active electrophilic species is the nitronium ion, $[O=N=O]^+$. This is formed in the typical nitrating mixtures by a reversible interaction of nitric

[1] Anhydrous nitric acid (sp g 1.50, sometimes called *white* fuming nitric acid) is a colorless liquid boiling at 86°; ordinary concentrated nitric acid (sp g 1.42) is the water-nitric acid azeotrope, bp 120°, containing 70 percent by weight of nitric acid; yellow fuming nitric acid contains 85–90 percent nitric acid with small amounts of oxides of nitrogen; red fuming nitric acid contains relatively large amounts of dissolved oxides of nitrogen.

[2] For a compound that is very sparingly soluble in the nitrating mixture (aromatic hydrocarbons, aryl halides, etc.) the rate of nitration may be governed by its rate of solution in the medium; hence good agitation hastens the reaction.

and sulfuric acids. Attack of the aromatic system by the nitronium ion,

$$HNO_3 + 2H_2SO_4 \rightleftharpoons [NO_2]^+ + [H_3O]^+ + 2[HSO_4]^-$$

$$NO_2^+ + H{-}C_6H_5 \rightleftharpoons [H{-}C_6H_5{-}NO_2]^+ \longrightarrow C_6H_5{-}NO_2 + H^+$$

followed by loss of a proton leads to the nitro derivative. The ease of nitration depends upon the nature of the substituents present; electron-releasing groups ($-OH$, $-NHCOCH_3$, $-CH_3$) facilitate the nitration and electron-withdrawing groups ($-NO_2$, $-CO_2H$, $-C{\equiv}N$) retard the reaction.

Since nitration is not a reversible reaction, the distribution of *ortho*, *meta*, and *para* isomers in the product is controlled by the *relative* rates of substitution at each position. In general, substituents may be divided into three broad categories:

| | |
|---|---|
| Activating, and *ortho/para* directive | $-OH$, $-OCH_3$, $-NHCOCH_3$, $-CH_3$, $-CH_2CO_2CH_3$ |
| Deactivating, and *ortho/para* directive | $-CH_2Cl$, $-Cl$, $-Br$, $-I$ |
| Deactivating, and *meta* directive | $-CCl_3$, $-COCH_3$, $-CO_2CH_3$, $-NO_2$, $-(NR_3)^+$ |

The observed kinetic order of the reaction (in organic solvents)[3] varies as the rate of nitration changes in relation to the rate of formation of the nitronium ion. For the more active compounds, in a large excess of nitric acid, the rate is independent of the concentration of the aromatic compound; the reaction is pseudo-zero order. Compounds of intermediate activity ($C_6H_6$, $C_6H_5{-}Cl$) show kinetic orders intermediate between pseudo-zero and pseudo-first order. For compounds that undergo nitration with difficulty ($C_6H_5{-}NO_2$, $C_6H_5{-}CO_2CH_3$) the rate is proportional to the concentration of the aromatic compound; the reaction is pseudo-first order.[3]

## 18a Nitrobenzene

In a 500-ml round-bottomed flask, place 4 ml of water[4] and add with shaking, a cooled mixture of 0.43 mole (35 g, 25 ml) of concentrated nitric acid and 55 g (30 ml) of concentrated sulfuric acid. Add in small portions, with vigorous swirling, 0.3 mole (23.5 g, 27 ml) of benzene. Cool the flask from time to time in running water so that the temperature is kept at 50–60°

[3] For a detailed discussion of aromatic substitution see Ingold, *Structure and Mechanism in Organic Chemistry*, Cornell University Press, Ithaca (1953) and Gould, *Mechanism and Structure in Organic Chemistry*, Holt, Rinehart and Winston, Inc., New York (1959).

[4] A small amount of water is added to decrease further nitration of nitrobenzene to *m*-dinitrobenzene.

and swirl the mixture vigorously.[2] When all the benzene has been added and the temperature no longer rises on vigorous shaking, attach a water-cooled reflux condenser and heat the flask in a water bath at 60° for 40 min with frequent shaking.

➤**CAUTION:** The vapor of nitrobenzene is poisonous and should not be inhaled. The liquid is a powerful skin poison. If the liquid is spilled upon the skin it should be removed by washing with a little ethanol, followed by soap and warm water.

Cool the mixture to room temperature and pour the contents of the flask into about 250 ml of cold water. Separate the layer of nitrobenzene from the acid solution and wash it once with 75–100 ml of water. Separate *carefully* from the water, dry the product with 8–10 g of anhydrous calcium chloride, and distill on a wire gauze, using an air-cooled condenser. Collect in a weighed bottle the portion boiling at 200–206° (uncor.). The yield is 25–30 g. If a residue remains in the distilling flask it is probably *m*-dinitrobenzene, which is formed when the temperature of nitration is not carefully controlled. Be careful to *avoid superheating the residue* since it may decompose explosively.

*Questions follow Experiment 18c.*

## 18b *m*-Dinitrobenzene

In a 125-ml Erlenmeyer flask mix 0.03 mole (3.6 g, 3 ml) of nitrobenzene and 8 ml of concentrated sulfuric acid. After the nitrobenzene has dissolved add 0.085 mole (5 ml) of concentrated nitric acid (70 percent) and mix by gentle swirling.[5] Heat is evolved and the mixture warms up to 60–70° in about 5 min. After the initial heat evolution has subsided, float the flask in 100 ml of boiling water contained in a 400-ml beaker. Remove the flask occasionally and swirl the contents carefully. After 15 min heating pour the reaction mixture into 100 ml of cold water, with stirring. Collect the crude *m*-dinitrobenzene by suction filtration and wash it *thoroughly* on the filter with two or three 20-ml portions of water. It is important to wash out the admixed nitric acid before recrystallizing the product from ethanol. For effective washing release the suction, mix the crystals thoroughly with the washing liquid, apply suction again and press the crystals firmly.

[5] With small quantities it is permissible to add the nitric acid (or other reactant) all at once, but with large amounts of material one of the reactants should be added gradually in small portions so that the heat evolution can be controlled and the reaction temperature kept within the proper limits (note the procedures used in Experiments 18a, c, d).

➤**CAUTION:** All nitro compounds are poisonous and must be handled carefully. If any nitro compound comes in contact with the skin it should be removed by washing with a little ethanol, followed by soap and water.

Crystallize the washed product from 25 ml of ethanol, reserving a very minute quantity of material to be used as seed crystals. Allow the filtrate to cool until a slight turbidity develops, then introduce a few tiny particles of the seed crystals; swirl the solution gently and set aside to cool slowly. After crystallization is complete, mix the semisolid mass thoroughly, filter with suction and press the crystals on the filter. Wash the crystals with two 5-ml portions of cold ethanol and allow them to dry in the air. The yield is about 3 g.

*Questions follow Experiment 18c.*

## 18c *p*-Bromonitrobenzene

In a 125-ml Erlenmeyer flask, place 0.17 mole (14 g, 10 ml) of concentrated nitric acid and add carefully 18 g (10 ml) of concentrated sulfuric acid. Cool the mixture to room temperature and add in 2–3 portions, 0.05 mole (8 g, 5.5 ml) of bromobenzene. Shake the reaction flask continuously and cool it in running water, if necessary, to keep the temperature between 50° and 60°. After all the bromobenzene has been added and the temperature no longer tends to rise from the heat of reaction, place the flask in a beaker of boiling water and heat for 0.5 hr, with occasional swirling.

➤**CAUTION:** All nitro compounds are poisonous and must be handled carefully. If any nitro compound comes in contact with the skin it should be removed by washing with a little ethanol, followed by soap and water.

Cool the flask to room temperature and pour the reaction mixture into about 100 ml of cold water. Collect with suction the crude *p*-bromonitrobenzene (containing some *o*-bromonitrobenzene), wash it with water, and press on the filter with a clean cork or flat glass stopper. To purify the product crystallize it from 50–60 ml of hot ethanol. The *ortho* isomer is more soluble and remains in solution in the ethanol, while the *para* isomer crystallizes on cooling the solution. The purification can be monitored by TLC (see Exercise 9). The yield of pure *p*-bromonitrobenzene is 5–7 g.

### Questions

**1.** Why is sulfuric acid used in nitration?
**2.** Mention several important reactions that are characteristic of aromatic compounds and which differentiate them from aliphatic compounds.

**3.** What is formed by the reduction of nitrobenzene in the presence of acids?

**4.** What is the structural formula of picric acid? TNT? Tetryl?

**5.** What position will be taken by the entering nitro group when the following substances are nitrated: phenol? acetanilide? toluene? benzaldehyde? benzonitrile? benzoic acid? phenylacetic acid?

## 18d Methyl *m*-Nitrobenzoate

Place 25 g (14.5 ml) of concentrated sulfuric acid in a 125-ml Erlenmeyer flask, cool the acid to 0°, and add 0.05 mole (6.8 g, 6.2 ml) of methyl benzoate, with swirling. While maintaining the internal temperature at 5–15°, by cooling as needed in an ice-water bath, add dropwise a cold mixture of 9 g (5 ml) of concentrated sulfuric acid and 0.085 mole (7 g, 5 ml) of concentrated nitric acid. Swirl the solution during the addition and for 10 min after all of the acid has been added.

➤**CAUTION:** All nitro compounds are poisonous and must be handled carefully. If any nitro compound comes in contact with the skin it should be removed by washing with a little ethanol, followed by soap and water.

Pour the reaction mixture, with stirring, onto about 50 g of cracked ice to precipitate the crude methyl *m*-nitrobenzoate (which contains an appreciable amount of the *ortho* isomer and a trace of *para*). Collect the product with suction and wash it thoroughly on the filter with two or three 15-ml portions of water, to remove nitric and sulfuric acids. For effective washing release the suction, mix the material thoroughly with the washing liquid, apply suction, and press the crystals firmly.

Wash the product finally with two 5-ml portions of *ice-cold* methanol, in the manner described above, and press the crystals thoroughly. Proper washing removes most of the more soluble *ortho* isomer. The crude product weighs about 6–7 g and melts at 74–76°. It may be purified by recrystallization from a small volume of hot methanol;[6] reserve a minute amount of material to be used as seed crystals. The recorded melting point of pure methyl *m*-nitrobenzoate is 78.5°.

*Questions follow Experiment 18c.*

[6] To avoid possibility of contamination through ester exchange it is good practice to avoid using for recrystallization, an alcohol different from that corresponding to the alkoxyl group of the ester.

EXPERIMENT

# 19

# Bromination of an Aromatic Hydrocarbon

In the presence of Lewis acid catalysts such as ferric bromide or aluminum halides, at slightly elevated temperatures, bromine reacts with benzene through an electrophilic *substitution* process. Hydrogen bromide is evolved and bromobenzene, dibromobenzenes (mainly the *para* isomer), and polybromobenzenes are formed. The reaction can be directed largely to a desired degree of substitution by control of the relative proportions of the reactants and of the experimental conditions (temperature, catalyst, etc.). The fact that bromine is a deactivating substituent, diminishing the rate of further substitution, is a favorable factor in controlling the reaction.

The bromination of benzene[1] under ordinary conditions is an ionic reaction. Ferric bromide, easily formed directly in the reaction mixture from iron and bromine, is probably the best catalyst. The mechanism of reaction appears to vary in detail depending upon the reactivity of the substrate and the conditions of the reaction. In a polar solvent such as glacial acetic acid, bromination may occur through attack of the aromatic system by the bromonium ion, $Br^+$, arising from interaction of $Br_2$ with the catalyst. But when

$$C_6H_6 \cdot Br_2 + FeBr_3 \longrightarrow [C_6H_6^{+}](\bar{Br}-Br \cdot FeBr_3)(H)$$

$$\updownarrow$$

$$C_6H_5Br + HBr + FeBr_3 \longleftarrow [C_6H_6^{+}](Br)(H) + Br-\bar{Fe}Br_3$$

[1] Other examples of bromination are given in Experiments 24 (*p*-bromoacetanilide) and 29b (bromo-*p*-acetotoluidide).

molecular bromine is used in the absence of a polar solvent, a more acceptable view is that the Lewis acid acts upon a preformed complex of $Br_2$ with the aromatic hydrocarbon to generate the reactive intermediate species. Irreversible loss of a proton from the aromatic ring leads to the brominated hydrocarbon and hydrogen bromide, with regeneration of the catalyst. Under free radical conditions bromine (and chlorine) attack the side chain of an aromatic hydrocarbon. Thus, toluene at the boiling point with ultraviolet illumination gives benzyl bromide, $C_6H_5{-}CH_2Br$.

Ring halogenation of alkylbenzenes usually gives mixtures of *ortho* and *para* derivatives, in which the *para* isomer predominates (65–80 percent). Since it is often difficult to separate the isomers, pure aryl halides of this type are often prepared by means of the Sandmeyer reaction from the corresponding amines (see Experiment 28) which can be obtained pure.

Direct iodination of aromatic hydrocarbons requires the use of a reagent that will convert molecular iodine to a more active electrophile (possibly $I^+$). Benzene, iodine, and nitric acid give iodobenzene in 85–87 percent yields.

$$2C_6H_6 + I_2 + HNO_3 \longrightarrow C_6H_5{-}I + 2H_2O + 2NO_2$$

More frequently iodo derivatives are prepared by the Sandmeyer reaction.

## Bromobenzene and *p*-Dibromobenzene

Fit a 500-ml round-bottomed flask with a Claisen adapter to which is attached a reflux condenser and a separatory funnel having a *tightly fitting* and well lubricated stopcock (see Figure 37, page 169). Provide the upper end of the condenser with a cork holding a delivery tube bent twice at right angles to form a U. Connect this by means of a short rubber tube to a 50–60-mm funnel inverted above 250–300 ml of water in a large beaker (see Figure 38, page 170), to absorb the hydrogen bromide formed during the bromination. Small amounts of benzene and bromine may be carried over with the acid.

In the 500-ml flask place 0.3 mole (24 g, 27 ml) of benzene and, as a catalyst, add 2–3 g of iron in the form of wire or filings (or $\frac{1}{2}$ g of anhydrous aluminum chloride). Carefully pour 0.25 mole (40 g, 13 ml) of bromine through a small funnel into the separatory funnel. Allow 1–2 ml of bromine to flow into the benzene, and if necessary warm the flask slightly to start the reaction. When the evolution of hydrogen bromide has started, drop in slowly the remainder of the bromine, so that the reaction proceeds quietly. The addition of the bromine should be carried out over a period of about an hour. After all of the bromine has been added, warm the mixture slightly until no red vapors are visible above the liquid.

➤**CAUTION:** Handle bromine carefully since the liquid produces painful burns and the vapors are obnoxious (hood). Use both hands in adjusting the stopcock to avoid loosening and consequent leakage. Treat bromine burns immediately with a liberal quantity of glycerol. Follow this by a thorough washing with water, dry the skin, and apply a healing ointment or salve.

It is advisable to keep a small beaker containing glycerol at hand when manipulating bromine.

Wash the product several times with water, then steam distill by passing steam into the mixture and attaching a good water-cooled condenser (Figure 22, page 64). A partial separation of bromobenzene and *p*-dibromobenzene is made during the steam distillation. Collect the first portion of the distillate until crystals of *p*-dibromobenzene separate in the condenser tube, then change the receiver and collect the second portion, until all of the *p*-dibromobenzene has passed over. Reject the material that does not steam distill.

➤**CAUTION:** During the steam distillation, when *p*-dibromobenzene starts to solidify in the condenser, watch carefully to avoid the formation of a crystalline mass that will completely obstruct the tube. If a solid mass collects in the tube, stop the flow of water through the condenser *momentarily* and drain the water from the condenser jacket. The heat from the vapors will then melt the crystals and the obstruction will be removed. As soon as this occurs, start the water again in the condenser jacket.

From the first portion of the steam distillate separate the heavy liquid, which is chiefly bromobenzene. Dry it with 3–4 g of anhydrous calcium chloride and distill on a wire gauze, using an air-cooled condenser. Reject any material distilling below 140°, and collect for redistillation the fraction boiling at 140–170°. Pour the residue from the distilling flask, while it is still hot, into a small beaker or a watch glass and save it for the isolation of *p*-dibromobenzene. Redistill the 140–170° fraction and collect the portion boiling at 150–160° as bromobenzene. If an appreciable residue remains in the flask save this for isolation of *p*-dibromobenzene. The yield of bromobenzene is about 20 g.

***p*-Dibromobenzene.** Filter the second portion of the steam distillate with suction and press the crystals as dry as possible. Combine these crystals with the residue from the distillation of the bromobenzene, and crystallize from hot ethanol with the addition of 1–2 g of decolorizing carbon. Use about 4 ml of ethanol per gram of material to be crystallized. Filter the hot solution through a fluted filter, cool the filtrate thoroughly, and collect the crystals with suction. The yield of *p*-dibromobenzene is 3–7 g.

***Reactions.*** Bromobenzene (and other aryl halides) may be characterized by conversion to crystalline nitro derivatives, such as *p*-bromonitrobenzene, mp 127° (Experiment 18c) or 1-bromo-2,4-dinitrobenzene, mp 72°.

Another method involves sulfonation (Experiment 20b), conversion of the arylsulfonate salt to the arylsulfonyl chloride, and reaction of the latter with ammonia to form a crystalline bromobenzenesulfonamide.

Aryl bromides are quite unreactive in typical replacement reactions but are used frequently for the preparation of aryl Grignard reagents (see Experiment 41), which are employed often in the laboratory for introduction of aryl groups.

## Questions

**1.** Illustrate by means of equations, *addition* and *substitution* of bromine.

**2.** Indicate what substances would be formed by the action of bromine on the following compounds: benzoic acid; nitrobenzene; chlorobenzene; naphthalene; toluene—in the presence of iron and absence of sunlight; toluene—at the boiling point in the presence of sunlight.

**3.** Write equations for the stepwise conversion of bromobenzene to *p*-bromobenzenesulfonamide.

EXPERIMENT

# 20

# Sulfonation of Aromatic Compounds

Sulfonation of aromatic compounds is usually effected by means of fuming sulfuric acid ($H_2SO_4 + SO_3$, sometimes called oleum). For very reactive compounds, such as phenols, concentrated sulfuric acid suffices. When a sulfonyl chloride is desired, this can be obtained directly by chlorosulfonation with chlorosulfonic acid, $ClSO_3H$,[1] an excellent sulfonating agent (see Experiment 25). The sulfonyl chlorides may be obtained also from arylsulfonate salts by reaction with phosphorus pentachloride.

The detailed mechanism of sulfonation is not as well established as that of other electrophilic aromatic substitution reactions. The attacking electrophilic species appears to be sulfur trioxide, an electrically neutral but electron-deficient molecule. Sulfur trioxide is present in the fuming sulfuric acid commonly used for sulfonations; in concentrated sulfuric acid it may be formed by interaction of two molecules of the acid. Attack of the aromatic

$$2H_2SO_4 \rightleftharpoons SO_3 + [H_3O]^+ + [HSO_4]^-$$

$$SO_3 + H{-}C_6H_5 \rightleftharpoons H{-}C_6H_5{-}SO_3 \rightleftharpoons [C_6H_5{-}SO_3]^- + H^+$$

system, through a series of *reversible* reactions, leads to the dissociated arylsulfonic acid. Evidence from studies with deuterated compounds discloses a pronounced isotope effect, indicating that the rate-determining step is the breaking of a C−H bond of the aromatic ring. Sulfonation, like Friedel-Crafts alkylations (but unlike nitration and halogenation), is a reversible

[1] Sulfuryl chloride ($SO_2Cl_2$) is not used as a sulfonating agent but in the presence of peroxide catalysts it is employed as a chlorinating agent for alkyl side chains (free radical mechanism). Sulfuryl chloride is a highly corrosive liquid but for laboratory preparations it is used often in preference to gaseous chlorine, because it permits simpler control of the amount of reactive chlorine used.

process and the distribution of isomeric substitution products varies widely with the experimental conditions.

The sulfonic acids are very strong acids and the simpler arylsulfonic acids are quite soluble in water. They can be isolated conveniently in the form of the sodium arylsulfonates, which are salted out by addition of sodium chloride. In the manufacture of dyes, sulfonic acid groups are introduced to increase the solubility of complex molecules, through formation of water-soluble sodium salts.

## 20a Sodium Benzenesulfonate

In a 250-ml round-bottomed flask, place 57 g (30 ml) of fuming sulfuric acid (sp g 1.88), containing about 7 percent of sulfur trioxide, and add with shaking 0.15 mole (11.7 g, 13.3 ml) of benzene, a few milliliters at a time. Wait until the first portion of benzene has reacted (and dissolved) before adding the next portion. Keep the temperature of the reaction mixture above 35° and below 50°, by occasional cooling in a dish of cold water. When the benzene has dissolved completely, cool, and pour the solution slowly with stirring into about 150 ml of water. Cool to room temperature and if necessary filter from any diphenylsulfone, ($C_6H_5-SO_2-C_6H_5$), that may separate. Partly neutralize the excess sulfuric acid by adding *carefully* 18 g of solid sodium bicarbonate, in small portions (*caution—foaming!*); then add 30 g of clean salt (sodium chloride) and warm until it dissolves. A little more water may be added if necessary. Transfer the warm solution to a beaker, and cool with stirring. Allow the mixture to stand for about 0.5 hr, in an ice bath, with occasional stirring.

When the sodium benzenesulfonate has separated completely, collect the crystals with suction and press them as dry as possible with a flat glass stopper or a clean cork. Moisten with 20–30 ml of a filtered, saturated salt solution, apply suction and press the material as dry as possible. Allow the sulfonate to dry by spreading it on a watch glass or large filter paper. If the product is washed with a little ethanol it will dry more rapidly. The yield is 12–14 g of the dry sodium salt, which contains some sodium chloride and other salts but is pure enough for most laboratory purposes.[2] These impurities can be removed by crystallization from hot ethanol but 15–18 ml of ethanol is necessary for each gram of sodium benzenesulfonate to be crystallized. Do not attempt to determine its melting point.

***Reactions.*** Arylsulfonic acids and their metallic salts may be characterized readily by conversion to crystalline salts of organic bases, such as the

[2] The quality of the crude product may be improved by dissolving it in a limited amount of hot water, adding sodium chloride, filtering the hot solution through a previously warmed Büchner funnel, and chilling the filtrate.

S-benzylthiouronium salts, which have suitable melting points.[3] Another method, more elegant but less convenient, involves reaction with phosphorus pentachloride (or phosphorus oxychloride) to form the arylsulfonyl chloride, which is treated with ammonia (or an amine) to furnish a crystalline arylsulfonamide.

**S-Benzylthiouronium Salts.** Dissolve 1 g of the sodium arylsulfonate in a minimum amount of water (warming if necessary) and prepare separately a concentrated solution of 1 g of S-benzylthiouronium chloride[4] in water. Mix the solutions together, shake well, and cool the mixture in an ice-water bath. If crystals do not form within a few minutes, scratch the inside of the tube with a glass rod to promote crystallization. Collect the product on a small suction filter, wash sparingly with cold water and recrystallize it from 50 percent aqueous ethanol (reserving a seed crystal for inoculation). The melting points[5] of several benzenesulfonates are: benzene-, 148°; *p*-toluene-, 181–182°; *p*-bromobenzene-, 170°; *p*-chlorobenzene-, 175°.

**Benzenesulfonyl Chloride and Benzenesulfonamide.** In a large dry test tube place 1 g of pulverized *dry* sodium benzenesulfonate and add 5 ml (10-molar excess) of phosphorus oxychloride (*caution—handle carefully!*). Mix the reactants thoroughly, attach a small condenser and reflux the mixture for 2 hr in a water bath. It is essential to shake the tube at frequent intervals during the heating. Cool the mixture and pour it cautiously, in small portions, into 50 ml of cold water in a small beaker. Swirl the mixture to facilitate hydrolysis of the excess phosphorus oxychloride. Decant off a portion of the water and use this to rinse out material remaining in the reaction tube. After 20 min allow the oily drops of benzenesulfonyl chloride to settle and carefully decant off the aqueous solution. To the residual crude sulfonyl chloride add about

[3] Donleavy, *J. Amer. Chem. Soc.*, **58**, 1004 (1936); Chambers and Watt, *J. Org. Chem.*, **6**, 376 (1941).

[4] S-Benzylthiouronium chloride is a relatively expensive reagent. Suitable material may be prepared readily from thiourea and benzyl chloride. In a flask fitted with a reflux condenser, a mixture of 25 g (16 ml) of benzyl chloride (*caution—lachrymator*), 15 g of thiourea and 40 ml of ethanol (or methanol) is warmed on a steam bath. Soon a vigorous exothermic reaction occurs and the thiourea dissolves completely. The pale-yellow solution is refluxed for 30 min, transferred while hot to a beaker, and cooled in an ice-water bath. The mass of white crystals is collected on a suction filter, washed with several 15-ml portions of cold ethyl acetate or ethanol, and pressed well. The dried product is stored in an amber bottle. The yield is 30–35 g of the salt, $[C_6H_5CH_2{-}S{-}C(NH_2){=}NH_2]^+Cl^-$. The crude product is satisfactory for the preparation of derivatives. It may be purified by recrystallization from ethanol or from 15–20 percent aqueous hydrochloric acid. The compound is dimorphic: stable form, mp 174–176°; metastable form, mp 142–145° (with reversion to the stable form).

[5] An extensive list of the melting points of S-benzylthiouronium arylsulfonates, and of arylsulfonamides, is found in McElvain, *The Characterization of Organic Compounds*, pages 275–276, The Macmillan Co., New York (1953).

10 ml of water, shake well, and pour in carefully 6 ml of concentrated aqueous ammonia. Swirl until all of the chloride has reacted, then add about 20 ml of water and heat to boiling to dissolve the sulfonamide. Add a pinch of decolorizing carbon, filter the hot solution, and allow it to cool. Collect the product with suction, wash it with cold water and press it dry. The recorded melting point of benzenesulfonamide is 156°.

### Questions

**1.** Write equations indicating the action of the following reagents on sodium benzenesulfonate: fusion with sodium hydroxide; fusion with sodium cyanide; phosphorus pentachloride.

**2.** How do you account for the fact that sulfanilic acid does not form a salt with hydrochloric acid?

**3.** Indicate a series of reactions which may be used to prove that the sulfur atom in an aromatic sulfonic acid is directly attached to a carbon atom of the benzene ring.

**4.** To what properties of the sulfonic acids do you attribute the fact that many synthetic dyes contain sulfonic acid groups?

**5.** What products are formed when naphthalene is sulfonated? nitrated?

## 20b Sodium *p*-Bromobenzenesulfonate

In a 250-ml round-bottomed flask, place 20 g (11 ml) of fuming sulfuric acid (sp g 1.88), containing about 7 percent of sulfur trioxide, and add 0.05 mole (8 g, 5.5 ml) of bromobenzene, in small portions and with constant swirling, so that the temperature does not exceed 100°. If the sulfonation proceeds slowly, warm the flask gently on a steam bath after all of the bromobenzene has been added, and shake the mixture until all of the bromobenzene has passed into solution. After cooling, pour the solution cautiously into about 45 ml of water. If a precipitate separates at this point, owing to the formation of a little dibromodiphenylsulfone, ($BrC_6H_4-SO_2-C_6H_4Br$), filter the warm solution with suction. Add 15 g of clean salt (sodium chloride) to the filtrate and warm on the steam bath until the salt dissolves.

Cool the solution rapidly with stirring and allow the sodium *p*-bromobenzenesulfonate to crystallize. Collect the crystals with suction and press them as dry as possible with a flat glass stopper or a clean cork. The product still contains appreciable amounts of sodium chloride and sulfuric acid. To effect a partial purification, pulverize it thoroughly, transfer to a large beaker, and mix well with 25 ml of a filtered saturated salt solution. Heat on a steam bath for a short while, cool, collect the product on a suction filter, and press it as dry as possible. Wash with a little ethanol and allow the salt to dry by spreading it on a watch glass or large filter paper. The yield is 10–12 g of the dry sodium salt. The product contains sodium chloride and other impurities but is pure enough for most laboratory purposes. These impurities can be removed by crystallization from hot ethanol (or methanol), but about 25 ml of ethanol is necessary for each gram of sodium *p*-bromobenzenesulfonate to be crystallized. Do not attempt to determine the melting point of this substance.

*Questions follow Experiment 20a.*

## 20c Sulfanilic Acid

In a 250-ml round-bottomed flask, place 0.15 mole (14 g, 14 ml) of freshly distilled aniline and add cautiously in small portions, 40 g (22 ml) of concentrated sulfuric acid. Attach an air-cooled reflux condenser and heat the flask in an oil bath at 180–190° for 4–5 hr. To test for completion of the reaction add one or two drops of the reaction mixture to 5–6 ml of 10 percent aqueous sodium hydroxide. If a clear solution results, the aniline has been completely sulfonated; if oily drops of unreacted aniline separate, the heating should be continued. In carrying out this test it is essential to use enough sodium hydroxide solution so that the test portion has a strongly alkaline reaction to litmus.

Cool the reaction mixture to room temperature and pour it with stirring into about 100 ml of cold water. The crude sulfanilic acid separates as a gray crystalline mass. Collect the product with suction, wash it with 25–30 ml of cold water, and crystallize from hot water, with the addition of 2–3 g of decolorizing carbon. The yield is 12–14 g.

Sulfanilic acid crystallizes from water with one molecule of water of crystallization. Take this into account in calculating the theoretical yield. Anhydrous sulfanilic acid may be obtained by crystallization of the hydrate from concentrated hydrochloric acid, or by heating the hydrate at 100–110° for a short while.

*Questions follow Experiment 20a.*

EXPERIMENT

# 21

# Side Chain Oxidation

Oxidation of methyl groups and other side chains is an important method of preparing aromatic carboxylic acids. Side chain oxidation also is a useful aid in the identification of aromatic hydrocarbons and their derivatives because the acids are crystalline solids that can be purified and characterized readily.

Practically all side chains with a carbon atom directly attached to the benzene ring may be oxidized to carboxylic acid groups. Thus, the xylenes are oxidized to the corresponding phthalic acids. Common reagents for oxidizing side chains in aromatic compounds are chromic acid mixture, nitric acid, potassium permanganate, and potassium ferricyanide. Since the oxidation of an alkyl side chain is initiated by attack of a C–H group adjacent to the ring, a *t*-butyl side chain is extremely resistant.

The presence of halogens, nitro, and sulfonic acid groups does not interfere with the oxidation of the alkyl group, but if hydroxyl or amino groups are present in the benzene ring, most oxidizing agents destroy the molecules completely. On the other hand, alkyl groups in the presence of alkoxy or acetamido groups can be satisfactorily oxidized to carboxylic acids.

Phthalic anhydride, a very important intermediate in the preparation of anthraquinone dyes and other chemicals, is prepared by the direct air oxidation of naphthalene in the presence of vanadium oxide as a catalyst. This procedure applied to benzene is a convenient commercial process for the preparation of maleic anhydride.

## 21a *p*-Nitrobenzoic Acid

In a 500-ml round-bottomed flask place 0.065 mole (8.9 g) of *p*-nitrotoluene, and a solution of 0.09 mole (27 g) of sodium dichromate crystals

($Na_2Cr_2O_7 \cdot 2H_2O$) in 60 ml of water. Add 70 g (38 ml) of concentrated sulfuric acid slowly, with constant swirling and *thorough mixing*, and attach a water-cooled reflux condenser. Thorough mixing is essential to avoid the danger of the reaction getting out of control during the heating. Heat the reaction mixture carefully until oxidation starts, then remove the flame until the vigorous ebullition subsides. When the mixture has ceased to boil from the heat of reaction, replace the flame under the flask and reflux the material vigorously for 2 hr (*caution—bumping!*). Cool the reaction mixture and pour it into 100–120 ml of cold water. Collect the precipitate of crude *p*-nitrobenzoic acid with suction and wash it on the filter with two 30-ml portions of water.

Grind the precipitate thoroughly in a mortar to break up the lumps, and then transfer it to a 600-ml beaker. Add 40–45 ml of about 5 percent sulfuric acid, made by adding 2 ml of concentrated sulfuric acid to 70 ml of water. Warm on a steam bath and stir thoroughly to extract the chromium salts as completely as possible from the *p*-nitrobenzoic acid. Cool, filter with suction, and wash the product with two 30-ml portions of water. Transfer the crude *p*-nitrobenzoic acid to a beaker, break up any lumps of the material, and treat it with 75–80 ml of 5 percent aqueous sodium hydroxide. The *p*-nitrobenzoic acid dissolves, and any unchanged *p*-nitrotoluene remains undissolved; chromium salts will be converted largely to chromium hydroxide. Add 2–3 g of decolorizing carbon, warm to 50° with stirring for about 5 min, and filter the alkaline solution with suction.[1] Precipitate the purified acid by pouring the alkaline solution with stirring, *into* 80–90 ml of about 15 percent sulfuric acid (prepared by adding 8 ml of concentrated sulfuric acid to 80 ml of water). Collect the purified acid with suction, wash it thoroughly with cold water, and dry. The yield is 7–8 g. The *p*-nitrobenzoic acid obtained in this way is sufficiently pure for most purposes. To obtain a product of high purity a small sample may be crystallized from a large volume of hot water, or from glacial acetic acid.

Since the melting point of *p*-nitrobenzoic acid is fairly high (about 240°) use preferably a bath of Crisco or of di-*n*-butyl phthalate. The purity of the acid may be checked also by determination of its neutralization equivalent, as described for benzoic acid in Experiment 35 but with aqueous ethanol as the solvent.

## Questions

**1.** What procedure may be used to arrest the oxidation of a methyl side chain at the aldehyde stage? (See Experiment 8.)

**2.** What acid is formed by oxidation of *p*-cymene (*p*-isopropyltoluene)? of naphthalene?

[1] The addition of 10 g, about 30 ml in volume, of a filter aid (Celite) greatly facilitates the filtration.

**3.** How may *p*-aminobenzoic acid be prepared from *p*-toluidine?

**4.** Compare the ease of oxidation of toluene, benzyl alcohol, and benzaldehyde.

**5.** If you were given a substance that might be either *o*-, or *m*-, or *p*-chlorotoluene (all three of which boil at about the same temperature), how would you identify the substance?

## 21b *o*-Nitrobenzoic Acid[2]

In a 500-ml round-bottomed flask prepare a solution of 0.057 mole (9 g) of potassium permanganate in 150 ml of warm water and add 0.025 mole (3.5 g, 3 ml) of *o*-nitrotoluene.[2] Attach an upright condenser and heat the flask over a wire gauze until the mixture boils vigorously (*caution—bumping!*). Swirl the mixture frequently to minimize risk of breaking the flask.[3] Continue the refluxing for 2–3 hr; the time may be divided between two laboratory periods if necessary.

Filter the hot solution through a fluted filter. If the solution is colored purple by residual permanganate, add a pinch of solid sodium bisulfite and 3 ml of concentrated hydrochloric acid; stir the solution thoroughly. Add more bisulfite if necessary to decolorize the excess permanganate, but avoid an excess.

To complete the precipitation of the nitrobenzoic acid, pour the solution *into* a well-stirred mixture of 10 ml of concentrated hydrochloric acid and 25 ml of water. Dissolve the nitrobenzoic acid by heating, add about $\frac{1}{2}$ g of decolorizing carbon, and filter the hot solution through a fluted filter. Collect the filtrate in a 250-ml Erlenmeyer flask. If crystals have separated during the filtration, redissolve them by heating, then cork the flask very *loosely*, and set it aside to cool. Collect the crystals with suction, wash them with a little water, and allow them to dry thoroughly. The yield is 2.5–3 g.

*Questions follow Experiment 21a.*

[2] *o*-Chlorobenzoic acid may be prepared by this procedure, by substituting 3.2 g (3 ml) of *o*-chlorotoluene for the *o*-nitrotoluene.

[3] For this reaction it is advantageous to use a mechanical stirrer if one is available. Another way of moderating the bumping is to provide the flask with a Claisen adapter containing a glass tube in the central arm, extending nearly to the bottom of the flask, through which a *very slow* stream of compressed air (or steam) can be introduced.

EXPERIMENT

# 22

# Aniline and Acetanilide

## Reduction of Aromatic Nitro Compounds

Aromatic nitro compounds can be reduced to primary arylamines by catalytic methods and by means of chemical reducing agents, such as tin and hydrochloric acid, stannous chloride and hydrochloric acid, iron or zinc and acetic acid, sodium sulfide, and others. The reducing agent and the solvent medium can be varied to suit the properties of the particular nitro compound and the arylamine. For laboratory preparations it is convenient to use tin and hydrochloric acid for reducing the simpler nitro compounds to amines, but on a large scale nitrobenzene and the nitrotoluenes are reduced by means of iron turnings and hot water in the presence of small amounts of hydrochloric acid.

If the full reducing power of tin and hydrochloric acid were utilized (conversion to stannic chloride), only one and one-half gram-atoms of tin

$$C_6H_5{-}NO_2 + 3Sn + 7HCl \longrightarrow [C_6H_5{-}NH_3]^+Cl^- + 3SnCl_2 + 2H_2O$$

$$C_6H_5{-}NO_2 + 3SnCl_2 + 7HCl \longrightarrow [C_6H_5{-}NH_3]^+Cl^- + 3SnCl_4 + 2H_2O$$

would be required to furnish the six equivalents of hydrogen needed to reduce one mole of nitro compound to the aryl amine. In practice this is not feasible and it is customary to use about two gram-atoms of tin per mole of the nitro compound. In the reaction mixture the amine is converted to a chlorostannite or chlorostannate salt, $(R{-}NH_3)_2SnCl_4$ or $(R{-}NH_3)_2SnCl_6$. Subsequent treatment with sodium hydroxide solution liberates the amine and forms water-soluble tin compounds, sodium stannite or stannate ($Na_2SnO_2$ or $Na_2SnO_3$).

The reduction of nitrobenzene occurs in steps, each of which requires two equivalents of hydrogen. With strong reducing agents in acid media, the process is carried to completion and gives the primary amine. Under milder conditions (zinc and ammonium chloride) the reduction stops at the stage of

$$C_6H_5{-}NO_2 \longrightarrow C_6H_5{-}N{=}O \longrightarrow C_6H_5{-}NHOH \longrightarrow C_6H_5{-}NH_2$$

the arylhydroxylamine, R−NHOH. In strong hydrochloric acid phenylhydroxylamine undergoes molecular rearrangement to *p*-chloroaniline, which is the principal product if the acid concentration of the reducing medium becomes too high. For this reason, and to control the vigorous reaction, the acid is added in small portions to the reaction mixture.

Sodium sulfide and hydrosulfide (NaSH) can be used to reduce nitro compound to arylamines. They are particularly useful for partial reduction of dinitro compounds to nitroanilines, as in the preparation of *m*-nitroaniline from *m*-dinitrobenzene (Experiment 30). With alkaline reducing agents, such as zinc and sodium hydroxide, reduction of nitrobenzene leads to compounds related to azobenzene, $C_6H_5{-}N{=}N{-}C_6H_5$, which arise through interaction of the intermediate reduction products (see Experiment 27).

Aromatic nitro compounds are reduced smoothly to amines by means of molecular hydrogen over a platinum oxide/platinum catalyst (Adams' catalyst) at low pressures, in ethanol or ethyl acetate; palladium and Raney nickel also are effective catalysts.[1] An interesting variation of direct catalytic hydrogenation is the use of hydrogen transfer in the presence of a catalyst. Upon heating nitro compounds with cyclohexene in the presence of palladium catalyst, the arylamine is formed and the cyclohexene dehydrogenated to benzene.[2] Similarly, hydrazine serves as a source of hydrogen in the presence of palladium or Raney nickel[3] and is converted to nitrogen.

Aniline can be prepared from chlorobenzene and ammonia, but the low reactivity of the aryl halide necessitates the use of vigorous conditions. Aniline is manufactured industrially by operating with a large excess of ammonia and a cuprous chloride catalyst at 200° under high pressure. Activated aryl halides, such as *p*-chloronitrobenzene, react with ammonia (and other basic reagents) under milder conditions. Occasionally primary arylamines are prepared from a substituted benzoic acid, by conversion to the amide and treatment of the amide with sodium hypochlorite or hypobromite (Hofmann reaction).

## 22a Aniline

In a 1-liter round-bottomed flask provided with an air-cooled condenser, place 0.2 mole (25 g, 21 ml) of nitrobenzene and 0.4 gram-atom (48 g)

[1] *Organic Syntheses*, Collective Volume **I**, 61, 463 (1941); **III**, 63 (1955).
[2] Braude, Linstead, and Wooldridge, *J. Chem. Soc.*, 3590 (1954).
[3] Balcomb and Furst, *J. Amer. Chem. Soc.*, **75**, 4334 (1953).

of mossy (or granulated) tin.[4] Shake the mixture thoroughly and add through the condenser tube, in small portions (not exceeding 10 ml each at the beginning), 120 g (100 ml) of concentrated hydrochloric acid. The reaction is exothermic and must be controlled by dipping the flask in a bath of cold water. Apply just enough cooling to permit the reaction mixture to boil quietly. When the initial reaction moderates, warm the mixture gently on a steam bath; allow it to cool slightly, add a second 10-ml portion of hydrochloric acid and continue the addition of acid in this manner.

➤**CAUTION:** Nitrobenzene and aniline are toxic and must be handled carefully. Avoid breathing their vapors. If either liquid comes in contact with the skin, wash it off promptly with a little ethanol, followed by thorough washing with soap and water.

After half of the hydrochloric acid has been added increase the portions to about 15 ml each; observe the same precautions to control the reaction. As the reduction proceeds the chlorostannite or chlorostannate salt of aniline may separate as a crystalline solid. Continue to shake the mixture *thoroughly*. After all of the acid has been added, warm the mixture on a steam bath for 20 min. Meanwhile prepare a solution of 75 g of solid sodium hydroxide in 120 ml of water and arrange an assembly for steam distillation, with a long water-cooled condenser (see Figure 22, page 64).

➤**CAUTION:** The reaction mixture must be watched carefully when first placed upon the steam bath. If it has been kept so cold in the beginning that the reaction was slowed down too much, it will now suddenly become so violent that the contents of the flask may be forced out through the condenser tube.

After the heating period, no nitrobenzene should remain; a few drops of the reaction mixture when poured into water should give a perfectly clear solution. If droplets of nitrobenzene are still present, complete the reduction by adding more tin and hydrochloric acid, and warming the reaction mixture. At this stage the experiment may be interrupted if necessary and allowed to stand overnight or longer.

Cool the mixture to room temperature and add *carefully* the previously prepared sodium hydroxide solution, with gentle swirling; cool the mixture to prevent loss of aniline. The mixture must be strongly alkaline (test with litmus) to insure complete liberation of the aniline, which separates as an oily liquid. Steam distill the mixture as rapidly as the cooling capacity of the

[4] Mossy tin is preferable because it has a greater surface per unit of weight and has less tendency to form a solid cake in the reaction mixture. With either form of tin vigorous shaking is essential to obtain complete reduction.

condenser permits, without loss of product. Continue the distillation for a short while after the distilling liquid has ceased to be turbid.[5]

Saturate the distillate with clean salt (sodium chloride), using about 25 g per 100 ml of liquid. Transfer the mixture to a separatory funnel and extract the aniline with two successive 25-ml portions of benzene.[6] Dry the combined benzene solutions with 8–10 g of anhydrous magnesium sulfate. Filter the dried liquid into a 125-ml distilling flask, add two small boiling chips, and *carefully* distill off the benzene, using a water-cooled condenser. Take adequate precautions against fire hazards. When the temperature of the distilling vapor reaches 85°, discontinue the distillation.

Replace the water-cooled condenser by a dry air-cooled condenser, add a fresh boiling chip, and distill the aniline. A small amount of benzene may still be present; take care during the initial heating to avoid igniting any benzene vapor. Collect in a dry weighed bottle the fraction boiling at 180–185°. The yield of aniline is 15–17 g. Pure aniline is colorless but becomes discolored on standing.

***Reactions.*** Aniline, and primary arylamines in general, are extremely useful intermediates for the synthesis of other aromatic compounds. Reaction with nitrous acid in the presence of mineral acids leads to diazonium salts that undergo many types of replacement reactions, arylation reactions, and also coupling reactions with aromatic amines and phenols to form azo dyes. Reduction of the diazonium group with sulfur dioxide or stannous chloride furnishes an arylhydrazine.

Aniline is acylated readily to form acetanilide, benzanilide, N-phenylurea and similar derivatives (see Experiment 12). Conversion to acetanilide is a preliminary step for the reactions of nitration, ring halogenation, and chlorosulfonation, illustrated in Experiments 23, 24, and 25.

## Questions

**1.** What side reaction occurs in the reduction of nitrobenzene if the concentration of hydrochloric acid in the medium is too high? Explain.

**2.** How could the presence of a small amount of nitrobenzene in a sample of aniline be detected and how could it be removed?

**3.** How may aniline be converted to benzoic acid? How may benzoic acid be converted to aniline?

[5] If acetanilide is desired, it is advantageous to acetylate the total distillate directly by the Lumière-Barbier method, without isolating the aniline. For this purpose, convert the aniline to the hydrochloride by adding a solution of 0.2 mole (17 ml) of concentrated hydrochloric acid in 100 ml of water and transfer the solution (which should have a volume of 450–500 ml) to a 1-liter flask. Follow the procedure outlined in the next section, using 30 g of sodium acetate crystals and 24 ml of acetic anhydride.

[6] If desired methylene chloride can be substituted for benzene. Methylene chloride is not flammable and has less tendency to form emulsions.

**4.** How may methylaniline and dimethylaniline be prepared?

**5.** Show how aniline may be converted to: (a) phenylisocyanate ($C_6H_5-N=C=O$); (b) phenylhydrazine; (c) *p*-bromophenylurea ($R-NHCONH_2$).

# 22b Acetanilide

## Acetylation of an Arylamine

In synthetic procedures primary and secondary arylamines often are converted to their acetyl derivatives as a protective measure, to reduce their susceptibility to oxidative degradation and to moderate their high reactivity in electrophilic substitution reactions (especially in halogenations). In some instances the amino group is acetylated to avoid an undesired reaction with another functional group or a reagent, such as $-COCl$, $-SO_2Cl$, $HNO_2$, etc.[7] At the end of a synthetic sequence the amino group can be regenerated readily by hydrolysis with acids or bases.

Arylamines (and aliphatic amines) may be acetylated by means of acetic anhydride or acetyl chloride, or by heating the amine with glacial acetic acid under conditions that permit removal of the water formed in the reaction. The last procedure is an economical one but requires a relatively long period of heating.

Acetyl chloride is disfavored for several reasons: half of the amine escapes acetylation through conversion to the hydrochloride; it reacts with extreme vigor and is a disagreeable reagent to manipulate; it is unsuitable for use with acid-sensitive compounds. Acetic anhydride does not suffer from these disadvantages and is the preferred acetylating agent. Moreover, in some instances the solution of the acetyl derivative in glacial acetic acid, that results from the acetylation, may be used for a subsequent reaction of the acetylated compound without the necessity of its isolation.

With primary amines acetylation with acetic anhydride may go further, if the temperature and reaction time are increased, to form a bis-acetyl derivative, $Ar-N(COCH_3)_2$. The latter can be hydrolyzed under mild conditions to the monoacetyl derivative.

The rate of hydrolysis of acetic anhydride is sufficiently slow to permit acetylation of amines to be carried out in buffered aqueous solutions (method of Lumière and Barbier). This is a general procedure that gives a product of high purity in good yield but it is not suitable for acetylation of the nitroanilines and other extremely weak bases. By a similar procedure, acetic

[7] Examples of syntheses involving protective acetylation are the nitration, bromination, and chlorosulfonation of acetanilide (Experiments 23, 24, and 25) and the halogenation of *p*-acetotoluidide (Experiment 29).

anhydride may be used to acetylate phenols in an aqueous alkaline solution (method of Chattaway, illustrated in Experiment 34, part G).

Acetanilide is one of the oldest synthetic medicinals (1886) and was used for many years as an antipyretic (fever-reducing) and analgesic (pain-relieving) drug, under the name Antifebrin. *p*-Ethoxyacetanilide, called Phenacetin, has similar activity and is less toxic than acetanilide. These drugs have the effect of reducing the oxygen-carrying capacity of the blood stream. The most widely used antipyretic and analgesic drug is acetylsalicylic acid, Aspirin, which is one of the least toxic medicinals of this type.

**(A) Acetylation in Water. The Lumière-Barbier Method.**[8] Dissolve 0.05 mole (4.7 g, 4.6 ml) of aniline in 125 ml of water to which 0.05 mole (4.2 ml) of concentrated hydrochloric acid has been added. If the solution is discolored add 2–3 g of decolorizing carbon, stir the solution for a few minutes and filter it with suction. Meanwhile prepare a solution of 0.06 mole (8 g) of sodium acetate crystals ($CH_3CO_2Na \cdot 3H_2O$) in 25 ml of water; if any insoluble particles are present, filter the solution.

Transfer the solution of aniline hydrochloride to a 500-ml flask and warm it to 50°. Add 0.06 mole (6.2 g, 6 ml) of acetic anhydride and swirl the contents to dissolve the anhydride. Add *at once* the previously prepared sodium acetate solution and mix the reactants thoroughly by swirling. Cool the reaction mixture in an ice bath and stir vigorously while the product crystallizes. Collect the crystals on a suction filter, wash with cold water, and allow them to dry. The yield is 5–5.5 g. The material obtained by this acetylation procedure is usually quite pure and of better quality than that prepared by the acetylation in acetic acid. If necessary, the product may be recrystallized from water, with addition of 1–2 g of decolorizing carbon.

**(B) Acetylation in Acetic Acid.** In a 125-ml Erlenmeyer flask dissolve 0.05 mole (4.7 g, 4.6 ml) of aniline in 10 ml of glacial acetic acid. To the solution add 0.06 mole (6.2 g, 6 ml) of acetic anhydride and mix well by swirling. The solution becomes warm from the heat of reaction. Add two boiling chips, attach a short air-cooled condenser tube, and boil the solution *gently* for 15 min to complete the acetylation. To hydrolyze the excess of acetic anhydride and any bis-acetyl derivative, add cautiously through the top of the condenser tube 5 ml of water, and boil gently for 5 min longer. Allow the reaction mixture to cool slightly and pour it *slowly*, with good stirring, into 70–75 ml of cold water. After allowing the mixture to stand for about 15 min

[8] Other aromatic amines (toluidines, xylidines, anisidines, phenetidines) may be acetylated by the general procedures given here. It is advantageous to use the Lumière-Barbier method if the sample of amine is discolored since this procedure affords a preliminary treatment with decolorizing carbon. Very weak bases (nitroanilines, dibromoanilines) are resistant and cannot be acetylated by the Lumière-Barbier method.

Phenacetin (*p*-ethoxyacetanilide) may be prepared conveniently by this method from *p*-phenetidine.

with occasional stirring, collect the crystals on a suction filter, and wash them with a little cold water. Recrystallize the crude product from hot water (about 20 ml per gram) with addition of 1–2 g of decolorizing carbon. The yield is 4–4.5 g.

**(C) Direct Acetylation with Acetic Acid.** This is the most economical method of effecting acetylation but the operation requires a longer time than methods A and B, which are most suited for small scale operations.

In a 200-ml round-bottomed flask place 0.2 mole (18.6 g, 18.2 ml) of aniline and 0.4 mole (24 g, 23 ml) of glacial acetic acid. Provide the flask with a short fractionating column fitted with a thermometer and connected with a condenser arranged for distillation. For the receiver use a small graduated cylinder. Add two boiling chips and heat the flask gently so that the solution boils quietly and the vapor does not rise into the column.

After 15 min increase the heating slightly so that the water formed in the reaction, together with a little acetic acid, distills over very slowly at a *uniform* rate; vapor temperature 104–105°. After about an hour, when 11–12 ml of distillate has collected, increase the heating so that the temperature of the distilling vapor rises to about 120°. Continue the distillation slowly for about 10 min longer, to collect an additional 2–3 ml of distillate (total volume, 13–14 ml), and then discontinue the heating. The distillate, consisting of 70–75 percent acetic acid, may be discarded.

As the reaction mixture will solidify upon cooling, pour it out at once, into about 350 ml of ice and water in a large beaker. Stir the aqueous mixture vigorously to avoid formation of large lumps of the product. Collect the acetanilide with suction, wash with a little cold water, and press it firmly on the filter. Crystallize the moist product from hot water (about 20 ml per gram) with addition of 1–2 g of decolorizing carbon. For filtration use a large fluted filter and a large funnel with a short, wide stem. Cool the filtrate rapidly with vigorous stirring to obtain small crystals. Allow the material to stand for about 10 min in an ice-water bath and then collect the crystals with suction. Wash the product with a small amount of cold water and spread it on a clean paper to dry. If the material is dark-colored it should be recrystallized. The yield is 17–18 g.

## Questions

**1.** Show a detailed mechanism for the reaction of aniline with acetic anhydride, including the possibility of a cyclic intermediate for elimination of acetic acid from the initial addition product.

**2.** What product would you expect to obtain by reaction of aniline with the mixed anhydride, acetic-formic anhydride ($HCO{-}O{-}COCH_3$)? Explain.

**3.** What may be the particular advantages of using ketene ($CH_2=C=O$, bp $-80°$) as an acetylating agent?

**4.** When acetic acid is used for acetylation of an amine why is it desirable to use an excess of the acid and to distill off the water formed in the reaction?

**5.** In the preparation of *p*-nitroaniline why is aniline converted to acetanilide before nitration?

**6.** What product is obtained from aniline and each of the following reagents: (a) 1-naphthylisocyanate; (b) succinic anhydride, followed by heating; (c) potassium cyanate + acetic acid (HNCO); (d) dimethylketene; (e) phenylisothiocyanate?

**7.** When trimethylamine and isobutyryl chloride are mixed in dry ether trimethylamine hydrochloride is precipitated but the amine is not acylated. What is the other product?

# EXPERIMENT 23

# Nitration of an Arylamine

When aniline nitrate or sulfate is nitrated at low temperature with concentrated nitric and sulfuric acids, the product consists of about 60 percent *m*-nitroaniline and 38 percent *p*-nitroaniline, with very little of the *ortho* isomer. The yield is not high since some of the aniline is lost through oxidation.

The formation of *m*-nitroaniline would be anticipated because the powerful inductive electron-withdrawal of the positively charged $-NH_3^+$ group brings about strong deactivation, especially at the *ortho* and *para* positions. The formation of a substantial amount of *p*-nitroaniline has sometimes been attributed to the nitration of free aniline (*ortho*/*para* directive effect) that is present in a small amount in equilibrium with the anilinium cation and would undergo nitration at a much faster rate. Recent experimental evidence casts doubt on this explanation since the *meta*/*para* ratio changes less than expected with increase in acidity of the nitrating medium. The observed *meta*/*para* ratio evidently represents the actual directive effect of the $-NH_3^+$ group. The extent of *para* nitration diminishes with successive introduction of methyl groups on the nitrogen atom; $C_6H_5-N(CH_3)_3^+$ gives 89 percent *meta* and 11 percent *para* nitro derivative.

To diminish the susceptibility to oxidation and avoid the *meta*-directive effect of salt formation, it is customary to convert an aromatic amine to its acetyl derivative before carrying out nitration. The acyl group is removed subsequently by hydrolysis with aqueous acid or alkali. Thus, aniline is converted to acetanilide, which on nitration in the usual way furnishes almost exclusively *p*-nitroacetanilide, and the latter gives *p*-nitroaniline upon hydrolysis.

Since only a small amount of *o*-nitroacetanilide is obtained by nitration of acetanilide, an indirect method is used to secure *o*-nitroaniline as the principal product of a series of reactions. Aniline is converted to sulfanilic acid, in which the reactive *para* position is blocked by the $SO_3H$ group; nitration of sulfanilic acid produces 4-amino-3-nitrobenzenesulfonic acid; hydrolysis of the latter by boiling with 60 percent sulfuric acid brings about elimination of the $SO_3H$ group and yields *o*-nitroaniline in a state of high purity.

*m*-Nitroaniline is prepared most conveniently by the controlled partial reduction of *m*-dinitrobenzene, using ammonium sulfide or sodium hydrosulfide (NaSH) as the reducing agent (see Experiment 30a).

The three nitroanilines are extremely weak bases but differ appreciably from one another in base strength: *meta* ≫ *para* > *ortho*. Mixtures of the nitroanilines can be separated by dissolving them in strong aqueous acid and precipitating successively the *ortho*, *para*, and *meta* isomers by progressive neutralization with dilute ammonia.

*p*-Nitroaniline is manufactured commercially by heating *p*-chloronitrobenzene with ammonia. The ammonolysis occurs more readily than with chlorobenzene, owing to the activating effect of the nitro group in the *para* position. *p*-Nitroaniline is used to prepare *p*-phenylenediamine and also for the production of a special type of azo dye, such as Para Red (an ingrain color). Cotton cloth may be soaked in a dilute alkaline solution of 2-naphthol (or similar coupling component), dried, and dipped into an ice-cold solution of diazotized *p*-nitroaniline; coupling takes place and the dye is formed within the pores of the cellulose fibres.

*p*-Nitroaniline is used as an intermediate for laboratory syntheses leading to compounds that cannot be obtained readily, or in a pure state, by direct substitution processes. Thus, diazotization of *p*-nitroaniline and subsequent treatment of the diazonium fluoborate with sodium nitrite (in the presence of copper powder) affords a route to *p*-dinitrobenzene. Similarly,

$$[O_2N-C_6H_4-\overset{+}{N}\equiv N][\overset{-}{B}F_4] + NaNO_2 \xrightarrow{Cu} O_2N-C_6H_4-NO_2 + NaBF_4$$

replacement of the diazonium group by $-C\equiv N$, using sodium cyanide and cuprous cyanide (the Sandmeyer reaction), furnishes *p*-nitrobenzonitrile. Neither of these compounds can be obtained by direct nitration because of the *meta*-directive effect of the $-NO_2$ and $-C\equiv N$ groups.

## *p*-Nitroaniline

***p*-Nitroacetanilide.** In a 125-ml Erlenmeyer flask dissolve 0.10 mole (13.5 g) of pure acetanilide in 15 ml of glacial acetic acid by warming gently. Cool the warm solution until crystals begin to form and then add slowly, with swirling 20 ml of ice-cold concentrated sulfuric acid. Prepare a nitrating mixture by adding 0.12 mole (10 g, 7 ml) of concentrated nitric acid to 10 ml of cold concentrated sulfuric acid; cool the solution to room temperature and transfer it to a small separatory funnel.

Cool the acetanilide solution to 5° in an ice bath, remove the flask from the bath and add the nitrating mixture slowly, drop by drop. Swirl the reaction mixture to obtain good mixing in the viscous solution and do not permit the temperature to rise above 20–25°. After all of the nitrating mixture has been added, allow the solution to stand at room temperature for about 40 min (but not longer than 1 hr) to complete the reaction. Pour the solution slowly with stirring into a mixture of 200 ml of water and 40–50 g of chipped ice. Collect the product with suction, press it firmly on the filter and transfer the filter cake to a beaker. Mix the crystals thoroughly with about 150 ml of water to form a thin paste, return them to the suction filter, and wash thoroughly with more water to remove the nitric and sulfuric acids. Press the material as dry as possible. The crude, moist *p*-nitroacetanilide is sufficiently pure to be used directly for hydrolysis to *p*-nitroaniline. The moist product is equivalent to about 12 g of dry material.

A small portion of the material may be purified by crystallization from 80 percent aqueous ethanol, with addition of a little decolorizing carbon. Since the melting point of *p*-nitroacetanilide is about 215–216°, a bath of Crisco or di-*n*-butyl phthalate should be used if its melting point is to be determined.

***p*-Nitroaniline.** In a 250-ml Erlenmeyer flask mix the moist, crude *p*-nitroacetanilide with 30 ml of water and 35 ml of concentrated hydrochloric acid. Heat the mixture slowly on a wire gauze, with occasional swirling, and boil *gently* for 30–40 min. The material gradually dissolves and an orange-colored solution is formed.[1] When the hydrolysis has been completed, add 60 ml of cold water and cool the mixture to room temperature. Crystals of the product may separate.

Pour the *p*-nitroaniline hydrochloride slowly with good stirring, into a mixture of 40 ml of concentrated aqueous ammonia, 150 ml of water and 50–60 g of chipped ice. The mixture must be distinctly alkaline at the end of the mixing; test with litmus, and add a little more ammonia if necessary. Collect the orange-yellow precipitate of *p*-nitroaniline with suction and wash it with cold water. Recrystallize the product from a large volume of hot water; about 100 ml of water will be required per gram of material. The yield is 6–7 g.

[1] If the reaction mixture has been boiled too vigorously, it may be necessary to add an additional quantity of hydrochloric acid to replace that lost by evaporation and to heat longer to complete the hydrolysis.

## Questions

**1.** Why is it necessary to carry out the nitration of acetanilide at relatively low temperature? To obtain *p*-nitroaniline, why is aniline not nitrated directly.

**2.** What products are formed by nitration of *p*-acetotoluidide? of *m*-acetotoluidide? of *m*-cresol? of *p*-toluic acid?

**3.** Would benzoic acid undergo nitration satisfactorily under the mild conditions used in this experiment? Explain.

**4.** Can *p*-nitroacetanilide be hydrolyzed by alkalies? What accessory product may be formed by the action of hot aqueous alkalies on *p*-nitroaniline? (Consider the activating effect of the nitro group.)

**5.** Outline a series of reactions for the preparation of *o*-nitroaniline, starting from aniline; of *m*-nitroaniline, starting with benzene.

**6.** Is *p*-nitroaniline a stronger or weaker base than aniline? Is *p*-nitrophenol a stronger or weaker acid than phenol? Explain.

# EXPERIMENT 24

# Bromination of an Arylamine

The ability of the amino group to permit electron-release into the aromatic ring renders the system extremely reactive toward electrophilic substitution, particularly at the *ortho* and *para* positions. Catalysts such as aluminum or ferric halides are quite unnecessary and direct halogenation even under mild conditions usually leads to introduction of halogen at all available *ortho* and *para* positions. Aniline and aqueous bromine react instantly to form 2,4,6-tribromoaniline.

Acylation of the amino group reduces the reactivity and allows halogenation to be controlled and also diminishes oxidative degradation reactions. Conversion of aniline to acetanilide permits facile bromination to give mainly *p*-bromoacetanilide, with a little of the *ortho* isomer. The acetyl group is removed readily by alkaline hydrolysis to form *p*-bromoaniline. This method is a general one that may be used to prepare chloro or bromo derivatives of the toluidines (see Experiment 29), and other arylamines.

## *p*-Bromoaniline

***p*-Bromoacetanilide.** In a 250-ml Erlenmeyer flask dissolve 0.1 mole (13.5 g) of acetanilide in 50 ml of glacial acetic acid and add slowly with shaking, 0.11 mole (17.5 g, 5.4 ml) of bromine, measured from a burette. Allow the orange-colored solution to stand at room temperature for about 10 min to complete the reaction. If the solution is warm, crystals may separate. At the end of this time pour the solution with stirring into 400–500 ml of cold water, and rinse the flask with a little water. The crude *p*-bromoacetanilide precipitates in white or yellow flocks. The mixture is sometimes colored with free bromine, which may be removed by adding a few drops of strong sodium bisulfite solution. Collect the crystalline precipitate with suction, wash it thoroughly with cold water, and press as dry as possible on the filter.

To obtain pure *p*-bromoacetanilide crystallize the crude product from hot ethanol. The yield is 10–15 g.

➤**CAUTION:** Handle bromine carefully since the liquid produces painful burns and the vapors are obnoxious (hood). Use both hands in adjusting the stopcock of the burette to avoid loosening it, with consequent leakage. Treat bromine burns immediately with a liberal quantity of glycerol. Follow this by a thorough washing with water, dry the skin, and apply a healing ointment or salve.

It is advisable to keep a small beaker containing glycerol at hand when manipulating bromine.

***p*-Bromoaniline.** Dissolve 0.05 mole (10 g) of pure *p*-bromoacetanilide in 20 ml of boiling ethanol contained in a 250-ml round-bottomed flask provided with a reflux condenser. Through the condenser tube, add a solution of 0.075 mole (5 g of 85 percent pure material) of potassium hydroxide in 6 ml of water and reflux the reaction mixture for 30–40 min. Dilute with about 75 ml of water and fit the flask with an adapter leading to a condenser set downward for distillation. Distill the mixture on a wire gauze and collect about 50 ml of distillate, which consists of ethanol and water. Pour the residual mixture of *p*-bromoaniline and aqueous solution of potassium acetate into about 100 ml of cold water and rinse the flask with a little water. The *p*-bromoaniline separates as an oil which crystallizes on standing. Collect the precipitate with suction and wash it thoroughly with water.

Dissolve the crude *p*-bromoaniline in a mixture of 100 ml of water, 100 ml of ethanol, and 5 ml of concentrated hydrochloric acid; add 1 g of decolorizing carbon, warm on a steam bath for a few minutes, and filter. Pour the filtrate *slowly* with stirring *into* a mixture of 40 ml of 10 percent aqueous sodium hydroxide and 60–70 g of clean ice. If performed properly, this procedure produces *p*-bromoaniline in crystalline form, and not as an oil that later solidifies. Collect the crystals with suction, wash thoroughly with water, and allow to dry. Press between filter papers and dry thoroughly before determining the melting point. The yield is 5–8 g. The product may be purified by crystallization from dilute ethanol with addition of a little decolorizing carbon, but a considerable amount is lost in purification.

***Reactions.*** *p*-Bromoaniline can be diazotized to form *p*-bromobenzenediazonium salts that are useful intermediates for introducing the *p*-bromophenyl group into other molecules, by means of Gomberg-Bachmann or Meerwein arylations (Experiment 32) and by the many variations of diazonium replacement reactions. An interesting example is the preparation of *p*-bromobenzaldehyde from the diazonium compound and formaldoxime.[1] *p*-Bromobenzonitrile, *p*-bromobenzenearsonic acid, and many other

[1] Jolad and Rajagopal, *Organic Syntheses*, **46**, 13 (1966).

compounds that cannot be obtained by direct bromination can be obtained through diazonium replacement reactions.

*p*-Bromoaniline can be converted readily into various acyl derivatives. Reaction with sodium cyanate in the presence of aqueous acetic acid leads to *p*-bromophenylurea.

$$Br{-}C_6H_4{-}NH_2 + NaNCO + CH_3CO_2H \longrightarrow Br{-}C_6H_4{-}NH{-}CO{-}NH_2 + CH_3CO_2Na$$

***p*-Bromophenylurea** *(Optional)*. The following procedure is a general method for the preparation of arylureas.[2]

In a 25 × 150-mm test tube dissolve 0.005 mole (0.9 g) of *p*-bromoaniline in 3 ml of glacial acetic acid and 6 ml of water at 35°. To this add slowly with vigorous shaking, a solution of 0.01 mole (0.65 g) of sodium cyanate [or potassium cyanate (0.8 g)] dissolved in 5 ml of water at 35° (*caution—frothing!*). A small rise in temperature occurs and a white precipitate forms. Allow the mixture to stand with occasional vigorous shaking for an hour or longer, add 2–3 ml of water and cool the mixture in an ice bath. Collect the product on a suction filter, wash it with a little cold water, and recrystallize it from 75 percent aqueous ethanol (~15 ml per gram). *p*-Bromophenylurea forms colorless prisms, reported mp 225–227°. The yield is about 0.6 g.

## Questions

**1.** Outline methods for the following syntheses: (a) *p*-bromoaniline from bromobenzene; (b) *m*-bromoaniline from benzene; (c) *p*-bromophenylhydrazine.

**2.** Write equations for specific examples illustrating the difference between *addition* and *substitution* of bromine.

**3.** What is the principal product of nitration of *p*-bromoacetanilide under mild conditions?

**4.** What products would be formed by the action of bromine on the following: benzoic acid? chlorobenzene? naphthalene? nitrobenzene? toluene, at the boiling point in the presence of ultraviolet light?

**5.** List the following in order of diminishing base strength: aniline, dimethylaniline, *p*-bromoaniline, *p*-nitroaniline, *m*-nitroaniline. Explain.

[2] Kurzer, *Organic Syntheses*, Collective Volume **IV**, 49 (1963).

EXPERIMENT

# 25

# Sulfanilamide

A convenient synthesis of sulfanilamide (*p*-aminobenzenesulfonamide) and related aminoarylsulfonamides makes use of *p*-acetamidobenzenesulfonyl chloride. It is not feasible to convert sulfanilic acid to the sulfonyl chloride having a free amino group, since *p*-aminobenzenesulfonyl chloride contains two functional groups that interact and lead to polymers of the type

$$\text{+NH}-C_6H_4-SO_2NH-C_6H_4SO_2\text{+}$$

Protecting the amino group of aniline by conversion to acetanilide permits direct chlorosulfonation with chlorosulfonic acid to obtain *p*-acetamido-

$$R-C_6H_4-H \xrightarrow{ClSO_3H} HCl + R-C_6H_4-SO_3H \xrightarrow{ClSO_3H} R-C_6H_4-SO_2Cl + H_2SO_4$$

benzenesulfonyl chloride. A slight excess over two moles of chlorosulfonic acid is used, since chlorosulfonation is a two step process. The sulfonic acid is formed in the first stage and converted to the sulfonyl chloride by reaction with a second molecule of chlorosulfonic acid.

Reaction of *p*-acetamidobenzenesulfonyl chloride with an excess of aqueous ammonia produces the corresponding sulfonamide. For reaction with aminothiazole, aminodiazole, and similar compounds, a tertiary base such as pyridine is added to combine with the hydrogen chloride formed.

$$CH_3CONH-C_6H_4-SO_2Cl + 2NH_3 \longrightarrow CH_3CONH-C_6H_4-SO_2NH_2 + NH_4Cl$$

$$CH_3CONH-C_6H_4-SO_2NH_2 + H_2O \xrightarrow{HCl} CH_3CO_2H + H_2N-C_6H_4-SO_2NH_2$$

Since a carboxylic amide is hydrolyzed more easily than a sulfonamide, it is a simple matter to de-acetylate the acetamido compound by selective hydrolysis under controlled conditions and secure the aminobenzenesulfonamide.

Hundreds of compounds related to sulfanilamide have been prepared.

Of these, two have proved of great therapeutic value against pneumococci, gonococci, and a variety of streptococcal infections. The two substances are sulfathiazole and sulfadiazine and are prepared in a similar manner to sulfanilamide, by substituting for ammonia the heterocyclic bases aminothiazole and aminodiazine.

$H_2N-C_6H_4-SO_2NH-C_3H_2NS$ (thiazole ring: S—CH=CH—N=C)

Sulfathiazole

$H_2N-C_6H_4-SO_2NH-C_4H_3N_2$ (pyrimidine ring)

Sulfadiazine

# p-Aminobenzenesulfonamide

**p-Acetamidobenzenesulfonyl Chloride.** In a *dry* 50-ml Erlenmeyer flask place 0.3 mole (35 g, 20 ml) of chlorosulfonic acid[1] ($Cl-SO_3H$) (*handle carefully—the acid causes severe burns if dropped on the skin*). Cool the acid to 10–15°, but *not below 10°*, in a water bath containing a few pieces of ice. Then add 0.05 mole (6.8 g) of finely powdered, *dry* acetanilide, in small portions and with good mixing, so that the temperature does not rise above 20°. After the acetanilide has dissolved (a few small particles may remain undissolved), place the flask in a 600-ml beaker containing just enough water to reach about to the level of the reaction mixture in the flask. Heat the water to 60–70° and maintain this temperature for an hour. Take care that the water level is maintained during the heating so that the bath does not go dry.

Pour the reaction mixture *slowly and carefully* in a thin stream onto a well-stirred mixture of 150 g of finely cracked ice and enough water to make stirring easy. *It is important to carry out this step cautiously to avoid spattering of the chlorosulfonic acid.* The reaction product separates as a white or pinkish-white gummy mass that soon becomes hard and can be broken up with a stirring rod. Break up any lumps that have formed, collect the product on a Büchner funnel, and wash it with several portions of cold water. The crude product, sucked as dry as possible, is used directly in the next step.[2]

[1] If a poor grade of chlorosulfonic acid is used the yield will be lowered. Discolored and impure specimens of the acid may be purified by distillation from an ordinary distilling flask (taking precautions to protect the distillate from moisture) provided the side tube is passed as far as possible into the condenser to avoid contact of the vapor and distillate with corks. The distillation should be carried out fairly rapidly and not more than 50 ml distilled in a single operation, since the corks are gradually attacked; rubber stoppers seem to be even less satisfactory than corks. If an all-glass distilling apparatus (with ground-glass joints) is available, this is ideal for the distillation. It is suitable to collect for use a fraction of chlorosulfonic acid boiling at 148–150° (760 mm).

[2] Moist *p*-acetamidobenzenesulfonyl chloride can be conveniently purified by dissolving in a mixture of acetone and benzene (1 : 1), separating the water, and allowing the solvent to evaporate until crystallization takes place. Obtained in this form the product keeps well, whereas the crude material even though dry cannot be preserved without considerable decomposition.

**p-Acetamidobenzenesulfonamide.** Place the crude, damp product from the first step in a 100-ml Erlenmeyer flask and add 25 ml of concentrated aqueous ammonia (28 percent). An immediate reaction ensues with the liberation of heat. Rub the mixture with a glass stirring rod until a smooth, thin paste is obtained and then heat at 70° for 30 min. Remove the flask to an ice bath and after cooling add dilute sulfuric acid until the mixture is acid to Congo red test paper.[3] After thorough chilling in the ice bath collect the product on a Büchner funnel, wash with cold water, and dry. The yield of crude product is 6–7 g. This material is sufficiently pure for the next step.

Recrystallize a small sample of the sulfonamide from hot water, using a little decolorizing carbon and take the melting point of the purified material. Test the solubility of the product in dilute acid and dilute alkali. Explain the result. The recorded melting point for pure *p*-acetamidobenzenesulfonamide is 219°.

**p-Aminobenzenesulfonamide.** Place the crude, dry *p*-acetamidobenzenesulfonamide obtained in the previous preparation in a 100-ml round-bottomed flask and add an amount of dilute hydrochloric acid (1 vol of concentrated acid to 2 vol of water) equal to *two ml* of dilute acid *per gram* of substance. Boil the mixture under reflux for 1 hr. Care should be taken at the start of the heating period to avoid charring of the initially pasty mixture.

Pour the solution into a 100-ml Erlenmeyer flask and dilute it with an equal volume of water. Add a little decolorizing carbon, heat the solution to boiling, and filter through a fluted filter into a clean 600-ml beaker. Add solid sodium carbonate in small portions with continuous stirring until the solution is just alkaline to litmus. During the neutralization the free amine separates as a white, crystalline precipitate. After thorough cooling in an ice bath, collect the product by suction filtration, wash with cold water, and dry. The yield is about 0.70–0.72 g per gram of *p*-acetamidobenzenesulfonamide.

Recrystallize the product from water, using about 12 ml of water per gram of sulfanilamide. Add a little decolorizing carbon (about 0.5 g) to the hot solution, boil for a few moments, and filter through a fluted filter paper in a short-stemmed funnel. The filter paper and funnel should be preheated by pouring boiling water through them just before filtering the solution, to prevent the material from crystallizing in the stem of the funnel. The sulfanilamide[4] separates on cooling as long, silky white needles. Test the solubility of sulfanilamide in dilute acid and dilute alkali. Explain the result.

The recorded melting point of pure *p*-aminobenzenesulfonamide is 163°.

[3] Congo red test paper may be prepared by dipping strips of filter paper into a solution of Congo red and allowing to dry. This indicator changes to a *blue* color with mineral acids, but is not affected by weak acids or acidic salts. Why is it necessary that the solution should be *strongly* acid at this point in the experiment?

[4] Sulfanilamide is a powerful therapeutic agent and must be taken only with the advice and supervision of a physician.

## Questions

**1.** Compare the ease of hydrolysis of benzenesulfonamide and benzamide; of benzenesulfonyl chloride and benzoyl chloride. Compare the behavior of sodium benzenesulfonate and sodium benzoate on fusion with sodium hydroxide.

**2.** What is saccharin? chloramine-T? How are they made commercially?

**3.** Why is it impractical to synthesize sulfanilamide by any process that would require *p*-aminobenzenesulfonyl chloride as an intermediate? What reaction would occur if *p*-aminobenzenesulfonyl chloride were formed momentarily in a synthetic process?

EXPERIMENT

# 26

# N-Phenylsydnone

## A Mesoionic Compound

The resonance structures of many compounds (vinyl ethers and halides, amides, and the like) involve one or more contributors that carry formally separated electrical charges. N-Phenylsydnone, a derivative of 1,2,3-oxadiazole, is a characteristic example of the special class of compounds described as mesoionic, in which *all of the important resonance structures* have a separation of charges. The name *sydnones* was given to them by their discoverers at the University of Sydney (Australia).

A general method of synthesis consists in eliminating the elements of water from the N-nitroso derivatives of N-arylaminoacetic acids (I) by means of acetic anhydride or thionyl chloride.[1] The sydnones undergo numerous

$$\text{Ar-NH-CH}_2\text{-CO}_2\text{H} + \text{HO-N=O} \longrightarrow \text{Ar-N(N=O)-CH}_2\text{-CO}_2\text{H} \quad \text{I}$$

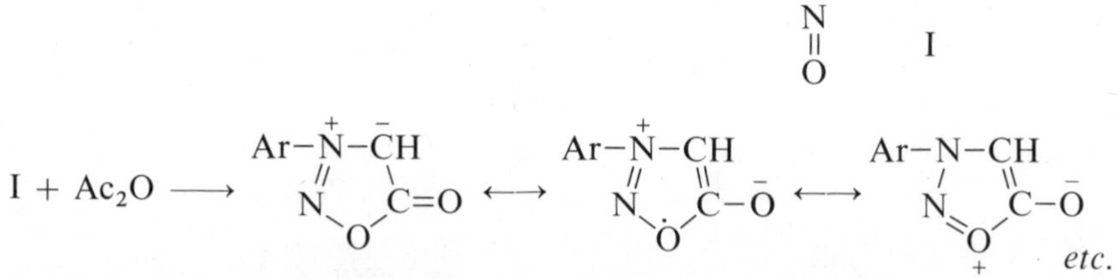

$$\text{I} + \text{Ac}_2\text{O} \longrightarrow \text{Ar-}\overset{+}{\text{N}}\text{-}\overset{-}{\text{CH}}\ (\text{N}\ \text{C=O}\ \text{O ring}) \longleftrightarrow \text{Ar-}\overset{+}{\text{N}}\text{-CH}\ (\text{N}\ \text{C-}\overset{-}{\text{O}}\ \dot{\text{O}}\ \text{ring}) \longleftrightarrow \text{Ar-N-CH}\ (\text{N}\ \text{C-}\overset{-}{\text{O}}\ \overset{+}{\text{O}}\ \text{ring}) \quad etc.$$

electrophilic substitutions and also 1,3-cycloaddition reactions (involving positions 2 and 4) with quinones, alkenes, and alkynes. The resulting adducts on loss of carbon dioxide furnish derivatives of pyrazoline or pyrazole ($C_3H_4N_2$; 1,2-diazole).

[1] The procedure given is that of Thoman and Voaden, *Organic Syntheses*, **45**, 97 (1965); references to other methods of preparation and to reactions of sydnones are given there.

**N-Phenylglycine.** In a 250-ml round-bottomed flask place 0.10 mole (9.5 g) of chloroacetic acid and 20 ml of water. *Neutralize* the acid by careful, slow addition of 10 percent aqueous sodium hydroxide solution (~40 ml; 0.1 mole), with shaking and cooling. To the solution of sodium chloroacetate add 0.11 mole (10.0 ml, 10.2 g) of aniline, adjust a reflux condenser and boil the mixture gently for 20 min. Cool the solution to room temperature in an ice bath, add 5 g of sodium hydroxide pellets and swirl the flask in the ice bath until all of the solid has dissolved. Extract the basic solution with two 50-ml portions of methylene chloride to remove the unreacted aniline. Acidify the aqueous solution by dropwise addition of concentrated hydrochloric acid until the pH is about 4 (Hydrion *or* Congo Red paper). If the product separates as an oil, induce crystallization by scratching with a glass rod.

Collect the crystals on a suction filter, wash them sparingly with ice-cold water, and spread them to dry in the air. The yield of crude product is 9–10 g. This may be used directly for the next step; if desired, it may be crystallized from 40 percent aqueous ethanol. Pure N-phenylglycine forms white crystals, mp 125–127°.

**N-Nitroso-N-Phenylglycine.** In a 250-ml Erlenmeyer flask prepare a slurry of 0.06 mole (9.1 g) of finely pulverized N-phenylglycine in 100 ml of water and cool the mixture in an ice-salt bath, with constant swirling, until the inner temperature falls below 0°. Without allowing the temperature to rise above 0° add dropwise over a period of 15–20 min, with good mixing, an ice-cold solution of 0.066 mole (4.6 g) of sodium nitrite in 30 ml of water. Filter the turbid, colored solution with suction on a large Büchner funnel. Transfer the filtrate to an Erlenmeyer flask and swirl the solution vigorously while adding 0.1 mole (9 ml) of concentrated hydrochloric acid. Continue to swirl the slurry of crystals for several minutes after all of the acid has been added. Collect the crystals on a suction filter, wash them with two 15-ml portions of cold water, and allow the product to dry *thoroughly* in the air. The yield is 7–9 g; reported mp 103–104°.

**N-Phenylsydnone.** Place the dried N-nitroso-N-phenylglycine in a 250-ml round-bottomed flask and add 0.5 mole (50 ml; a large excess) of acetic anhydride. Attach a reflux condenser fitted with a drying tube at the top, and heat the reaction mixture in a bath of boiling water for about 40 min, with occasional swirling. Cool the solution to 20–30° and pour it slowly, with vigorous stirring, into about 250 ml of water in a large beaker. After the excess acetic anhydride has been hydrolyzed collect the crystals on a suction filter and wash them thoroughly with several portions of cold water. After drying, the product weighs 5–6 g; reported mp 136–137°. On crystallization from boiling water the sydnone forms light-tan needles.

## Questions

**1.** Write equations for another method of converting aniline to N-phenylglycine.

**2.** Show the tautomeric forms and the resonating structures of indoxyl, $C_8H_7NO$, which is made commercially by cyclization of N-phenylglycine with sodium amide.

**3.** Devise structural formulas for two or more analogs of N-phenylsydnone in which the heterocyclic system has been altered. Show their resonance structures.

**4.** Write a structure for the cycloaddition of methyl propiolate ($H-C{\equiv}C-CO_2CH_3$) to N-phenylsydnone and show the formula of the substituted pyrazole that could be formed by elimination of $CO_2$ from the adduct. (The final product should be 3-carbomethoxy-1-phenylpyrazole.)

# EXPERIMENT 27

# Intermediate Reduction Products of Nitro Compounds

When nitrobenzene is reduced by means of zinc and aqueous ammonium chloride solution, the reaction proceeds through nitrosobenzene to phenylhydroxylamine but does not go on to aniline. Phenylhydroxylamine is of some interest because of its molecular rearrangements. When treated with

$$C_6H_5-\overset{+}{N}(-\overset{-}{O})=O \xrightarrow{2H} C_6H_5-N=O \xrightarrow{2H} C_6H_5-NH-OH$$

Nitrobenzene — Nitrosobenzene — Phenylhydroxylamine

$$HO-C_6H_4-NH_2 \xleftarrow{H_2SO_4} H-C_6H_4-NH-OH \xrightarrow{HCl} Cl-C_6H_4-NH_2$$

strong sulfuric acid it is converted to *p*-aminophenol; concentrated hydrochloric and hydrobromic acids effect similar transformations to give *p*-chloroaniline and *p*-bromoaniline, respectively. These are examples of a broad class of molecular rearrangements in which a substituent attached to nitrogen (or oxygen) is shifted to a carbon atom in the *ortho* or *para* position of the aromatic ring.[1]

In the presence of alkaline reagents, nitrosobenzene and phenylhydroxylamine interact readily to form azoxybenzene, an analog of nitrobenzene in which one of the oxygen atoms is replaced by the group $=N-C_6H_5$. This leads to a series of compounds designated as *bimolecular reduction products*,

$$C_6H_5-\overset{+}{N}(-\overset{-}{O})=N-C_6H_5 \xrightarrow{2H} C_6H_5-N=N-C_6H_5 \xrightarrow{2H} C_6H_5-NH-NH-C_6H_5$$

Azoxybenzene, pale yellow, mp 36° — Azobenzene, orange-red, mp 68° — Hydrazobenzene, colorless, mp 126°

[1] For a discussion of these rearrangements see Gould, *Mechanism and Structure in Organic Chemistry*, Holt, Rinehart and Winston, Inc., New York (1959).

that contain two benzene rings and two nitrogen atoms. With very mild alkaline reducing agents (sodium methoxide, sodium arsenite) the reduction of nitrobenzene proceeds merely to azoxybenzene. Alkaline sodium stannite solution (prepared from stannous chloride and excess sodium hydroxide) takes the reduction one step farther, to azobenzene. Zinc and aqueous sodium hydroxide (or magnesium and methanol) carry the reduction through azobenzene to hydrazobenzene.

Azoxybenzene and hydrazobenzene afford additional examples of rearrangements. With acid catalysts, azoxybenzene is converted to *p*-hydroxyazobenzene and hydrazobenzene undergoes rearrangement involving both rings, to form benzidine (4,4′-diaminobiphenyl).

$$C_6H_5-\overset{+}{N}(O^-)=N-C_6H_5 \xrightarrow{H^+} HO-C_6H_4-N=N-C_6H_5$$

$$C_6H_5-NH-NH-C_6H_5 \xrightarrow{H^+} H_2N-C_6H_4-C_6H_4-NH_2$$

Hydrazobenzene also exhibits another type of reaction, disproportionation or intermolecular oxidation/reduction. At temperatures somewhat above its melting point hydrazobenzene undergoes disproportionation to form azobenzene and aniline. This transformation parallels the conversion of hydrogen peroxide to molecular oxygen and water.

$$2C_6H_5-NH-NH-C_6H_5 \xrightarrow{\text{heat}} C_6H_5-N=N-C_6H_5 + 2C_6H_5-NH_2$$

## Hydrazobenzene

In a 500-ml round-bottomed flask, place 130 ml of ethanol, 0.15 mole (18.5 g, 15 ml) of nitrobenzene and a solution of 10 g of sodium hydroxide in

➤**CAUTION:** The vapor of nitrobenzene is poisonous and should not be inhaled. The liquid is a powerful skin poison. If the liquid is spilled on the skin it should be removed by washing with a little ethanol, followed by soap and warm water.

25 ml of water. Provide the flask by means of a rubber stopper with an air-cooled condenser tube about 70–80 cm in length and 10–12 mm in diameter. Heat the flask on a steam bath to about 70–75°, remove it from the steam

bath, wrap in a towel, shake *vigorously* and add about 5–6 g of zinc dust[2] from a quantity of 35 g (0.5 mole) that has been weighed previously. Shake the mixture *vigorously* until the reaction starts. If the reaction becomes violent, check it by dipping the flask in cold water.

When the reaction subsides, heat the flask on the steam bath, remove it, and add another portion of zinc dust in the same manner as before, and shake *vigorously*. Continue to warm and add portions of zinc dust until the solution changes from orange-red to *pale yellow* in color. About 35 g of zinc dust will be required; if necessary an additional quantity of 5–10 g of zinc dust may be used.

When the solution has become pale yellow,[3] add 30–40 ml of ethanol, heat to boiling, and filter it rapidly through a previously warmed Büchner funnel. Wash the sludge of zinc oxide with a small amount of hot ethanol. Transfer the filtrate to a large flask and cool it rapidly in an ice bath. Collect the crystals of hydrazobenzene with suction, working as rapidly as possible in order to avoid oxidation. Do not draw air through the filter cake. Wash the product with a little cold ethanol and press it firmly on the filter. The yield of crude product is 12–14 g.

Do not recrystallize the hydrazobenzene. Reserve 2 g of the crude product for the disproportionation experiment and divide the remainder into two equal parts (5–6 g each) to be used in the preparation of benzidine and azobenzene.

**The Benzidine Rearrangement.** Dissolve 5–6 g of the crude hydrazobenzene in 60 ml of ether and add this solution in small portions with shaking to 50 ml of ice-cold 20 percent hydrochloric acid (concentrated hydrochloric acid diluted with an equal volume of water) contained in a 250-ml flask. Stopper the flask and shake it thoroughly after each addition of the ethereal solution. Benzidine hydrochloride separates during the reaction. After all the hydrazobenzene has been introduced, add 30 ml of concentrated hydrochloric acid and allow the reaction mixture to stand for 0.5 hr in an ice bath. Collect the benzidine hydrochloride with suction, using a hardened filter paper. Wash the product with 12–15 ml of 20 percent hydrochloric acid, followed by two or three 10-ml portions of ether. The yield of benzidine hydrochloride is 6–8 g.

➤**CAUTION:** Handle benzidine carefully. If any comes in contact with the skin it should be removed by washing with a little ethanol, followed by soap and water. Benzidine is toxic—ingestion may cause nausea and liver or kidney damage. It is suspected of causing bladder-tumors.

[2] Commercial zinc dust is occasionally found to be nonreactive. It can be activated by swirling with dilute hydrochloric acid followed immediately by washing with water. The activated material must be used at once.

[3] If necessary the operation may be interrupted at this point; if this is done the flask should be fitted with a tight cork. The flask should be opened cautiously since a slight pressure may be developed.

If the free base is desired, dissolve the hydrochloride in 120 ml of warm water, filter the solution and cool rapidly to 20°. Pour the solution with stirring into a mixture of 30 ml of 10 percent aqueous sodium hydroxide and 100 g of ice, contained in a beaker. The benzidine separates in gray flocks. Collect the base with suction and wash it thoroughly with distilled water. The crude product may be crystallized from a very small amount of ethanol or from hot water (about 15 ml of ethanol, or 100 ml of water, will be required for each gram of benzidine). The yield is 2–3 g.

The benzidine rearrangement takes place readily with substituted hydrazo compounds only if the *p*-positions to the NH groups are free and if other hydrogens in the benzene rings are not unfavorably substituted. *o,o'*-Dimethoxyhydrazobenzene, and *o,o'*-dimethylhydrazobenzene, formed by alkaline reduction of *o*-nitroanisole and *o*-nitrotoluene, are converted smoothly to dianisidine (4,4'-diamino-3,3'-dimethoxybiphenyl) and tolidine (4,4'-diamino-3,3'-dimethylbiphenyl) which are important intermediates for the manufacture of substantive azo dyes for cotton (they dye directly without a mordant).

If one or both of the *p*-positions of the hydrazo compound is substituted, the rearrangement may proceed only half-way and yield an *o*- or *p*-amino derivative of diphenylamine (called a semidine).

$H_3C$–$C_6H_4$–NH–NH–$C_6H_4$–$CH_3$ $\xrightarrow[\text{warm}]{\text{HCl}}$ $H_3C$–$C_6H_4$–NH–$C_6H_3(NH_2)(CH_3)$

*p*-Hydrazotoluene — an *o*-semidine

**Oxidation to Azobenzene.** In a 125-ml Erlenmeyer flask dissolve 3 g of sodium dichromate dihydrate in 30 ml of water and add 2.5 ml of concentrated sulfuric acid, followed by 5 ml of glacial acetic acid. Swirl the solution to obtain good mixing and add 5–6 g of the crude hydrazobenzene. Warm the mixture gently for about 5 min, with continuous vigorous swirling. The azobenzene separates as an orange-red liquid that crystallizes when the mixture is cooled. Collect the product with suction and wash it thoroughly with water. After pressing firmly on the filter, crystallize the moist material from ethanol (3–4 ml per gram). The yield is about 4 g.

Azobenzene exists in geometrically isomeric forms: *cis*-, mp 70–71°; *trans*-, mp 68°. The more stable *trans* isomer is usually encountered.

**Disproportionation of Hydrazobenzene.** Place 2 g of the crude hydrazobenzene in a large test tube and warm it over a small luminous flame, while shaking the tube constantly, until the hydrazobenzene melts. Continue the heating carefully until the aniline produced in the reaction just begins to boil, then allow the tube to cool, with continued shaking. Crystals of azobenzene separate, mixed with oily drops of aniline. Add 15 ml of 10 percent aqueous acetic acid, stopper the tube, and shake vigorously to dissolve the aniline. Collect the azobenzene on a small suction filter and wash it with a little water. Crystallize the material from a small amount of ethanol.

The aniline in the dilute acetic acid filtrate may be characterized by conversion to acetanilide by the Lumière-Barbier method (Experiment 22a),[4] or to benzanilide by the Schotten-Baumann method (Experiment 12, part D).

## Questions

**1.** What reducing agents are used in the preparation of primary aromatic amines from nitro compounds?

**2.** Write the names and formulas of the substances which are formed by the reduction of nitrobenzene with zinc dust in neutral solution.

**3.** What is formed by the action of the following reagents on phenylhydroxylamine: concentrated hydrochloric acid? cold dilute sulfuric acid? potassium dichromate and dilute sulfuric acid? nitrous acid?

**4.** What relation does azobenzene bear to the azo dyes? Is azobenzene a dye?

**5.** Why is benzidine of particular importance in the preparation of azo dyes?

[4] For this purpose dilute the filtrate to a volume of 30 ml, shake with a little decolorizing carbon and filter the solution. Warm the filtrate to 40–50°, add 1 ml of acetic anhydride, shake well and add immediately a solution of 2 g of sodium acetate crystals in 5 ml of water.

EXPERIMENT

# 28

# Replacement of the Diazonium Group

## The Sandmeyer Reaction

The replacement of an aromatic primary amino group by other atoms or groups through the agency of the corresponding diazonium salts is an extremely important preparative reaction because of its simplicity and broad scope. The ease of formation and the reactivity of the diazonium salts is influenced by the character of the substituents in the aromatic nucleus and also by the nature of the anion of the salt. In typical replacement reactions the diazonium salts need not be isolated but are used in aqueous solution or suspension. In a solid dry state many diazonium salts undergo decomposition with explosive violence.

On warming in an aqueous acidic medium, most diazonium salts undergo mainly hydrolytic decomposition with formation of the corresponding phenol and nitrogen (see Experiment 30). A notable exception is the behavior of the diazonium iodides, which furnish readily the corresponding aryl iodides on heating in the presence of water.

Sandmeyer discovered that replacement of the diazonium group by chlorine, bromine, and the cyano group occurs smoothly and the yields and quality of product are improved greatly by introducing the corresponding cuprous salt into the reaction mixture. This was an important advance in the practical utilization of diazonium salts in organic syntheses. Subsequently Gattermann found that finely divided copper is also very effective in accelerating the replacement reactions.

In addition to the typical Sandmeyer reactions, the diazonium group may be replaced by a variety of other atoms and groups. Examples of the types of aromatic compounds that may be obtained in this way include the following: R–$OCH_3$, R–F, R–SCN and R–SH, R–$NO_2$, R–$AsO_3H_2$, R–$SbO_3H_2$, R–R, and R–R′.[1]

Mild reducing agents effect replacement of the diazonium group by hydrogen; this reaction is useful for the synthesis of *meta*-substituted derivatives of toluene (see Experiment 29). Arylhydrazines are obtained by reduction with sulfur dioxide or stannous chloride.

An important reaction of the diazonium salts, one in which both nitrogen atoms are retained in the product, is the coupling with arylamines and phenols to form azo dyes and azo indicators, such as methyl orange (see Experiment 31).

# *o*-Chlorotoluene[2]

**Preparation of Cuprous Chloride Solution.**[3] In a 1-liter round-bottomed flask, prepare a solution of 0.24 mole (60 g) of powdered copper sulfate crystals ($CuSO_4 \cdot 5H_2O$) and 18 g of sodium chloride in 200 ml of hot water. In a beaker prepare a solution of 14 g of sodium bisulfite and 9 g of solid sodium hydroxide in about 100 ml of water, and add this solution with swirling to the hot copper sulfate solution, over a period of 5–10 min. Cool the mixture to room temperature, allow the solid to settle, and decant off the liquid. Wash the precipitated cuprous chloride two or three times with water, by decantation. The cuprous chloride is obtained as a white powder that darkens on exposure to the air. Dissolve the cuprous chloride (as $HCuCl_2$) by adding 70 ml of concentrated hydrochloric acid and 25 ml of water. Cork the flask to minimize oxidation and place it in an ice bath.

**Diazotization of *o*-Toluidine.** In a 500-ml Erlenmeyer flask dissolve 0.2 mole (21.5 g, 21.5 ml) of *o*-toluidine in 30 ml of water and 20 ml of concentrated hydrochloric acid. When the amine has dissolved add a further quantity of 45 ml of concentrated hydrochloric acid and cool the mixture to 2–5° (internal temperature) in an ice bath. *o*-Toluidine hydrochloride separates as a finely divided crystalline precipitate. Meanwhile prepare a solution of 0.2 mole (14 g) of sodium nitrite in 40 ml of water and place the solution in a small separatory funnel.

[1] For a discussion of diazotization and replacement reactions of diazonium salts see Zollinger (translation by Nursten), *Diazo and Azo Chemistry*, Interscience Publishers, New York (1961) and Hodgson; "The Sandmeyer Reaction," *Chem. Revs.*, **40**, 251 (1947); Sidgwick, *The Organic Chemistry of Nitrogen*, third edition, revised by Millar and Springall, Clarendon Press, Oxford (1966), Chapter 16.

[2] *p*-Chlorotoluene or *m*-chlorotoluene may be prepared by the same procedure from *p*- or *m*-toluidine.

[3] An alternative procedure is to dissolve 20 g of commercial cuprous chloride in 70 ml of concentrated hydrochloric acid and 25 ml of water.

To the cold suspension of *o*-toluidine hydrochloride add 10–25 g of chipped ice, swirl the mixture, and introduce the sodium nitrite solution slowly, with the stem of the separatory funnel extending below the level of the liquid. This avoids loss of nitrous acid by surface decomposition into oxides of nitrogen; these will appear as brown fumes if the diazotization is not carried out properly. Add a small amount of ice from time to time and maintain the temperature of the reaction mixture at 2–5°. Addition of the nitrite solution requires about 10 min and a clear solution of *o*-toluene-diazonium chloride results. Allow the solution to stand for a few minutes (but not longer than 10 min) in the ice bath.[4] Proceed without delay to the next step.

**Conversion to *o*-Chlorotoluene.** While the diazotization is in progress, cool the cuprous chloride solution to 2–5° in an ice bath. Pour the cold diazonium solution slowly with shaking into the cold cuprous chloride solution. A double salt of the diazonium chloride and cuprous chloride,

$$[CH_3{-}C_6H_4{-}N{\equiv}N]^+CuCl_2^-$$

is precipitated and nitrogen is evolved slowly. Allow the mixture to stand at room temperature for about 10 min, then warm it gradually on a steam bath, with occasional swirling so that the nitrogen being evolved is disengaged from the semi-solid mass without loss of material through foaming. The cuprous complex begins to decompose with evolution of nitrogen and formation of chlorotoluene at about 15°.

Arrange an assembly for steam distillation, with a good water-cooled condenser (Figure 22, page 64). When the cuprous chloride complex has decomposed completely, steam distill the mixture until 75–80 ml of distillate has been collected; this will suffice to carry over all of the *o*-chlorotoluene. Separate carefully the layer of *o*-chlorotoluene and wash it with two 20-ml portions of cold 80 percent sulfuric acid (prepared by adding 35 ml of concentrated sulfuric acid to 10 ml of water). Shake the mixture thoroughly to extract *o*-cresol,[5] formed as a side product, and separate the layers carefully.[6] Wash the product twice with water to insure removal of sulfuric acid and dry it with 3–4 g of anhydrous calcium chloride.

[4] It is advisable to test for the presence of free nitrous acid (to insure complete diazotization and to avoid a large excess of nitrous acid), by placing a drop of the solution on potassium iodide-starch test paper. In the presence of nitrous acid, iodine is liberated and the starch is colored blue *immediately*. The test paper may be prepared by dipping strips of filter paper into 1 percent aqueous potassium iodide, then into colloidal starch solution (prepared readily from "instant starch" and cold water), and allowing them to dry.

[5] Sulfuric acid is used to remove the phenolic compound instead of sodium hydroxide solution because the latter tends to form troublesome emulsions.

[6] If in doubt about which layer is the organic halide, draw off carefully the lower layer and add one drop of it to 5 ml of water in a test tube. The organic halide will be insoluble in the water but the usual washing liquids (sulfuric acid, aqueous alkalies, etc.) will dissolve readily in the water.

Filter the liquid into a small dry distilling flask, attach an air-cooled condenser, add a boiling chip, and distill over a wire gauze. Collect the fraction boiling at 153–158°. If the distillation is pushed too far a yellow impurity will contaminate the product. Pure *o*-chlorotoluene is a colorless liquid. The yield is 12–16 g.

The chlorotoluenes may be characterized by oxidation with permanganate to the corresponding chlorobenzoic acids (see Experiment 21).

## Questions

**1.** Why is the diazotization carried out at low temperature? What side products would be formed if the solution were not strongly acidic?

**2.** Why is the Sandmeyer reaction used to prepare pure *o*- or *p*-chlorotoluene rather than direct ring chlorination of toluene?

**3.** Write a series of reactions, including the reagents and conditions used, for preparing the following compounds from *o*-toluidine: (a) *o*-toluic acid; (b) *o*-tolylhydrazine; (c) *o*-fluorobenzoic acid; (d) an azo dye.

EXPERIMENT

# 29

# Synthesis of a *meta*-Substituted Toluene

This sequence of reactions illustrates a general method that can be used to obtain a *meta*-substitution product that is not accessible by a direct substitution process, owing to the *ortho/para* directive effect of the group initially present in the aromatic system ($CH_3-$, $Cl-$, etc.). The method involves using the powerful directive influence of an acylated amino group to introduce the new substituent in the desired position and subsequently removing the amino group (replacement by hydrogen). Acetylation of the amino group diminishes reactivity of the aromatic system and reduces markedly the susceptibility to oxidative degradation.

In this synthetic sequence *p*-toluidine is acetylated by means of acetic anhydride and the resulting *p*-acetotoluidide is chlorinated by chlorine generated in the reaction mixture, from sodium chlorate and hydrochloric

$$NaClO_3 + 6HCl \longrightarrow 3Cl_2 + NaCl + 3H_2O$$

acid. For laboratory preparations this procedure is more convenient than the use of gaseous chlorine from a cylinder. In a similar way bromine can be generated from sodium bromate, sodium bromide and sulfuric acid, to avoid manipulating liquid bromine.

$$NaBrO_3 + 5NaBr + 6H_2SO_4 \longrightarrow 3Br_2 + 6NaHSO_4 + 3H_2O$$

2-Chloro-4-methylacetanilide obtained from the chlorination is subjected to hydrolysis to remove the acetyl group and the resulting amine is diazotized in acidic solution. Replacement of the diazonium group by hydrogen, with elimination of nitrogen, is effected by a mild reducing agent. Ethanol, in the presence of copper powder, is used for this purpose but gives rise to some extent to replacement of the diazonium group by an ethoxyl group. This side reaction is avoided by use of hypophosphorous acid, which

$$[R-N\equiv N]^+Cl^- + H_3PO_2 + H_2O \longrightarrow R-H + HCl + H_3PO_3 + H_2O$$

generally gives a higher yield and purer product[1] but is a relatively expensive reagent.

## 29a *m*-Chlorotoluene

**_p_-Acetotoluidide (4-Methylacetanilide).** Prepare a solution of *p*-acetotoluidide in glacial acetic acid, by refluxing for 30 min a mixture of 0.5 mole (53.5 g) of *p*-toluidine crystals, 160 g (152 ml) of glacial acetic acid, and 0.55 mole (56.5 g, 52 ml) of acetic anhydride. Use this solution directly for the next step without isolating the product. If a sample of the intermediate is desired, pour a small portion of the solution into water and crystallize the *p*-acetotoluide from 50 percent ethanol.

**Chloro-_p_-acetotoluidide (2-Chloro-4-methylacetanilide).** Remove the reflux condenser and cool the above solution in a bath of cold water, while swirling the solution vigorously, until the temperature has fallen to about 40°; this may cause a small portion of the *p*-acetotoluidide to separate in crystalline flakes. Without delay add 140 ml of concentrated hydrochloric acid and cool the mixture in an ice bath to 0–5°. While maintaining the temperature at 0–5° and swirling the reaction mixture, add dropwise from a separatory funnel a solution of 0.21 mole (22 g) of sodium chlorate ($NaClO_3$) in 50 ml of water. After all of the chlorate solution has been added allow the mixture to stand at room temperature (not above 20°) for an hour, with occasional shaking.

Pour the contents of the flask into a vigorously stirred mixture of 250 g of ice and 250 ml of water, to which 6–7 g of solid sodium bisulfite has been added. Disintegrate any large lumps of the precipitated 2-chloro-4-methylacetanilide and collect the product on a suction filter. Wash the crystals *thoroughly* with water and press them as dry as possible on the filter. The slightly moist crude product may be used directly for the next step.

If a specimen of this intermediate is desired a small sample may be recrystallized from 20 percent aqueous ethanol, with addition of decolorizing carbon. 2-Chloro-4-methylacetanilide melts at 115–116°.

**Chloro-_p_-toluidine Hydrochloride (2-Chloro-4-methylaniline Hydrochloride).** Transfer the chloro-*p*-acetotoluidide to a 1-liter round-bottomed flask, add 150 ml of ethanol, adjust a reflux condenser and warm the mixture until the solid has dissolved completely. Through the condenser tube add 125 ml of concentrated hydrochloric acid, introduce a boiling chip, and boil the mixture gently for 2–3 hr. During this time crystals of the amine hydrochloride

[1] Kornblum, *Organic Reactions*, **2**, 263–340 (1944).

may separate. Pour the hot reaction mixture into a large beaker and cool it thoroughly. Collect the crystals of the hydrochloride with suction, press them thoroughly and wash them with two 15-ml portions of cold ethanol. The yield is 53–60 g.

**2-Chloro-4-methylaniline.** Suspend the amine hydrochloride in about 200 ml of water in a large flask and add a solution of 20 g of solid sodium hydroxide in 100 ml of water. Swirl the mixture thoroughly for 10 min to insure complete liberation of the amine. The mixture must be strongly basic; test with litmus paper and add more alkali if necessary. Cool the mixture to 20°, transfer it to a separatory funnel, and separate carefully the crude chloro-*p*-toluidine. The yield is 42–50 g. It is unnecessary to purify the amine before going on to the next step.

If the pure amine is desired the crude product may be purified by a preliminary steam distillation, followed by drying over solid sodium hydroxide, and distillation. 2-Chloro-4-methylaniline distills at 108–110° at 10 mm pressure, or 220–225° at atmospheric pressure. The pure amine crystallizes on cooling and melts at 5–7°.

***m*-Chlorotoluene.** To a cold mixture of 200 ml of ethanol and 50 ml of concentrated sulfuric acid in a 2-liter round-bottomed flask, add the crude amine from the previous step (42–50 g). Cool the solution to 5° and while shaking thoroughly, add slowly from a separatory funnel a solution of 0.4 mole (28 g) of sodium nitrite in 50 ml of water. Keep the reaction mixture cold during the slow addition of the sodium nitrite solution and allow it to stand for 10 min after all the nitrite has been added, to insure complete diazotization. At this time a drop of the solution on starch-iodide test paper should give a blue color immediately, indicating free nitrous acid (see footnote, page 324).

To the *cold* solution add 8–9 g of powdered copper (copper bronze) that has been washed previously with a little ether. Fit the flask at once with a long reflux condenser and prepare an ice-water bath to cool the flask if the reaction becomes too vigorous.

Warm the flask *very cautiously* on a steam bath until a rapid evolution of gases begins (nitrogen and acetaldehyde), then chill at once in an ice bath to moderate the vigorous reaction (*caution—flammable vapors!*). When the reaction has subsided, again warm the flask gently and finally heat for 10 min on the steam bath. At the end of the reaction the color of the solution changes from reddish-brown to yellow. The success of this step depends upon controlling the evolution of gases to avoid loss of material through violent decomposition.

Add about 500 ml of water and steam distill the reaction mixture (Figure 22, page 64) as long as any oily drops pass over (about 125–150 ml of distillate). Separate the yellow organic liquid, wash it with two 25-ml portions of cold 80 percent sulfuric acid[2] (prepared by adding 40 ml of concentrated sulfuric acid to 14 ml of water). Wash the product finally with two 25-ml portions of water to remove all of the sulfuric acid. Dry the material with 3–4 g of anhydrous calcium chloride and filter the liquid into a small dry distilling flask. Distill, using an air-cooled condenser, and collect the *m*-chlorotoluene boiling at 156–162°. The yield is 20–25 g. Calculate the percentage yield on the basis of the original *p*-toluidine.

***Reactions.*** *m*-Chlorotoluene may be characterized by oxidation to *m*-chlorobenzoic acid, mp 158°, and also by conversion to various substitution products. The directive effects of the two substituents reinforce one another, consequently the introduction of a new group occurs mainly at the 4- or 6-position. Disubstitution products are usually the 4,6-derivatives, such as 3-chloro-4,6-dinitrotoluene, mp 91°.

Side chain chlorination, at elevated temperature and with chlorine and ultraviolet illumination (or with sulfuryl chloride and a peroxide catalyst), leads to *m*-chlorobenzyl chloride. Ring chlorination, in the presence of ferric or aluminum chloride, gives a mixture of 3,4- and 2,5-dichlorotoluenes.

*m*-Chlorotoluene is unaffected by hot aqueous alkalies but at high temperatures is converted to a mixture of *o*-, *m*-, and *p*-cresols. This anomalous behavior is attributed to the formation of a *benzyne* intermediate, formed by elimination of hydrogen chloride *before* introduction of the hydroxyl group. *p*-Chlorotoluene under similar conditions gives a mixture of *m*- and *p*-cresols, but no *o*-cresol:

$CH_3$–(benzyne) ⟶ $CH_3$–(ring)–OH (meta) and $CH_3$–(ring)–OH (para)

The benzyne mechanism is suppressed by the addition of copper-cuprous oxide catalyst, which permits the conversion of aryl halides to isomer-free phenols.[3] With this catalyst *p*-bromotoluene furnishes pure *p*-cresol.

## 29b *m*-Bromotoluene

The synthesis of *m*-bromotoluene from *p*-toluidine is an example of the general method outlined in the introduction to this experiment. The sequence

[2] A general procedure for determining which layer is the organic bromide is to separate the two layers carefully and add cautiously a few drops of one layer to 5 ml of water in a test tube. The organic bromide will be insoluble in the water but usual washing liquids (sulfuric acid, aqueous sodium bicarbonate, etc.) will be completely soluble.

[3] Gumprecht, *Organic Syntheses*, **48**, 96 (1968). For a review of benzyne chemistry see Wittig, *Z. Angew. chem. Intern. Ed.*, **4**, 731 (1965).

of steps is the same as that used for *m*-chlorotoluene and differs merely in the introduction of bromine into *p*-acetotoluidide instead of chlorine. Liquid bromine is used directly instead of generating the halogen in the reaction mixture. The handling of bromine requires special care.

**Bromo-*p*-acetotoluidide (2-Bromo-4-methylacetanilide).** Prepare a solution of 0.5 mole of *p*-acetotoluidide in glacial acetic acid as described in Experiment 29a. Remove the condenser and cool the solution while swirling the solution vigorously, until the temperature has fallen to 45–50°; this may cause a small portion of the *p*-acetotoluidide to separate in crystalline flakes. From a separatory funnel allow 0.52 mole (84 g, 27 ml) of bromine to drop slowly

➧CAUTION: Handle bromine carefully since the liquid produces painful burns and the vapors are obnoxious (hood). Use both hands in adjusting the stopcock of the separatory funnel to avoid loosening it, with consequent leakage. Treat bromine burns immediately with a liberal quantity of glycerol. Follow this by thorough washing with water, dry the skin, and apply a healing ointment or salve.

It is advisable to keep a small beaker containing glycerol at hand when manipulating bromine.

into the solution, over a period of about 20 min. Adjust the rate of addition of the bromine so that the temperature is held at 45–55°, and continue to stir or shake vigorously during the addition of the bromine. After allowing the reaction mixture to stand for 15 min with occasional shaking, pour it into a vigorously stirred mixture of 350 g of ice and 350 g of water, to which 6–7 g of solid sodium bisulfite has been added. Disintegrate any large lumps of the precipitated 2-bromo-4-methylacetanilide and collect the product with suction. Wash the crystals *thoroughly* with water and press them as dry as possible on the filter. Remove the crude product from the funnel and allow it to dry until its weight does not exceed 125 g before proceeding to the hydrolysis. It is unnecessary to purify the crude bromo-*p*-acetotoluidide at this stage.

If a specimen of this intermediate is desired, a small sample may be crystallized from ethanol. 2-Bromo-4-methylacetanilide melts at 117–118°.

**Bromo-*p*-toluidine Hydrochloride (2-Bromo-4-methylaniline Hydrochloride).** Prepare the hydrochloride using the same method and quantities of reagents used in the preparation of chloro-*p*-toluidine hydrochloride described in Experiment 29a. Wash the collected crystals with two 25-ml portions of chilled ethanol. The yield is 65–75 g.

**Bromo-*p*-toluidine (2-Bromo-4-methylaniline).** Convert the hydrochloride into the free amine using the same method and quantities of reagents specified in the preparation of 2-chloro-4-methylaniline described in Experiment 29a.

The yield is 55–62 g. It is unnecessary to purify the amine before proceeding to the next step.

If the pure amine is desired the crude product can be purified by drying over solid sodium hydroxide and distilling under diminished pressure. 2-Bromo-4-methylaniline distills at 120–122° under 30 mm pressure; the purified amine solidifies on cooling and melts at 25–26°.

***m*-Bromotoluene.** Deaminate the bromo-*p*-toluidine using the same procedure and quantities of reagents specified in the preparation of *m*-chlorotoluene described in Experiment 29a. After drying the product distill it, using an air-cooled condenser, and collect the *m*-bromotoluene boiling at 178–183°. The yield is 20–25 g. Calculate the percentage yield on the basis of the original *p*-toluidine.

## Questions

**1.** In the preparation of chloro-*p*-toluidine why is it necessary to acylate *p*-toluidine instead of chlorinating the amine directly?

**2.** Compare the behavior of aniline and *p*-acetotoluidide toward oxidizing agents.

**3.** Show the reactions that would occur if *o*-acetotoluidide were used in place of *p*-acetotoluidide in this experiment.

**4.** Outline a series of reactions suitable for the preparation of the following compounds in a pure state: *m*-chlorophenol from benzene; 3,5-dinitrotoluene from *p*-toluidine.

**5.** Give an explanation for the observation that *m*-chlorotoluene when aminated with sodamide ($NaNH_2$), produces a mixture of the three isomeric toluidines. (Consider the benzyne intermediate.)

EXPERIMENT

# 30

# Synthesis of a *meta*-Substituted Derivative of Aniline and of Phenol

Owing to the strong *ortho/para* directive influence of the phenolic hydroxyl group, *meta*-substituted derivatives of phenol and its homologs cannot be prepared by the usual substitution reactions. Nitration of phenol affords a direct route to *o*- and *p*-nitrophenol, which are separated easily since only the *ortho* isomer is volatile with steam. Likewise, *ortho*- and *para*-substituted anilines may be obtained by nitration or halogenation of acetanilide and subsequent removal of the acetyl group (see Experiments 23 and 24).

A good starting material for the synthesis of several *meta*-substituted derivatives of aniline and phenol is *m*-dinitrobenzene, which is obtained readily by further nitration of nitrobenzene (Experiment 18b) and is an inexpensive industrial intermediate. Powerful reducing agents, such as tin or iron in acidic medium, effect reduction of both nitro groups to form *m*-phenylenediamine but partial reduction to *m*-nitroaniline can be accomplished by controlled action of mild reducing agents. For this purpose sodium hydrosulfide (NaSH) or sodium polysulfide ($Na_2S_3$, etc.) is suitable.

When sodium polysulfide is used a quantity of elemental sulfur is precipitated with the *m*-nitroaniline, which must then be isolated by repeated extraction with hot aqueous hydrochloric acid. Usually some 4,4′-dinitroazoxybenzene is formed as a side product, owing to the high alkalinity of the reducing medium (see equation 1). Sodium hydrosulfide has the advantage of producing a purer product and little or no elemental sulfur is formed. The reduction process involves two or more concurrent reactions, corresponding approximately to equation 2.

$$R{-}NO_2 + 3Na_2S + 4H_2O \longrightarrow R{-}NH_2 + 6NaOH + 3S \qquad (1)$$

$$4R{-}NO_2 + 6NaSH + H_2O \longrightarrow 4R{-}NH_2 + 3Na_2S_2O_3 \qquad (2)$$

For small scale laboratory preparations a solution of sodium hydrosulfide in aqueous methanol can be prepared conveniently by treating aqueous sodium sulfide with sodium bicarbonate and precipitating the resulting sodium carbonate by addition of methanol.[1]

$$Na_2S + NaHCO_3 \longrightarrow NaSH + Na_2CO_3$$

Replacement of the amino group of *m*-nitroaniline by other groups is accomplished by diazotization, to form a *m*-nitrobenzenediazonium salt, followed by replacement of the diazonium function by the desired substituent. Upon hydrolysis in hot dilute sulfuric acid *m*-nitrophenol is produced. By means of appropriate replacement reactions *m*-bromonitrobenzene, *m*-nitrobenzenearsonic acid and many other *meta*-substituted nitro compounds

$$[O_2N{-}C_6H_4{-}N{\equiv}N]^+HSO_4^- + H_2O \xrightarrow[\text{heat}]{} O_2N{-}C_6H_4{-}OH + H_2SO_4 + N_2$$

may be obtained. Reduction of these nitro compounds, in turn, affords access to the corresponding amines: *m*-aminophenol, *m*-bromoaniline, *m*-arsanilic acid ($H_2N{-}C_6H_4{-}AsO_3H_2$), etc.

# 30a *m*-Nitroaniline

**Sodium Hydrosulfide Solution.**[2] In a 600-ml beaker dissolve 0.2 mole (48 g) of sodium sulfide crystals ($Na_2S \cdot 9H_2O$)[3] in 90 ml of water. To the solution, maintained at 15–20°, add 0.2 mole (17 g) of finely pulverized anhydrous sodium bicarbonate, in small portions, with good stirring. After the bicarbonate has dissolved completely add slowly 120 ml of methanol, with stirring and cooling. Precipitation of sodium carbonate monohydrate begins at once, with evolution of heat. After allowing the mixture to stand for 30 min filter off the sodium carbonate crystals with suction (save the filtrate and washings) and wash them with three 20-ml portions of cold methanol.

The combined filtrate and washings contain about 0.2 mole (11 g) of sodium hydrosulfide and a little dissolved sodium carbonate. The solution should be used promptly. If necessary the experiment may be interrupted at

[1] Hodgson and Ward, *J. Chem. Soc.*, 242 (1948).

[2] If technical sodium hydrosulfide dihydrate ("Sodium Sulfhydrate") is available this solution may be prepared merely by dissolving 19 g of the dihydrate in 110 ml of water and adding 150 ml of methanol. This is the simplest procedure and gives excellent results. Technical sodium hydrosulfide may be purchased from Matheson, Coleman and Bell (catalog No. SX585) or Will Scientific, Incorporated. Sodium hydrosulfide (NaSH) must not be confused with sodium hydrosulfite ($Na_2S_2O_4$).

[3] Sodium sulfide is available through chemical suppliers as the crystalline nonahydrate, $Na_2S \cdot 9H_2O$, and as the trihydrate, $Na_2S \cdot 3H_2O$, known as "fused flakes (60%)." In this preparation 26.5 g of the trihydrate and 24 ml of water may be used instead of 50 g of the nonahydrate.

the next step, after refluxing the mixture of *m*-dinitrobenzene and hydrosulfide solution.

***m*-Nitroaniline.** In a 1-liter round-bottomed flask dissolve 0.12 mole (20 g) of pure *m*-dinitrobenzene in 150 ml of hot methanol and add, with shaking, the previously prepared methanolic solution of sodium hydrosulfide. Adjust a reflux condenser, add a boiling chip, and boil the orange-brown mixture gently for 20 min.[4] Allow the mixture to cool slightly and connect the flask by a bent glass tube to a condenser arranged for distillation. Distill the mixture until most of the methanol has been removed (250–300 ml of distillate) and pour the residue, with stirring, into about 600 ml of cold water. Collect the yellow crystals of *m*-nitroaniline with suction, wash them thoroughly with water and press dry. The moist product weighs 18–20 g; this corresponds to about 13–14 g of dry *m*-nitroaniline.

The crude product (mp 108–112°) is usually sufficiently pure for conversion to *m*-nitrophenol. If desired the material may be purified by recrystallization from hot 75 percent aqueous ethanol, with addition of a little decolorizing carbon. Pure *m*-nitroaniline crystallizes in yellow needles, mp 113–114°.

➤**CAUTION:** Manipulate *m*-nitroaniline and *m*-nitrophenol carefully. These substances will discolor the hands and stain clothing.

***Reactions.*** *m*-Nitroaniline may be characterized by acylation of the amino group with acetic anhydride to form *m*-nitroacetanilide, mp 152–153°, or with benzenesulfonyl chloride to form N-(3-nitrophenyl)benzenesulfonamide, mp 136°.

Diazotization of the amino group leads to *m*-nitrobenzenediazonium salts that can be subjected to replacement reactions and also to coupling reactions with amines and phenols to form azo dyes.[5] The coupling reaction is discussed in Experiment 31.

The three nitroanilines are much weaker bases than aniline; the reduction of basic strength is greatest in the *ortho* isomer and least in the *meta* isomer. The basicity constants, $K_B$, have approximately the following relationship if aniline ($K_B = 4.2 \times 10^{-10}$) is assigned arbitrarily a value of 1000: *m*-nitro, 0.76; *p*-nitro, 0.024; *o*-nitro, 0.008.

[4] If the hydrosulfide solution was prepared from sodium sulfide and sodium bicarbonate, a small amount of sodium carbonate (1–2 g) may precipitate as a fine powder in the hot solution. The carbonate need not be removed as it will dissolve later.

[5] Reference books in this field are: Sidgwick, Taylor, and Baker, *The Organic Chemistry of Nitrogen*, third edition, revised by Millar and Springall, Clarendon Press, Oxford (1966), Chapter 16; Zollinger (translation by Nursten), *Diazo and Azo Chemistry*, Interscience Publishers, New York (1961).

## 30b *m*-Nitrophenol

In a 600-ml beaker place 0.1 mole (14 g) (dry basis) of finely pulverized *m*-nitroaniline and add slowly with stirring, a cold solution of 50 percent aqueous sulfuric acid prepared by adding 22 ml of concentrated sulfuric acid to 35 ml of water. Stir the mixture thoroughly and add 45–50 g of ice. Continue the stirring until the suspension of *m*-nitroaniline sulfate has become completely homogeneous.

To the well-stirred suspension of the amine sulfate add in small portions, over a period of 5 min,[6] a solution of 0.11 mole (6.8 g) of sodium nitrite in 20 ml of water. The mixture should be maintained at 5–10° during the diazotization by adding small pieces of ice as needed. After stirring for 10 min longer a drop of the solution should give a positive test for nitrous acid (immediate blue color) with starch-iodide paper.[7] If necessary add a *small* additional amount of sodium nitrite, not more than 0.5 g. Finally allow the crystals of the diazonium salt to settle undisturbed for 5 or 10 min. Decant most of the supernatant solution from the crystals into an Erlenmeyer flask. Proceed at once to the next step.

In a 1-liter round-bottomed flask, firmly supported by a clamp, place 50 ml of water and add carefully with shaking, 65 ml of concentrated sulfuric acid. Heat the diluted sulfuric acid to the boiling point ($\sim 60°$) and add slowly through a separatory funnel, the decanted solution of the diazonium salt (*caution—foaming!*). Adjust the rate of addition and the heating so that the hot mixture continues to boil vigorously. About 20 min is required for the addition.

After all the solution has been introduced remove the separatory funnel and add the moist diazonium sulfate crystals cautiously, in *small* portions, to the boiling reaction mixture. Control the rate of addition so that the evolution of nitrogen does not cause excessive foaming and loss of material. During the reaction the mixture becomes brown in color and some *m*-nitrophenol separates as a dark oil.

Maintain the reaction mixture at the boiling point for 15 min after all the solid has been added. Allow the mixture to cool to about 100° and pour it slowly, with vigorous stirring, into a large beaker cooled in an ice-water bath. During this operation effective stirring is essential to obtain small homogeneous crystals of the product. When the mixture has become thoroughly chilled, collect the crude *m*-nitrophenol on a suction filter, wash it

[6] The nitrite is added as rapidly as possible in order to prevent coupling of the diazonium salt with undiazotized amine, which would form the yellow, insoluble diazoamino compound. The strongly acidic medium also retards the coupling process (see Experiment 31).

[7] The presence of free nitrous acid indicates that sufficient reagent is present to diazotize all of the amine sulfate. It is customary to use 3–5 per cent excess of sodium nitrite over the calculated requirement. Excessive amounts of nitrous acid give rise to side reactions such as replacement of the diazonium group by $-NO_2$ and nitrosation of the *m*-nitrophenol.

with several small portions of ice-cold water and press dry. The yield of slightly moist, brown product at this stage is 8–10 g.

Dissolve the crude *m*-nitrophenol in hot 15 percent hydrochloric acid (prepared by adding 50 ml of concentrated acid to 80 ml of water), using about 10–12 ml of the solvent per gram of the moist material. Boil the solution with a small quantity of decolorizing carbon, then allow it to cool slightly before filtering. When the filtrate has cooled to about 50° pour the solution with vigorous stirring into a large beaker cooled in an ice-water bath. If the operation is done carefully the *m*-nitrophenol separates directly in pale yellow crystals. Chill the material thoroughly, collect the crystals on a suction filter, and wash them with small portions of ice-cold water. After drying, the purified product weighs 5–6 g. The recorded melting point of *m*-nitrophenol is 96°.

***Reactions.*** *m*-Nitrophenol can be transformed into derivatives through reactions of the phenolic hydroxyl function and of the nitro group, and also by introduction of additional substituents through direct substitution reactions.

Acetylation with acetic anhydride and sodium acetate gives *m*-nitrophenyl acetate, mp 55–56°; benzoyl chloride in the presence of aqueous sodium hydroxide (Schotten-Baumann method) gives *m*-nitrophenyl benzoate, mp 95°. Reduction with stannous chloride and hydrochloric acid or with sodium hydrosulfite leads to *m*-aminophenol.

Alkylation of *m*-nitrophenol in alkaline medium with dimethyl sulfate gives the methyl ether, *m*-nitroanisole; with chloroacetic acid in alkaline medium *m*-nitrophenoxyacetic acid, mp 154–156°, is formed.

## Questions

**1.** Suggest an explanation for the observation that *m*-nitroaniline is not attacked by boiling sodium hydroxide solution, whereas *o*- and *p*-nitroanilines are hydrolyzed to the corresponding nitrophenols and ammonia (a nucleophilic substitution reaction).

**2.** In the diazotization of an aromatic amine why does a strongly acidic medium disfavor coupling of the diazonium salt? What conditions are favorable for the coupling reaction with amines and phenols? (See Experiment 31.)

**3.** Give a series of reactions, including reagents and experimental conditions, for the preparation of the following compounds starting from a *m*-nitrobenzenediazonium salt: (a) *m*-fluoroaniline; (b) a diazoamino compound; (c) an arylarsonic acid; (d) 3,3′-dinitrobiphenyl.

**4.** Explain why *m*-nitrophenol is a stronger acid than phenol but weaker than *o*- and *p*-nitrophenol.

**5.** Give a series of reactions for the conversion of *m*-nitrophenol into the following: *m*-phenetidine ($C_2H_5O{-}C_6H_4{-}NH_2$); *m*-nitrophenyl N-phenylcarbamate ($O_2N{-}C_6H_4{-}O{-}CO{-}NHC_6H_5$).

# EXPERIMENT 31

# Diazonium Coupling Reactions

Methyl orange belongs to a class of dyes known as "azo colors," which contain the $-N{=}N-$ group linked to two aromatic nuclei. In addition to the azo group the dyes must contain salt-forming groups such as hydroxyl, amino, sulfonic acid or carboxyl groups (auxochromes) which usually intensify the color and at the same time enable the molecule to attach itself to the fabric, or combine with the mordant to form a lake. Two typical commercial azo dyes are shown below.

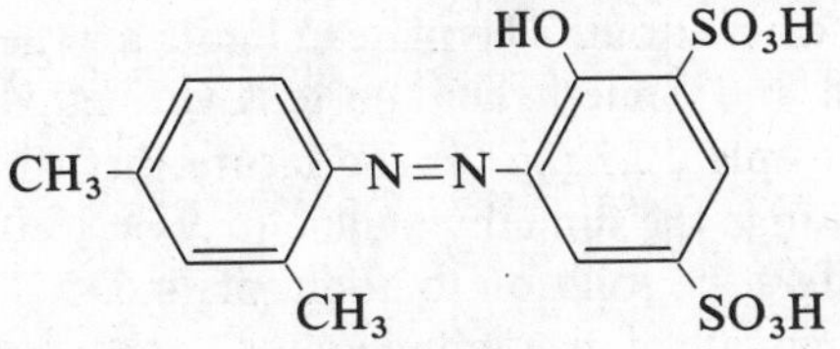

Ponceau 2R
(a brilliant scarlet)

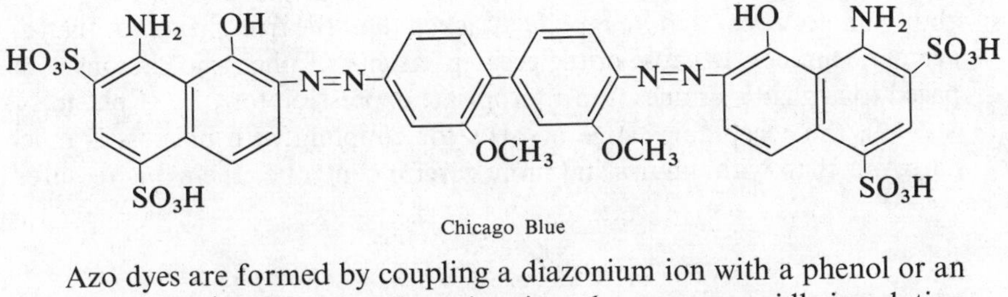

Chicago Blue

Azo dyes are formed by coupling a diazonium ion with a phenol or an aromatic amine.[1] Since many diazonium ions decompose rapidly in solution

[1] Reference books in this field are: Sidgwick, *The Organic Chemistry of Nitrogen*, third edition, revised by Millar and Springall, Clarendon Press, Oxford (1966), Chapter 16; Zollinger (translation by Nursten), *Diazo and Azo Chemistry*, Interscience Publishers, New York (1961).

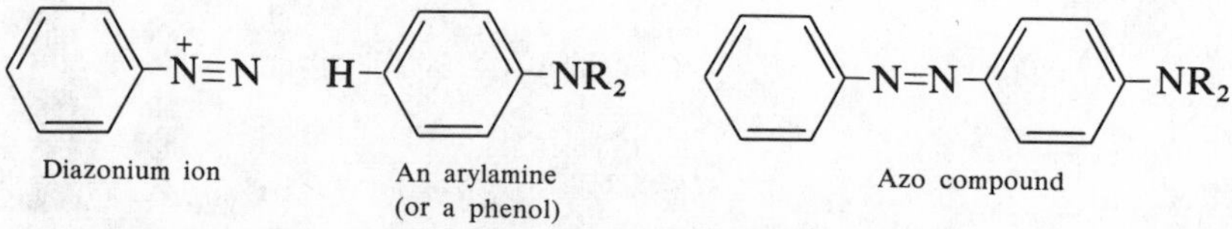

$C_6H_5\overset{+}{N}\equiv N$ — Diazonium ion

$H-C_6H_4-NR_2$ — An arylamine (or a phenol)

$C_6H_5-N=N-C_6H_4-NR_2$ — Azo compound

it is desirable that the coupling reaction be completed quickly. The rate at which a diazonium ion couples with an aromatic amine is proportional to the product of the concentrations of the diazonium ion and the free (unprotonated) amine. At high pH's the diazonium ion is converted into the unreactive diazoate anion and at low pH's the free amine is converted into the unreactive ammonium salt. Only at intermediate pH's will there be a

$$ArN_2^+ + H_2O \underset{}{\overset{K_1}{\rightleftharpoons}} ArN_2O^- + 2H^+$$

$$Ar\overset{+}{N}HR_2 \underset{}{\overset{K_2}{\rightleftharpoons}} ArNR_2 + H^+$$

sufficient concentration of both required species to give a significant coupling rate.

The pH dependence of the relative rate of coupling can be expressed quantitatively by solving the equilibrium equations for the concentrations of free amine and diazonium ion in terms of $[H^+]$ and $K_1$ and $K_2$. It is found that the rate is nearly constant for pH's between $\frac{1}{2}pK_1$ and $pK_2$ ($pK$ is defined as $-\log K$ by analogy to pH) and diminishes sharply for pH's outside this range. There exists a broad plateau of maximum coupling rate that is of great practical significance, for it is within the range of pH's defining this plateau that the coupling reaction should be carried out. The plateau limits for the coupling of *p*-diazobenzenesulfonate and dimethylaniline to give methyl orange are about pH 4.4 ($pK_2$) and ~pH 10.7 ($\frac{1}{2}pK_1$). To insure that the reaction solution is not outside this range the dimethylaniline may be converted into its acetate salt, which buffers the solution to about pH 4.6.

The influence of pH on the coupling rate of diazonium ions with phenols is qualitatively similar to the situation with amines. With phenols the *active* coupling species is the phenoxide anion, which, because of the greater availability of electrons, is more rapidly attacked than the free phenol (or than an aromatic amine). Because of the greater basicity of phenoxide anions compared to aromatic amines (the appropriate expression for $pK_2$ of phenols is ~7–10), the acceptable range of pH's for coupling with phenols is much narrower than with amines and more careful control of the pH is required.

# 31a Methyl Orange

**Diazotization.** In a 600-ml beaker, place 50 ml of a 5 percent solution of sodium carbonate, dilute it with water to about 100 ml, and add 0.06 mole (10.5 g) of anhydrous sulfanilic acid (or 11.5 g of the hydrate). Warm slightly on a steam bath and if the sulfanilic acid does not dissolve completely add an additional 5–10 ml of 5 percent aqueous sodium carbonate (do not add more than 10 ml). If necessary, filter the solution with suction to remove any undissolved residue. Carefully weigh out 0.05 mole (3.5 g) of sodium nitrite, dissolve it in about 20 ml of water, and add the nitrite solution to the sodium sulfanilate. Cool the solution in a slush of water and ice until the temperature is between 3–5°, then stir vigorously and add drop by drop, a solution of 7 g (6 ml) of concentrated hydrochloric acid diluted with about 10 ml of water. Do not allow the resulting diazonium solution to stand any longer than necessary; proceed at once to the next step.[2]

**Coupling.** To 0.05 mole (6 g, 6.3 ml) of dimethylaniline in a test tube, add 0.053 mole (3 g, 3 ml) of glacial acetic acid and mix thoroughly. To the diazonium salt solution add quickly, with thorough mixing, the dimethylaniline acetate and allow the mixture to stand with occasional stirring for 5–10 min. Finally make the solution alkaline by adding a solution of 7 g of solid sodium hydroxide in about 20 ml of water. This causes the deep red color to change to a yellowish orange. The methyl orange separates at once; it may be made to precipitate more completely by adding about 20 g of clean salt. Collect the precipitate with suction, using a hardened filter paper, and crystallize the impure product from hot water. Usually about 20–25 ml of hot water will be required for each gram of material to be crystallized. Cool the hot filtered solution, filter the crystals with suction, wash them with ethanol, and finally with ether. The yield is 10–13 g of purified methyl orange. Do not attempt to determine the melting point of this substance.

Dissolve a little methyl orange in water, add a few drops of dilute hydrochloric acid, then make alkaline again with dilute sodium hydroxide solution. Observe the color changes. The effect of acids and alkalies is probably represented by the following structural changes:

[2] It is advisable to test for the presence of free nitrous acid (to insure complete diazotization and to avoid a large excess of nitrous acid) by placing a drop of the solution on potassium iodide-starch test paper. In the presence of nitrous acid, iodine is liberated and the starch is colored blue *immediately*. The test paper may be prepared by dipping strips of filter paper into 1 percent aqueous potassium iodide, then into colloidal starch solution (prepared readily from "instant starch"), and allowing them to dry.

$^{-}O_3S-C_6H_4-N=N-C_6H_4-N(CH_3)_2$

Anion (alkaline solution) yellow

$H^+ \downarrow \quad \uparrow OH^-$

$^{-}O_3S-C_6H_4-NH-N=C_6H_4=N(CH_3)_2{}^+$

Inner salt (acid solution) red

Many types of organic molecules can be used as indicators. All these have the property of undergoing practically instantaneous change (or changes) in structure in going from acid to alkaline solution, or the reverse, generally within a narrow pH range.

## 31b Para Red[2]

In a 600-ml beaker dissolve 0.025 mole (1 g) of sodium hydroxide, 0.025 mole (9.5 g) of commercial trisodium phosphate ($Na_3PO_4 . 12H_2O$)[3] and 0.01 mole (1.4 g) of 2-naphthol in 200 ml of water in that order. Chill the solution by allowing it to stand in an ice bath for 5 min with intermittent stirring.

Prepare a solution of 0.01 mole of diazotized *p*-nitroaniline as described in Experiment 32 and pour it all at once into the chilled alkaline solution of 2-naphthol. Stir vigorously for a few minutes to insure complete reaction and then, after adding 5 ml of concentrated hydrochloric acid, raise the temperature to $\sim 30°$ on the steam bath and stir for 0.5–1 hr.

Collect the bright red dye with suction, using hardened filter paper. Allow the solid to dry completely and then extract the inorganic salts by stirring with 100 ml of water. Collect the solid with suction and wash with water until the filtrate is essentially free of chloride ion.[4] Wash the crystals with small portions of ethyl alcohol. After drying, the product weighs about 2.7 g. Do not attempt to determine the melting point of this substance.

[2] This preparation is based on the procedure described by Lucas and Pressman, *Principles and Practice in Organic Chemistry*, John Wiley and Sons, New York (1949), p. 384.

[3] The phosphate is added as a buffer.

[4] Acidify a 2-ml portion of the filtrate with 1 ml of concentrated nitric acid and add 3 ml of dilute (5–10 percent) aqueous silver nitrate.

## Questions

**1.** What effect will the following substituents have on the rate of coupling of the aryldiazonium ion, relative to the benzenediazonium ion: (a) *p*-nitro; (b) *m*-nitro; (c) *p*-methoxy; (d) 2,4-dimethyl?

**2.** What is a diazoamino compound? How may it be converted to an aminoaryl azo compound?

**3.** At pH 5–6, 7-amino-2-naphthol undergoes coupling at the 8-position, but at pH 10 the coupling occurs at the 1-position. Explain this behavior.

**4.** What compounds are formed by the reduction of methyl orange with strong reducing agents, such as sodium hydrosulfite ($Na_2S_2O_4$) or stannous chloride? Suggest a method for separating and characterizing the reduction products.

**5.** What is meant by a chromophore group? an auxochrome group? Give examples of each.

**6.** Define or explain the following terms: (a) direct or substantive dye; (b) mordant or adjective dye; (c) vat dye. Cite an example of each.

EXPERIMENT

# 32

# The Meerwein Arylation Reaction[1]

The low reactivity of aryl halides does not permit direct introduction of aryl groups by the conventional $S_N2$ nucleophilic displacement reactions that occur with typical alkyl halides. A method of effecting arylation known as the Gomberg-Bachmann reaction[2] involves the interaction of an aryl diazotate and an aromatic substrate (hydrocarbon, alkoxy or nitro compound, benzoic ester, thiophene, etc.):

$$[\text{Ar}-\overset{+}{\text{N}}\equiv\text{N}]\overset{-}{\text{X}} + \text{C}_6\text{H}_5-\text{R} \xrightarrow[\textit{or}\ \text{NaOAc}]{\text{NaOH}} \text{Ar}-\text{C}_6\text{H}_4-\text{R} + \text{N}_2 + \overset{+}{\text{Na}}\overset{-}{\text{X}}$$

A modification of the arylation process that has extended the scope of the reaction makes use of the N-nitroso derivatives of acetylated arylamines as a source of the diazotate. The nitroso derivatives are obtained by nitrosation

$$\text{R}-\text{C}_6\text{H}_4-\text{NH}-\text{Ac} \longrightarrow \text{R}-\text{C}_6\text{H}_4-\underset{\text{N=O}}{\underset{|}{\text{N}}}-\text{Ac} \longrightarrow \text{R}-\text{C}_6\text{H}_4-\text{N}=\text{N}-\text{OAc}$$

with nitrosyl chloride or oxides of nitrogen. This variation has led to improved yields and made accessible many compounds that could not have been obtained by the original method; for example, various terphenyls and quaterphenyls.

A typical Meerwein arylation[3] involves the reaction of an aryldiazonium chloride with an activated *aliphatic* double bond, in the presence of acetate buffer, usually in acetone solution and with cupric chloride as catalyst. The

[1] The name of Professor Hans Meerwein is associated also with the Meerwein-Ponndorf-Verley reduction and the Wagner-Meerwein rearrangement (p. 445).

[2] Gomberg and Bachmann, *J. Amer. Chem. Soc.*, **46**, 2339 (1924); Bachmann and Hofmann, *Organic Reactions*, **2**, 224 (1944); also, *Organic Syntheses*, Collective Volume **I**, 113 (1941).

[3] Meerwein, Büchner, and van Emster, *J. prakt. Chem.*, [2] **152**, 239 (1939); Rondestvedt, *Organic Reactions*, **11**, 189 (1960).

aliphatic reactants include $\alpha,\beta$-unsaturated acids, esters, and nitriles; maleic esters and imides; furfural; styrene, 1,3-dienes, and quinones.

$$Ar{-}N_2{-}Cl + CH_2{=}CH{-}CO_2H \xrightarrow[Cu^{++}]{NaOAc} Ar{-}CH{=}CH{-}CO_2H + N_2$$

$$Ar{-}N_2{-}Cl + CH_2{=}CH{-}Ar' \xrightarrow[Cu^{++}]{NaOAc} Ar{-}CH{=}CH{-}Ar' + N_2$$

$$Ar{-}N_2{-}Cl + C_6H_5{-}CH{=}CH{-}CO_2H \xrightarrow[Cu^{++}]{NaOAc} C_6H_5{-}CH{=}CH{-}Ar + N_2 + CO_2$$

In the present experiment the stilbene synthesis from cinnamic acid is illustrated, with *p*-nitroaniline as starting material for the aryldiazonium component. Nitro groups and halogen substituents in the diazonium components are relatively favorable but the yields are usually moderate. Even so the ready availability of suitable reactants often makes the method useful.

Much evidence points to an aryl free radical as the active arylating agent. In the Gomberg-Bachmann reactions the position of attack in the aromatic substrate is mainly *ortho* and *para*, regardless of the nature of the substituent present. Thus, *ortho*/*para* arylation is observed both with anisole and nitrobenzene, although these groups exert opposite directive influences in typical ionic (electrophilic) substitution reactions. Also, the arylation of acrylic acid and acrylonitrile in the $\beta$-position, a site of low electron density, is a strong argument against attack by an aryl cation.

The fact that cinnamic acid is attacked in the $\alpha$-position is probably due to more favorable resonance interaction of the unpaired electron in the intermediate adduct with the aryl group already present in the $\beta$-position.

$$Ar\cdot + C_6H_5{-}CH{=}CH{-}CO_2H \longrightarrow C_6H_5{-}\dot{C}H{-}\underset{\substack{|\\ Ar}}{CH}{-}CO_2H \longleftrightarrow \cdot C_6H_5{=}CH{-}\underset{\substack{|\\ Ar}}{CH}{-}CO_2H$$

A schematic representation of steps in a free radical mechanism for the reaction of a diazo acetate with acrylonitrile is:

$$C_6H_5{-}N{=}N{-}OAc \longrightarrow \underset{I}{C_6H_5\cdot} + N_2 + \cdot OAc$$

$$C_6H_5\cdot + CH_2{=}CH{-}CN \longrightarrow \underset{II}{C_6H_5{-}CH_2{-}\dot{C}H{-}CN}$$

$$II + Cu^{++} \longrightarrow \underset{III}{C_6H_5{-}CH_2{-}\overset{+}{C}H{-}CN} + Cu^{+}$$

$$III + \overset{-}{O}Ac \longrightarrow C_6H_5{-}CH{=}CH{-}CN + H{-}OAc$$

$$Cu^{+} + \cdot OAc \longrightarrow Cu^{++} + \overset{-}{O}Ac$$

The cupric chloride catalyst evidently acts as a radical chain-transfer agent that circumvents vinyl polymerization of the intermediate radical II.

The presence of diazo chloride and chloride ions in the reaction mixture could give rise to chlorine atoms, which would account for observed chlorination of the solvent, acetone, and for formation of chloro adducts (in certain instances). Other accessory products are formed by replacement of the diazo group by hydrogen and by chlorine.

## *p*-Nitrostilbene

**Diazotization of *p*-Nitroaniline.** Dissolve 0.05 mole (7 g) of *p*-nitroaniline by warming it in a small flask with 10 ml of concentrated hydrochloric acid diluted with 15 ml of water. Add an additional 5 ml of concentrated hydrochloric acid (total, 0.18 mole) and pour the solution into a 250-ml Erlenmeyer flask containing 40 g of chopped ice. Swirl the mixture vigorously to obtain a fine suspension of crystals of the hydrochloride. Before proceeding to the diazotization prepare the materials mentioned below that will be required for the arylation.

To the vigorously agitated suspension of *p*-nitroaniline hydrochloride, maintained at 5–10°, add as quickly as possible a cold solution of 0.06 mole (4 g, a 20 percent excess) of sodium nitrite in 10–12 ml of water. Swirl the flask until most of the amine hydrochloride dissolves (about 3 min) and allow it to stand a few minutes, until the diazotization is complete. Proceed at once to the next step.

***p*-Nitrostilbene from Cinnamic Acid.** In a 1-liter flask prepare a solution of 0.05 mole (7.5 g) cinnamic acid in 75 ml of acetone and a separate solution of 0.015 mole (2.5 g) of cupric chloride dihydrate in about 10 ml of water in a small beaker. Also weigh out 0.18 mole (25 g of the trihydrate) of sodium acetate crystals.

To the chilled cinnamic acid solution add the cold *p*-nitrobenzenediazonium chloride solution and the weighed quantity of sodium acetate. Swirl the solution to obtain good mixing and add the solution of cupric chloride (*caution—evolution of nitrogen begins at once*). Allow the reaction mixture to stand for 0.5 hr, with occasional swirling, and maintain the temperature at about +15° by adding a little ice when needed.[4] After this period the reaction mixture may be allowed to stand at room temperature.

Arrange the flask for steam distillation and steam distill the reaction mixture to remove acetone, small amounts of chloroacetone (*caution—lachrymator!*), nitrobenzene, and *p*-chloronitrobenzene. Decant off the hot green solution and dissolve the brown residue in 100 ml of methylene

[4] Temperature control is critical here. Above +20° the reaction will yield unwanted side-products; below +12° the reaction will cease. If the reaction does stop it can be restarted by addition of 0.1 g of sodium nitrite.

chloride. Wash the methylene chloride solution with water and then *carefully* (to avoid emulsions) with two 75-ml portions of 5 percent aqueous sodium bicarbonate. Finally wash the methylene chloride solution with two portions of water.

Separate the methylene chloride solution and filter it through a fluted filter directly into a distilling flask. Attach a condenser and distill off the solvent from a steam bath. Take up the solid residue in hot ethanol, add 0.5 g of decolorizing carbon, and filter the hot solution through a fluted filter. Chill the filtrate, collect the crystals with suction, and wash them with a little cold ethanol. The yield is 3–4 g. A purer product can be obtained by sublimation. The reported melting point of *trans-p*-nitrostilbene is 155°.

## Questions

**1.** Indicate a suitable combination of reactants that should serve to furnish each of the following compounds, in a pure condition, by means of a Gomberg-Bachmann arylation reaction: (a) 4-bromo-2′,4′-dimethylbiphenyl ($Br{-}C_6H_4{-}C_6H_3(CH_3)_2$); (b) *p*-terphenyl ($C_6H_5{-}C_6H_4{-}C_6H_5$); (c) phenanthrene (Pschorr synthesis).

**2.** Indicate a suitable synthesis of each of the following compounds by means of a Meerwein arylation reaction: (a) 2-*p*-nitrostyrylfuran ($O_2N{-}C_6H_4{-}CH{=}CH{-}C_4H_3O$), starting from aniline and furfural; (b) *p*-bromophenylsuccinic acid; (c) 1,4-diphenyl-1,3-butadiene.

**3.** Stilbene obtained by the Meerwein phenylation of cinnamic acid is the *trans* isomer, mp 125°. The *cis* isomer, mp +6°, is obtained by controlled partial hydrogenation of diphenylacetylene. Upon treatment of the *cis*- and *trans*-isomers with a reagent that effects *cis*-hydroxylation, the resulting hydrobenzoins ($C_6H_5{-}CHOH{-}CHOH{-}C_6H_5$) melt at 137° and 120°, respectively. What is the stereochemical character of these products?

EXPERIMENT

# 33

# A Modified Wittig Synthesis

The veritable Wittig syntheses of alkenes[1] are based upon the interaction of an alkylidene phosphorane (a phosphorus ylide, II) with an aldehyde or ketone. The requisite ylides can be obtained by the action of strongly nucleophilic reagents, such as phenyllithium or sodium hydride, upon appropriate quaternary phosphonium halides (I). Since the ylides are unstable they are

$$\underset{\text{I}}{[(C_6H_5)_3\overset{+}{P}-CH_2R]\overset{-}{X}} \xrightarrow{C_6H_5Li} \underset{\text{IIa}}{(C_6H_5)_3P{=}CH-R} \longleftrightarrow \underset{\text{IIb}}{(C_6H_5)_3\overset{+}{P}-\overset{-}{C}H-R}$$

$$\text{II} + (C_6H_5)_2C{=}O \longrightarrow C_6H_5-CH{=}C(C_6H_5)_2 + (C_6H_5)_3P{=}O$$

usually generated in the reaction mixture in the presence of the carbonyl compound. Advantages of the Wittig synthesis are: carbon-carbon bond formation occurs without production of isomeric alkenes, and acid-sensitive alkenes can be prepared because the reaction occurs under mild conditions in an alkaline medium.

If the organic halide ($R-CH_2X$) used for the formation of the quaternary phosphonium halide (I) is a highly reactive one, such as $C_6H_5-CH_2-Cl$, a simpler procedure may be used. An ester of phosphorous acid is converted by the Arbusov reaction[2] to a phosphonic ester (III), which reacts with carbonyl compounds, in the presence of a base, in the same way as an ylide.

$$(C_2H_5O)_3P + R-CH_2Cl \xrightarrow{\text{heat}} \underset{\text{III}}{(C_2H_5O)_2\overset{+}{\underset{|\atop O^-}{P}}-CH_2R} + C_2H_5Cl$$

$$\text{III} + (C_6H_5)_2C{=}O \xrightarrow{\text{NaOEt}} R-CH{=}C(C_6H_5)_2 + NaO-\underset{\|\atop O}{P}(OC_2H_5)_2$$

[1] Wittig and Geisler, *Ann.*, **580**, 44 (1953); Maercker, *Organic Reactions*, **14**, 270 (1965).
[2] Kosolapoff, *Organic Reactions*, **6**, 276 (1951).

From benzyl type halides and substituted benzaldehydes unsymmetrical stilbenes can be synthesized,[3] and with benzophenones triarylethylene derivatives are formed. $\alpha,\beta$-Unsaturated aldehydes such as crotonaldehyde and cinnamaldehyde furnish derivatives of 1,4-butadiene.

$$C_6H_5{-}CH{=}CH{-}CH{=}O + C_6H_5{-}CH_2{-}PO(OC_2H_5)_2 \xrightarrow{\text{NaOEt}} C_6H_5{-}CH{=}CH{-}CH{=}CH{-}C_6H_5$$

$\alpha$-Halogenated esters also will react with triethyl phosphite to form phosphonic esters that are useful intermediates for modified Wittig syntheses. Ethyl bromoacetate gives triethyl phosphonoacetate (IV), which can be deprotonated with sodium hydride to form the highly active anion (V). This ylide-like intermediate reacts with aldehydes and ketones to produce

$$(C_2H_5O)_3P + Br{-}CH_2CO_2Et \longrightarrow (C_2H_5O)_2\overset{+}{P}(\underset{-}{O}){-}CH_2CO_2Et \quad \text{IV}$$

$$\text{IV} + NaH \longrightarrow (C_2H_5O)_2P(\underset{-}{O}){=}CH{-}CO_2Et \ (\text{V}) \xrightarrow{R_2CO} R_2C{=}CH{-}CO_2Et$$

mono- and di-substituted acrylic esters. The Wittig synthesis of such compounds is often more satisfactory than a Reformatsky[4] sequence. The latter

$$R_2C{=}O + Br{-}CH_2CO_2Et \xrightarrow[I_2]{Zn} R_2C(OH){-}CH_2CO_2Et \xrightarrow{Ac_2O} R_2C{=}CH{-}CO_2Et$$

involves reaction of the ketone with ethyl bromoacetate and zinc, with iodine as catalyst, to produce a $\beta$-hydroxypropionic ester, which is dehydrated by means of acetic anhydride, potassium acid sulfate, and similar reagents.

## *p*-Methoxystilbene[5]

**Diethyl Benzylphosphonate.** In a 100-ml round-bottomed flask place 0.05 mole (9 ml, 8.3 g) of triethyl phosphite and 0.05 mole (5.8 ml, 6.3 g) of

[3] Seus and Wilson, *J. Org. Chem.*, **26**, 5243 (1961); Wadsworth and Emmons, *J. Amer. Chem. Soc.*, **83**, 1733 (1961).

[4] Shriner, *Organic Reactions*, **1**, 1, 11 (1942).

[5] The specific example of the modified Wittig synthesis given below may be varied at the second step by using *p*-chlorobenzaldehyde (0.05 mole, 7 g) to give *trans*-4-chlorostilbene, mp 129. Another interesting example is the use of cinnamaldehyde (0.05 mole, 6.6 g) to produce 1,4-diphenyl-1,3-butadiene (see Fieser, *Organic Experiments*, 2nd edition, Raytheon Education Company, Lexington, Mass., 122 (1968).

benzyl chloride (*caution—lachrymator!*). Attach a condenser and heat the mixture gently for 1 hr. When the temperature reaches 130–140° evolution

➤**CAUTION:** Avoid contact of phosphorus compounds with the skin. Wash off any spilled material thoroughly with soap and water.

of ethyl chloride (bp + 12°) begins. The internal temperature continues to rise and attains about 190° by the end of the hour. Allow the product to cool and dissolve it in 10 ml of dimethylformamide.

***p*-Methoxystilbene.** In a 125-ml Erlenmeyer flask place 0.052 mole (2.8 g) of sodium methoxide[6] and the solution of diethyl benzylphosphonate. Swirl

➤**CAUTION:** Handle sodium methoxide carefully. Any material spilled on the hands should be washed off promptly with a large quantity of water.

the mixture and add dropwise a solution of 0.05 mole (6.8 g) of *p*-methoxybenzaldehyde in 40 ml of dimethylformamide with intermittent cooling in an ice bath so that the temperature of the reaction mixture is maintained between 30 and 40°. Allow the reaction mixture to stand overnight or longer.

Pour the reaction mixture into about 50 ml of water, with stirring, and collect the product on a suction filter. After washing thoroughly with water, crystallize the material from ethanol. The recorded melting point of *p*-methoxystilbene is 136°. The yield is 6–7 g.

## Questions

**1.** Give a specific example of a reaction of an arsenite (or other arsenic$^{+3}$ compound) and one of a sulfite, analogous to the formation of a phosphonate from a phosphite.

**2.** Indicate an appropriate synthesis for each of the following compounds by two approaches—a Wittig synthesis and one other method (Grignard reaction, Meerwein arylation, Perkin reaction, Reformatsky reaction, and others): (a) 4-methoxy-4′-chlorostilbene, (b) methyl β-methylcinnamate [$C_6H_5-C(CH_3)=CH-CO_2CH_3$], (c) 4,4′-distyrylbenzene ($C_6H_5-CH=CH-C_6H_4-CH=CH-C_6H_5$), (d) 3,4-dimethoxy-α-methylcinnamic acid.

[6] Commercial sodium methoxide gives erratic results. If fresh reagent is not available it is advisable to use twice the stated amount.

Sodium methoxide sufficient for five preparations can be prepared by the procedure of Cason described in *Organic Syntheses*, **39**, 51 (1959). To 130 ml of anhydrous methanol contained in a 250 ml round-bottomed flask equipped with an upright condenser add *through* the condenser tube 6.0 g of clean sodium cut in small pieces. In order to keep the reaction under control one piece of sodium should be allowed to react completely before another is added. After all of the sodium has reacted the excess methanol is removed by distillation, first at atmospheric pressure and then under an aspirator vacuum using a heating bath maintained at 150°. The resulting free-flowing sodium methoxide can be stored in a desiccator for several weeks.

# EXPERIMENT 34

# Phenols

## 34a Phenol

Pure phenol is a white crystalline solid, mp 42°. The solid, or concentrated solutions of phenol, cause painful burns if allowed to come in contact with the skin. For the following tests use a 5 percent aqueous solution of phenol prepared by the laboratory instructor.

**(A) Acidity of Phenol.** Test an aqueous solution of phenol with red and blue litmus paper, Congo red paper, and Hydrion indicator paper. Carry out parallel tests with dilute aqueous solutions (about 5 percent) of benzyl alcohol, oxalic acid, and acetic acid. Arrange these compounds in the order of increasing acid strength.

Congo red is insensitive to weak acids and changes to the acid form (blue) only in the presence of relatively strong acids, pH 3–5.

**(B) Ferric Chloride Enol Test.** To 5 ml of ethanol in a test tube add 1–2 drops of a 5 percent aqueous solution of phenol and a few drops of a 3 percent aqueous solution of ferric chloride. Shake well and observe.

Carry out the enol test with the following substances instead of phenol: (a) a crystal of salicylic acid; (b) a crystal of 2-naphthol; (c) a crystal of catechol (*o*-dihydroxybenzene); (d) a drop of acetone or of allyl alcohol ($CH_2{=}CHCH_2OH$). Record your observations and comment on the results.

Many phenols and related compounds form colored coordination compounds with ferric iron, in which six molecules of a monohydric phenol are combined with one atom of iron in the form of a complex anion. Aliphatic enols (ethyl acetoacetate, acetylacetone) give a positive test. Some phenols do not give a color reaction.

**(C) Oxidation Test.** To 5 ml of 0.5 percent aqueous potassium permanganate solution add 2 ml of a 5 percent aqueous solution of phenol. Warm the solution and shake vigorously. Is phenol oxidized by the permanganate solution?

Carry out a similar test using 2–3 drops of anisole (methoxybenzene) instead of the phenol solution, and compare the result with that obtained with phenol.

**(D) Bromination: Tribromophenol.** In a 125-ml Erlenmeyer flask place 4 ml of a 5 percent aqueous solution of phenol (0.002 mole) and 20 ml of water. In a separatory funnel place 35 ml of 3 percent aqueous bromine solution (bromine water—0.007 mole) and allow it to flow dropwise into the phenol solution, while swirling the flask to obtain good mixing. Continue to add the bromine solution until a yellow color persists in the solution after thorough mixing. Collect the precipitate on a small suction filter and wash it *thoroughly* with two successive 10-ml portions of aqueous sodium bisulfite solution (5 percent). Finally, wash the material with two 10-ml portions of water and press it as dry as possible on the filter.

Dissolve the crude product in 20 ml of hot ethanol, by warming on a steam bath, and filter the hot solution through a fluted filter into a clean 125-ml Erlenmeyer flask. To the hot filtrate add 40 ml of hot water (75°), mix thoroughly, and set the solution aside to cool undisturbed. After the product has crystallized, collect the purified 2,4,6-tribromophenol by suction filtration, wash it with 5 ml of cold 50 percent aqueous ethanol, and spread it on a clean filter paper to dry. Determine the melting point.

The precipitate formed by reaction of bromine water with aqueous phenol is a tetrabromo quinonoid derivative (2,4,4,6-tetrabromo-2,5-hexadienone), which is converted to 2,4,6-tribromophenol by treatment with sodium bisulfite.

**(E) Nitrosation: *p*-Nitrosophenol/*p*-Benzoquinone Monoxime.** In a 125-ml Erlenmeyer flask place 20 ml of a 5 percent aqueous solution of phenol (0.01 mole); 5 ml of 10 percent sodium hydroxide solution (0.013 mole); and 0.013 mole (0.9 g) of pulverized sodium nitrite.[1] Cool the solution to 5–7° in an ice-water bath and add dropwise, while swirling the liquid, 8 ml of cold 25 percent sulfuric acid.[2] Allow the reaction mixture to stand in the cooling bath, with occasional swirling, for about an hour. The nitroso compound separates slowly as the reaction progresses. Collect the crystals with suction and wash them thoroughly with cold water. Recrystallize the material at once from hot water, with addition of a little decolorizing carbon, and chill the

[1] It is advantageous to add also 2–3 ml of 10 percent sodium bisulfite solution to counteract the deleterious effect of traces of nitrate present as impurity in the sodium nitrite.

[2] Sulfuric acid of approximately 25 percent concentration (by weight) may be prepared by adding 2 ml of concentrated acid to 11 ml of water.

filtrate rapidly with stirring. Collect the purified product with suction, press it well on the filter, and dry it between filter papers. *p*-Nitrosophenol melts with vigorous decomposition in the range 120–126°. The observed melting point varies with the purity of the sample and the rate of heating. The substance becomes discolored on standing.

*p*-Nitrosophenol and *p*-benzoquinone monoxime have a tautomeric relationship. Nitrosation of phenol and oximation of *p*-benzoquinone (with

$$HO-C_6H_4-N{=}O \rightleftharpoons O{=}C_6H_4{=}N-OH$$

one equivalent of hydroxylamine) lead to the same product. Oxidation of this compound produces *p*-nitrophenol; reaction with hydroxylamine furnishes *p*-benzoquinone dioxime.

***Oxidation to p-Nitrophenol (Optional).*** For each gram of *p*-nitrosophenol to be oxidized use 8 ml of dilute nitric acid, prepared by dissolving 2 ml of concentrated nitric acid (70 percent) in 6 ml of water. Warm the diluted acid to 40° and add 1 g of *p*-nitrosophenol in small portions, with shaking. The solution becomes dark red in color and soon sets to a mass of light brown needles. Recrystallize the crude product from hot water. The purified *p*-nitrophenol is almost colorless; mp 114°.

**(F) Benzoylation by the Schotten-Baumann Method: Phenyl Benzoate.** In a 125-ml Erlenmeyer flask place 20 ml of a 5 percent aqueous solution of phenol (0.01 mole) and 20 ml of 10 percent sodium hydroxide solution. Add 0.0085 mole (1.2 g, 1 ml) of benzoyl chloride (*caution—irritating vapor!*) to the sodium phenoxide solution, being careful not to allow the benzoyl chloride to touch the lip or neck of the flask. Mix the contents of the flask by swirling gently and warm the flask by dipping it into a beaker of hot water (60°). Continue to warm and swirl the mixture until the disagreeable odor of benzoyl chloride can no longer be detected (*test cautiously!*).

After all the benzoyl chloride has reacted, chill the flask in cold water and swirl the contents to promote crystallization of the product. Collect the crystals by suction filtration and wash them thoroughly with several 10-ml portions of water. Transfer the moist product to a 25-ml Erlenmeyer flask and dissolve it in a small amount of hot methanol, using a steam bath or a bath of hot water. Filter the hot solution through a small filter and allow the filtrate to stand undisturbed to crystallize. Collect the crystals of purified phenyl benzoate by suction filtration and spread them out to dry on a filter paper; mp 68–69°.

*Questions follow Experiment 34c.*

## 34b 2-Naphthyl Acetate

The hydroxyl group of a phenol cannot be esterified (acylated) by direct interaction of a phenol and a carboxylic acid under the conditions used commonly for esterification of primary and secondary alcohols. Phenolic compounds are converted to acetates by means of acetyl chloride or acetic anhydride. Acetylation with acetic anhydride can be accomplished by heating, usually with an acid or base as catalyst (sulfuric or phosphoric acid; pyridine or sodium acetate). It can also be done in an aqueous medium by the method of Chattaway,[3] in a manner similar to the Lumière-Barbier acetylation of amines (Experiment 22).

In a small flask dissolve 0.007 mole (1 g) of 2-naphthol ($\beta$-naphthol) in 5 ml of 10 percent sodium hydroxide solution, add 10–12 g of finely crushed ice, and 0.018 mole (1.9 g, 2 ml) of acetic anhydride. Cork the flask firmly and shake the reaction mixture vigorously. The reaction is complete in about 10 min and the product separates in colorless crystals. Continue to shake the mixture occasionally until the excess anhydride has been hydrolyzed. Collect the crystals with suction, wash them with cold water, and press dry on the filter. Recrystallize the material from warm 70 percent aqueous methanol. The yield is 1.0–1.2 g; mp 70–71°.

The acetylation can be carried out equally well by treating 1 g of 2-naphthol with 4 ml of acetic anhydride and *one* drop of sulfuric acid, and warming the mixture gently for 5 min. The solution is cooled to room temperature, diluted with 40 ml of water, and shaken vigorously to promote crystallization of the product. The crystals are collected with suction and purified as described above.

*Questions follow Experiment 34c.*

## 34c Acetylsalicylic Acid (Aspirin)

Derivatives of salicylic acid have been used in medicine for many years. Salicylic acid occurs in nature in the form of esters in a variety of glycosides and essential oils. The methyl ester is present in oil of wintergreen and in many other fragrant oils from flowers, leaves and bark.

The sodium salt of salicylic acid is prepared commercially by heating

[3] Chattaway, *J. Chem. Soc.*, 2495 (1931). This method is applicable to a variety of phenols, including bifunctional phenols (catechol, resorcinol, hydroquinone) but salicylic acid is not acetylated under these conditions.

sodium phenoxide with carbon dioxide at 150° under slight pressure (the Kolbe synthesis). If potassium phenoxide is used in place of sodium phenoxide and the reaction carried out at 180–200°, the carboxyl function is introduced into the *para* instead of the *ortho* position; this affords a practical synthesis of *p*-hydroxybenzoic acid.

Salicylic acid is used for the manufacture of medicinal compounds, artificial oil of wintergreen, and certain dyes. Aspirin is used as an analgesic (to relieve pain) and antipyretic (to reduce fever). The phenyl ester of salicylic acid, known as Salol, is used as an intestinal antiseptic.

Other compounds related to salicylic acid are the corresponding aldehyde and primary alcohol, salicylaldehyde and saligenin, which also occur in nature. Salicylaldehyde may be prepared, together with the *para* isomer, by the action of chloroform on phenol in the presence of excess alkali (the Reimer-Tiemann reaction). The reactive intermediate is *dichlorocarbene*, $CCl_2$, formed by stepwise abstraction of a proton and a chloride anion from chloroform.

**Acetylsalicylic Acid.** In a 50-ml Erlenmeyer flask place 0.01 mole (1.4 g) of pure salicylic acid. Add 0.03 mole (3.1 g, 3.3 ml) of acetic anhydride in such a way as to wash down any material adhering to the walls of the flask, and then 5 drops of syrupy (85 percent) phosphoric acid. Heat the flask for 5 min on a steam bath or in a beaker of water heated to 85–90°. Remove the flask from the bath and without allowing it to cool, add 2 ml of water in one portion. The excess acetic anhydride decomposes vigorously and the contents of the flask come to a boil (*caution—hot acid vapors!*).

When the decomposition is complete add 20 ml of water and allow the flask to stand at room temperature until crystallization begins. The crystallization can be hastened by occasionally scratching the walls of the flask at the surface of the solution with a glass stirring rod. When crystals begin to appear place the flask in an ice bath, add 10–15 ml of cold water, and chill thoroughly until crystallization is complete. Collect the product on a Büchner funnel and press it firmly to remove the mother liquor. Wash the material with cold water and allow it to dry thoroughly. The yield is 1.5–1.6 g.

Purify the crude product in the following way (see solvent-pairs, page 85). In an Erlenmeyer flask dissolve the *thoroughly dried* material in 20–25 ml of ether by warming gently and stirring. To the solution (if not clear, filter into a clean flask) add 20 ml of petroleum ether (bp 30–60°), stopper the flask, place in an ice bath, and allow to cool undisturbed for about an hour. Collect the product on a Büchner funnel, wash with a little petroleum ether, and spread it on a clean paper to let traces of solvent evaporate.

Dissolve a few milligrams of the pure product in about 1 ml of methanol and add a drop of ferric chloride solution. Perform the same test on salicylic acid.

## Questions

**1.** What are the two principal industrial methods of manufacturing phenol? Write equations for the reactions involved, starting from benzene.

**2.** Indicate suitable reactions, including the reagents and conditions used, for the conversion of phenol to: (a) phenetole; (b) salicylic acid; (c) 2,4-Dichlorophenoxyacetic acid (2,4-D); (d) *o*-anisidine; (e) an azo dye.

**3.** What is meant by the *phenol-coefficient* of an antiseptic?

**4.** The equilibrium constant for the esterification of phenol and acetic acid, $K_E$, is roughly 0.01, in contrast to $K_E$ values of 2–4 for esterification of typical primary and secondary alcohols. Calculate the percentage conversion of phenol to phenyl acetate, at equilibrium, starting from an equimolar mixture of phenol and acetic acid (see Experiment 10).

**5.** *o*-Nitrophenol and *o*-hydroxybenzaldehyde (salicylaldehyde) are volatile with steam but the *meta* and *para* isomers are not. The *ortho* isomers also are less soluble in water and more soluble in organic solvents than the *meta* and *para* isomers. Give an explanation for these distinctive properties of the *ortho* isomers.

# EXPERIMENT 35

# The Cannizzaro Reaction

Aromatic aldehydes, in common with aliphatic aldehydes, undergo addition reactions of the carbonyl group leading to cyanohydrins, acetals, oximes, phenylhydrazones, and similar derivatives (see Experiment 9). They also undergo certain reactions generally considered to be characteristic of aromatic aldehydes as distinct from typical aliphatic aldehydes. Examples of such reactions are the Cannizzaro reaction, the benzoin condensation (Experiment 37), and the Perkin reaction (Experiment 39). These particular reactions are attributable mainly to the circumstance that the aldehyde function is attached to a tertiary carbon atom (one that bears no hydrogen atoms). For this reason an aromatic aldehyde cannot undergo reactions requiring the presence of an active methylene group in the alpha position (of its own molecule) but is limited to reactions that arise from the aldehyde group itself.

$$C_6H_5-\underset{\displaystyle H}{\overset{\displaystyle O^-}{\underset{|}{\overset{|}{C}}}}-OH + O{=}CH-C_6H_5 \longrightarrow C_6H_5-\overset{\displaystyle O}{\overset{\|}{C}}-OH + \bar{O}-CH_2-C_6H_5$$

$$\downarrow$$

$$C_6H_5-\overset{\displaystyle O}{\overset{\|}{C}}-O^- + HO-CH_2-C_6H_5$$

In the presence of strong alkalies, benzaldehyde (like formaldehyde) undergoes disproportionation to form the corresponding primary alcohol and a salt of the carboxylic acid: the Cannizzaro reaction.[1] The process involves addition of hydroxyl ion to the carbonyl group of one molecule and transfer of hydride anion from the adduct to a second molecule of benz-

[1] For a discussion of the Cannizzaro reaction see Geissman, *Organic Reactions*, **2**, 94 (1944).

aldehyde, accompanied by proton interchange to form the benzoate anion and benzyl alcohol. If the reaction is effected under anhydrous conditions with the sodium derivative of benzyl alcohol ($NaOCH_2C_6H_5$) as catalyst the product is the ester, benzyl benzoate. Aluminum alkoxides in catalytic amount, under anhydrous conditions, convert aromatic and aliphatic aldehydes to esters (the Tishchenko reaction).

For more efficient conversion of an aromatic aldehyde to the corresponding alcohol one employs a *crossed* Cannizzaro reaction, with formaldehyde to serve as the donor of hydride ion. An excess of formaldehyde is used and the aromatic aldehyde is transformed almost entirely to the alcohol; surplus formaldehyde is converted to potassium formate and methanol. Formaldehyde may be used in this manner with furfural and with tertiary aliphatic aldehydes.

$$H{-}CH{=}O + C_6H_5{-}CH{=}O + KOH \longrightarrow H{-}CO_2K + C_6H_5{-}CH_2{-}OH$$

Benzaldehyde differs from aliphatic aldehydes in its behavior toward ammonia. Three molecules of the aldehyde react with two molecules of ammonia to form a crystalline *hydramide*, hydrobenzamide (mp 101–102°): $C_6H_5{-}CH{=}N{-}CH(C_6H_5){-}N{=}CH{-}C_6H_5$. Another difference is that aromatic aldehydes do not form cyclic trimers (1,3,5-trioxane derivatives), such as those obtained from formaldehyde and acetaldehyde.

## 35a Benzyl Alcohol[2]

In a small beaker dissolve 0.27 mole (18 g of 85 percent pure solid) of solid potassium hydroxide in 18 ml of water and cool the solution to about 25°. Place 0.2 mole (21 g, 20 ml) of benzaldehyde in a 125-ml Erlenmeyer flask (or narrow-mouthed bottle) and to it add the potassium hydroxide solution. Cork the flask firmly and shake the mixture thoroughly until an emulsion is formed. Allow the mixture to stand for 24 hr or longer. At the end of this period the odor of benzaldehyde should no longer be detectable.

To the mixture add just enough distilled water to dissolve the precipitate of potassium benzoate. Shake the mixture thoroughly to facilitate solution of the precipitate. Extract the alkaline solution with three or four 20-ml portions of ether (*caution—flammable solvent!*) to remove the benzyl alcohol and traces of any unconverted benzaldehyde.[3] Combine the ethereal extracts for isolation of benzyl alcohol and reserve the aqueous solution to obtain the benzoic acid.

[2] In planning his laboratory schedule the student should observe that this experiment requires materials to be mixed and allowed to stand for 24 hr or longer.

[3] Benzene is less satisfactory than ether for the extraction because it tends to form troublesome emulsions with the alkaline solution.

**Benzyl Alcohol.** Concentrate the ethereal solution of benzyl alcohol by distillation from a steam bath, using a water-cooled condenser, until the volume of the residual liquid has been reduced to 15–20 ml. Cool the liquid, transfer it to a small separatory funnel (using 2–3 ml of ether to rinse the distilling flask), and shake it thoroughly with two 5-ml portions of 20 percent aqueous sodium bisulfite to remove any benzaldehyde. Wash the ether solution finally with two 10-ml portions of water and dry it with 3–4 g of anhydrous magnesium sulfate. Filter the solution into a small dry distilling flask and carefully distill off the ether (*avoid fire hazards!*). Attach a short air-cooled condenser and distill the benzyl alcohol, by heating the flask directly with a luminous flame kept in motion. Collect the material boiling at 200–206°. The yield is 4–5 g.

***Reactions.*** Benzyl alcohol may be characterized by reaction with 1-naphthyl isocyanate to form the N-arylcarbamic ester (*naphthylurethan*) or

$$C_{10}H_7{-}N{=}C{=}O + C_6H_5{-}CH_2OH \longrightarrow C_{10}H_7{-}NH{-}CO{-}OCH_2C_6H_5 \text{ (mp } 134°)$$

$$O_2N{-}C_6H_4{-}CO{-}Cl + C_6H_5{-}CH_2OH \xrightarrow{C_5H_5N} O_2N{-}C_6H_4{-}CO{-}OCH_2C_6H_5 \text{ (mp } 85°)$$

by treatment with *p*-nitrobenzoyl chloride, in the presence of pyridine, to obtain the crystalline *p*-nitrobenzoic ester.

**Benzyl 1-Naphthylcarbamate** *(Optional).* Aryl isocyanates react with water to form sparingly soluble diarylureas (R−NH−CO−NH−R). It is essential to use an anhydrous sample of the alcohol and a well dried test tube.

Place 1 g (1 ml) of benzyl alcohol in a dry test tube, add 0.5 ml of 1-naphthyl isocyanate (*caution—irritating vapor!*), cork the tube loosely, and heat in a beaker of boiling water for 5 min. Cool the tube and scratch the inside wall with a glass rod to induce crystallization. Dissolve the product in 5 ml of ligroin, bp 100–120° (*flammable solvent!*), filter from any insoluble material, and allow the hot filtrate to cool. Collect the crystals, recrystallize them from ligroin; mp 134°.

**Benzyl *p*-Nitrobenzoate** *(Optional).* The *p*-nitrobenzoyl chloride needed for this experiment may be prepared in the following way. In a large dry test tube place 0.5 g of finely pulverized, dry *p*-nitrobenzoic acid, 2 ml of thionyl chloride and *one drop* of pyridine or dimethylformamide.[4] Attach a small condenser, add a boiling chip, and boil the mixture in a water bath for 0.5 hr. During this time the acid should react and dissolve completely. Continue to boil the solution for 5 min long. Detach the condenser and replace it with a cork holding a 15-cm length of glass tubing, bent to form a 90°

[4] Dimethylformamide is a particularly effective catalyst for the conversion of carboxylic and sulfonic acids to the acid chlorides using thionyl chloride.

angle. Connect the glass tube to a water pump (not an oil pump!) and apply suction to distill off the excess thionyl chloride. Warm the test tube gently to hasten the process. Disconnect the suction line and cool the test tube until the crude *p*-nitrobenzoyl chloride crystallizes (mp 69–72°).

To the crude acid chloride add 1 g (1 ml) of benzyl alcohol and 1 ml of dry pyridine. Heat the mixture for 10 min in a bath of boiling water, with occasional shaking. Cool the mixture, add 10–15 ml of 5 percent sodium bicarbonate solution, and shake it vigorously to dissolve any *p*-nitrobenzoic acid. Collect the product with suction and wash it thoroughly with water. Recrystallize the crude benzyl *p*-nitrobenzoate from ethanol and take the melting point; recorded mp 85°.

## 35b Benzoic Acid

To free the acid pour the aqueous solution of potassium benzoate (from which the benzyl alcohol has been extracted) *into* a vigorously stirred mixture of 40 ml of concentrated hydrochloric acid, 40 ml of water, and 40–50 g of chipped ice. Test the mixture with indicator paper to make sure that it is strongly acidic. Collect the benzoic acid with suction and wash it once with cold water. Crystallize the product from hot water, collect the crystals and allow them to dry thoroughly. The yield is about 8 g.

Aromatic and aliphatic carboxylic acids generally are characterized by conversion to crystalline amides.[5] Benzoic acid may be converted to benzamide, mp 130°, or benzanilide, mp 160°. For this purpose the acid usually is converted by means of thionyl chloride to the acid chloride (essentially as described above for *p*-nitrobenzoyl chloride) and this is treated with ammonia or an arylamine to obtain the desired amide.

**Neutralization Equivalent of an Acid** *(Optional)*. Transfer an accurately weighed sample of thoroughly dry benzoic acid (about 0.200 g) to a 250-ml Erlenmeyer flask and wash the material down the sides of the flask with about 50 ml of water. Warm the mixture, if necessary, to dissolve the acid[6] and add *two drops* of phenolphthalein indicator solution. Titrate the acid with standardized sodium hydroxide solution, about 0.1 *N*, until a faint

[5] A valuable aid in the identification of an unknown organic acid is the determination of its equivalent weight (neutralization equivalent), by titration with standard base. This determination may be used also to check the purity of a sample of a known acid.

[6] It is not essential to have an acid completely dissolved but the titration may be slow if the acid is sparingly soluble in water. Ethanol may be added to increase the solubility. Ethanol or ethanol-benzene may be used as solvent instead of water, but bromothymol blue should be used as indicator since phenolphthalein gives an indistinct end point. When ethanol or other organic solvent is used, it is necessary to run a blank determination on the solvent, employing the same amount of indicator solution that is added for titration of the acid.

*permanent* pink coloration remains after swirling the solution. Calculate the neutralization equivalent (equivalent weight) by means of the following relationship:

$$\text{Neutralization Equivalent} = \frac{\text{Weight of acid (g)} \times 1000 \text{ ml}}{\text{Volume of base (ml)} \times \text{Normality factor}}$$

The result should check the theoretical value within about 1 percent. With a thoroughly purified and dried sample of acid, accurate weighing, and superior manipulative skill the result may check within 0.3 percent.

## Questions

**1.** Write equations for the preparation of benzaldehyde from: (a) benzene; (b) toluene; (c) benzoic acid.

**2.** Benzaldehyde forms two stereoisomeric oximes, mp 35° and 130°. How may their configurations be determined?

**3.** Write equations for the reaction of benzaldehyde with the following reagents: (a) methanol (+hydrogen chloride catalyst); (b) semicarbazide; (c) *p*-tolylmagnesium bromide, followed by dilute acid; (d) sodium cyanide and ammonium chloride, followed by hydrolysis (the Strecker reaction); (e) aluminum isopropoxide (the Meerwein-Pondorff reaction).

**4.** Compare the aldol condensation and the Cannizzaro reaction from the standpoint of the structure of the aldehyde involved.

**5.** Acetaldehyde, when treated with an excess of formaldehyde in the presence of a basic catalyst, furnishes pentaerythritol, $C(CH_2{-}OH)_4$ (mixed aldol + crossed Cannizzaro reaction). Write equations, stepwise, for the reactions involved.

**6.** When an equimolecular mixture of benzaldehyde and cyclohexanone is treated with semicarbazide hydrochloride and sodium acetate, and the reaction mixture is worked up *within a few minutes* the product is cyclohexanone semicarbazone. But if the reaction mixture is allowed to stand overnight or longer, the product is benzaldehyde semicarbazone! Can you account for this difference? (Consider rates of reaction *vs* equilibria.)

EXPERIMENT

# 36

# Mixed Aldol Condensations

In its simplest form the aldol condensation is a self-addition involving two molecules of the same aldehyde or ketone and results in the formation of a new carbon-carbon bond joining the carbonyl carbon of one molecule to the $\alpha$-position of the second. To react in this way it is necessary that the molecule

$$CH_3{-}CH_2{-}CH{=}O + H{-}\underset{\displaystyle CH_3}{\underset{|}{C}}H{-}CH{=}O \xrightarrow[\text{or } H^+]{OH^-} CH_3{-}CH_2{-}\underset{\displaystyle OH}{\underset{|}{C}}H{-}\underset{\displaystyle CH_3}{\underset{|}{C}}H{-}CH{=}O$$

possess a reactive carbonyl function *and* one or more reactive hydrogens in the $\alpha$-position.

An aromatic aldehyde has no hydrogen atoms in the $\alpha$-position but is capable of participating in a mixed (crossed) aldol condensation with another aldehyde or ketone that can furnish an active methylene group. Usually the

$$C_6H_5{-}\overset{\displaystyle O}{\overset{\|}{C}}H + CH_3{-}\overset{\displaystyle O}{\overset{\|}{C}}H \xrightarrow{OH^-} C_6H_5{-}\overset{\displaystyle OH}{\overset{|}{C}}H{-}CH_2{-}\overset{\displaystyle O}{\overset{\|}{C}}H$$

$$\downarrow$$

$$C_6H_5{-}CH{=}CH{-}\overset{\displaystyle O}{\overset{\|}{C}}H + H_2O$$

mixed aryl aldol undergoes dehydration spontaneously to form the $\alpha,\beta$-unsaturated aldehyde or ketone. The reaction is known as the Claisen-Schmidt condensation.

Aldol-type condensations of aromatic aldehydes extend to other reactants containing an active methylene group—such as malonic ester, acid anhydrides, nitriles, nitroalkanes, and similar compounds.

Either acids or bases will catalyze the aldol condensation but basic catalysts are generally preferred. Dilute aqueous or ethanolic sodium hydroxide, sodium ethoxide, and secondary amines (diethylamine or piperidine) are effective catalysts. The first step in the process is the formation of the

$$CH_3{-}CH{=}O \underset{H_2O}{\overset{OH^-}{\rightleftharpoons}} :\overset{-}{C}H_2{-}CH{=}O \longleftrightarrow CH_2{=}CH{-}\overset{-}{O}$$

$$C_6H_5{-}CH{=}O + :\overset{-}{C}H_2{-}CH{=}O \underset{OH^-}{\overset{H_2O}{\rightleftharpoons}} C_6H_5{-}\underset{\displaystyle OH}{\underset{|}{C}H}{-}CH_2{-}CH{=}O$$

enolate anion of the active methylene component by action of the base. The resulting carbanion combines with the carbonyl reactant and proton interchange with the solvent leads to the mixed aldol, which then undergoes dehydration to the $\alpha,\beta$-unsaturated compound.

The kinetics of aldol-type condensations vary with the character of the reactants and the experimental conditions. In the self-condensation of acetaldehyde under typical conditions, with aqueous sodium hydroxide as catalyst, the enolization step is slow (and irreversible) and the second step is very fast. In the condensation of benzaldehyde and acetophenone, with sodium ethoxide as catalyst, the enolization of acetophenone is fast (and reversible) and combination of the carbanion with benzaldehyde is the slow step.

Mixed aldol condensations furnish intermediates for synthetic procedures used to obtain aromatic compounds having a variety of functional groups in the side chain and also structures having several aromatic rings attached to an aliphatic system. A few examples of such transformations are mentioned below in the discussion of reactions of benzalacetophenone.

## Benzalacetophenone

**The Claisen-Schmidt Condensation.** In a 125-ml Erlenmeyer flask place 25 ml of 10 percent aqueous sodium hydroxide (0.063 mole), 15 ml of ethanol, and 0.05 mole (6 g, 6 ml) of acetophenone. Cool the mixture in an ice-water bath, shake well, and add 0.05 mole (5.3 g, 5 ml) of benzaldehyde. Maintain the internal temperature at 25–30° and shake the mixture vigorously from time to time for a period of 1–2 hr. Then chill the reaction mixture in an ice bath for 30 min or longer.

It is desirable, but not essential, to induce crystallization of the product by inoculating the reaction mixture with a few small crystals of benzalacetophenone about 30 min after the reactants are mixed.

➤**CAUTION:** Benzalacetophenone should be handled carefully since it acts as a skin irritant—some individuals are especially sensitive to it.

Collect the product with suction and wash it thoroughly with water, until the washings are neutral to litmus. Finally wash the crystals with 5–6 ml of ice-cold ethanol and press dry on the filter. Reserve a few seed crystals, and recrystallize the crude product from warm ethanol (using 4–5 ml of solvent per gram). The yield of pale-yellow crystals is 7–8 g; mp 56–57°. This substance exists in at least five polymorphic modifications.[1]

***Reactions.*** Benzalacetophenone may be characterized by addition of bromine and by epoxidation, to form crystalline adducts, and also by means of carbonyl derivatives, such as the oxime or phenylhydrazone. Reaction with hydroxylamine hydrochloride furnishes an oxime (I), mp 115–116°,

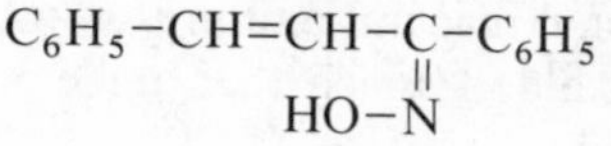

$$\underset{\text{I}}{C_6H_5{-}CH{=}CH{-}\underset{\underset{HO-N}{\|}}{C}{-}C_6H_5} \qquad \underset{\text{II}}{C_6H_5{-}CH{=}CH{-}CO{-}NH{-}C_6H_5}$$

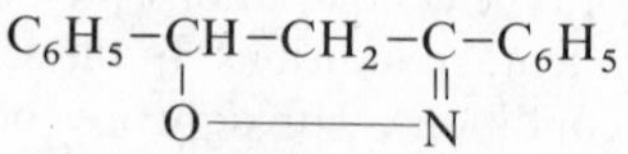

$$\underset{\text{III}}{C_6H_5{-}\underset{\underset{O}{|}}{CH}{-}CH_2{-}\underset{\underset{N}{\|}}{C}{-}C_6H_5} \quad (\text{O}{-}\text{N bonded})$$

which gives the anilide of cinnamic acid (II) when subjected to the Beckmann rearrangement. On treatment with cold sulfuric acid the oxime is cyclized to form 3,5-diphenyl-isoxazoline (III). The phenylhydrazone undergoes a similar cyclization to form 1,3,5-triphenylpyrazoline.

The conjugated system present in benzalacetophenone (and similar $\alpha,\beta$-unsaturated carbonyl compounds) gives rise to 1,4-addition reactions with a number of reagents. These include hydrogen cyanide, amines, Grignard reagents, and compounds having a reactive methylene group (nitromethane, methyl cyanoacetate, dimethyl malonate). A basic catalyst such as sodium methoxide is usually employed for the reactions. With aluminum chloride or sulfuric acid as catalyst, benzene undergoes addition to benzalacetophenone to form $(C_6H_5)_2CH{-}CH_2{-}CO{-}C_6H_5$.

**Benzalacetophenone Dibromide** *(Optional)*. Dissolve 1 g of dry benzalacetophenone in 4 ml of carbon tetrachloride and cool the solution in an ice-water bath. Add gradually, with shaking, 4 ml of a 20 percent solution of bromine in carbon tetrachloride (20 g of bromine per 100 ml of solution). After allowing the mixture to stand for 10 min, with shaking, collect the crystals with suction and wash them with two small portions of ethanol. Recrystallize the material from hot ethanol. The product obtained is the $\alpha$-dibromide, mp 156–157°. The stereoisomeric $\beta$-dibromide, mp 108–109°,

[1] The one usually obtained is the $\alpha$-form, leaflets, in purest form melting at 58–59°. Others are $\beta$-form, prisms or needles, mp 56–57°; $\gamma$-form, mp 48°; and two additional forms melting at 28° and 18°. The unusual crystalline forms have been obtained by inoculation with foreign seed crystals of different crystal habit.

is formed in small amounts in the reaction and is retained in solution in the mother liquors. The yield is 1.5 g.

Treatment of the dibromide with potassium acetate eliminates hydrogen bromide to form the α-bromo derivative, $C_6H_5{-}CH{=}C(Br){-}CO{-}C_6H_5$ which is a useful synthetic intermediate. Treatment of this compound with excess of a secondary amine leads to the α-amino derivative (IV), from which the 1,2-diketone (benzylphenylglyoxal, V) is obtained by acidic hydrolysis.

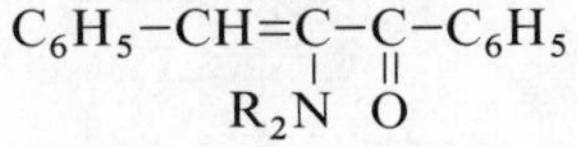

$$C_6H_5{-}CH{=}\underset{R_2N}{\underset{|}{C}}{-}\underset{O}{\underset{\|}{C}}{-}C_6H_5$$

IV

$$C_6H_5{-}CH{=}\underset{HO}{\underset{|}{C}}{-}\underset{O}{\underset{\|}{C}}{-}C_6H_5 \rightleftharpoons C_6H_5{-}CH_2{-}\underset{O}{\underset{\|}{C}}{-}\underset{O}{\underset{\|}{C}}{-}C_6H_5$$

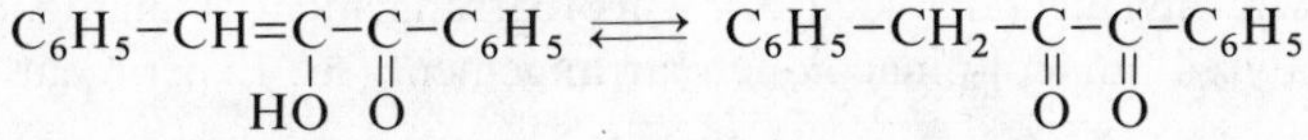

V

The keto-enol tautomers of this compound have been isolated in pure crystalline form:[2] keto, mp 35–36°; enol, mp 67°.

The α-bromo derivative when treated with two moles of sodium methoxide in methanol gives the β-methoxy derivative (VI), from which the 1,3-diketone (dibenzoylmethane, VII) is produced by acidic hydrolysis.[3]

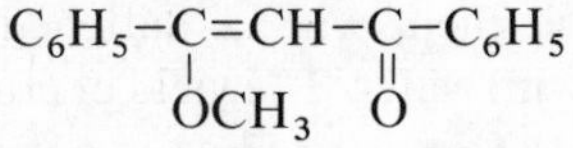

$$C_6H_5{-}\underset{OCH_3}{\underset{|}{C}}{=}CH{-}\underset{O}{\underset{\|}{C}}{-}C_6H_5$$

VI

$$C_6H_5{-}\underset{OH}{\underset{|}{C}}{=}CH{-}\underset{O}{\underset{\|}{C}}{-}C_6H_5 \rightleftharpoons C_6H_5{-}\underset{O}{\underset{\|}{C}}{-}CH_2{-}\underset{O}{\underset{\|}{C}}{-}C_6H_5$$

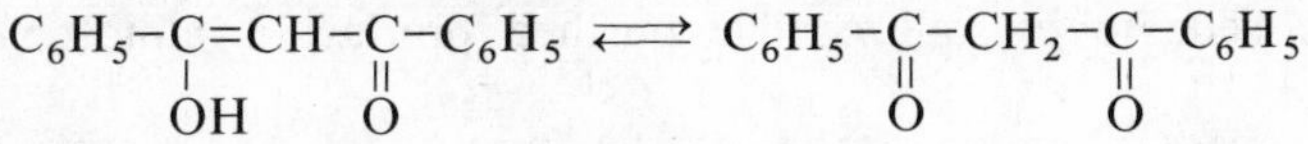

VII

Dibenzoylmethane has been isolated only as the enol form (pale yellow), which exists in two modifications: stable, mp 77–78°; metastable, mp 70–71°.

**Epoxidation of Benzalacetophenone** *(Optional)*. In a 125-ml Erlenmeyer flask place 2 g of benzalacetophenone, 25 ml of ethanol, and 1.5 ml of 30 percent aqueous hydrogen peroxide solution. Swirl the mixture and add *dropwise* 4 ml of 5 percent aqueous sodium hydroxide solution. Cool the reaction mixture so that the temperature does not rise above 30°. After a short while the epoxide separates as a mass of colorless crystals. Collect the product with suction and wash it with a little cold ethanol. The yield is 1.6–1.8 g. The product is usually quite pure, mp 89–90°. It may be recrystallized from 7–8 ml of ethanol. The recorded melting point of the epoxide is 90°.

[2] H. Moureu, *Ann. chim.* (*Paris*), [10] **14**, 303 (1930).
[3] *Organic Syntheses*, Collective Volume **I**, 205 (1941).

The epoxide undergoes two interesting transformations. On warming with alkali it is converted to the enol form of benzylphenylglyoxal (V). Upon treatment with sulfuric acid, in glacial acetic acid, the epoxide (VIII) is isomerized through a carbonium ion intermediate, so that the phenyl group in the $\beta$-position is shifted to the $\alpha$-position. The product is the enol form of

$$\text{V} \xleftarrow{\text{OH}^-} \underset{\text{VIII}}{C_6H_5\text{-}\overset{}{CH}\text{-}CH\text{-}CO\text{-}C_6H_5}\ (\text{epoxide, O bridging the two CH}) \xrightarrow{\text{H}^+} \underset{\text{IX}}{H\text{-}\underset{\displaystyle OH}{C}{=}\overset{\displaystyle C_6H_5}{C}\text{-}CO\text{-}C_6H_5}$$

$\alpha$-benzoylphenylacetaldehyde (IX). This rearrangement is analogous to the acid-catalyzed pinacol-pinacolone rearrangement (see Experiment 42).

**$\alpha$-Benzoylphenylacetaldehyde** *(Optional)*. In a 125-ml Erlenmeyer flask dissolve 1 g of benzalacetophenone oxide (VIII) in 10 ml of glacial acetic acid and add a cold solution of 6 ml of concentrated sulfuric acid in 6 ml of glacial acetic acid. The solution becomes warm and develops a yellow color, which darkens to a reddish brown. Swirl the solution from time to time and after about 30 min add 60–65 ml of cold water. The product separates in yellow, oily droplets that crystallize slowly. Collect the crystallized globules on a small suction filter, crush them with a flattened glass rod (not a metal spatula), and wash them with water. The yield of crude product is about 0.5 g. For purification dissolve the crude product in a very small amount of warm acetone, add water dropwise until a slight turbidity develops, and set the solution aside to crystallize. The purified product is almost colorless; mp 111–112°.

The enol is a moderately strong acid and dissolves in aqueous sodium carbonate solution. In ethanolic solution it gives a violet-brown color with ferric chloride solution. With cupric acetate it forms a chartreuse-colored copper chelate derivative.

## Questions

**1.** What product would be formed by mixed aldol condensation of benzaldehyde with acetaldehyde? with acetone (in excess)?

**2.** Write equations, stepwise, showing the mechanism of addition of hydrogen chloride to benzalacetophenone.

**3.** What products are formed by reaction of benzalacetophenone with the following reagents: (a) phenylmagnesium bromide, followed by water and dilute acid; (b) diethylamine; (c) phenylhydrazine, followed by cyclization with sulfuric acid?

**4.** Write projection formulas for the stereoisomeric forms of benzalacetophenone and its dibromide.

**5.** Write a stepwise mechanism for the acid-catalyzed rearrangement of benzalacetophenone epoxide (formula VIII) to the enol form of benzoylphenylacetaldehyde (formula IX).

EXPERIMENT

# 37

# The Benzoin Condensation

In the benzoin condensation two molecules of an aromatic aldehyde react to form a new carbon-carbon bond by joining the carbon atoms of the carbonyl functions. The product is an $\alpha$-hydroxy ketone (an acyloin). Cyanide *ion* is a specific catalyst for the reaction and functions by addition to the carbonyl group to form a cyanohydrin *anion* (step 1). In the adduct the acidity of the

$$C_6H_5{-}CH{=}O \underset{(1)}{\overset{CN^-}{\rightleftharpoons}} C_6H_5{-}\underset{H}{\overset{^-O}{C}}{-}C{\equiv}N \underset{(2)}{\rightleftharpoons} C_6H_5{-}\overset{OH}{\underset{\cdot\cdot}{C}}{-}C{\equiv}N \longleftrightarrow C_6H_5{-}\overset{OH}{C}{=}C{=}N^-$$

$$\xrightarrow[(3)]{+C_6H_5-CH=O}$$

$$C_6H_5{-}\overset{O}{\overset{\|}{C}}{-}\overset{OH}{CH}{-}C_6H_5 \underset{(5)}{\rightleftharpoons} C_6H_5{-}\underset{C\equiv N}{\overset{^-O}{C}}{-}\overset{OH}{CH}{-}C_6H_5 \underset{(4)}{\rightleftharpoons} C_6H_5{-}\underset{C\equiv N}{\overset{OH}{C}}{-}\overset{O^-}{CH}{-}C_6H_5$$

C$-$H bond is enhanced (being in the $\alpha$-position to a nitrile group) and by proton exchange the isomeric carbanion is formed (step 2). The latter adds to a second molecule of the aldehyde (step 3); proton interchange and loss of cyanide ion (steps 4 and 5) lead to the benzoin. The rate determining step in the case of benzaldehyde appears to be step 3.

Substituted benzaldehydes, such as the tolualdehydes and methoxybenzaldehydes (and also furfural) form the corresponding *symmetrical* benzoins. Two different aldehydes can react to form *unsymmetrical* or mixed benzoins, such as anisbenzoin or benzfuroin.

$$R{-}CH{=}O + R'{-}CH{=}O \xrightarrow{CN^-} R{-}CO{-}CH(OH){-}R' \text{ or } R{-}CH(OH){-}CO{-}R'$$

 Although two isomeric mixed benzoins and two symmetrical benzoins could be formed in these reactions, it is not uncommon to obtain a single mixed benzoin as the principal product.[1]

Some substituted benzaldehydes, such as *p*-chloro- and *p*-dimethylaminobenzaldehyde, react poorly or not at all to form simple benzoins, but will form mixed benzoins in conjunction with another aldehyde. *p*-Chlorobenzaldehyde and *p*-nitrobenzaldehyde have good carbonyl reactivity but the nucleophilic activity of the carbonium ion needed for step 3 is reduced (or almost completely lost for the *p*-nitro compound), through the electron-attracting effect of the substituent. *p*-Chlorobenzaldehyde will form a mixed benzoin by reaction with an appropriate partner.

*p*-Dimethylaminobenzaldehyde fails to form a simple benzoin because the electrophilic activity of its carbonyl group (needed for step 3) is greatly reduced through electron-release by the *p*-amino group. With benzaldehyde it forms a mixed benzoin: $(CH_3)_2N-C_6H_4-CO-CH(OH)-C_6H_5$.

Benzoin and similar α-hydroxy aldehydes and ketones (including the simple sugars) are capable of existing in a tautomeric ene-diol structure. Under anhydrous conditions benzoin can be converted to the dipotassium salt of the ene-diol form. By the action of strong bases, the less stable isomer of a mixed benzoin can be converted to the more stable isomer, through the

$$R-CO-CH(OH)-R' \underset{H_2O}{\overset{OH^-}{\rightleftharpoons}} \begin{matrix} HO & & R' \\ & C=C & \\ R & & OH \end{matrix} \underset{OH^-}{\overset{H_2O}{\rightleftharpoons}} R-CH(OH)-CO-R'$$

enolate that is common to both. The diacetate of the ene-diol of benzoin (diacetoxystilbene) exists in two stereoisomeric forms: *trans*-, mp 155°; *cis*-, mp 118–119°. Saponification of either form produces benzoin.

The conversion of benzoin to benzil and the use of these compounds as intermediates for syntheses are discussed in the experimental section.

## 37a Benzoin

In a 250-ml round-bottomed flask place 0.03 mole (1.5 g) of sodium cyanide and dissolve it in 15 ml of water. To the cyanide solution add 30 ml of ethanol and 0.15 mole (16 g, 15 ml) of pure benzaldehyde. Add a small boiling chip, attach an upright condenser, and reflux the solution gently on a steam bath for 30 min.

➤**CAUTION:** Sodium cyanide is extremely poisonous and must be handled with the greatest care. Wash the hands thoroughly after using it. Do not handle sodium cyanide if you have an open cut on the hand. Take care not to acidify a solution containing cyanide.

Pour all filtrates containing cyanide directly into the drain pipe of the sink and wash the material away with a large volume of water.

[1] Ide and Buck, "The Synthesis of Benzoins," *Organic Reactions*, **4**, 269 (1948).

Remove the flask from the steam bath, detach the condenser, and swirl the solution gently as it cools, to promote crystallization. After cooling the flask in an ice-water bath, collect the product with suction and wash the crystals thoroughly, using two 15-ml portions of *cold* 50 percent ethanol and several portions of water. Press the crystals as dry as possible and spread them to dry in the air. The yield is 11–13 g (dry weight).

The product may be used without careful drying or recrystallization, for the preparation of derivatives or for conversion to benzil and benzilic acid. Benzoin may be purified, with loss of 10–15 percent, by recrystallization from methanol (12 ml per gram of benzoin) or from ethanol (8 ml per gram).

***Reactions.*** Benzoin and other acyloins (α-hydroxyketones) may be characterized through reactions of the secondary alcohol function (acetylation, oxidation) or of the carbonyl group (formation of oximes, hydrazones, semicarbazones). Benzoin, like the keto sugars, is converted to an osazone by warming with excess phenylhydrazine. By reduction with sodium amalgam or sodium borohydride ($NaBH_4$) benzoin is converted to *meso*-hydrobenzoin (I); with zinc and hydrochloric acid the reduction product is desoxybenzoin (II).

$$\underset{\text{I}}{C_6H_5{-}CH(OH){-}CH(OH){-}C_6H_5} \qquad \underset{\text{II}}{C_6H_5{-}CH_2{-}CO{-}C_6H_5}$$

**Benzoin Acetate** *(Optional)*. In a test tube place 1 g of benzoin, 1 ml of glacial acetic acid and 1 ml of acetic anhydride. Introduce *one drop* of concentrated sulfuric acid and shake well to assure good mixing; the mixture becomes warm and the benzoin dissolves. Heat the tube in a beaker of boiling water for 5–10 min (not longer), allow the solution to cool slightly, and pour it carefully into 25 ml of water. Collect the crystals with suction, wash them thoroughly with water, and press dry. Recrystallize the product directly from methanol or ethanol. The recorded melting point of benzoin acetate is 82–83°.

**Benzoin α-oxime** *(Optional)*. In a small Erlenmeyer flask place 1 g of benzoin, 5 ml of ethanol, 1 ml of a 35 percent aqueous solution of hydroxylamine hydrochloride, and 2 ml of 30 percent aqueous sodium hydroxide. Add a small boiling chip, attach a short condenser, and reflux the mixture gently for 1.5–2 hr. Pour the warm reaction mixture, with stirring, into 50 ml of cold water to which 2 ml of concentrated sulfuric acid has been added. Collect the product on a suction filter, wash it well with water and press dry. The oxime prepared in this way contains about 10 percent of the β-isomer (oximino hydroxyl in *syn*-configuration relative to the $-CHOH-C_6H_5$ group). Recrystallize the crude product from ethanol, or, after thoroughly drying the material, from benzene. The recorded melting points are: α-oxime, 152–155°; β-oxime, 99°.

The acetates of these oximes exhibit stereospecific behavior toward cold 5 percent aqueous sodium hydroxide:[2] the α-oxime acetate is cleaved readily to benzaldehyde, benzonitrile, and sodium acetate; the β-oxime acetate is not cleaved but merely hydrolyzed to regenerate the β-oxime and sodium acetate. The α-oxime forms chelate derivatives with metal ions such as $Cu^{++}$, and has usefulness as an analytical reagent.[3]

## 37b Benzil

α-Hydroxyketones (acyloins) are oxidized to the corresponding 1,2-diketones by cupric oxide and other cupric compounds, which are thereby reduced to the cuprous state (see Fehling's test for reducing sugars, Experiment 16). In the present procedure[4] cupric acetate is used in catalytic amount (less than 1 percent of the stoichiometric requirement) and is continuously re-oxidized from the reduced (cuprous) state by ammonium nitrate. The latter is reduced to ammonium nitrite, which decomposes in the reaction mixture into nitrogen and water. Benzoin may be oxidized to benzil also by means of nitric acid.

**(A) Oxidation by Cupric Salts.** In a 100-ml round-bottomed flask place 0.04 mole (8.5 g) of benzoin, 25 ml of glacial acetic acid, 0.05 mole (4 g) of pulverized ammonium nitrate, and 5 ml of a 2 percent solution of cupric acetate.[5] Add one or two small boiling chips, attach a reflux condenser, and heat the flask gently on a wire gauze, with occasional swirling. As the reactants dissolve, evolution of nitrogen begins. Boil the green solution for 1.5 hr to complete the reaction. Cool the solution to 50–60° and pour it into 40 ml of ice water, with stirring. After crystallization of the benzil is complete, collect the crystals on a suction filter and wash them thoroughly with water. Press the product as dry as possible on the filter. The yield is 7.5–8 g (dry weight). Benzil obtained in this way is usually sufficiently pure for conversion to derivatives or to benzilic acid. It may be purified by recrystallization from methanol or 75 percent aqueous ethanol.

***Test for the Presence of Benzoin.*** Dissolve a few crystals of the product in 1 ml of ethanol and add a drop of sodium hydroxide solution. A purple coloration develops if any unoxidized benzoin is present. The color fades on shaking the solution with air but reappears on standing undisturbed.

[2] Blatt and Barnes, *J. Amer. Chem. Soc.*, **56**, 1149 (1934); **58**, 1900 (1936).

[3] Welcher, *Organic Analytical Reagents*, Van Nostrand Co., Inc., New York (1947), Vol. III, 239.

[4] Weiss and Appel, *J. Amer. Chem. Soc.*, **70**, 3666 (1948).

[5] The catalyst solution may be prepared by dissolving 2.5 g of cupric acetate monohydrate in 100 ml of 10 percent aqueous acetic acid, stirring well, and filtering to remove any basic copper salts that have precipitated.

**(B) Oxidation by Nitric Acid** *(Alternative Procedure).* In a 250-ml Erlenmeyer flask place 0.04 mole (8.5 g) of benzoin and 40 ml of glacial acetic acid. While swirling the flask to obtain good mixing, add 0.34 mole (28 g, 20 ml) of concentrated nitric acid. Heat the reaction mixture on a steam bath, in a hood, for 2 hr. Cool the flask in an ice-water bath, add 150 ml of water, mix thoroughly, and allow the yellow precipitate of benzil to settle. Collect the product with suction and wash it thoroughly with water to remove nitric acid (test with moist litmus paper). Press the crystals as dry as possible with a clean cork or flat glass stopper. The yield is 7–8 g (dry weight). Benzil obtained in this way is usually sufficiently pure for conversion to derivatives or to benzilic acid. It may be purified by recrystallization from methanol or 75 percent aqueous ethanol.

Test the product for the presence of unoxidized benzoin as described in part A.

***Reactions.*** 1,2-Diketones may be characterized by conversion to a variety of crystalline derivatives, such as the mono- and dioximes, hydrazones, and semicarbazones. The monosemicarbazone undergoes cyclization readily upon heating, or treatment with cold aqueous alkali, to give a triazine derivative (I).

$$\begin{array}{l} \quad\quad N{-\!\!-}NH \\ C_6H_5{-}C \quad\quad C{=}O \\ C_6H_5{-}C{=}O \quad H_2N \end{array} \longrightarrow$$

$$\begin{array}{c} C_6H_5{-}C{=}N{-}NH \\ C_6H_5{-}C{=}N{-}C{=}O \\ \text{I} \end{array} \rightleftharpoons \begin{array}{c} C_6H_5{-}C{=}N{-}N \\ C_6H_5{-}C{=}N{-}C{-}OH \end{array}$$

Pyrazine derivatives are formed by reaction of 1,2-diketones with aliphatic and aromatic 1,2-diamines; the latter give quinoxalines (benzopyrazines, II) that may be used for purposes of identification. Benzil under-

$$\begin{array}{c} C_6H_5{-}C{=}N \\ C_6H_5{-}C{=}N \\ \text{II} \end{array}\!\!\!\bigcirc \qquad \begin{array}{c} C_6H_5{-}C{-\!\!-}N \\ C_6H_5{-}C{-}NH{-}C{-}C_6H_5 \\ \text{III} \end{array}$$

$$\begin{array}{c} C_6H_5{-}C{-\!\!-}C{-}C_6H_5 \\ C_6H_5{-}C{-}C({=}O){-}C{-}C_6H_5 \\ \text{IV} \end{array}$$

goes other cyclization reactions, such as the formation of triphenylimidazole (lophine, III) on treatment with ammonia and benzaldehyde, and conversion to tetraphenylcyclopentadienone (tetracyclone, IV) by reaction with dibenzyl ketone.[6]

The most interesting reaction of benzil, discovered by Justus Liebig in 1838, is its transformation to benzilic acid (V) by heating with strong alkalies (Experiment 38). The mechanism of this rearrangement involves addition of hydroxyl ion to one of the carbonyl groups, followed by a cycle of intramolecular shifts leading to the stable benzilate anion. An analogous re-

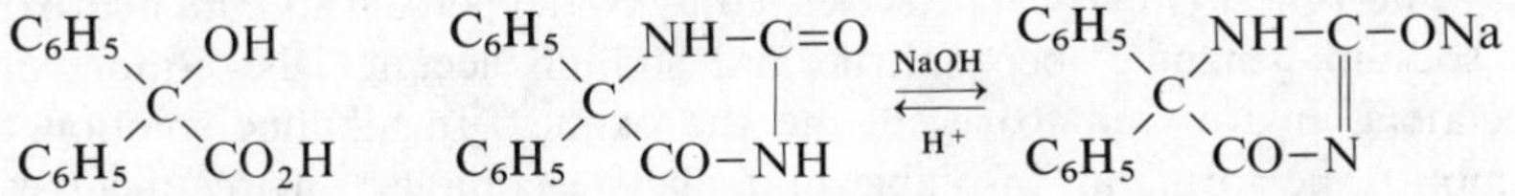

$$(C_6H_5)_2C(OH)CO_2H \qquad (C_6H_5)_2C\begin{matrix}NH-C=O\\ | \\ CO-NH\end{matrix} \underset{H^+}{\overset{NaOH}{\rightleftharpoons}} (C_6H_5)_2C\begin{matrix}NH-C-ONa\\ \| \\ CO-N\end{matrix}$$

V VI Dilantin Sodium

arrangement occurs when benzil is warmed with urea and alkali; the principal product is 5,5-diphenylhydantoin (VI). The sodium salt of this hydantoin is an anticonvulsant medicinal, *Dilantin Sodium*, used in the treatment of epilepsy.

**Benzil Monohydrazone** *(Optional)*. Dissolve 1 g of benzil in 5 ml of warm ethanol and add dropwise, with shaking, 1 ml of a 25 percent aqueous solution of hydrazine hydrate (5-molar solution of $NH_2-NH_2$). Boil the solution for 5–10 min in a water bath, add 10 ml of 50 percent aqueous ethanol, and allow the reaction mixture to cool. Collect the crystals, wash them with cold 50 percent ethanol and dry in the air. The product may be recrystallized from hot ethanol. The recorded melting point of the monohydrazone is 149–150° (dec).

On heating the monohydrazone at the melting point, nitrogen is evolved and desoxybenzoin ($C_6H_5-CO-CH_2-C_6H_5$) is formed.[7] Oxidation with mercuric oxide converts the monohydrazone to phenylbenzoyldiazomethane, which loses nitrogen at 100° and undergoes rearrangement to give diphenylketene,[8] $(C_6H_5)_2C=C=O$. Benzil dihydrazone may be obtained by heating benzil with an excess of hydrazine hydrate in ethylene glycol. The dihydrazone upon oxidation with mercuric oxide gives 1,2-diphenyl-*bis*-diazoethane, which loses nitrogen to form diphenylacetylene, $C_6H_5-C\equiv C-C_6H_5$.[7]

**Benzil α-Monoxime** *(Optional)*. In a test tube place 1 g of finely pulverized benzil, 5 ml of ethanol and 1 ml of a 35 percent aqueous solution of hydroxylamine hydrochloride. Cool the mixture in an ice bath and add dropwise, with shaking, 2 ml of 30 percent aqueous sodium hydroxide solution. Allow the

[6] *Organic Syntheses*, Collective Volume **III**, 806 (1955).

[7] Curtius and Thun, *J. prakt. Chem.*, [2] **44**, 161, 186 (1891); Schlenk and Bergmann, *Ann.*, **463**, 76 (1928).

[8] *Organic Syntheses*, Collective Volume **III**, 356 (1955).

tube to stand in the cooling bath, with occasional shaking, until a drop of the reaction mixture when placed in water is *almost* completely soluble (2–3 hr is required). Pour the reaction mixture into 20 ml of water, filter the solution and acidify the filtrate with 20 percent sulfuric acid. Collect the product on a suction filter, wash it with water and recrystallize from 30 percent ethanol or from methanol. The recorded melting point of the α-monoxime is 137–138°; it corresponds in configuration to benzoin α-oxime (oximino hydroxyl in *anti*-configuration relative to the $-CO-C_6H_5$ group). The stereoisomeric β-monoxime melts at 105–108°.

The α-monoxime acetate is cleaved by cold 5 percent sodium hydroxide to sodium benzoate, benzonitrile and sodium acetate; the β-monoxime acetate is hydrolyzed to regenerate the oxime.[2] In alkaline solution the α-monoxime forms an insoluble blue chelate complex on treatment with ferrous sulfate; the β-monoxime does not form such complexes. The α-monoxime dissolves in 10 percent aqueous potassium hydroxide to form a deep orange solution; the β-monoxime forms a canary yellow solution.

**Benzil Monosemicarbazone** *(Optional)*. In a 50-ml Erlenmeyer flask dissolve 1 g of benzil in 20 ml of warm ethanol and cool the solution rapidly, with swirling, to obtain a fine suspension of small crystals. To this add a solution of 0.5 g of semicarbazide hydrochloride and 1.5 g of sodium acetate crystals ($CH_3CO_2Na \cdot 3H_2O$) in 5 ml of water. Cork the flask firmly and shake the mixture vigorously. Allow the flask to stand for several days at room temperature, then add 15–20 ml of water, and cool in an ice bath. Collect the product with suction, wash it with water, and press dry on the filter. The yield is 0.5 g. The crystals melt with decomposition in the range 170–175°, when heated rapidly.

To effect cyclization of the product, dissolve a small portion in 5 percent aqueous sodium hydroxide, with shaking, and then acidify the solution with glacial acetic acid. Collect the yellow crystals of 5,6-diphenyl-1,2,4-triazin-3-one and wash them with water. This compound sinters at 190° and melts at 224–226°.

**5,5-Diphenylhydantoin: Dilantin** *(Optional)*. In a small Erlenmeyer flask place 1 g of benzil, 0.5 g of urea, 15 ml of ethanol, and 3 ml of 30 percent aqueous sodium hydroxide. Attach an upright condenser, add a boiling chip, and boil the mixture gently for 2 hr. Cool the reaction mixture, add 25 ml of water, and filter the solution to remove a sparingly soluble side product. Acidify the filtrate with hydrochloric acid, collect the product on a suction filter, and wash it thoroughly with water. This hydantoin may be recrystallized from ethanol. The yield is 0.7–0.8 g. Do not attempt to determine the melting point (recorded mp 286–295°). *Warning: Dilantin is a powerful drug!*

## Questions

**1.** Compare the mechanism of the benzoin condensation with that of the aldol condensation and the Cannizzaro reaction, including the role of the reagents employed.

**2.** Write formulas for the products formed by the action of excess phenylmagnesium bromide on the following compounds, after treating the reaction mixture with water and mineral acid: (a) benzoin; (b) benzil; (c) mandelonitrile (benzaldehyde cyanohydrin).

**3.** Write configurational formulas for the stereoisomers of benzil dioxime.

**4.** Application of the benzoin condensation to an equimolar mixture of benzaldehyde and 4-methoxybenzaldehyde (anisaldehyde) furnishes the compound $CH_3O{-}C_6H_4{-}CO{-}CH(OH){-}C_6H_5$ (mp 106°). (a) Suggest a method for establishing definitely that the carbonyl group is the assigned position rather than adjacent to the phenyl group; (b) Devise a synthesis that would furnish the isomeric methoxybenzoin (mp 89–90°).[1]

**5.** When benzil is heated with urea and alkali to prepare 5,5-diphenylhydantoin (formula VI), a sparingly soluble compound of the molecular formula $C_{16}H_{14}N_4O_2$ is produced in small amount by a side reaction. Devise a structural formula for this compound. (The side reaction does not involve rearrangement of benzil.)

# EXPERIMENT 38

# The Benzilic Acid Rearrangement

When benzil is warmed with strong alkalies it is converted into a salt of α-hydroxydiphenylacetic acid (benzilic acid). The 1,2-molecular rearrangement is initiated by addition of hydroxyl ion to the diketone (step 1), followed by transfer of the aryl group with its bonding electrons (carbanion rearrangement) to the adjacent carbon atom (step 2). By concurrent proton interchange the stable benzilate anion is formed. It has been established by means

$$\begin{array}{c} C_6H_5-C{=}O \\ | \\ C_6H_5-C{=}O \end{array} \underset{(1)}{\overset{OH^-}{\rightleftharpoons}} \begin{array}{c} O^- \\ | \\ C_6H_5-C-OH \\ | \\ C_6H_5-C{=}O \end{array} \underset{(2)}{\longrightarrow} (C_6H_5)_2C(O^-)-\overset{O}{\overset{\|}{C}}-OH \underset{(3)}{\longrightarrow} (C_6H_5)_2C(OH)-\overset{O}{\overset{\|}{C}}-O^-$$

of oxygen-exchange with $O^{18}$-labeled water that step 1 is reversible and faster than step 2. With methoxide ion in methanol, a similar rearrangement occurs and the product is the methyl ester of benzilic acid.[1]

The benzilic acid rearrangement extends to many substituted diaryl 1,2-diketones but not to simple aliphatic analogs (such as $CH_3-CO-CO-CH_3$). The latter undergo complex aldol condensations under the influence of alkaline reagents.

Cyanide ion in ethanolic solution, in catalytic amounts, causes a rapid and complete cleavage of benzil at the central carbon-carbon bond; the products are ethyl benzoate and benzaldehyde. The mechanism of the cleavage

$$\begin{array}{c} C_6H_5-C{=}O \\ | \\ C_6H_5-C{=}O \end{array} + C_2H_5OH \xrightarrow{CN^-} \begin{array}{c} C_6H_5-C{=}O \\ | \\ OC_2H_5 \end{array} + \begin{array}{c} C_6H_5-C{=}O \\ | \\ H \end{array}$$

[1] For a discussion of the benzilic acid and related rearrangements see Gould, *Mechanism and Structure in Organic Chemistry*, Holt, Rinehart and Winston, Inc., New York (1959) and Selham and Eastman, *Quarterly Reviews* (London), **14**, 221 (1960).

is uncertain but it is likely that the process involves addition of cyanide ion to a hemi-ketal formed by interaction of the diketone with ethanol.

## Benzilic Acid

**(A) From Benzil.** In a 100-ml flask dissolve 0.075 mole (5 g of 85 percent pure solid) of potassium hydroxide pellets in 10 ml of water, add 15 ml of ethanol and mix well by swirling. To the solution add 0.024 mole (5 g) of pure benzil (a bluish black coloration is developed), attach an upright condenser and reflux the solution on a steam bath for 10–15 min. Transfer the contents of the flask to a small beaker or a porcelain dish and cover with a watch glass. Allow the reaction mixture to stand for several hours, preferably overnight, until crystallization of the potassium salt of benzilic acid is complete. Collect the crystals on a suction filter and wash them sparingly with ice-cold ethanol. The ethanolic mother liquor will furnish a small additional quantity of potassium benzilate if allowed to stand overnight.

Dissolve the potassium salt in about 150 ml of water and add to the solution, with stirring, *two drops* of concentrated hydrochloric acid. A reddish-brown, slightly sticky precipitate is formed. Add a small amount of decolorizing carbon and filter off the solid material. If the procedure has been performed successfully the filtrate will be colorless or only faintly yellow in color. Pour the clear filtrate slowly, with stirring, into a solution of 8 ml of concentrated hydrochloric acid in 50 ml of water. Collect the precipitated benzilic acid with suction, wash it thoroughly with water to remove chlorides, and press dry. The crude product is usually light pink or yellow in color and weighs 4–4.5 g. Crystallize the material from hot water, or from benzene (about 6 ml per gram), with addition of a little decolorizing carbon. The yield of purified benzilic acid is 3.5–4 g.

**(B) From Benzoin** *(Alternative Procedure)*. In a small porcelain dish prepare a solution of 0.175 mole (7 g) of solid sodium hydroxide and 0.01 mole (1.5 g) of sodium bromate (or 1.7 g of potassium bromate) in 15 ml of water. To the warm solution add in portions, 6–6.5 g of the slightly moist benzoin obtained in Experiment 37 (this amount corresponds to about 0.025 mole of dry benzoin). During and after the addition of the benzoin, heat the reaction mixture on a steam bath and stir it constantly. The mixture should not be heated above 90–95° since higher temperatures favor decomposition to form benzohydrol (diphenylmethanol). From time to time add small portions of water (in total about 12–15 ml) to keep the mixture from becoming too thick. Continue the heating and stirring until a small test portion is completely or almost completely soluble in water. This usually requires 1.5–2.0 hr.

Dilute the reaction mixture with 60 ml of water and allow it to stand, preferably overnight. Filter the solution to remove the oily or solid side product (benzohydrol). To the filtrate add slowly, with stirring, sufficient 40 percent sulfuric acid (prepared by adding 5 ml of concentrated sulfuric acid to 15 ml of water) to reach a point just short of the liberation of bromine.[2] Usually about 17 ml of the sulfuric acid is required. Collect the benzilic acid with suction, wash it well with water, and press dry. The yield is 4.5–5 g and the product is usually quite pure. Benzilic acid may be crystallized from hot water, or from benzene (about 6 ml per gram).

***Reactions.*** Benzilic acid may be characterized by acylation of its alcohol group or through reactions of the carboxyl group (esterification, amide formation). Oxidation with dichromate mixture converts it to benzophenone and carbon dioxide; reduction with hydriodic acid leads to diphenylacetic acid, $(C_6H_5)_2CH-CO_2H$.

The tertiary alcohol group also enters readily into condensation reactions with aromatic hydrocarbons and phenols; for example, benzilic acid reacts with benzene in the presence of stannic chloride as catalyst to give triphenylacetic acid.

Benzilic acid is converted by warming with phosphorus pentachloride (2 moles) into diphenylchloroacetyl chloride but a different reaction occurs upon treatment with thionyl chloride ($SOCl_2$): benzophenone and carbon monoxide are produced, together with sulfur dioxide and hydrogen chloride. Thionyl chloride (3 moles) in carbon tetrachloride solution, or phosphorus oxychloride ($POCl_3$) under mild conditions, converts benzilic acid to diphenylchloroacetic acid.

**Acetylbenzilic Acid (α-Acetoxydiphenylacetic Acid)** *(Optional)*. In a large test tube place 0.5 g of benzilic acid, 1 ml of glacial acetic acid, and 1 ml of acetic anhydride. Add *one drop* of concentrated sulfuric acid, mix well, and heat the tube in a bath of boiling water for 2 hr. Cool the solution to 25° and add 5 ml of water, dropwise, with shaking. Allow the reaction mixture to stand overnight or longer to permit crystallization of the acetyl derivative. Collect the product with suction, wash it well with water, and press dry on the filter. The air-dried material is a monohydrate, mp 96–98°. Prolonged drying in a vacuum desiccator over sulfuric acid is necessary to obtain anhydrous acetylbenzilic acid, mp 104–105°.

**Methyl Benzilate** *(Optional)*. In a large diameter test tube place 0.5 g of benzilic acid, 5 ml of methanol and 0.5 ml of concentrated sulfuric acid. Addition of the acid causes the development of a red color that disappears on shaking the tube. Attach a short condenser, add a boiling chip, and reflux the solution for 30 min. Cool the reaction mixture and pour it in small por-

[2] To minimize the danger of passing the end point it is advisable to set aside in a test tube, 8–10 ml of the filtrate and add sulfuric acid to the remainder until a *trace* of bromine is liberated. This is removed by adding the small portion of the solution from the test tube.

tions into 20 ml of 5 percent aqueous sodium carbonate solution (*caution—foaming!*). Chill the mixture in an ice bath, collect the crystals with suction, and wash them well with water. The recorded melting point of methyl benzilate is 74–75°.

**Reduction to Diphenylacetic Acid** *(Optional)*. In a large diameter test tube place 10 ml of glacial acetic acid, 0.6 g of red phosphorus (*handle carefully*) and 0.2 g of iodine crystals. Mix the materials by swirling and allow the tube to stand for 5 min; then add 5–6 drops of water and 3.5 g of benzilic acid. Attach a short condenser, add a boiling chip, and boil the mixture gently without interruption for 2 hr. Filter the hot solution (*caution—acid vapors!*) to remove the excess of red phosphorus and pour the warm filtrate, with stirring, into 40 ml of cold water in which 1 g of sodium bisulfite has been dissolved (to remove free iodine). Collect the precipitated diphenylacetic acid on a suction filter, wash it with water, and press dry on the filter. Purify the material by recrystallization from warm 50 percent ethanol (6–7 ml per gram), with addition of decolorizing carbon. The yield is 2.5–3 g.

## Questions

**1.** Account for the difference in type of reagent used to bring about the benzilic acid and the pinacol-pinacolone rearrangements.

**2.** What products are formed in the following reactions:

(a) benzilic acid + toluene (+$SnCl_4$ catalyst)?

(b) benzilic amide + NaOBr + NaOH (Hofmann reaction)?

(c) methyl benzilate + excess $C_6H_5{-}MgBr$ (followed by $H_2O$ + acid)?

(d) methyl benzilate + ammonia (in methanol)?

(e) methyl benzilate + phenyl isocyanate ($C_6H_5{-}N{=}C{=}O$)?

**3.** In the presence of a trace of sulfuric acid, benzilic acid reacts with acetone to form a crystalline product, mp 48°, having the molecular formula $C_{17}H_{16}O_3$. Suggest a structural formula for this compound.

**4.** In the reduction of benzilic acid to diphenylacetic acid by hydriodic acid, formed from iodine, phosphorus, and water, only a small amount of iodine (0.1 gram-atom) is needed for reducing a large amount of benzilic acid (1 mole). Explain why this is possible.

EXPERIMENT

# 39

# The Perkin Reaction

The Perkin reaction is an aldol-type condensation in which an aldehyde of aromatic type reacts with an acid anhydride in the presence of a mild base, such as the potassium salt of the acid corresponding to the anhydride, potassium carbonate, or a tertiary amine. The role of the base is to promote enolization of the acid anhydride (step 1). The sequence of steps, illustrated for benzaldehyde and acetic anhydride, involves addition of the enolate to the aldehyde in the usual aldol manner (step 2); O-acyl exchange within the

$$CH_3{-}CO{-}O{-}CO{-}CH_3 + CH_3{-}CO_2^- \underset{(1)}{\rightleftharpoons} CH_3{-}CO_2H + :\bar{C}H_2{-}CO{-}O{-}CO{-}CH_3$$

$$C_6H_5{-}CH{=}O + :\bar{C}H_2{-}CO{-}O{-}CO{-}CH_3 \underset{(2)}{\longrightarrow} C_6H_5{-}\underset{}{\overset{O^-}{\overset{|}{C}}}H{-}CH_2{-}CO{-}O{-}CO{-}CH_3$$

$$\downarrow (3)$$

$$C_6H_5{-}CH{=}CH{-}CO{-}O^- + CH_3{-}CO_2H \underset{(4)}{\longleftarrow} C_6H_5{-}\overset{O-CO-CH_3}{\overset{|}{C}}H{-}CH_2{-}CO{-}O^-$$

resulting adduct to give the more stable $\beta$-acetoxypropionate anion (step 3); and finally, $\alpha,\beta$-elimination of acetic acid to produce the anion of cinnamic acid ($\beta$-phenylacrylic acid).

If the salt used is not derived from the same acid as the anhydride a mixture of two different arylacrylic acids may be formed, because an exchange reaction between salt and anhydride can occur in the reaction mixture. When one of the components has appreciably greater $\alpha$-methylene activity than the

$$R{-}CH_2{-}CO_2K + CH_3{-}CO{-}O{-}CO{-}CH_3 \overset{100^\circ}{\rightleftharpoons} R{-}CH_2{-}CO{-}O{-}CO{-}CH_3 + CH_3{-}CO_2K$$

 other, the product will consist mainly or entirely of the arylacrylic acid derived from that component. Thus, benzaldehyde, acetic anhydride, and a salt of phenylacetic acid furnish α-phenylcinnamic acid, and little or no cinnamic acid.

Homologs of acetic anhydride having two hydrogens in the α-position furnish α-substituted cinnamic acids. If only one hydrogen is present in the α-position, as in isobutyric anhydride, $(CH_3)_2CH{-}CO{-}O{-}CO{-}CH(CH_3)_2$, the normal α,β-elimination reaction (step 4) cannot occur. The products consist of the acyl derivative of the β-hydroxy acid (I) *and* the unsaturated

$$\underset{\text{I}}{C_6H_5{-}\underset{\displaystyle O{-}CO{-}CH(CH_3)_2}{\underset{|}{C}H}{-}C(CH_3)_2{-}CO{-}O^-} \longrightarrow C_6H_5{-}CH{=}C(CH_3)_2 + CO_2 + {}^-O{-}CO{-}CH(CH_3)_2$$

hydrocarbon formed by a concerted decarboxylation and α,β-elimination. Even with mono-substituted derivatives of acetic anhydride the side reaction may become important, especially at higher reaction temperatures, and reduce the yield of the substituted cinnamic acid.

Substituted benzaldehydes containing alkyl, halogen, or nitro groups take part readily in the Perkin reaction. Alkoxy derivatives usually give lower yields (25–30 percent) and *p*-dimethylaminobenzaldehyde gives very unsatisfactory results. The condensation may be carried out with cinnamaldehyde ($C_6H_5{-}CH{=}CH{-}CH{=}O$) and with furfural.[1]

An interesting extension of the Perkin reaction, called the Erlenmeyer synthesis, involves the reaction of aromatic aldehydes with acyl derivatives of aminoacetic acid $C_6H_5CO{-}NH{-}CH_2{-}CO_2H$ or $CH_3CO{-}NH{-}CH_2{-}CO_2H$ to furnish derivatives of α-aminocinnamic acid. These are employed as intermediates for the synthesis of derivatives of phenylalanine, phenylpyruvic acid, and related compounds (see Experiment 51b).

Aromatic and aliphatic aldehydes undergo condensation readily with malonic acid, and with malonic and cyanoacetic esters, in the presence of secondary amines as catalysts (the Knoevenagel reaction). Cinnamic acids may be prepared conveniently by heating an aromatic aldehyde with malonic acid and an amine catalyst, since the intermediate benzylidenemalonic acid undergoes decarboxylation in the reaction mixture.[2]

## Cinnamic Acid

In a 100-ml round-bottomed flask place 0.06 mole (6 g) of *freshly fused*

[1] For a review of the Perkin reaction and the use of cinnamic acid derivatives in syntheses, see *Organic Reactions*, **1**, 210 (1942).

[2] For examples of condensation with malonic acid see footnote 1, and also *Organic Syntheses*, Collective Volume **III**, 425 (1955); **31**, 35 (1951); **33**, 62 (1953).

and pulverized potassium acetate,[3] 0.16 mole (16 g, 15 ml) of acetic anhydride, and 0.1 mole (10.5 g, 10 ml) of freshly distilled benzaldehyde.[4] Mix the materials thoroughly and provide the flask with an air-cooled reflux condenser. Heat the reaction mixture in an oil bath at 155–160° for 15 min, then raise the temperature of the bath to 165–170° and continue the heating for 1.5–2.0 hr.

Since the mixture will solidify upon cooling, pour it out while hot into a 1-liter round-bottomed flask and rinse the reaction flask with two 100-ml portions of boiling water to remove all of the product. Add a further quantity of 300 ml of water and provide the flask with an unpacked fractionating tube connected to a water-cooled condenser. Distill the mixture over a wire gauze until the unchanged benzaldehyde has been removed by steam distillation (about 50 ml of distillate will suffice). The distillate may be discarded.[5]

Add enough water to the contents of the flask to bring the volume to about 600 ml, introduce 2–3 g of decolorizing carbon, and boil the solution gently for about 5 min. Filter the hot solution through a fluted filter into a large flask or beaker, heat the filtrate to the boiling point and make it strongly acid by addition of 6–8 ml of concentrated hydrochloric acid. Cool the solution, with stirring, in an ice bath. After the cinnamic acid has crystallized completely, collect the crystals with suction and wash them with cold water. Allow the acid to dry completely before taking the melting point and weighing the product. If the product is not sufficiently pure it should be crystallized from hot water. The yield is 6–7 g.

Cinnamic acid exhibits geometrical isomerism;[6] the common form is the *trans* isomer, mp 133°. The *cis* isomer exists in three polymorphic modifications: *allo*-, mp 68°; Liebermann's *iso*-cinnamic acid, mp 57–58°; and Erlenmeyer's *iso*-cinnamic acid, mp 40°. The polymorphs of the *cis* isomer

[3] This is prepared by melting ordinary potassium acetate in a porcelain dish and heating *gently*, with occasional stirring, until no more water vapor is evolved and the salt is completely fluid. Allow the melt to cool, grind the solid quickly in a mortar, and transfer it to a tightly corked bottle until needed.

Instead of fused potassium acetate, potassium carbonate may be used to produce potassium acetate in the reaction mixture, but it is necessary to control the foaming that results from the evolution of carbon dioxide. Place 5 g of anhydrous potassium carbonate in the flask and add 18 ml of acetic anhydride, 1 ml of glacial acetic acid, and 10.5 g of benzaldehyde. Adjust an air-cooled condenser and warm the mixture *cautiously* over a wire gauze, with constant shaking, until the foaming has subsided. Transfer the flask to an oil bath and heat the mixture at 155–160° for 15 min, then at 175–180° for 2 hr, and proceed as directed above to work up the product.

[4] 2-Furanacrylic acid ($C_4H_3O{-}CH{=}CH{-}CO_2H$) may be prepared by the same procedure, if benzaldehyde is replaced by an equivalent quantity of furfural (9.6 g, 8 ml).

[5] If an appreciable amount of benzaldehyde is recovered, separate it carefully and measure its volume. This material may be taken into account in calculating the yield.

[6] The designations *cis* (same) and *trans* (across) refer to the orientation of the two substituents about the double bond. With three substituents this nomenclature is ambiguous, leading to different designations for the same compound depending on the assignment of substituent priorities. An unambiguous nomenclature has been proposed (J. Amer. Chem. Soc., **90**, 509 (1968)) by the Chemical Abstracts Service.

may be interconverted by inoculation of a solution with seed crystals of another form.

***Reactions.*** Cinnamic acid may be characterized by conversion to the anilide (mp 152°) in the usual way—formation of the acid chloride by reaction with thionyl chloride, followed by treatment with aniline. Addition of bromine gives a crystalline dibromide (mp 200°). With a mild base, such as sodium acetate, the dibromide is converted by dehydrohalogenation and decarboxylation into β-bromostyrene, $C_6H_5-CH=CH-Br$. The latter has an odor resembling hyacinths and is used to a limited extent in perfumes.

Treatment of the dibromide with two equivalents of a strong base gives the *cis* and *trans* isomers of α-bromocinnamic acid (mainly the *cis* form). The *trans* isomer undergoes further dehydrohalogenation much more readily than the *cis* isomer, to form phenylpropiolic acid ($C_6H_5-C{\equiv}C-CO_2H$); decarboxylation occurs as a side reaction and some phenylacetylene is formed also.

***Cis* and *trans*-α-Bromocinnamic Acids**[7] *(Optional)*. Dissolve 3 g of dry cinnamic acid in 35 ml of hot carbon tetrachloride in a 125-ml Erlenmeyer flask and cool the solution rapidly to obtain a suspension of fine crystals. From a separatory funnel add slowly, while swirling the mixture vigorously, 16 ml of a 20 percent solution of bromine in carbon tetrachloride (20 g of bromine per 100 ml of solution). It is desirable to carry out the reaction under strong illumination. After all of the bromine solution has been added, allow the mixture to stand for 15 min with occasional swirling. Chill the flask in an ice bath, collect the dibromide with suction, and wash it with 5–6 ml of cold carbon tetrachloride. Spread the crystals on a clean paper to allow residual solvent to evaporate. The yield of crude α,β-dibromophenylpropionic acid is 5–6 g. This material is pure enough to use directly in the next step. A small sample may be purified by crystallization from warm ethanol. Do not attempt to determine the melting point (~200–205°).

In a 125-ml Erlenmeyer flask place 5 g of the cinnamic acid dibromide and 20 ml of ethanol. While swirling the mixture, add a solution of 2 g of solid potassium hydroxide in 5 ml of water. Adjust a small reflux condenser and boil the mixture gently for 10 min, with constant shaking. Transfer the material to a small distilling flask and distill off the ethanol (and some water) until the vapor temperature rises to about 90°. Pour the residual solution and any precipitate into a beaker and add sufficient water to bring the volume to about 50 ml. Add dropwise, with good stirring, just enough glacial acetic acid to *neutralize* the solution (test with litmus paper). Stir the material thoroughly and add a solution of 1 g of barium chloride dihydrate in 5 ml of water, to precipitate the barium salt of *trans*-α-bromocinnamic acid. After allowing the mixture to stand for a short while, collect the precipitate on a small suction

[7] The designations *cis* and *trans* refer to the orientation of the phenyl and carboxylic acid groups.

filter and wash it with 5–6 ml of water (reserving the filtrate and washings for isolation of the *cis* isomer). Suspend the damp barium salt in 10 ml of water, heat the mixture to boiling, and add 1 ml of concentrated hydrochloric acid. After cooling in an ice bath collect the crystals of the *trans* isomer on a small filter and wash them with a little water. The yield is about 0.3–0.4 g. The recorded melting point of *trans*-α-bromocinnamic acid is 131–132°.

Treat the filtrate and washings from the barium salt precipitation with 5 ml of concentrated hydrochloric acid and stir the mixture thoroughly. Cool the mixture, collect the crystals of the *cis* isomer with suction, and wash them with a little water. The yield is 2–2.5 g. The acid may be purified by crystallization from carbon tetrachloride. The recorded melting point of *cis*-α-bromocinnamic acid is 120–121°.

## Questions

**1.** Write equations for the reaction of benzaldehyde with the following: (a) diethyl malonate; (b) acetone, in excess; (c) acetaldehyde. How could cinnamic acid be obtained from each of these products?

**2.** How may the following compounds be prepared from cinnamic acid: (a) β-phenylpropionic acid; (b) cinnamoyl chloride; (c) *p*-aminocinnamic acid?

**3.** What acid would be obtained by heating benzaldehyde with propionic anhydride and potassium propionate?

**4.** In 1868 Wm. H. Perkin, Sr., synthesized coumarin, which is now manufactured commercially for use in artificial vanilla-type flavoring materials. What is coumarin and how may it be prepared (starting from phenol)?

**5.** Discuss the stereochemistry of the addition of bromine to ordinary (*trans*) cinnamic acid, and dehydrohalogenation of the dibromide to α-bromocinnamic acid(s), and further dehydrohalogenation of the α-bromocinnamic acids to phenylpropionic acid.

EXPERIMENT

# 40

# Reduction by Sodium Borohydride

By appropriate choice of reagent and reaction conditions a diaryl ketone can be reduced to a diarylcarbinol, a pinacol (see Experiment 42), a diarylmethane, or a tetra-arylethane.

Reduction with zinc dust in the presence of ethanol and alkali is the conventional method of converting benzophenone to benzohydrol.[1] This gives good yields and is well-suited to moderate scale laboratory preparations. For small scale syntheses sodium borohydride ($NaBH_4$) is a particularly convenient reducing agent. It has extremely high reducing capacity—one mole is able to reduce four moles of a ketone.

$$4(C_6H_5)_2C{=}O + \overset{+}{Na}\overset{-}{B}H_4 \longrightarrow \overset{+}{Na}\overset{-}{B}[OCH(C_6H_5)_2]_4 \xrightarrow{H_2O} 4(C_6H_5)_2CH{-}OH$$

Since sodium borohydride decomposes at an appreciable rate in water[2] or methanol it is desirable to effect reactions in ethanol or 2-propanol. In these media at room temperature sodium borohydride will reduce aldehydes and ketones to the corresponding primary and secondary alcohols. It is quite selective and does *not* reduce nitriles, nitro compounds, carboxylic acids or esters, or lactones. Lithium aluminohydride ($LiAlH_4$) is a somewhat stronger reagent and will reduce esters, lactones, and amides.[3]

In aprotic solvents such as dioxane and 1,2-dimethoxyethane (glyme), sodium borohydride will reduce acid chlorides to alcohols. Secondary and tertiary halides are reduced at 50° by 4-molar solutions of sodium borohydride in a 65 volume percent solution of diglyme and 1-molar aqueous

[1] Wiselogle and Sonneborn, *Organic Syntheses*, Collective Volume **I**, 90 (1941).

[2] The decomposition is slowed markedly by addition of alkali.

[3] Surveys of the uses of these reducing agents are found in Fieser and Fieser, *Reagents for Organic Synthesis*, pp. 1049–1055 (for $NaBH_4$); pp. 581–600 (for $LiAlH_4$); John Wiley and Sons, Inc., New York, 1967. See also, for $LiBH_4$, W. G. Brown, *Organic Reactions*, **6**, 649 (1951).

sodium hydroxide. Other uses of sodium borohydride are replacement of the diazonium group by hydrogen, and reduction of ozonides to the corresponding alcohols. Thus, the ozonide of oleic acid furnished 1-nonanol and 9-hydroxynonanoic acid.

$$C_8H_{17}-CH=CH-(CH_2)_7-CO_2H \xrightarrow{O_3} \text{ozonide} \xrightarrow{NaBH_4}$$
$$C_8H_{17}CH_2OH + HO-CH_2-(CH_2)_7CO_2H$$

# Benzohydrol

In a 100-ml round-bottomed flask place 0.03 mole (5.5 g) of benzophenone and add a slurry of 0.015 mole (0.6 g, a large excess) of sodium borohydride in 25 ml of 2-propanol. Add two boiling chips and reflux the mixture for 0.5 hr on a steam bath. Allow the solution to cool; no harm is done if it stands overnight or longer.

To decompose the boric ester complex add 30 ml of 10 percent aqueous sodium hydroxide and swirl the reaction mixture vigorously until the precipitate has dissolved completely. Break up any resistant lumps carefully with a stirring rod. Transfer the alkaline solution to a separatory funnel with the aid of 30 ml of water. Extract the benzhydrol by shaking with two successive 40-ml portions of methylene chloride. Combine the extracts, transfer them to a distilling flask and carefully distill off the methylene chloride (traces of water in the methylene chloride will steam-distill). On cooling and standing the residue will crystallize to give a nearly quantitative yield of almost pure benzhydrol, mp 68–69°. If desired the product may be recrystallized from 60 percent water-methanol.

***Reactions.*** Benzohydrol has a reactive hydroxyl group: resonance involving the two aryl groups of the benzohydryl cation facilitates rupture of the C−OH bond in the transition state. It is converted easily into dibenzohydryl ether merely by boiling with dilute mineral acids and reacts readily with hydrogen chloride to give diphenylchloromethane. For characterization benzohydrol may be converted to the acetate (mp 41–42°), benzoate (mp 88°), or 1-naphthylcarbamate (mp 136°).[4]

Benzohydrol reacts directly with some active methylene compounds and 1,4-quinones. With ethyl acetoacetate it gives the α-benzohydryl derivative, I, which produces 4,4-diphenyl-2-butanone on warming with dilute alkali. With 1,4-naphthoquinone in glacial acetic acid and a little sulfuric acid it introduces a benzohydryl group in the 2-position, II.

[4] This derivative may be prepared by essentially the same procedure given for benzyl alcohol (page 357), using 0.6 g of benzohydrol and 0.3 ml of 1-naphthylisocyanate.

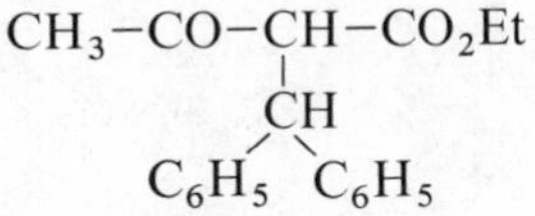

$$\begin{array}{c} CH_3{-}CO{-}\underset{\displaystyle\underset{\displaystyle C_6H_5 \quad C_6H_5}{CH}}{\underset{|}{CH}}{-}CO_2Et \end{array}$$

I

[2-(diphenylmethyl)-1,4-naphthoquinone: naphthoquinone ring bearing $-CH(C_6H_5)_2$]

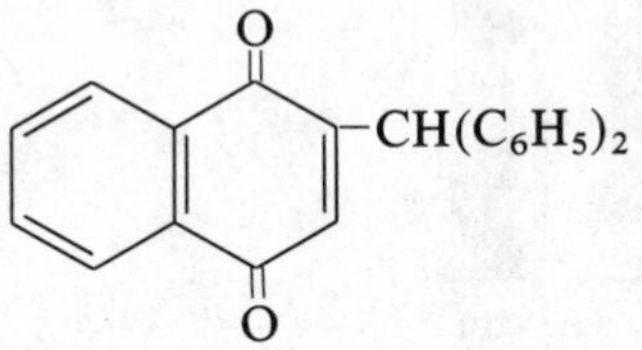

II

lemon yellow, mp. 168°

## Questions

**1.** Reaction of the Lewis acid $BH_3$ with a terminal alkene occurs so that hydrogen is added at the 2-position (to form $R{-}CH_2CH_2BH_2$) instead of the 1-position, as observed with acids such as hydrogen chloride (which forms $R{-}CHCl{-}CH_3$). Give an explanation for this behavior.

**2.** Indicate syntheses for the following compounds: (a) 4-methylbenzohydrol, starting from benzene and toluene; (b) 3,3′-dibromobenzohydrol, starting from bromobenzene.

EXPERIMENT

# 41

# Triphenylcarbinol

Triphenylmethane and its derivatives may be synthesized conveniently by means of the Friedel-Crafts reaction or the Grignard reaction. Hydroxyl and amino derivatives, used in the manufacture of triphenylmethane dyes, are obtained by condensation of aromatic aldehydes and diaryl ketones with phenols and arylamines in the presence of acid catalysts.

In the presence of anhydrous aluminum chloride, chloroform reacts stepwise with benzene (in excess) to form benzylidene chloride, benzohydryl chloride, and finally triphenylmethane. With carbon tetrachloride the end product is triphenylchloromethane (also called trityl chloride). The successive

$$CHCl_3 \xrightarrow[AlCl_3]{C_6H_6} C_6H_5{-}CHCl_2 \xrightarrow{C_6H_6} (C_6H_5)_2CHCl \xrightarrow{C_6H_6} (C_6H_5)_3CH$$

$$CCl_4 \xrightarrow[AlCl_3]{C_6H_6} C_6H_5{-}CCl_3 \xrightarrow[20^\circ]{C_6H_6} (C_6H_5)_2CCl_2^- \xrightarrow[70^\circ]{C_6H_6} (C_6H_5)_3CCl$$

steps of arylation require increasingly vigorous conditions and it is not possible to introduce a fourth aryl group by the Friedel-Crafts reaction. Tetraphenylmethane has been obtained, in low yield, by heating trityl chloride with phenylmagnesium bromide. Trityl chloride is extremely reactive and is hydrolyzed rapidly by cold water to form triphenylcarbinol (triphenylmethanol).

Triphenylcarbinol can be synthesized readily by the Grignard reaction. Phenylmagnesium bromide is prepared by direct reaction of bromobenzene with metallic magnesium in the presence of anhydrous diethyl ether (or tetrahydrofuran). Usually a crystal of iodine is added to aid in starting the

$$C_6H_5{-}Br + Mg + (C_2H_5)_2O \longrightarrow C_6H_5{-}Mg{-}Br \cdot 2(C_2H_5)_2O$$

 reaction, which must be carried out with carefully purified reagents and under anhydrous conditions. A small amount of biphenyl, $C_6H_5-C_6H_5$, is formed as an accessory product through coupling of the aryl groups (Wurtz-Fittig reaction).

The tertiary alcohol, triphenylcarbinol, may be obtained by reaction of phenylmagnesium bromide with any one of several reagents: (a) with dimethyl or diethyl carbonate; (b) with an ester of benzoic acid; (c) with benzophenone. These reactants require, respectively, three, two, and one mole(s) of the Grignard reagent, as shown in the sequence of steps (a, b, c) starting

$$CH_3O-CO-OCH_3 + C_6H_5-MgBr \xrightarrow{a} C_6H_5-CO-OCH_3 + CH_3O-MgBr$$

$$C_6H_5-CO-OCH_3 + C_6H_5-MgBr \xrightarrow{b} C_6H_5-CO-C_6H_5 + CH_3O-MgBr$$

$$C_6H_5-CO-C_6H_5 + C_6H_5-MgBr \xrightarrow{c} (C_6H_5)_3C-O-MgBr$$

from dimethyl carbonate. It is not feasible to arrest the reaction at an intermediate stage because the reactivities toward phenylmagnesium bromide decrease in the sequence: benzophenone > methyl benzoate > dimethyl carbonate, corresponding to the diminishing electrophilic activity of the carbonyl group of these reactants. This sequence is the reverse of the reactivities observed in the Friedel-Crafts arylation of carbon tetrachloride: $CCl_4 > C_6H_5-CCl_3 > (C_6H_5)_2CCl_2$. The relative reactivities in this series correspond to the diminishing electrophilic activity of the corresponding carbonium ions: $[CCl_3]^+ > [C_6H_5-CCl_2]^+ > [(C_6H_5)_2CCl]^+$.

Triphenylcarbinol, like other tertiary alcohols, is not acetylated by reaction with acetyl chloride but is converted to triphenylchloromethane. The colorless carbinol dissolves in cold concentrated sulfuric acid to form a yellow solution of the halochromic salt, containing the relatively stable triphenylcarbonium ion, $(C_6H_5)_3C^+$. On dilution with water the carbinol is regenerated.

Triphenylcarbinol is the parent structure of the color bases of triphenylmethane dyes, such as Malachite Green and Crystal Violet. It is related in a

$$(CH_3)_2N-C_6H_4-C(Ar)(OH)-C_6H_4-N(CH_3)_2 \underset{OH^-}{\overset{H^+}{\rightleftharpoons}} (CH_3)_2\overset{+}{N}{=}C_6H_4{=}C(Ar)-C_6H_4-N(CH_3)_2$$

$$HO-C_6H_4-C(Ar)(-O-CO-)C_6H_4 \underset{H^+}{\overset{OH^-}{\rightleftharpoons}} O{=}C_6H_4{=}C(Ar)-C_6H_4-CO_2^-$$

similar way to the rhodamine dyes, prepared from *m*-aminophenols and phthalic anhydride, and to the phthalein and sulfonphthalein acid-base indicators (see Experiment 50). The colored compounds are salts of amino or hydroxyl derivatives that have quinonoid structures of the types illustrated above. Reduction of these compounds gives the colorless leuco bases, which are derivatives of triphenylmethane.

# Triphenylcarbinol

**Grignard Synthesis of a Tertiary Alcohol.** In Grignard reactions it is essential that the reagents be free from ethanol and water, and the apparatus perfectly clean and dry. Rubber stoppers should not be used because they contain extractable, deleterious sulfur compounds.

➤**CAUTION:** It is advisable to have a bath of ice and water at hand during this preparation as the reaction may start suddenly with vigorous ebullition of the ether. Take care that no flame is nearby.

**Preparation of Phenylmagnesium Bromide Solution.** In a 250-ml round-bottomed flask provided with a Claisen adapter bearing an addition funnel and a vertical condenser, place 0.1 mole (2.4 g) of magnesium turnings (*carefully weighed*). Introduce directly into the flask a mixture of 3 g (2 ml) of bromobenzene, 7 ml of *absolute* ether[1] and a small crystal of iodine. If a reaction does not start at once, warm the flask gently in a bath of warm water. *After the reaction has started*, as evidenced by disappearance of the iodine color, appearance of turbidity and spontaneous boiling, add 70 ml of *absolute* ether. For the success of the experiment it is essential that the reaction begin before the main portions of the ether and bromobenzene are added.

Place in the addition funnel 13.5 g (9 ml) of bromobenzene (a total of 0.105 mole) and allow it to flow drop by drop into the previously activated reaction mixture, at such a rate that the ether refluxes without external heating.

[1] Ether used as the solvent must be carefully purified to remove impurities (ethanol and water) that are present in some technical grades of ether. The procedure used in the preparation of absolute, anhydrous ether consists in: (1) washing with strong calcium chloride solution; (2) preliminary drying with calcium chloride; (3) drying with phosphorus pentoxide; (4) distillation, with precautions to protect the product from atmospheric moisture. In some laboratories absolute ether is dried over metallic sodium instead of phosphorus pentoxide.

Technical ether of high quality, such as that from Carbide and Carbon Chemicals Co., contains no ethanol or water, and in fact is satisfactory for direct use in this experiment provided it has not been exposed to air, from which it absorbs moisture. It is advisable to store this ether over phosphorus pentoxide or metallic sodium.

After all of the halide has been added, reflux the mixture gently for 30 min on a steam bath. Do not heat the material so vigorously that ether vapors traverse the condenser. The reaction is complete when the magnesium has dissolved; some dark particles of impurities will remain undissolved. Remove the heating bath and proceed without delay to the next step.

**Reaction of Phenylmagnesium Bromide with Methyl Benzoate.**[2] Cool the reaction flask containing the Grignard reagent to 15–20° and place in the addition funnel a solution of 0.05 mole (7 g, 6.5 ml) of pure methyl benzoate[2] in about 25 ml of *absolute* ether. Allow the methyl benzoate solution to flow slowly into the Grignard reagent, with swirling, and cool the flask from time to time to control the reaction. The bromomagnesium derivative of the carbinol separates as a white precipitate. After all of the methyl benzoate has been added, allow the mixture to stand at room temperature for 30 min or longer. Meanwhile arrange a 1-liter flask for steam distillation (see Figure 22, page 64).

Pour the contents of the flask as completely as possible into a mixture of about 50 g of ice, 100 ml of water, and 5–6 ml of concentrated sulfuric acid, contained in a 500-ml flask. Add 4–5 ml of strong sodium bisulfite solution to remove any free iodine. Shake the mixture thoroughly to complete the decomposition of the magnesium derivative and rinse the reaction flask with the acid mixture to remove material that adheres to the wall of the flask. Add about 75 ml of *ordinary* ether to aid in extracting the carbinol completely. Separate the ethereal layer, wash it with two 25-ml portions of 5–10 percent sulfuric acid, and finally with water.

Transfer the ethereal solution to a 1-liter round-bottomed flask, that will be used later for steam distillation. Connect the flask with a condenser arranged for distillation and distill off the ether from a steam bath (*caution—flammable vapor!*). To the residue in the flask add about 50 ml of water, arrange the apparatus for steam distillation, and pass in a rapid current of steam. Residual bromobenzene and methyl benzoate are removed readily; biphenyl, formed in a side reaction in preparing the Grignard reagent, comes over slowly. Triphenylcarbinol is not volatile and remains in the distillation flask. Continue the distillation until about 400 ml of distillate has been collected, which will suffice to remove most of the biphenyl. Open wide the drain of the steam trap and *disconnect the steam entry to the distillation flask* before shutting off the current of steam, otherwise the contents of the flask may be sucked back into the drain.

Cool the distillation flask and collect the crude product on a suction filter. Break up any lumps of the material, wash it with water, and press as dry as possible. Crystallize the product from a small amount of hot ethanol, with addition of a little decolorizing carbon. Chill the filtrate in an ice bath

[2] In place of methyl benzoate, 8 g (8 ml) of pure ethyl benzoate may be used.

before collecting the crystals of the purified carbinol. The yield is 7–8 g. Pure triphenylcarbinol is colorless and melts at 162°.

## Questions

**1.** Write equations for the action of phenylmagnesium bromide on the following compounds, including hydrolysis of the reaction mixture with dilute acid: (a) carbon dioxide; (b) ethanol; (c) oxygen; (d) *p*-tolunitrile; (e) ethyl formate.

**2.** How may the following compounds be prepared from phenylmagnesium bromide: (a) 1,2-diphenylethanol; (b) benzaldehyde; (c) benzyl alcohol; (d) benzopinacol?

**3.** Indicate a series of reactions for the conversion of triphenylcarbinol to hexaphenylethane. Cite evidence showing that hexaphenylethane undergoes dissociation (homolysis) to form triphenylmethyl radicals. Why is this radical more stable than a free methyl radical?

**4.** Write equations for the synthesis of a typical triphenylmethane dye, starting from simple aromatic compounds. Explain the terms leuco base and color base.

EXPERIMENT

# 42

# The Pinacol-Pinacolone Rearrangement

Ketones may be reduced by means of amalgamated magnesium, or a mixture of magnesium and iodine (magnesious iodide), to form bimolecular reduction products called pinacols, of the type $R_2C(OH)-C(OH)R_2$. Aryl ketones, such as benzophenone, may be reduced photochemically by exposing a solution of the ketone in ethanol or 2-propanol to ultraviolet illumination.

$$2\,CH_3-CO-CH_3 + 2\,H(Mg-Hg) \longrightarrow \underset{OH}{(CH_3)_2C}-\underset{OH}{C(CH_3)_2}$$

$$2\,C_6H_5-CO-C_6H_5 + CH_3CHOHCH_3 \xrightarrow{h\nu} \underset{OH}{(C_6H_5)_2C}-\underset{OH}{C(C_6H_5)_2} + CH_3COCH_3$$

The reduction is a one-electron process that involves a free radical intermediate.

When heated with strong acids, or a catalyst such as iodine, substituted 1,2-diols of the pinacol type undergo dehydration and rearrangement to ketones (the pinacol-pinacolone rearrangement). The transformation involves protonation of one of the hydroxyl groups (I), followed by elimination of water to give a carbonium ion intermediate (II). Migration of an alkyl or

$$\underset{\text{I}}{\underset{{}^{+}\underset{H}{OH}}{R_2C}-\underset{OH}{CR_2}} \longrightarrow \underset{\text{II}}{\overset{+}{R_2C}-\underset{OH}{CR_2}} \longrightarrow \underset{\text{III}}{R_3C-\underset{{}^{+}OH}{\overset{\|}{C}}-R}$$

aryl group from the adjacent carbon atom, with its bonding electrons, leads to the protonated form of the pinacolone (III). Carbonium ion rearrangements of the pinacol type are encountered with 1,2-halohydrins, 1,2-amino-alcohols, α-hydroxyaldehydes, and similar structures. The mechanism of the rearrangement has been studied extensively.[1]

Studies of the rearrangement of a series of symmetrical pinacols (type IV) have disclosed that substituents exert a marked effect on the selectivity of migration of the aryl group. The *migration aptitudes* of a few substituted

$$\underset{\text{IV}}{R{-}C_6H_4{-}\overset{\overset{\displaystyle OH}{|}}{\underset{\underset{\displaystyle C_6H_5}{|}}{C}}{-}\overset{\overset{\displaystyle OH}{|}}{\underset{\underset{\displaystyle C_6H_5}{|}}{C}}{-}C_6H_4{-}R} \qquad \underset{\text{V}}{(R{-}C_6H_4)_2\overset{\overset{\displaystyle OH}{|}}{C}{-}\overset{\overset{\displaystyle OH}{|}}{C}(C_6H_5)_2}$$

groups, relative to $C_6H_5-$ are: *p*-methoxy-, 500–1000; *p*-methyl-, 15–18; *p*-chloro-, 0.7; *m*-methoxy-, 0.3. The rearrangement of unsymmetrical pinacols (type V) does not involve selectivity of migration but is governed by selectivity in the formation of one of the two possible carbonium intermediates (VI and VII). In this situation the ease of formation of the carbonium ion is the dominant factor and the effect of substituents is different from that observed with the symmetrical pinacols.

$$\underset{\text{VI}}{(R{-}C_6H_4)_2\overset{+}{C}{-}\overset{\overset{\displaystyle OH}{|}}{C}(C_6H_5)_2} \qquad \underset{\text{VII}}{(R{-}C_6H_4)_2\overset{\overset{\displaystyle OH}{|}}{C}{-}\overset{+}{C}(C_6H_5)_2}$$

# Benzopinacol

**(A) Photochemical Reduction.** Place 0.05 mole (9.1 g) of benzophenone in a 100-ml round-bottomed flask, add 70 ml of 2-propanol (isopropyl alcohol) and dissolve the solid by warming on a steam bath. To the solution add *one drop* of glacial acetic acid,[2] and sufficient 2-propanol to fill the flask almost completely. Stopper the flask firmly with a good cork that fits tightly and projects about half its length into the neck of the flask. Wire the cork firmly to the neck of the flask by means of soft copper wire.

Invert the flask, support it firmly by means of a condenser or burette clamp, and expose it to direct sunlight. Benzophenone is activated by absorption of light in the near ultraviolet region, that is partially transmitted through

[1] Collins, The Pinacol Rearrangement, *Quarterly Reviews* (London), **14**, 357 (1960); Gould, *Mechanism and Structure in Organic Chemistry*, Holt, Rinehart and Winston, Inc., New York (1959); Hine, *Physical Organic Chemistry*, McGraw-Hill Book Co., Inc., 2nd edition, New York (1962).

[2] The acid is added to neutralize traces of alkali from the glass vessel. Alkali is deleterious because it catalyzes cleavage of the pinacol to benzophenone and benzohydrol ($C_6H_5{-}CHOH{-}C_6H_5$).

ordinary glass. As the reduction progresses benzopinacol separates in colorless, dense crystals. By occasional tapping and swirling, the crystals may be made to settle in the neck of the flask. Five or six days' exposure to moderate sunlight will furnish an abundant crop of crystals. Collect these with suction and, if necessary, return the filtrate for further exposure to complete the reaction. Under favorable conditions the yield will attain at least 90 percent. The product is quite pure (mp 186–188°) and may be used directly for conversion to benzopinacolone.

**(B) Reduction by Magnesious Iodide**[3] *(Alternative Procedure)*. The reagents and apparatus for this preparation must be dried carefully. In a 100-ml round-bottomed flask place 0.063 mole (1.5 g) of powdered magnesium, 10 ml of anhydrous ether and 15 ml of dry benzene (*caution—flammable solvents!*). Attach a short reflux condenser and warm the flask gently on a water bath. Through the condenser tube add 0.018 mole (4.5 g) of iodine crystals, in portions, at a rate that keeps the solution boiling vigorously. About half the magnesium dissolves and the supernatant solution is almost colorless.

Cool the reaction mixture to room temperature, remove the condenser, and add a solution of 0.03 mole (5.5 g) of benzophenone in 8–10 ml of warm benzene. A heavy white precipitate is formed. Cork the flask firmly and shake the mixture vigorously until the precipitate dissolves, with the formation of a deep red solution. About 10 min shaking is needed; do not shake the mixture longer than necessary, or the iodomagnesium pinacolate will precipitate on the surface of the residual magnesium.

Allow the excess magnesium to settle and decant the solution through a fluted filter into a 250-ml Erlenmeyer flask. Extract the residual magnesium with a mixture of 10 ml of ether and 15 ml of benzene, and filter the extract into the flask. To the filtrate add a solution of 8 ml of concentrated hydrochloric acid in 20 ml of water, and a small pinch of sodium bisulfite (to remove any free iodine). Shake the mixture vigorously to decompose the pinacolate and draw off the aqueous layer. Wash the organic layer with two 10-ml portions of water to remove acid. Transfer the organic layer to a small distilling flask and distill off about three-fourths of the solvent mixture (*caution—flammable vapor!*), using a water-cooled condenser. Pour the residual solution into a small beaker and rinse the distilling flask with 5–10 ml of ethanol. After cooling thoroughly collect the crystals with suction and wash them with a little cold ethanol. The yield is 4–4.5 g.

## Benzopinacolone

In a 100-ml round-bottomed flask place 0.08 mole (3 g) of benzopinacol, 15 ml of glacial acetic acid, and a *small* crystal of iodine. Attach a short

[3] Gomberg and Bachmann, *J. Amer. Chem. Soc.*, **49**, 236 (1927).

reflux condenser and reflux the solution for 10 min. Allow the solution to cool slightly, add 15 ml of ethanol, swirl the mixture thoroughly, and allow it to cool. Collect the crystals with suction and wash them with cold ethanol to remove iodine. Benzopinacolone forms colorless crystals, mp 179–180°. The yield is 2.5–2.7 g.

## Questions

**1.** Indicate a stepwise mechanism for the alkali-catalyzed cleavage of benzopinacol.[2] Benzopinacolone is cleaved by heating with aqueous alkali; what are the products?

**2.** Write equations for the reaction of *p*-tolylmagnesium bromide (in excess) on the following compounds, including hydrolysis of the reaction mixture with dilute acid: (a) $C_6H_5{-}CO{-}CO{-}C_6H_5$ (benzil); (b) methyl ester of benzilic acid; (c) dimethyl oxalate.

**3.** Write the structures of the isomeric pinacolones that could be obtained by pinacol rearrangement of symmetrical 4,4′-dimethylbenzopinacol. This reaction actually furnishes one of the pinacolones to the extent of more than 90 percent. How may its structure be determined?

# EXPERIMENT 43

# Benzoquinone

Controlled oxidation of 1,2- and 1,4-dihydroxybenzene (catechol and hydroquinone) leads to *o*- and *p*-benzoquinone. The quinones possess a strong chromophore arising from extended conjugation of the carbonyl groups, for which the generic term *quinonoid structure* is used. The ring system has lost

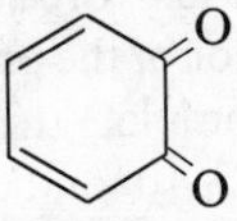

*o*-Benzoquinone
red

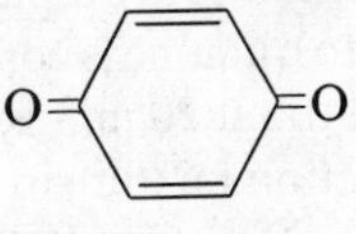

*p*-Benzoquinone
yellow

its aromatic character; the quinones are highly active α,β-unsaturated ketones and their typical behavior involves 1,4-addition reactions. They are oxidizing (dehydrogenating) agents and can be reduced to their colorless, benzenoid precursors.

On a small scale *p*-benzoquinone may be prepared conveniently by oxidation of hydroquinone with dichromate or potassium bromate in acid solution. Commercially *p*-benzoquinone is made by oxidation of aniline with manganese dioxide and aqueous sulfuric acid. 2-Methylbenzoquinone is prepared in a similar way from *o*-toluidine.[1] The yields in the amine oxidations are not high but the starting materials are readily available and cheap.

Direct oxidation of naphthalene and 2-methylnaphthalene with chromic acid in acetic acid gives the 1,4-naphthoquinones. Chromic acid oxidation of anthracene and phenanthrene gives the 9,10-quinones. An excellent general procedure for the synthesis of anthraquinones is the ring closure of *o*-benzoyl-

[1] For a review see Cason, "Synthesis of Quinones by Oxidation," *Organic Reactions*, **4**, 305 (1948).

benzoic acids (see Experiment 44) obtained by Friedel-Crafts acylations of aromatic compounds with phthalic anhydride.

## *p*-Benzoquinone

In a 125-ml Erlenmeyer flask place 0.05 mole (5.5 g) of hydroquinone and a warm solution of 3 ml of concentrated sulfuric acid in 50 ml of water. Dissolve the hydroquinone completely by gentle warming and then cool the solution at once in an ice bath, with vigorous swirling, to obtain a suspension of fine crystals.

➤**CAUTION:** Benzoquinone is extremely volatile and is toxic. Its vapor is dangerously irritant to the eyes and the solid is a skin irritant. Manipulate quinone carefully to minimize contact with the vapor and the crystals.

Remove the flask from the cooling bath and add in small portions, with swirling, a solution of 0.024 mole (7.1 g) of sodium dichromate dihydrate in 10 ml of water. Cool the reaction mixture as needed to maintain its temperature at 20–25°. During the early stages of oxidation a greenish-black precipitate of quinhydrone (a 1:1 complex of quinone and hydroquinone) separates. As the oxidation to quinone is completed the color of the precipitate becomes yellowish-green. About 20 min is required to complete the oxidation.

Cool the reaction mixture in an ice bath to 5–10° and collect the crystals of quinone with suction. Wash them with two or three 5-ml portions of ice-cold water, press firmly on the filter, and suck them as dry as possible. Purify the crude product by sublimation, using one of the methods shown in Figures 31 and 32 (pages 92, 93). Transfer the sublimed crystals quickly to a dry, weighed sample tube. The yield is about 3 g. The bright yellow crystals of quinone become discolored upon standing.

***Reactions.*** Benzoquinone undergoes 1,4-addition reactions with halogen acids, with alcohols, and with amines. The initial adducts undergo isomerization to form substituted hydroquinones; hydrogen chloride, for example, gives chlorohydroquinone. Acetic anhydride, in the presence of acid catalysts, adds in the same way and the intermediate diacetoxy compound is acetylated further to yield the triacetate of hydroxyhydroquinone (1,2,4-trihydroxybenzene).

The addition of alcohols and amines leads to alkoxy- and aminoquinones. The intermediate hydroquinones are dehydrogenated by benzoquinone, because it has a higher oxidation potential than the alkoxy- and aminoquinones. This oxidation does not occur with the halogenated hydroquinones, since the halogenated quinones are stronger oxidizing agents than benzoquinone.

Quinone forms a monoxime and a dioxime. The monoxime is identical with *p*-nitrosophenol (see Experiment 34a, Part E). Oximation is done with hydroxylamine hydrochloride, since free hydroxylamine reduces the quinone.

Benzoquinone acts as a dienophile in the Diels-Alder reaction (see Experiment 55). With one molecule of 1,3-butadiene it gives 5,8,9,10-tetrahydro-1,4-naphthoquinone; a second molecule of the diene can be added to obtain octahydro-9,10-anthraquinone.[2] Dehydrogenation of the latter affords a route to anthraquinone.

+Diene → +Diene →

**Hydroxyhydroquinone Triacetate** *(Optional).* In a test tube place 1 g. of benzoquinone and 3 ml of acetic anhydride. Mix the reactants thoroughly and add *one* drop of concentrated sulfuric acid. The temperature of the reaction mixture should not be allowed to rise above 50°. After 5 min shaking pour the solution into 15 ml of cold water and rinse the tube with 4–5 ml of water. Collect the crystals with suction and crystallize them from 4–5 ml of ethanol. The yield is about 1 g. Hydroxyhydroquinone triacetate forms colorless crystals, mp 96–97°. The product becomes discolored on standing in the air.

## Questions

**1.** Write equations for the reaction of benzoquinone with the following: (a) hydrogen chloride; (b) sulfurous acid ($SO_2 + H_2O$); (c) hydroxylamine hydrochloride (one equivalent).

**2.** What is quinhydrone? Give examples of other addition compounds of similar type.

**3.** Suggest a method, starting from *o*-toluidine, for the preparation of

[2] For a review of Diels-Alder reactions of quinones see Butz and Rytina, *Organic Reactions*, **5**, 136 (1949).

2-methyl-1,4-naphthoquinone. What is the structural relationship of this quinone to the anti-hemorrhagic (blood-clotting) factors, vitamins $K_1$ and $K_2$?

**4.** What product(s) would you expect to obtain by treating hydroquinone diacetate with excess aluminum chloride (Fries rearrangement)?

EXPERIMENT

# 44

# Polycyclic Quinones

Anthracene and phenanthrene can be oxidized directly at the 9,10-positions by means of chromic acid to give the corresponding quinones. These condensed polycyclic quinones are more stable to oxidation than the simple benzoquinones and naphthoquinones. In the oxidation reaction some complex products are formed by coupling and disproportionation of the reactive

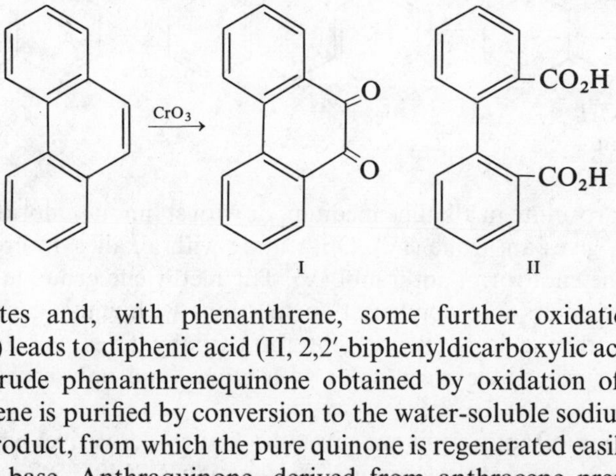

$CrO_3$ →

I

$CO_2H$

$CO_2H$

II

intermediates and, with phenanthrene, some further oxidation of the quinone (I) leads to diphenic acid (II, 2,2′-biphenyldicarboxylic acid).

The crude phenanthrenequinone obtained by oxidation of technical phenanthrene is purified by conversion to the water-soluble sodium bisulfite addition product, from which the pure quinone is regenerated easily by treatment with base. Anthraquinone, derived from anthracene present as an impurity in technical phenanthrene, is a less reactive carbonyl compound and does not form a sodium bisulfite adduct.

Phenanthrenequinone bears a strong resemblance to the 1,2-diketone, benzil, and undergoes a benzilic acid rearrangement upon warming with concentrated aqueous alkalies; the product is 9-hydroxyfluorene-9-carboxylic acid.

A general method for the preparation of anthraquinone and its derivatives consists in the cyclization of *o*-benzoylbenzoic acids, which are obtained readily by the reaction of phthalic anhydride with benzene, toluene, chlorobenzene, and similar aromatic compounds, in the presence of anhydrous aluminum chloride. The cyclization can be effected by means of concentrated

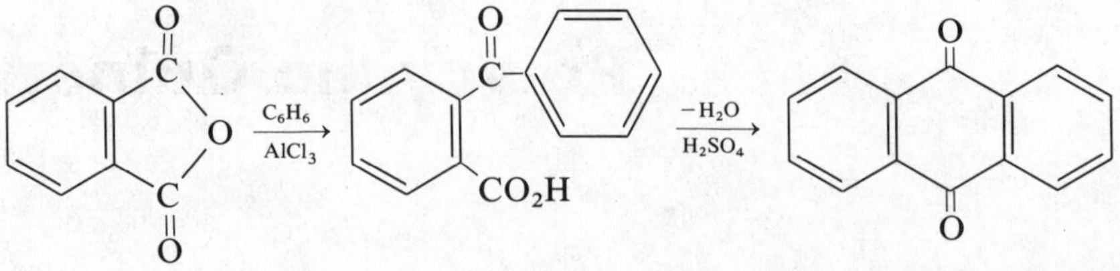

$$\text{phthalic anhydride} \xrightarrow[AlCl_3]{C_6H_6} o\text{-}C_6H_5CO\text{-}C_6H_4\text{-}CO_2H \xrightarrow[H_2SO_4]{-H_2O} \text{anthraquinone}$$

sulfuric acid or polyphosphoric acid. Derivatives of anthraquinone are used in the manufacture of important vat dyes (*Indanthrene Brown*, *Caledon Jade Green*, etc.), that are unusually stable to light and to washing.

Anthraquinone is reduced easily to anthrahydroquinone (III), which dissolves in aqueous alkalies to form a deep red solution; air or mild oxidizing agents regenerate the quinone. Reduction of anthraquinone with

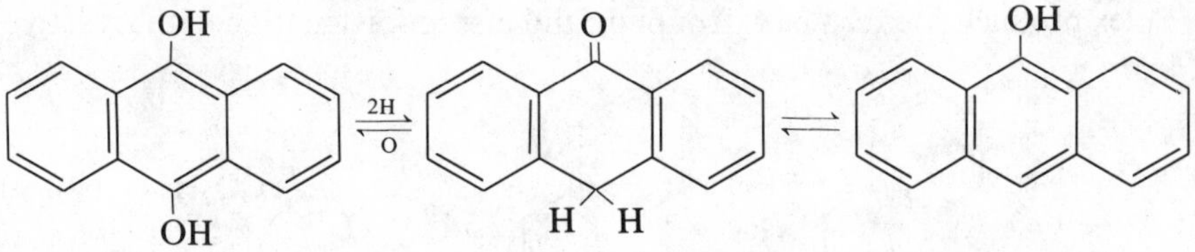

$$\underset{\text{III}}{\text{anthrahydroquinone}} \underset{O}{\overset{2H}{\rightleftharpoons}} \underset{\text{IV}}{\text{anthrone}} \rightleftharpoons \underset{\text{V}}{\text{anthranol}}$$

sodium hydrosulfite in alkaline medium, or with stannous chloride in glacial acetic acid, gives anthrone (IV). On heating with alkalies anthrone is converted to the enol form, anthranol (V). The methylene group at position 5 in anthrone enters into condensation reactions with active carbonyl compounds. Anthrone is a test reagent for carbohydrates: in sulfuric acid solution it gives a blue-green color with mono- and polysaccharides, and with glycosides.

Nitration of anthraquinone gives the 1-nitro derivative, which can be reduced to 1-aminoanthraquinone. Sulfonation at 140–150° with fuming sulfuric acid gives anthraquinone-2-sulfonic acid. Ammonolysis of the sodium salt of the sulfonic acid furnishes 2-aminoanthraquinone.

Many derivatives of anthraquinone occur in nature as pigments and active principles of plants, fungi, lichens, and insects. An example is emodin, the cathartic principle of cascara sagrada and of rhubarb, which is 1,3,8-trihydroxy-6-methylanthraquinone.

# 44a Anthraquinone

In a 250-ml round-bottomed flask place 0.022 mole (5 g) of anhydrous *o*-benzoylbenzoic acid[1] and 25 ml of concentrated sulfuric acid. Heat the flask on a steam bath and swirl the mixture until the solid dissolves. By means of a clamp support the flask firmly in a steam bath; place a towel or cloth around the bath and flask (to reduce heat loss), and heat the material at 100° for 30 min. With a medicine dropper or a pipette, add 5 ml of water, dropwise, with swirling. The product will begin to separate. Remove the flask from the steam bath and allow it to cool. Dilute the reaction mixture with about 150 ml of water and transfer it to a 600-ml beaker. Add enough chipped ice to bring the total volume to 300 ml and stir the mixture thoroughly. Collect the product by suction filtration, preferably on a hardened (*Shark Skin*) filter paper, and wash it well with water. To remove any unchanged starting material wash the product carefully with 10 ml of concentrated aqueous ammonia diluted with 50 ml of water, followed by a washing with water. To facilitate drying, wash the quinone finally with a little ice-cold acetone, and then spread it on a clean paper. Do not determine the melting point (284–286°). The yield is 4–4.5 g. A small sample may be purified by sublimation, at 230–250°, using the apparatus shown in Figure 31 (page 92).

**Reduction to Anthrone** *(Optional)*. In a 500-ml flask place 0.02 mole (4.1 g) of anthraquinone, 100 ml of water, 0.07 mole (12.2 g) of sodium hydrosulfite dihydrate ($Na_2S_2O_4 \cdot 2H_2O$), and a solution of 0.12 mole (5 g) of sodium hydroxide in 10 ml of water. Adjust a reflux condenser and boil the mixture gently for 40–50 min. The quinone is reduced first to anthrahydroquinone, which dissolves to give a deep red solution of the sodium salt. Further reduction gives the pale yellow, sparingly soluble anthrone.

Cool the reaction mixture, collect the product with suction, and wash it thoroughly with water. Dissolve the moist product in hot ethanol and filter the solution through a fluted filter. Chill the filtrate and collect the purified material. Concentrate the mother liquor and obtain a second crop of crystals. The yield is 3–3.5 g. Pure anthrone forms pale yellow crystals, mp 150–155° (dec.).

*Questions follow Experiment 44b.*

[1] The anhydrous acid melts at 127–128°; it forms a monohydrate, mp 94–95°. The pure anhydrous acid can be purchased from chemical supply firms.

## 44b Phenanthrenequinone

In a 500-ml Erlenmeyer flask place 30 ml of glacial acetic acid and 100 ml of water, and add carefully 60 ml of concentrated sulfuric acid. Swirl the solution, add 0.03 mole (6 g) of technical 90 percent phenanthrene, and heat the mixture to 95° (internal temperature) in a bath of boiling water. Prepare a solution of 0.12 mole (36 g) of sodium dichromate dihydrate in 25 ml of warm water and add this in portions, with swirling, to the hot suspension of phenanthrene. Watch the temperature of the reaction mixture carefully to observe the onset of a strongly exothermic reaction. When this occurs stop the addition of the oxidizing solution, remove the flask from the heating bath, and swirl the mixture vigorously. The temperature will rise to 110–120°. As soon as the temperature begins to fall, resume addition of the dichromate solution and do not allow the temperature to drop below 85–90°. Dip the flask in the boiling water bath, when required, to maintain the desired temperature. After all of the oxidizing agent has been added, heat the mixture for 30 min in the boiling water bath, with frequent swirling.

Cool the reaction mixture and pour it into a well-stirred mixture of about 300 ml of water and 100 g of chipped ice. Break up any lumps of the product and collect the crude quinone on a large suction filter. Wash the crystals thoroughly with water, until the green chromous sulfate has been removed. Transfer the moist product to a 500-ml Erlenmeyer flask, add 65 ml of ethanol, swirl the mixture vigorously, and add a solution of 0.16 mole (30 g) of sodium bisulfite in 60 ml of water. The yellow color of the quinone is discharged as the bisulfite adduct is formed. Allow the mixture to stand for about 20 min, with occasional swirling, to complete the reaction. Add water until the flask is nearly filled, cork it firmly, and shake the mixture thoroughly to dissolve the sodium bisulfite addition compound. Filter the mixture through a fluted filter to remove anthraquinone and other impurities. After washing the insoluble residue with 20 ml of 50 percent aqueous ethanol it may be discarded.

Place the filtrate and washings in a large beaker and add saturated aqueous sodium carbonate in small portions (*caution—foaming!*), with good stirring, until the solution is distinctly alkaline. Collect the phenanthrenequinone on a suction filter, wash it thoroughly with cold water, and press as dry as possible on the filter. Place the damp product in a 250-ml round-bottomed flask, add 55–60 ml of glacial acetic acid, and heat the mixture under a short reflux condenser to dissolve the quinone. Cool the solution to room temperature and allow the material to stand for 20 min or longer, with occasional swirling before collecting the orange crystals of the purified product. Wash the crystals thoroughly with cold water and allow them to dry in the air. The yield is about 3 g. The recorded melting point of phenanthrenequinone is 206°.

## Questions

**1.** What is the source of technical anthracene and phenanthrene? What methods are used to obtain these hydrocarbons in a state of high purity?

**2.** Write equations for a synthesis of 2-methylanthraquinone: (a) starting from naphthalene and toluene; (b) starting from naphthalene and aliphatic compounds, and making use of the Diels-Alder reaction.

**3.** Write equations for the reaction of phenanthrenequinone with the following reagents: (a) hot aqueous potassium hydroxide solution; (b) *o*-phenylenediamine; (c) hydrogen peroxide, in glacial acetic acid.

EXPERIMENT

# 45

# Derivatives of 1,1-Diphenylethane

In the presence of concentrated sulfuric acid and similar acidic catalysts, aliphatic and aromatic aldehydes and ketones undergo condensation reactions with aromatic hydrocarbons, arylamines, and phenols to furnish arylated derivatives of methane, ethane, propane, etc. Thus, acetaldehyde reacts with two molecules of benzene to form 1,1-diphenylethane, $CH_3{-}CH(C_6H_5)_2$, which is the parent structure of an important group of insecticides.

The first of the diarylethane insecticides to be introduced was 1,1-di(*p*-chlorophenyl)-2,2,2-trichloroethane, called DDT, which is prepared by condensation of chloral (trichloroacetaldehyde) with two molecules of chlorobenzene in the presence of sulfuric acid containing a small amount of

$$Cl_3C{-}CH{=}O + 2H{-}C_6H_4{-}Cl \xrightarrow{H_2SO_4} Cl_3C{-}CH(C_6H_4{-}Cl)_2 + H_2O$$

dissolved sulfur trioxide. Other examples of insecticides of this type are *Methoxychlor*, which is 1,1-di(*p*-methoxyphenyl)-2,2,2-trichloroethane, and DDD, which is 1,1-di(*p*-chlorophenyl)-2,2-dichloroethane.

Another example of this general reaction is the condensation of acetone with two molecules of phenol, in the presence of sulfuric acid, to produce 2,2-*bis*(*p*-hydroxyphenyl)propane. This compound, called industrially *Bisphenol-A*, is a starting material for the manufacture of epoxy resins.

$$CH_3{-}CO{-}CH_3 + 2H{-}C_6H_4{-}OH \xrightarrow{H_2SO_4} HO{-}C_6H_4{-}\overset{\overset{\large CH_3}{|}}{\underset{\underset{\large CH_3}{|}}{C}}{-}C_6H_4{-}OH + H_2O$$

# DDT

In a small separatory funnel place 15 ml of concentrated sulfuric acid, add 0.12 mole (20 g) of chloral hydrate crystals, $Cl_3C{-}CH(OH)_2$, and shake thoroughly until the crystals have liquefied. The reaction is endothermic. Draw off the sulfuric acid layer and shake the chloral layer with a fresh 15-ml portion of concentrated sulfuric acid. Separate the sulfuric acid layer carefully, pour the anhydrous chloral[1] into a small dry flask, and mix it with 0.25 mole (28 g, 26 ml) of dry chlorobenzene. Cork the flask tightly until needed for the next step.

Fit a 500-ml round-bottomed flask with a rubber stopper bearing a thermometer reaching nearly to the bottom of the flask, and a curved piece of glass tubing (to serve as gas exit tube). Provide a dish of water for cooling the reaction mixture. In the reaction flask place 130 g (70 ml) of concentrated sulfuric acid (95 percent), add 80 g (42 ml, containing 0.15 mole of sulfur trioxide) of 15 percent oleum (fuming sulfuric acid),[2] and cool until the internal temperature has fallen to 30°. Add about 10 ml of the freshly prepared solution of anhydrous chloral and chlorobenzene, and swirl the material vigorously. *The success of the reaction depends upon obtaining good contact of the immiscible liquids.* The temperature will rise slightly but should be maintained at 30–35° by occasional cooling, if necessary, in the water bath. Add the remainder of the chloral-chlorobenzene solution in portions and continue to mix the reactants thoroughly by swirling. After the addition of the chloral-chlorobenzene has been completed, maintain the internal temperature at 30–35° for 2 hr longer and shake the mixture frequently. During this time the product separates as a fine crystalline or semi-solid precipitate. If necessary, continue to warm the reaction mixture at 30–35°, with shaking, until a precipitate has formed.

Transfer the reaction mixture to a separatory funnel, add 60 ml of methylene chloride[3] and *mix well* to insure complete extraction of the product by the organic solvent. Draw off the sulfuric acid layer carefully and wash the organic layer twice with water. Wash it finally with 5 percent

[1] The crude anhydrous chloral is obtained as a colorless, slightly turbid liquid that absorbs moisture avidly to regenerate the crystalline hydrate. The crude product is sufficiently pure to use directly for the preparation of DDT. If pure chloral were desired the material could be distilled through a short column and the distillate protected from moisture.

[2] Other suitable combinations of acid and oleum are: 150 g of concentrated sulfuric acid and 60 g of 20 percent oleum, *or* 160 g of concentrated sulfuric acid and 55 g of 25 percent oleum.

[3] If methylene chloride is not available carbon tetrachloride may be used, but a larger amount of solvent (100 ml) is required.

aqueous sodium bicarbonate to insure complete removal of acid (test with litmus paper). Separate carefully and dry the organic layer by shaking with 5–6 g of anhydrous magnesium sulfate. Filter through a small plug of cotton and distill off the solvent on a steam bath in a hood. Allow the syrupy residue of crude DDT (containing admixed chlorobenzene and *o,p*-isomer of DDT) to cool, and stir with a glass rod until it solidifies to a mass of white crystals.

Triturate the crystals with cold 80 percent aqueous ethanol to remove small quantities of chlorobenzene, collect the crystals on a suction filter, and wash them with cold 80 percent aqueous ethanol. Purify the material by crystallization from 95 percent ethanol; about 8–10 ml of solvent will be required per gram of DDT. The yield of purified product is 15–20 g. Pure DDT[4] forms colorless crystals melting at 107–108°.

## Questions

**1.** How is chloral hydrate prepared commercially?

**2.** What product would be formed if benzaldehyde were used instead of chloral in this experiment?

**3.** How may the following compounds be prepared by means of a condensation reaction similar to that used in this experiment: (a) 4,4′-bis(dimethylamino)triphenylmethane (the leuco base of Malachite Green); (b) 4,4′-dimethoxydiphenylmethane?

[4] Pesticides like DDT and other organic chloro compounds have made possible greatly increased agricultural productivity but objections have been raised to their continued and widespread use. They are persistent materials and their residues accumulate in the soil and may contaminate food products.

EXPERIMENT

# 46

# Enamine Synthesis of a Diketone

Michael additions and many other acylation and alkylation reactions of less reactive carbonyl compounds involve the formation and reaction of an intermediate enolate anion. The scope of these reactions is limited by three

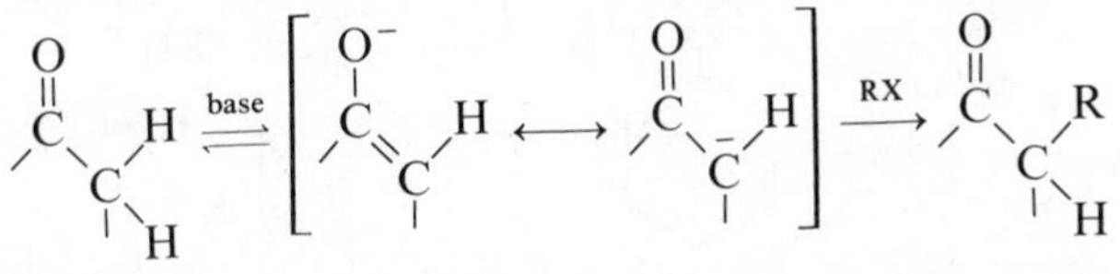

factors. In order to form the enolate anion it is necessary to use a strong base like sodium amide, triphenylmethide ion or *t*-alkoxides, which may react competitively with the acylating or alkylating reagent. A second complication is the unwanted base catalyzed aldol condensation of aldehydes and the self condensation of ketones (I). Finally, owing to the rapid equilibration of the intermediate anion with the product as it is formed, further

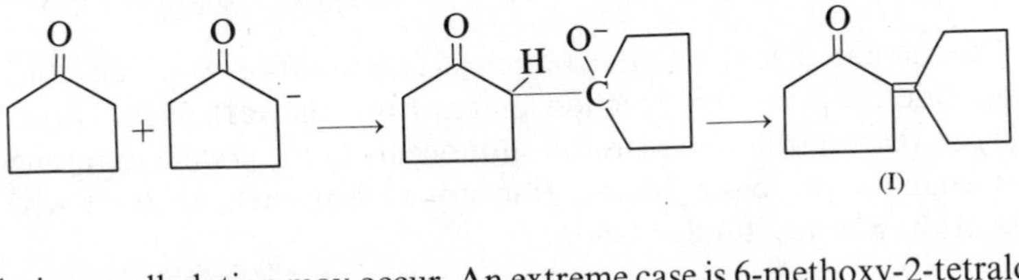

(I)

acylation or alkylation may occur. An extreme case is 6-methoxy-2-tetralone, which on attempted mono-methylation gives largely a one to one mixture of dimethylated ketone and unreacted starting material.

In 1954 Stork[1] introduced a new method for the acylation and alkylation of aldehydes and ketones via enamines (II), derived by condensation with a secondary amine. The enamine undergoes preferential attack at carbon as would the enolate anion but being neutral it does not condense with itself nor exchange a proton with the product to form a reactive intermediate. The enamines are formed readily by refluxing a mixture of the amine and the carbonyl compound in a solvent such as dioxane, acetonitrile, toluene or ethyl alcohol. The water formed in the reaction is removed by azeotropic distillation using an apparatus such as is shown in Figure 42. After acylation or alkylation the product is hydrolyzed by heating with water.

$$R_2NH + R{-}CO{-}CH_2{-}R' \rightleftharpoons H_2O + \left[ R_2N{-}\underset{R}{C}{=}\underset{R'}{CH} \longleftrightarrow R_2\overset{+}{N}{=}\underset{R}{C}{-}\underset{R'}{\overset{-}{C}H} \right]$$

(II)

$$\text{II} \xrightarrow{CH_3{-}X} R_2\overset{+}{N}{=}\underset{R}{C}{-}\underset{R'}{CH}{-}CH_3 \xrightarrow[H_2O]{} O{=}\underset{R}{C}{-}\underset{R'}{CH}{-}CH_3$$

With the enamine synthesis it is possible to prepare compounds that would otherwise require circuitous routes.

$$\text{cyclohexanone} \longrightarrow \text{enamine} \xrightarrow{Br{-}CH(CH_3){-}CO_2Et} \text{2-}[CH(CH_3)CO_2Et]\text{cyclohexanone}$$

$$\text{enamine} \xrightarrow{Cl{-}C(=O){-}R} \text{2-}[C(=O){-}R]\text{cyclohexanone} \xrightarrow{OH^-} HO{-}C(=O){-}(CH_2)_5{-}C(=O){-}R$$

The enamine also can undergo Michael type addition to $\alpha,\beta$-unsaturated carbonyl derivatives. This reaction differs from the earlier acylation and alkylation reactions in that a proton shift occurs in the adduct to regenerate a new enamine, which in ethanol (but not in benzene) can react with an excess of the $\alpha,\beta$-unsaturated ketone.

[1] Stork, Terrell, and Szmuszkovicz, *J. Amer. Chem. Soc.*, **76**, 2029 (1954); Stork, Brizzolara, Landesman, Szmuszkovicz, and Terrell, *ibid.*, **85**, 207 (1963).

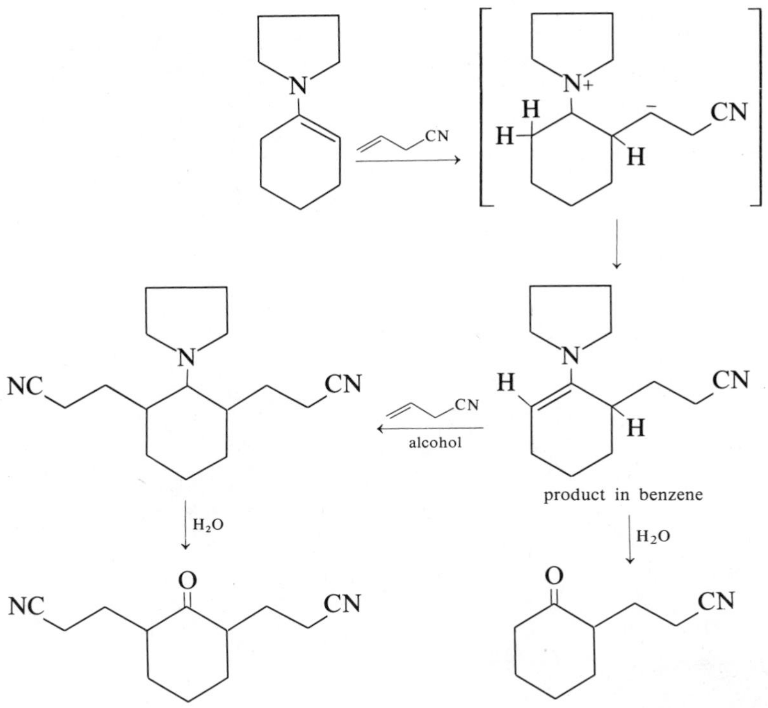

# 2-Acetylcyclohexanone

**N-1-Cyclohexenylpyrrolidine.** Arrange the assembly shown in Figure 42 using a 250-ml round-bottomed flask to hold the reagents and a 25-ml flask to serve as a water trap. In the 250-ml flask place 0.10 mole (9.8 g, 9.8 ml) of cyclohexanone,[2] 0.11 mole (7.8 g, 9.2 ml) of pyrrolidine and 0.1 g of *p*-toluenesulfonic acid. Attach the flask to the apparatus and add slowly about 60 ml of toluene through the condenser, so that the 25-ml flask and side-arm will be filled with toluene and the reaction flask slightly less than half-filled. Attach a drying tube filled with calcium chloride to the condenser and heat the reagents under reflux for 1.5 hr. During this time the water produced in the reaction will steam distill and collect as a lower layer in the trap.

Allow the reaction mixture to cool slightly, then replace the adapter, attached water trap and condenser, by an assembly for distillation. Remove the excess pyrrolidine by distilling the mixture until the thermometer reaches 105–108°. The residual enamine can be used for the next step without further purification.[3]

[2] The material prepared in Experiment 8a is suitable.

[3] Somewhat better yields are obtained if the cyclohexenylpyrrolidine (bp 106° at 13 mm) is purified by distillation before acylation.

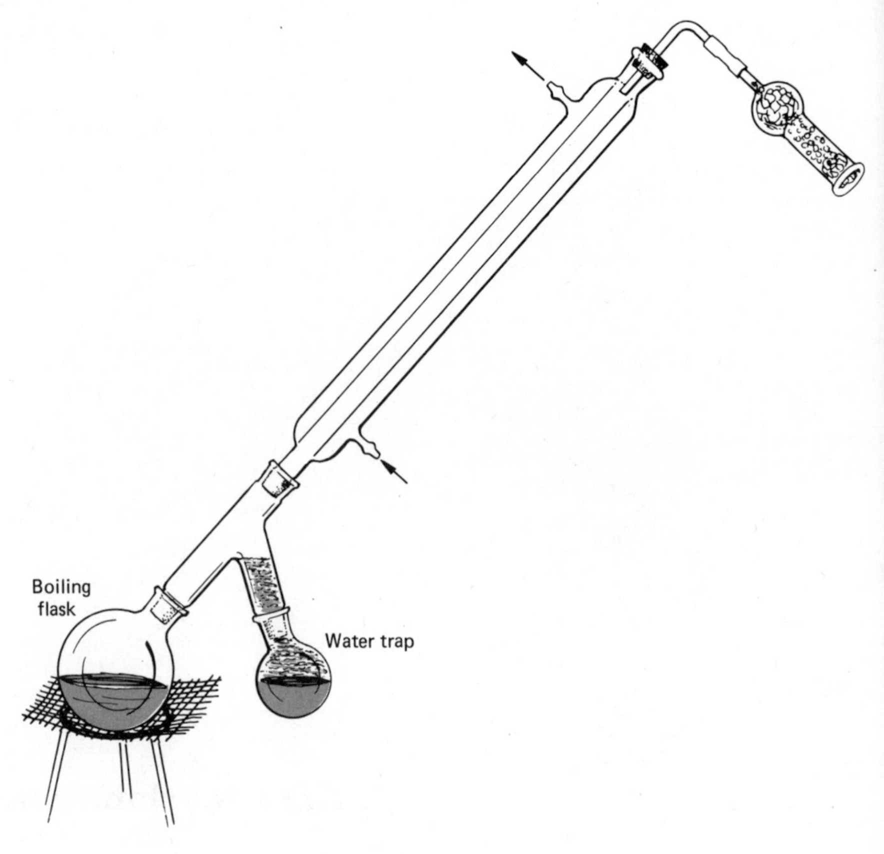


**Figure 42.** Water Separator

**2-Acetylcyclohexanone.** Add a solution of 0.11 mole (11.2 g, 10.4 ml) of acetic anhydride in 20 ml of toluene to the enamine, at room temperature. Mix the reagents well, attach a drying tube to the flask and set it aside for 24 hr or longer.

Add 10 ml of water to the reaction flask, arrange a condenser, and reflux the mixture for 0.5 hr. Wash the cooled toluene solution with water, then with 5 percent hydrochloric acid, and finally with water. Dry the solution with 2–3 g of anhydrous calcium chloride and remove the toluene by distillation. Distill the residual liquid under reduced pressure; bp 97–104° at 12–14 mm. The yield of 2-acetylcyclohexanone is 5.6–8.4 g.

## Questions

**1.** It is found that enamines prepared from pyrrolidine are more reactive than those from piperidine. Explain.

**2.** With unsymmetrical ketones the enamine synthesis introduces the alky or acyl group into the least substituted position. Explain.

**3.** Suggest a possible reason why use of ethanol as a solvent leads to dialkylated products in the reaction of enamines with $\alpha,\beta$-unsaturated carbonyl compounds.

**4.** Devise syntheses for the following molecules that make use of the enamine reaction: (a) 2-carbomethoxycyclopentanone; (b) pimelic acid; (c) 1,15-pentdecanedioic acid.

# EXPERIMENT 47

# Reactions of Ethyl Acetoacetate

Compounds in which two carbonyl groups are attached to the same carbon atom, such as the β-diketones ($R-CO-CH_2-CO-R$) and β-ketonic esters ($R-CO-CH_2-CO_2C_2H_5$), exhibit the phenomenon of tautomerism or dynamic isomerism. The classical example is ethyl acetoacetate; the liquid ester at room temperature is an equilibrium mixture containing about 92 percent of the ketonic isomer and 8 percent of the enol form. The individual isomers have distinct properties and can be isolated in a pure state, but either

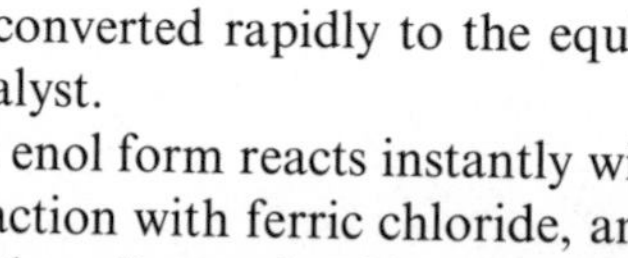

$$CH_3-\overset{\overset{O}{\|}}{C}-CH_2-\overset{\overset{O}{\|}}{C}-OC_2H_5 \qquad CH_3-\overset{\overset{O-H\cdots O}{|}}{C}{=}\underset{H}{C}-\overset{\|}{C}-OC_2H_5$$

mp −39°
bp 41°/2 mm

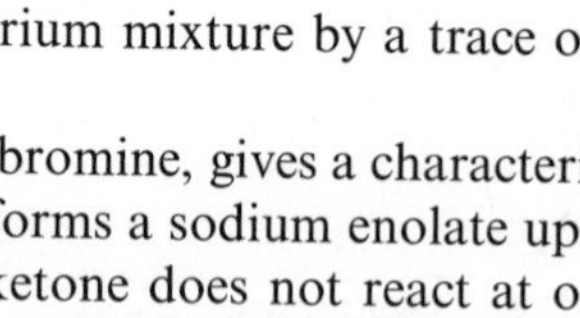

bp 33°/2 mm

form is converted rapidly to the equilibrium mixture by a trace of acid or base catalyst.

The enol form reacts instantly with bromine, gives a characteristic enol color reaction with ferric chloride, and forms a sodium enolate upon treatment with sodium ethoxide. The pure ketone does not react at once with bromine and does not immediately give a color reaction with ferric chloride but will show these reactions slowly as it undergoes isomerization to the enol form. The keto form reacts with carbonyl reagents to give typical derivatives such as the oxime and semicarbazone.

The enol form of a β-ketonic ester or β-diketone is stabilized by intramolecular proton-bonding to form a cyclic (chelate) structure. Metals such as copper, berylium, and aluminum form stable covalent chelate compounds with β-ketonic esters and β-diketones, that are soluble in organic solvents.

Thus, ethyl acetoacetate forms a green crystalline cupric chelate complex, mp 192°, that is soluble in ether, ethanol, and benzene. The enol form of

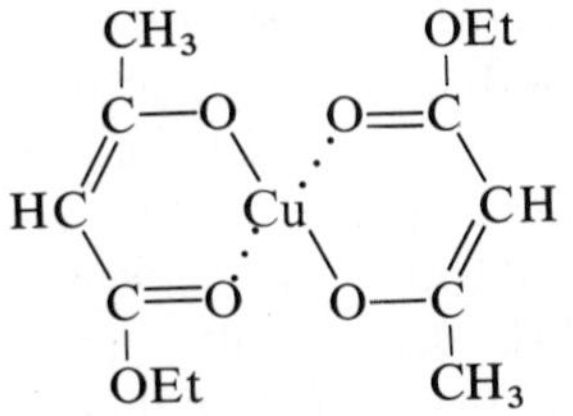

III

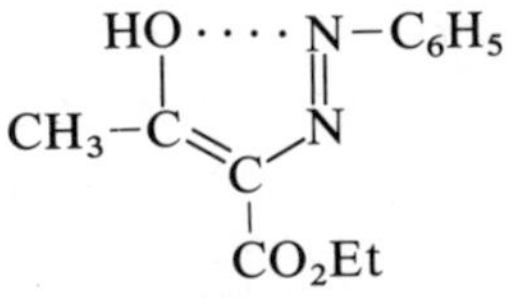

IV

ethyl acetoacetate undergoes coupling with aromatic diazonium salts to form azo derivatives such as IV.

Saponification of β-ketonic esters with cold dilute sodium hydroxide gives the sodium salt of the acid. The free β-ketonic acids upon warming undergo decarboxylation to form the corresponding ketones. This reaction forms the basis for a useful general method of synthesizing ketones from C-alkylated derivatives of ethyl acetoacetate (see Experiment 48). A competing reaction involves hydrolytic cleavage at the β-position to form carboxylic acids. This type of cleavage is the principal reaction if the β-ketonic ester is saponified with hot concentrated sodium hydroxide solution. With C-alkylated derivatives, one molecule of acetic acid and one molecule of a C-alkylated acetic acid are produced.

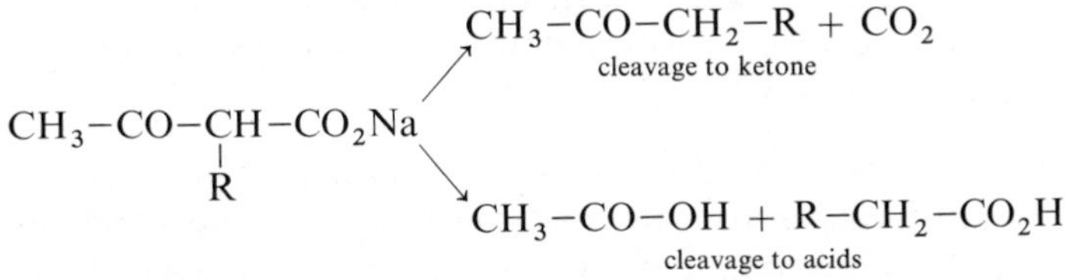

$$CH_3{-}CO{-}\underset{\displaystyle R}{\underset{|}{CH}}{-}CO_2Na \nearrow CH_3{-}CO{-}CH_2{-}R + CO_2 \quad \text{(cleavage to ketone)}$$

$$CH_3{-}CO{-}\underset{\displaystyle R}{\underset{|}{CH}}{-}CO_2Na \searrow CH_3{-}CO{-}OH + R{-}CH_2{-}CO_2H \quad \text{(cleavage to acids)}$$

**(A) Ferric Chloride Enol Test.** In a test tube dissolve one drop of ethyl acetoacetate in 5 ml of ethanol and add a few drops of a 3 percent solution of ferric chloride. Shake well and observe.

Carry out the ferric chloride test with one or more of the following substances in place of ethyl acetoacetate: (a) a crystal of salicylic acid or of catechol; (b) one or two drops of a 5 percent aqueous solution of phenol; (c) a drop of acetone or acetophenone. Record your observations and comment on the results (see Experiment 34, part B).

**(B) Bromine Test.** In a test tube dissolve 4–5 drops of ethyl acetoacetate in 5 ml of carbon tetrachloride and add a 2 percent solution of bromine in carbon tetrachloride, with shaking, until a *faint* bromine color persists for about 1 min. Allow the tube to stand for 5 min and observe again. Test to determine if the solution will now decolorize more of the bromine solution. Explain the result.[1]

**(C) Semicarbazone Formation.** In a 50-ml Erlenmeyer flask dissolve 1 g of semicarbazide hydrochloride and 1.5 g of sodium acetate crystals in 10 ml of warm water (50–60°). Add a solution of 1 g (1 ml) of ethyl acetoacetate in 10 ml of ethanol, cork the flask firmly, and shake the mixture vigorously for a few minutes. Allow the mixture to stand until the semicarbazone has separated completely. Collect the crystals by suction filtration, wash them with a little cold water, and spread them to dry on a clean filter paper. The recorded melting point of the semicarbazone is 129°.

**(D) Ketonic Cleavage.** In a 250-ml round-bottomed flask place 5 g (5 ml) of ethyl acetoacetate, 30 ml of water, and 20 ml of 10 percent aqueous sodium hydroxide solution. Attach a reflux condenser, add a boiling chip, and boil the solution gently for 20 min. Cool the flask to about 50° and remove the condenser. Fit the flask with a short unpacked fractionating tube, arrange a condenser for distillation, and distill the solution until about 20 ml of distillate has been collected. Use a 50-ml Erlenmeyer flask as receiver.

To the distillate add 1 g of semicarbazide hydrochloride and 1.5 g of sodium acetate crystals, cork the flask firmly, and shake the mixture vigorously. After a short while crystals of acetone semicarbazone begin to separate. Chill the mixture thoroughly in an ice bath, collect the crystals with suction, and wash them with two small portions of ice-cold water. Allow them to dry and take the melting point. The recorded melting point of acetone semicarbazone is 190°.

**(E) Copper Chelate Derivative.** In a 50-ml flask dissolve 2 g of copper sulfate crystals in 20 ml of water and add dilute aqueous ammonia, with shaking, until the precipitate of cupric hydroxide just redissolves. While swirling the deep blue solution add dropwise a solution of 1 g of ethyl acetoacetate in 5 ml of ethanol. Cork the flask firmly and shake the mixture thoroughly until green crystals of the chelate derivative have separated. Collect the product with suction, wash it well with water, and allow it to dry. When thoroughly dried the material may be crystallized from benzene. The recorded melting point of the copper derivative is 192°.

## Questions

**1.** Write equations for the preparation of ethyl acetoacetate: (a) from ethyl acetate, by means of the Claisen ester condensation; (b) from ketene dimer.

[1] A quantitative experiment to determine the equilibrium constant for the enolization of ethyl acetoacetate, by a modified Kurt Meyer bromine titration, has been described by Ward, *J. Chem. Educ.*, **39**, 95 (1962).

**2.** What products are formed by condensation of : (a) ethyl benzoate and ethyl propionate; (b) ethyl benzoate and acetophenone; (c) ethyl acetate and ethyl oxalate? What ketones would be formed by cleavage of the products obtained in reactions (a) and (b)?

**3.** The oxime and the phenylhydrazone of ethyl acetoacetate undergo cyclization upon heating, with elimination of ethanol. What are the products formed in these reactions?

**4.** What functional groups other than $-CO-R$ and $-CO_2R$ have a strongly activating effect upon a methylene group to which they are directly attached?

EXPERIMENT

# 48

# The Acetoacetic Ester Synthesis of Ketones

The acetoacetic ester synthesis affords a general method of preparing derivatives of acetone of the type $CH_3{-}CO{-}CH_2R$, $CH_3{-}CO{-}CHR_2$ and $CH_3{-}CO{-}CH(R){-}R'$. The steps involved in the synthesis are: (a) conversion of ethyl acetoacetate into its sodium derivative by means of sodium ethoxide in absolute ethanol; (b) C-alkylation of sodio-acetoacetic ester by reaction with an alkyl halide; (c) saponification of the C-alkylated acetoacetic ester with dilute aqueous alkali, and ketonic cleavage. The ketonic

$$\text{(a)}\quad \begin{matrix} CH_3{-}CO{-}CH_2{-}CO_2Et \\ \upharpoonleft\downharpoonright \\ CH_3{-}\underset{\displaystyle OH}{\underset{|}{C}}{=}CH{-}CO_2Et \end{matrix} \xrightarrow{\text{NaOEt}} Na^+\left[CH_3{-}\overset{\displaystyle O}{\overset{\|}{C}}{-}\ddot{C}H{-}\overset{\displaystyle O}{\overset{\|}{C}}{-}OEt\right]^-$$

$$\text{(b)}\ \xrightarrow{C_4H_9{-}Br} CH_3{-}\overset{\displaystyle O}{\overset{\|}{C}}{-}\underset{\displaystyle C_4H_9}{\underset{|}{C}H}{-}\overset{\displaystyle O}{\overset{\|}{C}}{-}OEt \rightleftharpoons CH_3{-}\overset{\displaystyle OH}{\overset{|}{C}}{=}\underset{\displaystyle C_4H_9}{\underset{|}{C}}{-}\overset{\displaystyle O}{\overset{\|}{C}}{-}OEt$$

$$\text{(c)}\ \xrightarrow{5\%\ NaOH} CH_3{-}\overset{\displaystyle O}{\overset{\|}{C}}{-}\underset{\displaystyle C_4H_9}{\underset{|}{C}H}{-}\overset{\displaystyle O}{\overset{\|}{C}}{-}\bar{O}Na^+ \longrightarrow \underset{\text{ketonic cleavage}}{CH_3{-}CO{-}CH_2{-}C_4H_9 + CO_2}$$

$$\xrightarrow{20\text{–}30\%\ NaOH} \underset{\text{cleavage to acids}}{CH_3{-}CO_2Na + C_4H_9{-}CH_2{-}CO_2Na}$$

cleavage occurs in warm, dilute alkaline solution or upon acidifying to form the free $\beta$-ketonic acid. A side reaction, involving cleavage at the $\beta$-position, occurs to a minor extent and produces acetic acid and an alkylacetic acid.

If the ester is saponified with concentrated alkali the cleavage to acids becomes the main reaction. This modification can be used for the preparation of alkylacetic acids but is less satisfactory than the malonic ester synthesis of acids (Experiment 49).

For the synthesis of dialkyl derivatives of acetone, the monoalkylated acetoacetic ester is converted to the sodium derivative, a second alkyl group is introduced (which may be different from the first group), and the dialkyl derivative is subjected to ketonic cleavage. Usually a small amount of the dialkylacetoacetic ester is formed during preparation of the monoalkyl

$$CH_3-CO-\underset{R}{\underset{|}{CH}}-CO_2Et \xrightarrow[R'-Br]{NaOEt} CH_3-CO-\underset{R}{\underset{|}{\overset{R'}{\overset{|}{C}}}}-CO_2Et \xrightarrow{NaOH} CH_3-CO-\underset{R}{\underset{|}{CH}}-R' + CO_2$$

derivative, because some of the sodium derivative of the alkylacetoacetic ester is present in equilibrium with ethyl sodio-acetoacetate and undergoes further alkylation. This side reaction is disfavored by using more than one mole of ethyl acetoacetate per mole of sodium ethoxide.

Alkylation of the sodium derivatives proceeds more readily with primary alkyl halides than with secondary halides. Tertiary alkyl groups cannot be introduced in the customary way because the tertiary alkyl halides are converted to alkenes under the conditions used. Owing to the low reactivity of aryl halides, aryl groups cannot be introduced but primary aralkyl halides of the type $Ar-CH_2Cl$ and $Ar-CH_2-CH_2Cl$ react satisfactorily in the usual alkylation procedure.

Dihalides such as 1,3-bromopropane may be used, with two moles of sodio-acetoacetic ester, to synthesize diketones of the type illustrated by nonane-2,8-dione: $CH_3-CO-CH_2-(CH_2)_3-CH_2-CO-CH_3$. $\alpha$-Chloro esters such as $Cl-CH_2CO_2Et$ may be used to obtain ketonic acids.

## 48a Ethyl *n*-Butylacetoacetate

**Ethyl Sodio-acetoacetate.** In a perfectly dry 500-ml round-bottomed flask provided with a reflux condenser, place 150 ml of high grade absolute ethanol.[1] In a small beaker containing kerosene or xylene weigh out 0.3 mole

[1] The grade of absolute ethanol used in this experiment is important since the yield of ethyl *n*-butylacetoacetate is decreased markedly by small amounts of water. For this experiment where the presence of methanol is not objectionable, a very anhydrous product may be obtained conveniently by the use of freshly prepared magnesium methoxide as a drying agent.

Provide a dry 1-liter round-bottomed flask with a reflux condenser fitted with a calcium

(6.9 g) of clean metallic sodium. Through the condenser tube add the metallic sodium in small pieces, at such a rate that the ethanol is maintained in ebullition. After all the sodium has dissolved, add 0.3 mole (39 g, 38 ml) of pure ethyl acetoacetate to the warm solution of sodium ethoxide, with shaking. Proceed rapidly without interruption to the next step.

**Ethyl *n*-Butylacetoacetate.** To the clear solution of ethyl sodio-acetoacetate add 0.33 mole (46 g, 36 ml) of *pure* *n*-butyl bromide, and heat the reaction mixture on a steam bath for 6–10 hr. Place a calcium chloride tube at the top of the condenser to exclude atmospheric moisture. The apparatus should be assembled so that the violent bumping that may occur when sodium bromide precipitates will not cause breakage. When the reaction is complete a drop of the solution will not show an alkaline reaction when tested with moist red litmus paper, provided an excess of sodium was not used. (Explain.)

If convenient, it is advantageous to allow the mixture merely to stand for a week at room temperature in a *tightly stoppered* round-bottomed flask, instead of refluxing for 6–10 hr.

Connect the flask by means of a wide delivery tube to a condenser set for distillation and distill off as much ethanol as possible by heating on a steam bath. Cool the contents of the flask to 20° and add 250 ml of water and 3 ml of concentrated hydrochloric acid. Transfer the mixture to a separatory funnel, draw off the aqueous layer of sodium bromide, and wash the organic layer with water. Dry the product with 8–10 g of anhydrous magnesium sulfate. Transfer the liquid to a Claisen flask of suitable size and distill under diminished pressure (see page 50).[2]

Collect a first fraction which distills below 100° at 25 mm pressure. This contains ethanol, water, *n*-butyl bromide, unchanged ethyl acetoacetate, and occasionally a trace of ethyl caproate. (Explain.) Collect the second fraction, which should consist chiefly of ethyl *n*-butylacetoacetate, from 120–130° at

---

chloride tube, and place in the flask 2.4 g of magnesium turnings and 20 ml of absolute methanol. Usually a vigorous reaction occurs spontaneously and the magnesium dissolves. If necessary, warm the mixture gently or add a small crystal of iodine to start the reaction. When the magnesium has dissolved, a white paste of magnesium methoxide results. Add 200 ml of commercial absolute ethanol (99–99.5 percent) and reflux the mixture for 4 hr. Remove the reflux condenser and distill the absolute ethanol directly into the flask in which it is to be used, *taking precautions to protect the distillate from atmospheric moisture.* It is advisable to reject the first 15–20 ml of distillate before collecting the main fraction.

Another convenient laboratory procedure for producing a high grade of absolute ethanol is by means of sodium ethoxide and diethyl phthalate. Dissolve 1.5 g of metallic sodium in 200 ml of commercial absolute ethanol, add 6 g of diethyl phthalate, reflux the mixture for 1 hr with the condenser protected from moisture with a calcium chloride tube. Distill off the ethanol through a short column, protecting the distillate from moisture.

[2] It is not essential to purify the crude product by vacuum distillation before proceeding to the ketonic cleavage, but it is good practice to do so. The crude ethyl *n*-butylacetoacetate may be treated directly with 5 percent aqueous sodium hydroxide as directed in the first paragraph of the following section.

25 mm.[3] The yield is 32–38 g. A third fraction may be collected which on redistillation will yield a small additional quantity of ethyl *n*-butylacetoacetate. The residue in the distilling flask contains ethyl di-*n*-butylacetoacetate (bp 140–145° at 25 mm).

To obtain perfectly pure ethyl *n*-butylacetoacetate it would be necessary to refractionate the main portion of the product and collect the material over a smaller boiling range. For the preparation of methyl *n*-pentyl ketone further purification is unnecessary.

## 48b Methyl *n*-Pentyl Ketone

**Ketonic Cleavage of an Alkylacetoacetic Ester.** ***Sodium n-Butylacetoacetate.*** In a 500-ml separatory funnel (or bottle) place 250 ml of 5 percent aqueous sodium hydroxide solution and add the ethyl *n*-butylacetoacetate (32–38 g) obtained in the preceding experiment. Stopper the container and shake thoroughly for about 0.5 hr. Allow the reaction mixture to stand at room temperature with occasional shaking for at least 3–4 hr, but not longer than two days. During this treatment the ethyl *n*-butylacetoacetate is saponified and passes into solution as the sodium salt of *n*-butylacetoacetic acid. A small oily layer of ethyl di-*n*-butylacetoacetate may remain undissolved; if the reaction mixture has been allowed to stand longer than 24 hr some of the ketone may also separate.

**Methyl *n*-Pentyl Ketone.** Transfer the solution of sodium *n*-butylacetoacetate to a 1-liter flask provided with a Claisen adapter fitted with a small separatory funnel and a distillation adapter connected to an efficient condenser set for distillation. Through the separatory funnel add slowly with shaking, a cold solution of 15 ml of concentrated sulfuric acid in 30 ml of water. During the addition of the acid a vigorous evolution of carbon dioxide occurs. When this has subsided, heat the reaction mixture slowly to the boiling point and distill slowly until 120–130 ml of distillate has been collected. At this time the methyl *n*-pentyl ketone should have distilled completely. The distillate contains the ketone, water, ethanol, and small amounts of acetic and caproic acids. Discard the residue.

To the distillate add small pieces of solid sodium hydroxide until the solution is distinctly alkaline (test with litmus). Redistill the solution until 75–80 percent of the material has passed over, and discard the residue. Separate the ketone layer in the distillate and redistill the aqueous layer until about one-third of the material has passed over. Separate the ketone layer and redistill the aqueous layer as before. Repeat this procedure as long as any appreciable amount of ketone is obtained in the distillate.

[3] The approximate boiling points of ethyl *n*-butylacetoacetate under various pressures are: 138°/50 mm; 132°/40 mm; 124°/30 mm; 116°/20 mm; 106°/15 mm.

Wash the combined portions of the ketone four times with one-third its volume of 35–40 percent aqueous calcium chloride. Dry over 4–5 g of anhydrous magnesium sulfate or Drierite and distill. Collect the methyl *n*-pentyl ketone from 146–152°. The yield is 10–15 g. If an appreciable low-boiling fraction is obtained it may be redried and redistilled (or used for the preparations described below).

**Semicarbazone of Methyl *n*-Pentyl Ketone** *(Optional)*. This experiment illustrates the conversion of a ketone to a crystalline derivative suitable for identification and characterization.

Dissolve 1 g of the ketone in 10 ml of 60 percent ethanol and pour this slowly, with shaking, into a solution of 1.2 g of semicarbazide hydrochloride and 1.8 g of sodium acetate crystals in 15 ml of water, contained in a small Erlenmeyer flask. Close the flask firmly with a cork and shake vigorously. The reaction takes place rapidly and white crystalline flocks of the semicarbazone soon fill the solution. Collect the crystals with suction and wash them with several small portions of cold water. Allow to dry and take the melting point. The product may be purified by recrystallization from 60 percent ethanol. The recorded melting point of this semicarbazone is 127°.

**2,4-Dinitrophenylhydrazone** *(Optional)*. This crystalline derivative may be prepared from methyl *n*-pentyl ketone by the same procedure used for cyclohexanone (see Experiment 9, part E). The recorded melting point of this dinitrophenylhydrazone is 89°.

## Questions

**1.** Explain what is meant by the term tautomerism (dynamic isomerism) and cite the experimental evidence for the existence of tautomeric forms of ethyl acetoacetate.

**2.** Write equations showing the mechanism for the formation of ethyl acetoacetate by the action of metallic sodium, or sodium ethoxide, on ethyl acetate (Claisen ester condensation).

**3.** In the ketonic cleavage of ethyl *n*-butylacetoacetate what side reaction leads to the formation of acetic and *n*-caproic acid? How are they removed in the purification of the ketone? How can methyl *n*-pentyl ketone be regenerated from the semicarbazone?

**4.** How could methyl *n*-pentyl ketone be prepared by means of a Grignard reaction?

**5.** How may the following compounds be prepared by means of the acetoacetic ester synthesis: (a) 3-methyl-2-pentanone; (b) 4-methyl-2-hexanone; (c) 2,6-heptanedione; (d) 4-oxo-3-methylpentanoic acid ($\gamma$-methyllevulinic acid)?

EXPERIMENT

# 49

# The Malonic Ester Synthesis of Acids

Diethyl malonate has a very reactive methylene group and resembles ethyl acetoacetate (Experiments 47 and 48) in its ability to form a sodium derivative when treated with sodium ethoxide. Unlike acetoacetic ester, it does *not* give an enol color reaction with ferric chloride, since the pure liquid ester does not contain an appreciable amount of the enol form (less than 0.1 percent as against 8 percent for acetoacetic ester). Malonic ester can be used for the synthesis of derivatives of acetic acid of the types $R{-}CH_2{-}CO_2H$, $R_2CH{-}CO_2H$ and $R'{-}CH(R){-}CO_2H$, in a manner similar to the acetoacetic ester synthesis of ketones.

The malonic ester synthesis of organic acids involves the following steps: (a) conversion of diethyl malonate to the sodium derivative by treatment with sodium ethoxide in absolute ethanol; (b) C-alkylation of the sodium derivative by means of an alkyl halide; (c) saponification of the alkylmalonic ester, followed by acidification and decarboxylation of the

$$\text{(a)}\quad EtO_2C{-}CH_2{-}CO_2Et \longrightarrow Na^+\left[EtO{-}\overset{\overset{\displaystyle O}{\|}}{C}{-}\ddot{C}H{-}\overset{\overset{\displaystyle O}{\|}}{C}{-}OEt\right]^-$$

$$\text{(b)}\quad \xrightarrow{C_4H_9{-}Br} NaBr + EtO{-}\overset{\overset{\displaystyle O}{\|}}{C}{-}\underset{\underset{\displaystyle C_4H_9}{|}}{CH}{-}\overset{\overset{\displaystyle O}{\|}}{C}{-}OEt$$

$$\text{(c)}\quad \xrightarrow[H_2O]{NaOH} C_4H_9{-}CH(CO_2Na)_2 \xrightarrow{H^+} C_4H_9{-}CH(CO_2H)_2 \xrightarrow{heat} C_4H_9{-}CH_2CO_2H + CO_2$$

alkylmalonic acid. Generally the alkylmalonic acids are not isolated because they are quite soluble in water and usually they can be decarboxylated by heating in strongly acid aqueous solutions. Dialkylmalonic esters, from which dialkylacetic acids may be obtained by the usual procedure, can be prepared by introducing a second alkyl group into a mono-substituted malonic ester.

In the alkylation reactions primary alkyl halides afford better yields than the secondary halides. Tertiary alkyl halides are converted mainly to alkenes in the basic reaction medium. The use of dihalides of the type $Br-(CH_2)_n-Br$, with one mole of malonic ester and two equivalents of sodium ethoxide, leads to cycloalkane-1,1-dicarboxylic esters having 3-, 4-, 5-, and 6-membered rings when $n$ is 2 to 5. Cyclization does not occur to an appreciable extent when $n$ is larger than 5. When two moles of sodio-malonic ester are used with these dihalides, one obtains linear tetracarboxylic esters of the type $(EtO_2C)_2CH-(CH_2)_n-CH(CO_2Et)_2$, that may be converted to dicarboxylic acids by the customary procedures.

Aryl halides, owing to their low reactivity, cannot be used for the synthesis of arylmalonic esters but the aryl derivatives can be obtained by other general methods that are applicable to the preparation of mono-substituted malonic esters. For example, phenylmalonic ester may be prepared from

$$C_6H_5-CH_2-CO_2Et + EtO-CO-OEt \xrightarrow{NaOEt} C_6H_5-CH(CO_2Et)_2 + EtOH$$

$$\begin{matrix} C_6H_5-CH_2-CO_2Et \\ + EtO_2C-CO_2Et \end{matrix} \xrightarrow{NaOEt} \begin{matrix} C_6H_5-CH-CO_2Et \\ | \\ CO-CO_2Et \end{matrix} \xrightarrow{heat} \begin{matrix} C_6H_5-CH-CO_2Et \\ | \\ CO-OEt \end{matrix} + C\equiv O$$

ethyl phenylacetate either by condensation with diethyl carbonate (Claisen ester condensation), or by condensation with ethyl oxalate to form phenyloxaloacetic ester, which undergoes loss of carbon monoxide (decarbonylation) upon heating.

Malonic esters undergo a cyclic condensation reaction with urea to form 2,4,6-trioxopyrimidines, known as barbituric acids. The 5,5-disubstituted barbituric acids are used in medicine as sedatives and hypnotics. These compounds are acidic (p$K$ ~7–8) and form water-soluble sodium salts.

$$O=C\begin{matrix} / NH_2 \\ \backslash NH_2 \end{matrix} + \begin{matrix} EtO_2C \quad C_2H_5 \\ \backslash \; / \\ C \\ / \; \backslash \\ EtO_2C \quad C_6H_5 \end{matrix} \xrightarrow[heat]{NaOEt} O=C\begin{matrix} / NH-CO \\ \backslash NH-CO \end{matrix}\begin{matrix} C_2H_5 \\ C \\ C_6H_5 \end{matrix} + 2C_2H_5-OH$$

*Barbital* (*Veronal*) is 5,5-diethylbarbituric acid; *Pentobarbital* (*Nembutal*) is the 5-ethyl-5-(2′-pentyl) analog. *Phenobarbital*, used in the treatment of epilepsy, is 5-ethyl-5-phenylbarbituric acid.

# 49a Ethyl *n*-Butylmalonate

**Ethyl Sodio-malonate.** In a perfectly dry 1-liter round-bottomed flask provided with a reflux condenser, place 250 ml of high-grade absolute ethanol.[1] In a small beaker containing kerosene or xylene weigh out 0.5 mole (11.5 g) of clean metallic sodium. Through the condenser tube add the metallic sodium in small pieces, at such a rate that the ethanol is maintained in ebullition. After all the sodium has dissolved allow the sodium ethoxide solution to cool slightly (to about 50°). To the warm solution add with shaking 0.55 mole (88 g, 80 ml) of pure ethyl malonate during the course of 5–10 min. Proceed at once to the next step.

**Ethyl *n*-Butylmalonate.** To the clear solution of ethyl sodio-malonate add gradually from a separatory funnel, 0.5 mole (68.5 g, 54 ml) of *pure n*-butyl bromide, while shaking the flask to insure thorough mixing of the reactants. When the addition of the butyl bromide has been completed, add two small boiling chips and reflux the reaction mixture for 4–5 hr on a steam bath. Place a calcium chloride tube at the top of the condenser to exclude atmospheric moisture. The apparatus should be assembled so that the violent bumping that may occur when sodium bromide precipitates will not cause breakage.

Connect the flask by means of a wide delivery tube to a condenser set for distillation and distill off as much ethanol as possible by heating on a steam bath. Cool the contents of the flask to 20° and add 200 ml of water and 5 ml of concentrated hydrochloric acid. Transfer the mixture to a separatory funnel and draw off the aqueous solution of sodium bromide. Wash the ethyl *n*-butylmalonate with a little water and dry with 10–15 g of anhydrous magnesium sulfate. Transfer the liquid to a Claisen flask of suitable size and distill under diminished pressure[2] (see page 50).

Collect first a low-boiling portion (ethanol, water, and *n*-butyl bromide) which distills below 100° at 25 mm pressure. Collect next an intermediate fraction (unchanged ethyl malonate) up to 115° at 20–25 mm pressure, and finally collect the main fraction of ethyl *n*-butylmalonate from 115–135° at

[1] The grade of absolute ethanol used in this experiment is important since the yield of ethyl *n*-butylmalonate is decreased considerably by small amounts of water. For this experiment, suitable absolute ethanol may be prepared by the methods given in Experiment 48, footnote 1, using 300 ml of commercial absolute alcohol.

[2] It is desirable but not essential to distill the product under diminished pressure; ethyl *n*-butylmalonate distills with only slight decomposition at 235–240° under 760 mm pressure if the distillation is carried out rapidly from a small flask.

20–25 mm pressure.[3] This weighs 65–75 g. If a large intermediate fraction is obtained this may be redistilled to secure an additional quantity of ethyl *n*-butylmalonate.

## 49b *n*-Caproic Acid

**Decarboxylation of a Substituted Malonic Acid.** ***n*-Butylmalonic Acid.** In this experiment use the entire quantity of ethyl *n*-butylmalonate obtained in the preceding preparation. In the first step use 1 g of solid potassium hydroxide (a 10 percent excess) for each gram of ester, and in the subsequent acidification use 2 g (1.1 ml) of concentrated sulfuric acid for each gram of potassium hydroxide.

In a 1-liter round-bottomed flask place the solid potassium hydroxide and add an equal weight of distilled water. Shake thoroughly until the potassium hydroxide has dissolved completely. While shaking the flask vigorously add the ethyl *n*-butylmalonate dropwise to the hot solution, from a separatory funnel. Usually a vigorous reaction occurs and much of the ethanol formed boils off. During this process the flask should be held in a towel or clamp, and care must be taken to avoid spattering of the strong alkali upon the skin. To minimize this danger the ester must be added slowly and the flask shaken thoroughly.

After all the ester has been added adjust a reflux condenser, add two small boiling chips, and boil the mixture for 2 or 3 hr to insure complete saponification. Remove the condenser, add about 150 ml of water, and set the condenser for distillation. Distill the mixture until 130–140 ml of distillate has been collected, to insure complete removal of ethanol. Discard the distillate.

Cool the reaction mixture to 20° and adjust a reflux condenser. Through the condenser tube add a *cold* solution of the requisite amount of concentrated sulfuric acid (see above) dissolved in twice its volume of water. Perform this addition slowly with shaking in order to prevent excessive foaming. During the addition of the sulfuric acid the reaction mixture becomes warm and may reflux spontaneously. The solution now contains *n*-butylmalonic acid, potassium acid sulfate, and sulfuric acid.

***n*-Caproic Acid.** Add a small boiling chip to the acid solution and reflux gently for 4–6 hr (longer refluxing may improve the yield slightly). Cool to 20°, separate the upper layer of organic acid, and extract the mother liquor with three 150-ml portions of benzene. Combine the original acid layer and the benzene extracts, wash with 20–25 ml of water, and dry with 10–15 g of anhydrous magnesium sulfate.

[3] Ethyl *n*-butylmalonate boils at 130–135° under 20 mm pressure. To obtain pure ethyl *n*-butylmalonate it would be necessary to refractionate the product but for many purposes (such as the preparation of *n*-caproic acid) further purification is unnecessary.

Separate from the drying agent and distill off the benzene and last traces of water through a short fractionating tube, until the vapors reach a temperature of approximately 100°. Transfer the residue to a small distilling flask and distill carefully on a wire gauze, using an air-cooled condenser. Collect the portion boiling from 198–206° in a dry, tared bottle. The yield is 20–25 g. Low-boiling material may be dried again and redistilled.

If desired, the final distillation may be carried out under diminished pressure, in a very small Claisen flask. The following are the boiling points of *n*-caproic acid under diminished pressure: 99°/10 mm, 111°/20 mm, 119°/30 mm, 125°/40 mm, 130°/50 mm.

***n-Caproanilide (Optional).*** Organic acids are often identified and characterized by conversion to crystalline amides (anilides, *p*-toluidides, etc.) or to solid esters (*p*-nitrobenzyl, phenacyl esters, etc.). Anilides may be obtained conveniently by forming the acid chloride by means of thionyl chloride, and treating with aniline.

In a large test tube mix 1 g of *n*-caproic acid with 2 g (1.5 ml) of *pure* thionyl chloride (sulfurous oxychloride, $SOCl_2$) and a few drops of dimethylformamide, attach a small reflux condenser, and boil gently for 30 min. After cooling, dissolve the liquid in 10 ml of *dry* benzene, and pour slowly into a solution of 2 g (2 ml) of aniline in 20 ml of *dry* benzene. Warm gently on a steam bath for a few minutes to complete the reaction. Decant the benzene solution of the anilide into a separatory funnel, wash thoroughly with two 5-ml portions of 5 percent hydrochloric acid, then with 10 ml of 5 percent aqueous sodium hydroxide, and finally with pure water. Distill off the solvent from a small distilling flask and recrystallize the crude anilide from dilute ethanol, with addition of decolorizing carbon. Dry the anilide thoroughly and take the melting point.[4]

## Questions

**1.** How may malonic ester (diethyl malonate) be prepared, starting from acetic acid?

**2.** Why is the yield of butylmalonic ester diminished by the presence of water in any of the reagents used?

**3.** When butylmalonic ester is heated with aqueous alkali, why is the saponification not necessarily complete as soon as the oily layer of the ester has disappeared?

**4.** Write equations for the preparation of the following compounds, starting from malonic ester: (a) 2-methylpentanoic acid; (b) cyclobutanecarboxylic acid; (c) $HO_2C{-}(CH_2)_5{-}CO_2H$ (pimelic acid); (d) 5,5-diethylbarbituric acid.

**5.** What methods other than alkylation of malonic ester may be used to prepare mono-substituted malonic esters? Illustrate with specific examples.

[4] The reported melting points of various anilides are: *n*-caproanilide, 95°; isocaproanilide, 110°; *n*-valeranilide, 63°; isovaleranilide, 109°; *n*-butyranilide, 92°; isobutyranilide, 105°.

# EXPERIMENT 50

# Phthaleins

Phthalic anhydride undergoes a stepwise condensation with two molecules of phenol, in the presence of acid catalysts such as sulfuric acid or zinc chloride, to form phenolphthalein. The phthaleins are derivatives of triphenylcarbinol and are related structurally to the triphenylmethane dyes. In neutral or acidic solutions phenolphthalein exists in the colorless, lactone form (I). In basic solutions, in the range pH 8.3–10, it is converted to the red

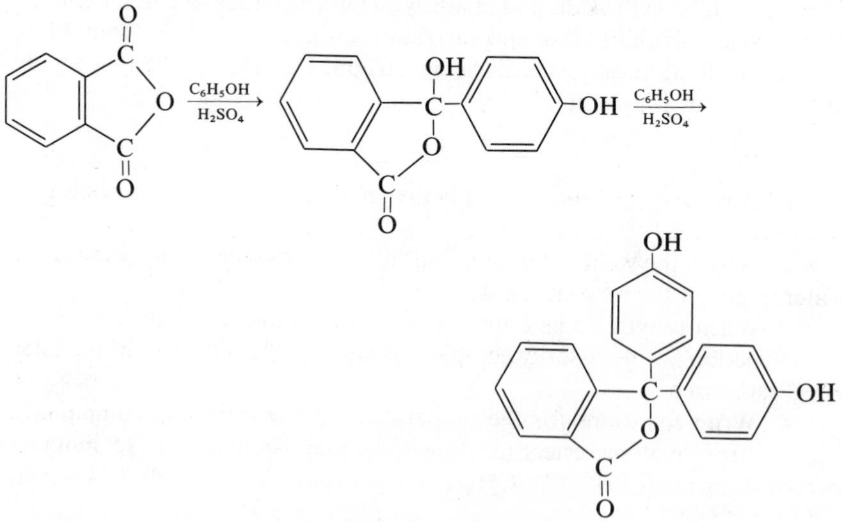

$\xrightarrow[H_2SO_4]{C_6H_5OH}$ $\xrightarrow[H_2SO_4]{C_6H_5OH}$

I

dianion (II); in very strongly alkaline solutions, hydroxyl ion is taken up at the central carbon atom (destroying the quinonoid structure) and the resulting trianion (III) is colorless. Similar phthaleins can be obtained by using *o*-cresol or thymol instead of phenol. If phthalic anhydride is replaced by the anhydride of *o*-sulfobenzoic acid, sulfonphthaleins are produced. Both types of phthaleins are useful acid-base indicators.

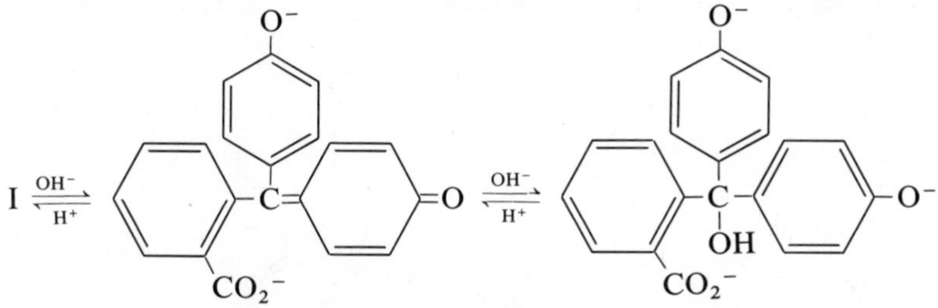

Lactone (colorless) — II Dianion (red) — III Trianion (colorless)

Condensation of phthalic anhydride with resorcinol (*m*-dihydroxybenzene) furnishes fluorescein (IV), which is a modified phthalein containing

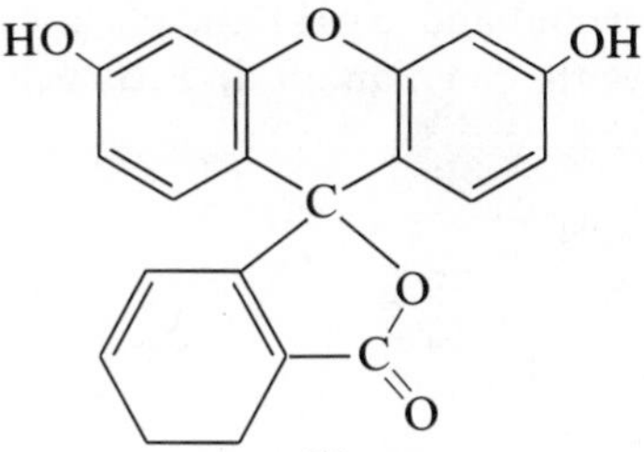

IV 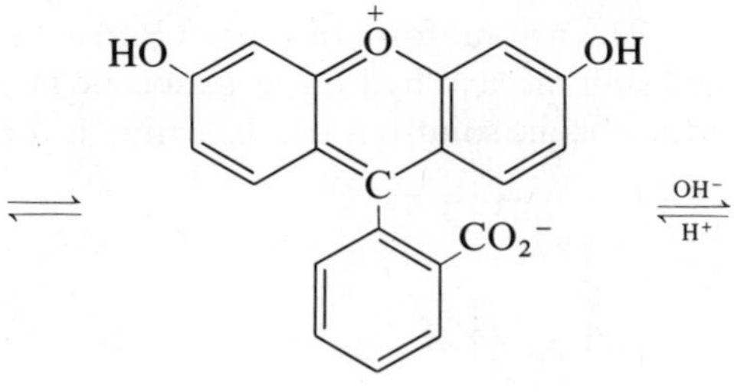 — V 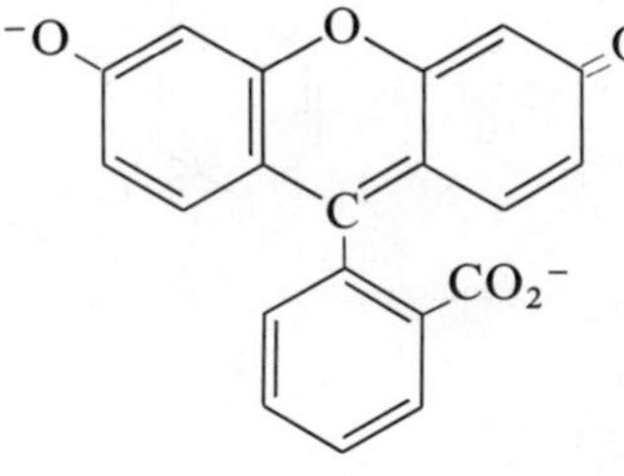 — VI

an oxygen heterocycle of the dibenzopyran type. In neutral and acidic solutions fluorescein is yellow and may exist as the lactone (IV) or the oxonium zwitterion (V). It dissolves in alkalies to give the orange dianion (VI) that exhibits an intense green fluorescence. Bromination of fluorescein takes place readily in the phenolic rings and gives a dibromo derivative and a tetrabromo derivative (eosin). The sodium salt of eosin dyes silk and wool a brilliant red. Sodium eosin is used also as a biological stain and in the manufacture of red ink. The analogous tetraiodo derivative of fluorescein, erythrosin, is used as a red coloring matter for food products and as a photographic sensitizer.

The sulfonphthaleins are excellent pH indicators since they are moderately soluble in water and give brilliant color changes over narrow pH ranges. The parent compound, phenolsulfonphthalein, is formed by condensation of *o*-sulfobenzoic anhydride with phenol.

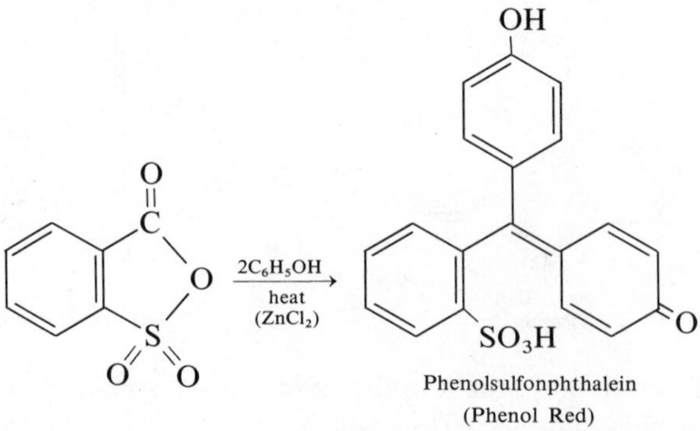

$\xrightarrow[\text{heat } (ZnCl_2)]{2C_6H_5OH}$

Phenolsulfonphthalein
(Phenol Red)

The neutral form of Phenol Red is yellow but above pH 8 the phenolic and sulfonic acid hydrogens dissociate to give the red dianion. In extremely strong acidic solutions the red cation is formed.

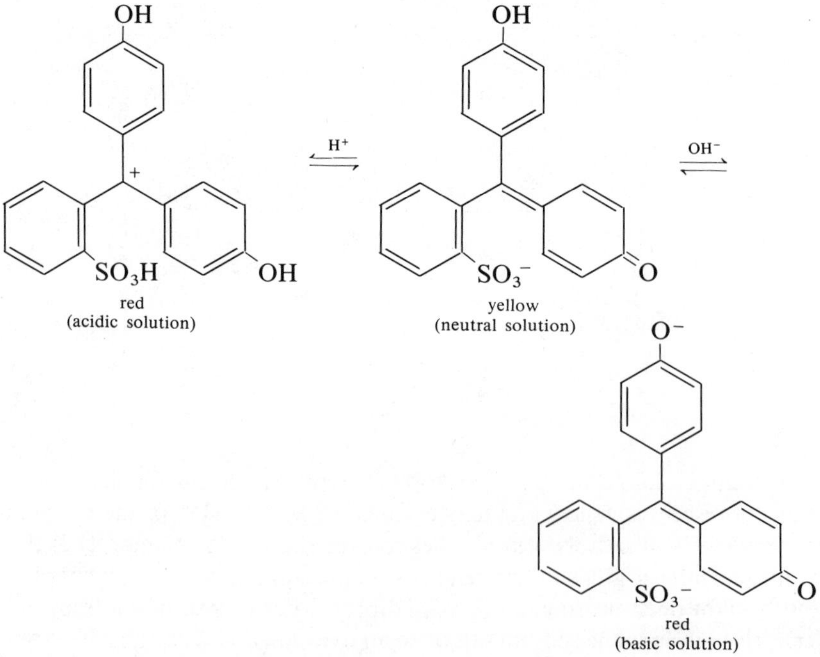

red
(acidic solution)

yellow
(neutral solution)

red
(basic solution)

If electron-withdrawing groups are added to the phenol rings both color changes occur at lower pH's; electron-releasing groups shift the color changes to higher pH's. The pH changes with different groups can be understood in terms of the relative stability of the cationic and anionic forms of the indicator.

The general method for preparing sulfonphthaleins is to heat *o*-sulfobenzoic anhydride with the appropriate phenol in the presence of a Lewis acid. The preparation of *o*-cresolsulfonphthalein described below is a typical procedure. Bromine substituents can be introduced by direct bromination of the sulfonphthaleins in acetic acid solution.

An alternate procedure for preparing sulfonphthaleins is to use the imide of *o*-carboxybenzenesulfonic acid (insoluble Saccharin) in place of *o*-sulfobenzoic anhydride. Substitution of the imide requires condensation catalysts such as sulfuric acid and more vigorous reaction conditions.

| SULFONPHTHALEIN INDICATORS | | pH AT COLOR CHANGE | ACID COLOR | BASE COLOR |
|---|---|---|---|---|
| Cresol Red | *o*-Cresolsulfonphthalein | 1.0–2.0 | Red | Yellow |
| Thymol Blue | Thymolsulfonphthalein | 1.2–2.8 | Red | Yellow |
| Meta Cresol Purple | *m*-Cresolsulfonphthalein | 1.2–2.8 | Red | Yellow |
| Bromophenol Blue | 3′,3″,5′,5″-Tetrabromo-phenolsulfonphthalein | 3.0–4.7 | Yellow | Blue |
| Bromocresol Green | 3′,3″,5′,5″-Tetrabromo-*m*-cresolsulfonphthalein | 3.8–5.4 | Yellow | Blue |
| Bromocresol Purple | 5′,5″-Dibromo-*o*-cresol-sulfonphthalein | 5.2–6.8 | Yellow | Purple |
| Bromothymol Blue | 3′,3″-Dibromothymol-sulfonphthalein | 6.0–7.6 | Yellow | Blue |
| Phenol Red | Phenolsulfonphthalein | 6.6–8.0 | Yellow | Red |
| Cresol Red | *o*-Cresolsulfonphthalein | 7.0–8.8 | Yellow | Red |
| Meta Cresol Purple | *m*-Cresolsulfonphthalein | 7.4–9.0 | Yellow | Purple |
| Thymol Blue | Thymolsulfonphthalein | 8.0–9.6 | Yellow | Blue |

## 50a Fluorescein and Eosin

**Fluorescein.** Grind together in a mortar 0.05 mole (7.4 g) of phthalic anhydride and 0.11 mole (12.1 g) of resorcinol, and transfer the mixture to a 500-ml round-bottomed flask. Heat the material to 180° (internal temperature) in an oil bath and, while stirring the mixture, add in small portions 5 g of pulverized fused zinc chloride. Take precautions to avoid contact of the anhydrous zinc chloride with atmospheric moisture, by placing the pulverized solid in a specimen tube and adding small portions from the tube.

➤**CAUTION:** Handle fluorescein and its colored derivatives with great care to avoid spilling the materials on the hands, clothing, laboratory desk and sink. The stains are extremely difficult to remove. Pour colored solutions *directly into the drain pipe of the sink.*

Continue to heat the mixture at 180–190° and stir the mass until it solidifies completely. The time required varies from 45–75 min. Remove the flask, cool, and add 150 ml of water and 8 ml of concentrated hydrochloric acid. Support the flask at an angle of about 45°, as in steam distillation, and pass water-free steam through a trap, or from a laboratory generator (see Figures 22 and 23, pages 64, 66) into the mixture for about 30 min, until the solid mass is loosened from the flask.

With a glass rod, *carefully* complete the disintegration of the mass and collect the material with suction. Crush the crude product in a mortar, transfer it to a suction filter, and wash thoroughly with water. Allow the product to dry in the air, or better, in a drying oven at 80–100°. The yield of crude fluorescein is 12–13 g. Do not attempt to determine its melting point (dec 290°).

The crude fluorescein is sufficiently pure for most purposes, and may be used directly for the preparation of eosin. Fluorescein may be purified by dissolving in warm alkalies and precipitation with acids, or better by conversion into the diacetate and hydrolysis of the pure diacetate.

**Fluorescein Diacetate.** Place 0.006 mole (2 g) of pulverized, dry fluorescein in a small flask and add 0.2 mole (21 g, 20 ml) of acetic anhydride and 1 ml of concentrated sulfuric acid. Warm the mixture *carefully* on a wire gauze with a small flame until solution is complete. Filter the hot solution through a small fluted filter and cool the filtrate. Stir with a glass rod and allow the cold solution to stand for about 30 min to allow complete separation of the crystals of the diacetate. If the product does not crystallize readily from the solution, scratch the sides of the beaker with a glass rod or introduce a seed crystal of the diacetate.

Collect the crystals with suction and wash them with a little ether to remove the adhering mother liquor. Crystallize from about 10 ml of hot acetic anhydride, collect the purified fluorescein diacetate with suction, and wash it with a small amount of ether. Allow the crystals to dry in the air. The yield is 1.0–1.2 g of purified product, which is practically colorless. The recorded melting point of the diacetate is 200°.

**Eosin (Tetrabromofluorescein).** In a 125-ml Erlenmeyer flask treat 0.015 mole (5 g) of pulverized fluorescein with 20 ml of glacial acetic acid, and add a solution of 0.075 mole (12 g, 3.8 ml) of bromine dissolved in 15 ml of glacial acetic acid. Heat the mixture (preferably in a hood) on a steam bath or in a bath of boiling water for 1 hr. The fluorescein dissolves as the soluble dibromo derivative is formed, and finally a crystalline precipitate of the tetrabromo derivative separates from the hot solution.

Pour the solution and the crystalline precipitate into 300–350 ml of cold water to which 6–8 ml of a saturated solution of sodium bisulfite has been added. Stir the mixture thoroughly and break up any lumps of the product. Collect the crude eosin with suction and wash it thoroughly with water.

Suspend the crude eosin in 50 ml of water and dissolve it as the sodium salt, by adding dilute aqueous sodium hydroxide (about 15–20 ml of a 10 percent solution is required). Pour the alkaline solution with stirring, into 300 ml of boiling water containing 5–7 ml of concentrated hydrochloric acid, and boil the mixture for a few minutes after all of the alkaline solution has been added. Cool the mixture and collect the purified eosin with suction. Wash thoroughly with water and allow to dry in the air. Eosin prepared in this way is a bright orange-colored powder. The yield is 8–9 g.

**Sodium Salt of Eosin.** In a mortar, grind thoroughly 6 g of dry eosin and 1 g of anhydrous sodium carbonate. Place the mixture in a 125-ml flask and moisten with 5 ml of ethanol. Add exactly 5 ml of water and warm gently on a steam bath, with stirring, until the evolution of carbon dioxide ceases. Add 25 ml of ethanol, heat to boiling, and filter the hot solution through a fluted filter. Upon cooling and allowing to stand, the sodium salt of eosin crystallizes from the solution in the form of green-red iridescent needles. The powdered crystals are dark-red in color. The sodium salt sometimes separates very slowly from the solution; it is advantageous to cool the solution and to scratch the sides of the flask with a glass rod or to introduce a seed crystal. Collect the crystals with suction and wash them with a little ethanol. The yield is 4–5 g.

## 50b *o*-Cresol Red

In a 15-ml test tube place 1 g of freshly fused and pulverized zinc chloride, 0.0055 mole (1 g) of *o*-sulfobenzoic anhydride and 0.015 mole (1.6 g) of *o*-cresol. Mix the reactants thoroughly with a stirring rod and then heat the mixture in a heating bath at 145° (bath temperature) for 1 hr. At the end of the heating period transfer the deep-red mass to a 250-ml Erlenmeyer flask with the aid of several small portions of 10 percent aqueous sodium hydroxide (use a total of about 50 ml of solution). Acidify the resulting wine-colored solution with concentrated hydrochloric acid until the product precipitates as dark crystals (about 12 ml of acid). During the addition of acid zinc salts will precipitate and then redissolve.

If the material in the flask remains liquid from unreacted *o*-cresol boil the solution in the hood until the volume has been reduced by a half. The unreacted *o*-cresol will steam distill and the product should solidify. If any oily material remains add more water and repeat the steam distillation.

Collect the iridescent, green crystals with suction and allow them to air dry. The yield is about 1–1.5 g. Do not attempt to determine the melting point.

## Questions

**1.** Write equations showing a method of preparing a triphenylmethane dye, such as Malachite Green or Crystal Violet.

**2.** What is meant by the term *leuco base*? Write the structural formula of leuco-eosin (formed by the action of zinc dust on eosin in the presence of sodium hydroxide solution).

**3.** Explain the color change that occurs when a phthalein, such as phenolphthalein or fluorescein, is treated with sodium hydroxide solution.

**4.** Explain the fact that fluorescein diacetate does not give a similar color change when treated with sodium hydroxide solution.

**5.** Compare the structure of phenolphthalein with the structure of phenol red and comment on the colors of these two indicators at high pH (dianion forms).

**6.** Phenolphthalein is colorless at pH 6 but phenol red is yellow. Suggest a reason for this difference in color.

**7.** Why does bromophenol blue change color at lower pH than phenol red?

**8.** What color would be expected for the tetrabromophenolphthalein dianion (high pH form)?

EXPERIMENT

# 51

# Hippuric Acid

Hippuric acid is normally present in considerable quantities in the urine of herbivorous animals but is present in very much smaller quantities in normal human urine. The average excretion of hippuric acid is about 0.7 g per day for an adult man. Hippuric acid is formed in the body by synthesis from benzoic acid and glycine. Fruits (especially plums, prunes, and cranberries) increase the hippuric acid content of the urine on account of their relatively high content of benzoic acid. The output of hippuric acid is decreased in fevers and in certain kidney disorders where the synthetic activity of the kidneys is diminished.

In order to increase the amount of hippuric acid in the urine so that it may be isolated conveniently, sodium or ammonium benzoate may be taken by mouth, either mixed with foodstuffs or in the form of an aqueous solution. The urine will then be found to contain much more hippuric acid than is normally present.

In the laboratory hippuric acid and other acylamino acids may be prepared conveniently by acylation of an amino acid with an aroyl chloride in the presence of aqueous sodium hydroxide solution, by the Schotten-Baumann method. Acetyl derivatives may be prepared by acetylation with acetic anhydride in aqueous medium, the Lumière-Barbier method. In this way glycine gives aceturic acid.[1]

In the presence of acetic anhydride and sodium acetate, aromatic aldehydes react with hippuric acid, or aceturic acid, to give azlactones (substituted oxazolones, I). The latter upon mild alkaline hydrolysis give α-acylaminocinnamic acids (II), which by reduction and hydrolysis furnish

[1] Herbst and Shemin, *Organic Syntheses*, Collective Volume **II**, 11 (1943).

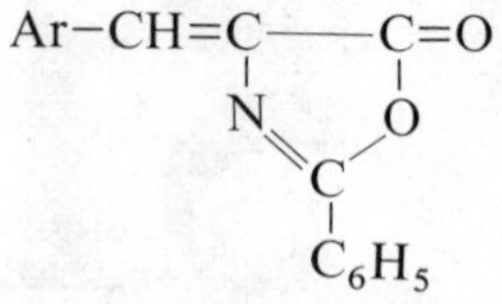

$$\text{Ar-CH=C}\overset{}{\text{—}}\text{C=O (ring: N, O, C–}C_6H_5\text{)} \quad \text{I}$$

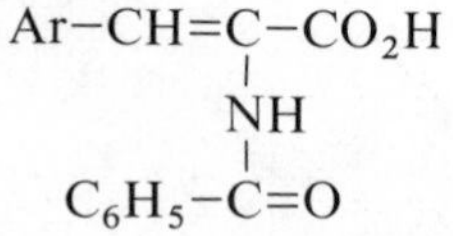

$$Ar{-}CH{=}C(NH{-}C(=O){-}C_6H_5){-}CO_2H \quad \text{II}$$

$$Ar{-}CH_2{-}CH(NH_2){-}CO_2H \quad \text{III}$$

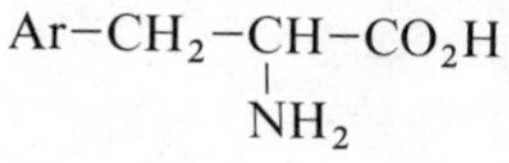

α-amino acids (III), such as phenylalanine and tyrosine. This sequence is known as the Erlenmeyer amino acid synthesis.[2]

## 51a Synthesis in the Human Body

**Ingestion of Benzoates.**[3] Before retiring at night, ingest 5 g of pure sodium benzoate (or ammonium benzoate), by dissolving the solid in about half a glassful of water and taking the solution by mouth. Collect the overnight urine voided the next morning and isolate the hippuric acid as described in the next section. The urine should be preserved from decomposition by adding about 10 ml of toluene if the isolation of hippuric acid is not to be started within the next 12 hr. If toluene is used, it should be separated before proceeding to the next step.

**Isolation of Hippuric Acid (Method of Roaf).** Measure the volume of urine, and for every 100 ml add 25 g of pulverized solid ammonium sulfate and 1.5 ml of concentrated sulfuric acid. Stir thoroughly until the ammonium sulfate has dissolved, and set aside for several hours in a cool place or in an ice bath until the hippuric acid has separated as completely as possible. It is advantageous to allow the solution to stand in a cool place overnight, since the hippuric acid separates slowly.

Collect the crystals with suction and wash with a little ice water. Transfer the crystals to a beaker and purify by crystallization from about 50 ml of hot water, with the addition of 1–2 g of decolorizing carbon. To aid in the removal of colored impurities it is advantageous to boil the hot solution for about 10 min with the decolorizing agent before filtering the solution. Allow the filtrate to stand for at least an hour, and cool in an ice bath before filtering the crystals of purified hippuric acid. Calculate the percentage yield of hippuric acid isolated, based upon the amount of benzoate ingested.

### Questions

**1.** What is the source of the glycine used in the body for the synthesis of hippuric acid?

[2] Carter, *Organic Reactions*, **3**, 205 (1946); Gillespie and Snyder, *Organic Syntheses*, Collective Volume **II**, 489 (1943).

[3] Students who are vegetarians or who have dietary or kidney disorders should not undertake this experiment.

**2.** How may hippuric acid be prepared in the laboratory from glycine?
**3.** What products are formed by hydrolysis of hippuric acid with alkali?
**4.** Suggest an explanation for the fact that benzoic acid is converted to hippuric acid in the body.
**5.** Compare in a general way, the synthetic processes that occur in animals, with the synthetic processes that occur in plants (especially in the green plants).

## 51b Benzoylation of Glycine

In a 250-ml Erlenmeyer flask dissolve 0.1 mole (7.5 g) of glycine in 0.3 mole (105 ml) of 10 percent aqueous sodium hydroxide. Add 0.11 mole (15.5 g, 13 ml) of benzoyl chloride (*caution—irritating vapor!*), stopper the flask firmly and swirl the mixture vigorously until the benzoyl chloride passes into solution. Cautiously release the internal pressure from time to time by carefully removing the stopper. The reaction mixture *must remain alkaline*; test with litmus and if necessary add more sodium hydroxide solution.

➤CAUTION: The vapors of benzoyl chloride are very irritating to the eyes. The flask in which this material is stored should not be rinsed out into the sink until all traces of benzoyl chloride have been destroyed with a mixture of ammonia and ethanol.

When there is no longer an odor of benzoyl chloride (sniff cautiously), transfer the solution to a beaker and rinse the flask with a small amount of water. If any precipitate is present in the alkaline solution remove this by filtering with suction (discard the precipitate and save the solution). Precipitate the crude hippuric acid (contaminated with benzoic acid) by pouring the clear alkaline solution with vigorous stirring *into* a mixture of 150 ml of concentrated hydrochloric acid and 50 g of ice, contained in a large beaker. Cool if necessary, collect the crystals with suction, and press them as dry as possible on the filter. Wash the crystals with about 10 ml of cold water, and again press them as dry as possible. To remove most of the benzoic acid, wash the crystals with several small portions of ether (*caution—flammable solvent!*) in the following way. Release the suction and carefully loosen the crystalline cake with a spatula or a flattened glass rod. Add the ether in small amounts and stir into a paste. Continue to add small quantities of ether and mix the solvent thoroughly with the crystals. When a thorough mixing has been obtained, apply suction and press the crystals as free as possible of ether. Repeat this procedure two or three times with fresh portions of ether. It is of great importance in washing crystals to stop the suction and to break up the crystalline cake so that the whole mass of crystals will actually come into contact with the washing liquid.

Crystallize the partly purified hippuric acid from hot water, with the addition of 1–2 g of decolorizing carbon, and allow the crystals to dry. The yield is 6–9 g. The melting point of hippuric acid is 187–188°.

**2-Phenyl-4-benzal-5-oxazolone (I)** *(Optional)*. In a large test tube (25 × 125 mm) place 0.02 mole (2.1 g, 2 ml) of benzaldehyde, 0.02 mole (3.6 g) of dry, pulverized hippuric acid, 0.02 mole (1.6 g) of pulverized, *fused* sodium acetate, and 0.06 mole (6.5, 6 ml) of acetic anhydride. Warm the tube gently with *constant shaking*, in the luminous flame of a burner. The mixture at first is almost solid but slowly becomes fluid and turns yellow. After the mixture has liquefied completely, stopper the tube *loosely* with a cork and heat it in a beaker of boiling water for 1 hr. During this time the azlactone begins to separate. Remove the tube from the heating bath and add *cautiously*, 15 ml of ethanol, with cooling and shaking.

After chilling thoroughly collect the crystals with suction and wash them with two 5-ml portions of ice-cold ethanol, followed by a similar washing with hot water. The crude product after drying weighs about 3 g and is fairly pure; mp 165–166°. The azlactone may be recrystallized from 5–6 ml of benzene but need not be purified for the next step.

**α-Benzoylaminocinnamic Acid (II)** *(Optional)*. In a 125-ml Erlenmeyer flask place 0.01 mole (2.6 g) of the azlactone, 25 ml of water and 5 ml (0.012 mole) of 10 percent aqueous sodium hydroxide. Heat the mixture just barely to the boiling point and swirl it occasionally until all of the azlactone has dissolved. Add a little water if necessary to maintain the original volume. Do not prolong the heating unnecessarily or boil the mixture vigorously as this may cause further hydrolysis to the arylpyruvic acid ($Ar{-}CH_2COCO_2H$).

Filter the hot solution and pour the warm filtrate into 12 ml of concentrated hydrochloric acid diluted with about 30 g of ice. Collect the colorless crystals of α-benzoylaminocinnamic acid with suction and wash them with a little cold water. The product is quite pure but melts with decomposition over a wide range (~215–225°).

By reduction with hydriodic acid and phosphorus this product furnishes phenylalanine (III).[2]

## Questions

**1.** How may glycine be prepared from acetic acid? from formaldehyde?

**2.** Alanine ($CH_3{-}CH(NH_2){-}CO_2H$), like the typical natural amino acids, occurs in proteins in an optically active form. How could synthetic *dl*-alanine be resolved into *d*- and *l*-enantiomorphs?

**3.** Show the steps in the conversion of an azlactone to an arylpyruvic acid, $Ar{-}CH_2{-}CO{-}CO_2H$, by vigorous hydrolysis with sodium hydroxide solution.

**4.** What is a peptide linkage? Illustrate by writing the formulas of glycylalanine and alanylglycine.

**5.**. How could you distinguish between the two isomeric dipeptides mentioned in question 4?

**6.** A useful acylating agent for the preparation of peptide derivatives is prepared by treatment of an acylamino acid, $R{-}CO{-}NH{-}CH_2CO_2H$, with triethylamine and ethyl chloroformate ($Cl{-}CO_2Et$). Show the structure of this type of acylating agent and explain how it reacts with an amino ester, $H_2N{-}CHR'{-}CO_2Et$.

EXPERIMENT

# 52

# The Skraup Synthesis

Quinoline is a nitrogen heterocycle, 2,3-benzopyridine, having a condensed aromatic-type ring system similar to naphthalene. Quinoline and its derivatives can be prepared from arylamines by a variety of ring closure reactions. A simple and convenient method is the Skraup synthesis, in which quinoline is obtained by heating aniline with glycerol and sulfuric acid in the presence of a mild oxidizing agent. Nitrobenzene is a particularly suitable oxidizing agent because it is reduced in the reaction mixture to aniline, which becomes available for the formation of an additional quantity of quinoline.[1] The reaction is strongly exothermic and may become violent unless moderated by the addition of a little ferrous sulfate. Boric acid is introduced to lessen side reactions brought about by the hot sulfuric acid.

The Skraup synthesis involves a sequence of several steps:[2] (a) dehydration of glycerol by means of hot sulfuric acid to produce acrolein; (b) 1,4-addition of aniline to the conjugated system of acrolein; (c) ring closure of the aniline-acrolein adduct, with elimination of water, to form dihydroquinoline; (d) dehydrogenation of the dihydro derivative by nitrobenzene to furnish quinoline (and aniline). At the completion of the reaction quinoline and aniline are present as their acid sulfate salts, together with unused nitrobenzene. The latter is removed readily by steam distillation from the strongly acidic mixture. Following this, the mixture is made strongly alkaline and a second steam distillation is performed to obtain the quinoline and aniline.

[1] The aniline formed in the reaction from nitrobenzene must be taken into account in calculating the theoretical yield. The stoichiometry of the reaction works out so that two molecules of aniline and one molecule of nitrobenzene, with sufficient glycerol and sulfuric acid, can furnish three molecules of quinoline.

[2] Manske and Kulka, "The Skraup Synthesis of Quinolines," *Organic Reactions*, **7**, 59 (1953).

$$HOCH_2{-}CH(OH){-}CH_2OH \xrightarrow[H_2SO_4]{-H_2O} [HOCH_2{-}CH{=}CH{-}OH] \xrightarrow[H_2SO_4]{-H_2O} CH_2{=}CH{-}CH{=}O$$

$$C_6H_5NH_2 + O{=}CH{-}CH{=}CH_2 \longrightarrow C_6H_5NH{-}CH_2{-}CH{=}CH{-}OH \xrightarrow{-H_2O} \text{1,2-dihydroquinoline}$$

$$\text{1,2-dihydroquinoline} \; (+\, C_6H_5NO_2) \longrightarrow \text{quinoline} \; (+\, C_6H_5NH_2)$$

Although quinoline (bp 237°) and aniline (bp 185°) differ appreciably in boiling point, it is difficult to achieve a complete separation by fractional distillation. Chemical methods based upon differences in reactivity are more effective and more convenient. By warming the mixture with an arylsulfonyl chloride, aniline is converted to the arylsulfonanilide and unaltered quinoline can be removed by extraction with dilute acid.[3] Another chemical method consists in diazotizing the aniline by means of sodium nitrite and aqueous sulfuric acid, hydrolyzing the diazonium salt to form phenol, and removing the phenol by extraction with aqueous alkali.

The Skraup synthesis can be applied to a number of ring-substituted anilines (toluidines, anisidines, etc.) for the preparation of quinoline derivatives having substituents in the benzene ring. Thus, *p*-toluidine (with *p*-nitrotoluene instead of nitrobenzene) yields 6-methylquinoline, and *o*-toluidine (with *o*-nitrotoluene) gives 8-methylquinoline. If the nitro compound corresponding to the amine is not available, arsenic acid may be used as oxidizing agent in its place.

Quinoline derivatives with alkyl groups in the pyridine ring can be synthesized from aniline and various aliphatic compounds. Aniline and acetaldehyde on heating with hydrochloric acid yield quinaldine (2-methylquinoline). Quinoline, quinaldine, and lepidine (4-methylquinoline) react with alkyl halides to give quaternary ammonium salts used as starting materials for the preparation of an important class of dyes, the cyanine and isocyanine dyes. These are excellent sensitizers for the preparation of panchromatic photographic plates.

One of the most important medicinal alkaloids, quinine, so widely used in the treatment of malaria, is a quinoline derivative. Quinine and cinchonine

[3] Quinoline may be separated from aniline also by conversion to the sparingly soluble zinc chloride double salt; see Gattermann-Wieland, *Die Praxis des organischen Chemikers*, 40th edition, Walter de Gruyter and Co., Berlin (1961), page 318.

are the principal alkaloids of a related group of natural bases occurring in the bark of the cinchona tree. The name quinoline was derived from its relation to the quinine (cinchona) alkaloids.

## Quinoline

Introduce into a 1-liter round-bottomed flask, in the sequence mentioned, 0.025 mole (7 g) of finely-powdered ferrous sulfate crystals (heptahydrate), 0.2 mole (19 g, 19 ml) of aniline, and 0.12 mole (15 g, 12.5 ml) of nitrobenzene. Dissolve 0.2 mole (12 g) of boric acid ($H_3BO_3$) in 0.82 mole (75 g, 60 ml) of anhydrous glycerol by heating gently, cool the solution to 20°, and pour it into the reaction flask. Mix the materials thoroughly by swirling and add 0.65 mole (65 g, 35 ml) of concentrated sulfuric acid. It is important to have the reactants very thoroughly mixed before applying heat.

➛**CAUTION:** Nitrobenzene, aniline, and quinoline are toxic substances and must be handled carefully. Avoid breathing their vapor or spilling the liquids. If any of these materials comes in contact with the skin, wash it off promptly with a little ethanol and cleanse the skin thoroughly with soap and water.

Attach a reflux condenser, add a boiling chip, and heat the mixture on a wire gauze over a small flame, with frequent swirling, until the boiling point is reached. Withdraw the burner until the vigorous reaction begins to subside, then continue to heat the mixture with the flame adjusted so the material refluxes gently for a period of 4–5 hr. Longer heating, up to 8–10 hr, will improve the yield appreciably. During the heating period arrange apparatus for steam distillation to be attached to the flask.

After allowing the reaction mixture to cool slightly, dilute it with an equal volume of water, and steam distill until all unreacted nitrobenzene has been removed (discard the distillate). Cool the contents of the flask and add *slowly* and *cautiously*, from a separatory funnel, a cold solution of 75 g of sodium hydroxide in 125 ml of water. During the neutralization cool the flask frequently in cold water. Steam distill the strongly alkaline mixture until no more oily drops appear in the distilling liquid (usually about 650 ml of distillate will be collected). Discard the mixture remaining in the distillation flask.

Separate the oily layer of crude quinoline in the distillate and extract the aqueous portion with two 40-ml portions of benzene.[4] Combine the

[4] Instead of removing aniline by means of toluenesulfonyl chloride as described in the experiment, it can be eliminated by the diazotization method, in the following way. Separate the crude quinoline in the steam distillate and dissolve it in a solution of 24 g (13 ml) of concentrated sulfuric acid in 15 ml of water. Cool the solution and add 10 percent aqueous sodium

crude quinoline and the benzene extracts, add 4 g of *p*-toluenesulfonyl chloride (or benzenesulfonyl chloride), and heat the solution for 0.5 hr on a steam bath. Pour the mixture into 75 ml of iced water and add concentrated hydrochloric acid until Congo red test paper shows an acid reaction.[5] Usually about 10–15 ml of acid is required. Separate the aqueous layer (which contains the quinoline hydrochloride) and wash it with two 10-ml portions of benzene. Save the aqueous layer and discard the benzene washings.

Render the aqueous solution strongly alkaline by adding 40 percent aqueous sodium hydroxide (prepared by dissolving 10 g of sodium hydroxide in 15 ml of distilled water), with stirring. Remove the quinoline by extracting the mixture twice with benzene, using 50 ml for the first extraction and 25 ml for the second. Dry the benzene extracts with 5–10 g of anhydrous magnesium sulfate, separate the drying agent, and distill off the benzene under ordinary pressure. Transfer the residual quinoline to a small flask[6] and distill under diminished pressure; bp 110–115°/14 mm, 118–120°/20 mm, 132°/40 mm. The yield is 16–17 g.

## Questions

**1.** Write all of the steps involved in the Skraup synthesis and explain the function of the nitrobenzene and of the ferrous sulfate.

**2.** What is the probable function of the boric acid?

**3.** Explain the use of *p*-toluenesulfonyl chloride in removing unchanged aniline (see Experiment 12).

**4.** How was the unchanged nitrobenzene removed from the crude quinoline?

**5.** What product would result if the Skraup synthesis were carried out using: 2-naphthylamine? *o*-phenylenediamine (1,2-diaminobenzene)?

nitrite solution dropwise, with stirring, until a drop of the solution gives an *immediate* blue color with starch-iodide test paper. After standing for 0.5 hr, make alkaline with 40 percent aqueous sodium hydroxide (*caution—with cooling*) and steam distill. When the distillate is clear, discard the reaction mixture in the flask, separate the quinoline in the distillate and extract the aqueous solution with three 25-ml portions of benzene. Combine the quinoline and benzene extracts, dry over 5–10 g of anhydrous magnesium sulfate, separate from the drying agent, distill off the benzene at ordinary pressure and distill the quinoline, preferably under diminished pressure.

[5] Congo red test paper may be prepared by dipping strips of filter paper into a solution of Congo red and allowing to dry. Congo red shows a *blue* color with mineral acids; it is not affected by weak acids or acidic salts. Quinoline hydrochloride shows an acid reaction to litmus but not to Congo red.

[6] Quinoline may be distilled without appreciable decomposition at atmospheric pressure (bp 236–240°, cor.), using an ordinary distilling apparatus and an air-cooled condenser.

EXPERIMENT

# 53

# Fumaric Acid from Furfural

## Oxidation with Vanadium Pentoxide Catalyst

Vanadium pentoxide ($V_2O_5$)[1] is a valuable catalyst for oxidations in the vapor phase and in liquid media. Air oxidation of naphthalene or *o*-xylene in the vapor phase at 350° affords the commercial routes to phthalic anhydride; in a similar way, benzene or crotonaldehyde can be oxidized to maleic anhydride. In liquid phase oxidations vanadium pentoxide has been used in conjunction with aqueous nitric acid or sodium chlorate (as in the present experiment). With 50 percent nitric acid cyclohexanol can be oxidized to adipic acid in 70 percent yield and cyclopentanone to glutaric acid in 80–85 percent yield.[2] In 90 percent aqueous acetic acid as the medium anthracene can be oxidized to anthraquinone in 90 percent yield by sodium chlorate and vanadium pentoxide.[3]

**Preparation of Vanadium Pentoxide Catalyst.** Suspend 1 g of finely ground C.P. ammonium metavanadate[1] in 10 ml of water and add slowly 1.5 ml of concentrated hydrochloric acid. Allow the reddish-brown, semi-colloidal precipitate to settle thoroughly (preferably overnight) and decant the supernatant solution. Add 10 ml of water, shake thoroughly, and then allow to

[1] Vanadium pentoxide and ammonium metavanadate can be purchased from Will Scientific, Incorporated (Rochester, N.Y. 14603). If the catalyst is to be prepared from ammonium metavanadate as described above, it should be noted that this process requires several days' standing and should be started in ample time ahead. The procedures for the oxidation and the preparation of the catalyst are essentially those of Milas [*Organic Syntheses*, Collective Volume **II**, 302, note 2 (1943)].

[2] Foster, *Organic Syntheses*, Collective Volume I, 19; Allen and Ball, *ibid.*, 290 (1941).

[3] Underwood and Walsh, *ibid.*, II, 554 (1943).

stand at room temperature for two or three days. This treatment renders the precipitate granular and easier to filter. Collect the precipitated vanadium oxide on a suction filter and wash several times with cold 5 percent aqueous sodium chloride to remove hydrochloric acid. Do not wash with pure water or the oxide may become colloidal. Dry the product at 120° for 12 hr, grind to a fine powder and dry again at 120° for 12 hr. The yield of catalyst is about 0.5 g, sufficient for two students.

**Oxidation of Furfural.** Fit a 500-ml round-bottomed flask with a Claisen adapter bearing a reflux condenser and a small separatory funnel. In the flask place a solution of 0.42 mole (45 g) of sodium chlorate ($NaClO_3$) in 100 ml of water, add 0.2–0.3 g of vanadium pentoxide catalyst,[1] and heat the flask to 70–75° in a water bath. From a quantity of 0.2 mole (19 g, 18 ml) of furfural (a technical grade is satisfactory) measure 2–3 ml and add this small portion to the oxidizing solution. Shake the flask and *wait for the initial oxidation to occur* before adding more furfural. The induction period varies from 5–20 min (occasionally longer).[4]

From the separatory funnel add the remainder of the furfural *drop by drop*, in small quantities, with swirling after each addition. During the reaction the temperature of the solution rises to 100–105°. The rate of addition should be such that the vigorous reaction is maintained and all of the furfural is introduced in 50–60 min. If the aldehyde is added too rapidly the reaction becomes violent. It is important to insure that each portion of the aldehyde has reacted before adding the next, but the oxidation must not be slowed down too much or the yield of fumaric acid is reduced.

After the addition of furfural is completed, heat the reaction mixture at 70–75° for 3 hr with occasional shaking, and then allow the mixture to stand overnight. Collect the crystalline fumaric acid with suction, wash it sparingly with cold water, and press thoroughly on the filter. Crystallize it from hot water and allow the crystals to dry in the air. Do not attempt to determine the melting point of fumaric acid unless special apparatus is available for high temperatures; fumaric acid melts in a sealed tube (to avoid sublimation) at 282–284°. The yield is 8–10 g.

A small additional amount of crude fumaric acid (1–2 g) may be obtained by adding 5 ml of concentrated hydrochloric acid to the filtrates, concentrating to 70 ml and cooling to 10–15°.

[4] It is essential *for safety* and for the success of the preparation to insure that the catalyst is active and that the first portion of furfural has reacted, before proceeding further. Occasionally specimens of vanadium pentoxide have little or no catalytic activity; if such material is encountered the experiment should be abandoned, since a sudden violent oxidation may occur if unreacted furfural is allowed to accumulate in the oxidizing mixture.

## Questions

1. How is furfural prepared commercially? Does it resemble aliphatic or aromatic aldehydes in its general reactions? Explain.
2. Write equations for two important commercial examples of vapor phase oxidations using vanadium pentoxide catalyst.
3. What type of isomerism is shown by maleic and fumaric acids? Cite experimental evidence used in establishing their configurations.

# EXPERIMENT 54

# Wagner-Meerwein Rearrangements

An implicit guiding principle used in predicting products of organic reactions is that minimum structural change occurs in the reaction. Important exceptions to the principle are those carbonium ion reactions in which an alkyl or aryl group adjacent to the developing positive charge migrates to the positive carbon atom and gives rise to products with rearranged carbon skeletons (Wagner-Meerwein rearrangement).

$$R-\underset{R}{\overset{R}{C}}-\underset{H}{\overset{H}{C}}-X \longrightarrow \left[R-\underset{R}{\overset{R}{C}}-\underset{H}{\overset{H}{C^{+}}}\right] \longrightarrow \left[R-\underset{+}{\overset{R}{C}}-C\begin{smallmatrix}H\\-H\\R\end{smallmatrix}\right] \longrightarrow \text{products}$$

The confusion that can arise from an unsuspected Wagner-Meerwein rearrangement is amplified in many of the reactions of bicyclic molecules because the rearranged products may retain the same bicyclic structure and differ from the non-rearranged products only in the position of substituents. The commercial conversion of camphene to isobornyl acetate is a classic example of a reaction of a bicyclic molecule proceeding with a Wagner-Meerwein rearrangement.[1]

[1] In studying this reaction the student is urged to prepare ball and stick models of each stage. It is difficult to see from two dimensional drawings how the transfer of one bond causes the movement of so many groups.

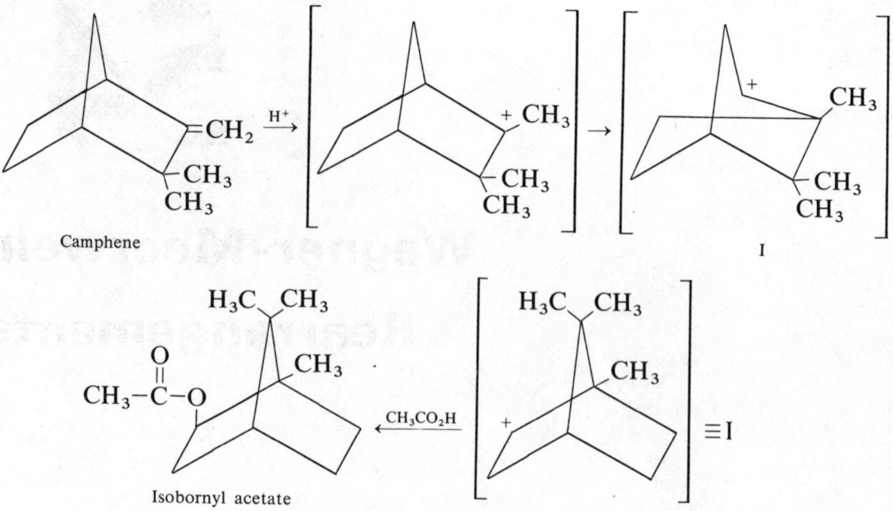

Camphene $\xrightarrow{H^+}$ [carbonium ion] $\rightarrow$ [I]

Isobornyl acetate $\xleftarrow{CH_3CO_2H}$ [carbonium ion] $\equiv$ I

The product formed is a secondary acetate rather than a tertiary acetate that might have been expected from the structure of camphene.

The remaining steps of the synthesis of camphor do not involve Wagner-Meerwein rearrangements since neither the saponification of isobornyl

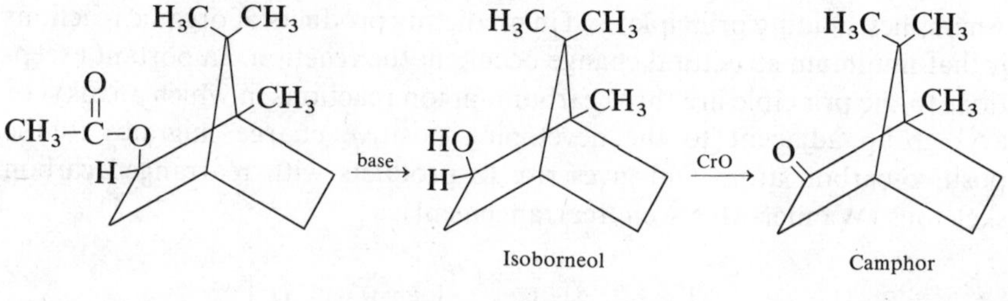

Isobornyl acetate $\xrightarrow{base}$ Isoborneol $\xrightarrow{CrO}$ Camphor

acetate to isoborneol nor the chromic acid oxidation of isoborneol to camphor produces a carbonium ion in the bicyclic ring.

## Camphor from Camphene

**Isobornyl Acetate.** In a 250-ml flask dissolve 0.25 mole (30 g) of camphene in 1.25 mole (75 g, 70 ml) of glacial acetic acid and add a solution of 1 ml of concentrated sulfuric acid in 2 ml of water. Warm the flask on a steam bath at 90–95° for 15 min with frequent swirling. Add 50 ml of water, mix well, and allow to cool.

Transfer the material to a separatory funnel and rinse the flask with a little water. Separate the ester layer, wash it well first with water and then with 10 percent aqueous sodium carbonate and dry it over calcium chloride. The crude product is suitable for conversion to isoborneol without further purification. The yield is about 35 g. If a purer material is desired, distill the crude product under reduced pressure and collect the fraction boiling between 106–107° at 15 mm.

**Isoborneol.** In a 250-ml round-bottomed flask add 0.125 mole (24 g) of isobornyl acetate to a solution of 0.15 mole (10 g of 85 percent pure solid) potassium hydroxide in 50 ml of ethanol and 15 ml of water, and heat the mixture under reflux on a steam bath for 1 hr. Pour the solution slowly onto 100 g of ice contained in a 600-ml beaker. Stir the mixture for several minutes until the isoborneol solidifies. Collect the solid on a Büchner funnel, wash it well with cold water and press it dry. The yield of crude isoborneol is about 20 g. The crude product is sufficiently pure for oxidation to camphor.

**Camphor.** In a 500-ml round-bottomed flask prepare a solution of 0.1 mole (15 g) of isoborneol in 25 ml of dry acetone and cool the solution in a beaker of ice-water to 15–25°. Add to the solution dropwise 0.054 mole (20 ml) of Jones' reagent.[2] Allow the reaction mixture to stand for 0.5 hr after the addition has been completed.

Add 200 ml of water to the distilling flask, attach a water-cooled condenser and distill about 50 ml of distillate. This first portion of distillate contains most of the organic impurities and should be discarded. Continue the distillation into a chilled receiver until no more product is collected. If the camphor collects in the condenser to the point where there is danger that the condenser may become plugged,[3] the distillation should be interrupted and the accumulation of camphor pushed out with a long glass rod.[4] The solid product is collected on a Büchner funnel and pressed dry. The yield is about 10 g. Camphor may be purified by sublimation (see Figure 32, page 93). Pure camphor has a melting point of 176°.[5]

[2] Jones' reagent is prepared by dissolving 27 g of chromium trioxide in 23 ml of concentrated sulfuric acid, followed by cautious dilution with water to 100 ml.

[3] A warning that this dangerous state is being approached comes when the camphor is not completely condensed during passage through the condenser.

[4] An alternate isolation procedure that avoids this hazard is to stop the distillation after the organic impurities have been removed and collect the solidified camphor by filtration. The product should be washed well with cold water.

[5] Because camphor has an unusually high change of melting temperature with pressure the observed melting point in an open capillary is several degrees below the melting point taken in a sealed tube. For the same reason, the melting point is particularly sensitive to traces of impurities and the melting point of mixtures of camphor with known amounts of foreign materials is useful in determining molecular weights (Rast method).

## Questions

**1.** How many asymmetric carbon atoms are present in camphor? How many optically active forms of camphor exist?

**2.** Explain why dehydration of borneol with aqueous acid gives camphene.

**3.** Explain why optically active *exo*- and *endo*-bicyclo[2.2.1]heptanol-2 (norborneol) undergo racemization when treated with acids.

# EXPERIMENT 55

# The Diels-Alder Diene Synthesis

One of the most interesting synthetic reactions of unsaturated compounds is the 1,4-addition of a conjugated diene to a molecule containing an active ethylenic or acetylenic bond (the dienophile), to form an adduct having a six-membered unsaturated ring. This cyclization process, known as the Diels-Alder reaction or diene synthesis,[1] is of exceedingly broad scope and has been applied to syntheses of important medicinal products, insecticides, terpene derivatives and intermediates for the manufacture of industrial chemicals.

$$CH_2{=}CH{-}C(CH_3){=}CH_2 + EtO_2C{-}CH{=}CH{-}CH_3 \longrightarrow \text{Adduct}$$

Diene | Dienophile | Adduct

Diene reactants include alkyl, halogen, and alkoxy derivatives of 1,3-butadiene, and also cyclic 1,3-dienes such as cyclopentadiene, 1,3-cyclohexadiene, and some terpenes (α-terpinene, α-phellandrene). Furan and a number of furan derivatives, and the inner ring of anthracene, also participate in the reaction as dienes. In general, the presence in the diene of substituents that facilitate electron-release favors the reaction.

The most typical dienophiles are α,β-unsaturated carbonyl compounds and nitriles, in which the β-carbon atom is activated by conjugation with an

[1] For reviews of the Diels-Alder reaction, see Kloetzel, *Organic Reactions*, **4**, 1 (1948); Holmes, **4**, 60 (1948); Butz and Rytina, **5**, 136 (1949). Dienophiles have been reviewed by Velluz, *Substances Naturelles de Synthèse*, **5**, 141 (Masson et Cie, Paris, 1959).

electron-withdrawing group. Examples are acrolein, *p*-benzoquinone[2] and 1,4-naphthoquinones, maleic anhydride, esters of acetylenedicarboxylic acid, acrylic esters and acrylonitrile. Under vigorous conditions vinyl ethers and halides, and even ethylene and acetylene, can be made to act as dienophiles toward the more reactive dienes.

The Diels-Alder reaction is a stereospecific *cis-cis* type of addition: the diene is obliged to assume a *cis* conformation to permit ring closure and it is observed that the configuration of substituents (*cis* or *trans*) in the dienophile is retained in the adduct. Cyclic dienes such as cyclopentadiene form adducts with maleic derivatives or quinones, in which the ring system for the dienophile has a *trans* (*endo*) relationship to the methylene bridge.

Active dienophiles are useful reagents for detecting the presence of a conjugated diene system and for analytical purposes. Diels-Alder adducts have been of value in establishing the structure of 1,3-dienes and in characterizing known dienes. N-Phenylmaleimide (maleanil, I) is a convenient reagent for identification purposes since it forms crystalline adducts that can be isolated and purified readily.

$$CH_2{=}CH{-}CH{=}CH_2 + \underset{\text{I}}{HC{=}CH(CO)_2N{-}C_6H_5} \longrightarrow \text{adduct } N{-}C_6H_5$$

The central ring of anthracene functions as a diene in Diels-Alder reactions even though its two double bonds participate formally in an aromatic structure. A striking example is the condensation of *p*-benzoquinone with anthracene to yield the bridged polycyclic system, II, containing two benzene rings and a cyclohexene-1,4-dione ring.[3] The dione ring corresponds structurally to the unstable diketone tautomer of hydroquinone. Treatment of the adduct with either acids or bases transforms it rapidly to the more stable, aromatized hydroquinone form, III, which is dihydroxytriptycene. The hydroxyl functions of III can be removed by several means[4] to yield the highly symmetric triptycene, IV.

Benzyne, $C_6H_4$, is a synthetically efficient dieneophile since with it a benzene ring can be introduced in a single step. Benzyne is extremely unstable (half-life of less than $10^{-4}$ sec in the gas phase) and is formed sluggishly

[2] The use of benzoquinone for the synthesis of naphthoquinones and anthraquinones is mentioned in Experiment 43, page 395.

[3] E. Clar, *Ber.*, **64**, 1676 (1931).

[4] P. D. Bartlett, M. J. Ryan, and S. G. Cohen, *J. Amer. Chem. Soc.*, **64**, 2649 (1942); A. C. Craig and C. F. Wilcox, Jr., *J. Org. Chem.*, **24**, 1619 (1959).

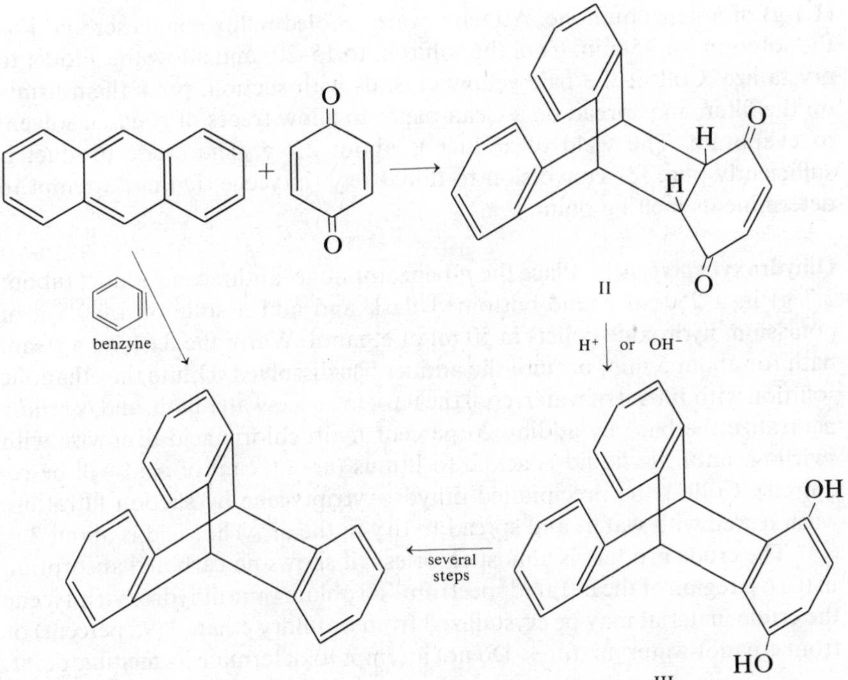

when conventional elimination reagents and reactions are used. Substances that yield benzyne at a practical rate under normal reaction conditions tend to be unstable and potentially explosive. An ingenious solution to the dual requirement of ready formation from relatively stable precursors was developed by L. Friedman[5] who introduced the use of benzenediazonium-2-carboxylate hydrochloride. In alcoholic solutions containing propylene oxide the hydrogen chloride of this stable double salt irreversibly cleaves the epoxide to yield propylene chlorohydrin and the remaining benzene-diazonium-2-carboxylate (which is a violent explosive when isolated in the dry state) decomposes spontaneously *in situ* by evolution of nitrogen and carbon dioxide to yield benzyne. If the cleavage and decomposition are carried out in the presence of a diene the transient benzyne intermediate is trapped and a good yield of Diels-Alder product results.

## 55a Dihydroxytriptycene

***p*-Benzoquinone–Anthracene Adduct.** In a 100-ml round-bottomed flask dissolve in 10 ml of xylene 0.01 mole (1.8 g) of pure anthracene and 0.01 mole

[5] L. Friedman, private communication. References to earlier methods for generating benzyne can be found in *J. Amer. Chem. Soc.*, **85**, 1549 (1963). For a general review see G. Wittig, *Z. Angew. Chem. Internat. Ed.*, **4**, 731 (1965).

(1.1 g) of *p*-benzoquinone. Attach a water-cooled reflux condenser and boil the solution for 45 min. Cool the solution to 15–20° and allow the adduct to crystallize. Collect the pale yellow crystals with suction, press them firmly on the filter, and spread on a clean paper to allow traces of residual solvent to evaporate. The yield of adduct is about 2.7 g. The crude product is sufficiently pure for conversion to dihydroxytriptycene. Do not attempt to determine its melting point.

**Dihydroxytriptycene.** Place the *p*-benzoquinone–anthracene adduct (about 2.7 g) in a 250-ml round-bottomed flask and add a solution of 0.5 g of potassium hydroxide pellets in 50 ml of ethanol. Warm the flask on a steam bath for about 5 min, or until the adduct has dissolved. Dilute the ethanolic solution with 100 ml of water, cool the flask in an ice-water bath, and *carefully* neutralize the base by adding 20 percent hydrochloric acid dropwise with swirling, until the liquid is acidic to litmus (about 2 ml of acid will be required). Collect the precipitated dihydroxytriptycene by suction filtration, wash it well with water, and spread to dry in the air. The yield is about 2 g.

The crude product is almost colorless; it shows no carbonyl absorption in the 6 $\mu$ region of the infrared spectrum. To obtain pure dihydroxytriptycene the crude material may be crystallized from ordinary ethanol (95 percent) or from ethanol-water mixtures. Do not attempt to determine its melting point.

## 55b Triptycene

**Benzenediazonium-2-carboxylate hydrochloride.** In a 100-ml beaker contained in an ice bath place a solution of 0.01 mole (1.4 g) of anthranilic acid in 15 ml of 95 percent ethanol. To the chilled solution add dropwise 1 ml of concentrated hydrochloric acid, followed by 0.02 mole (2.4 ml, 2.1 g; a large excess) of *n*-butyl nitrite.[6] Swirl the mixture for 10 min, then add 15 ml of

➤**CAUTION:** Alkyl nitrites are powerful heart stimulants and the vapor should not be inhaled; *n*-butyl nitrite is decomposed on exposure to air and light and should be stored in a cool place in a tightly closed container protected from light.

diethyl ether (*flammable!*) and continue swirling for an additional 5 min. Collect the crystals by suction filtration and wash them with a little ether (about 20 ml). The yield is 1.4–1.6 g. The hydrochloride is not explosive but it flashes on ignition and should not be stored in the desk.

[6] Sufficient *n*-butyl nitrite can be prepared readily by the method described in *Organic Syntheses*, Collective Volume II, 108 (1943), using one-hundredth of the quantities specified. The crude nitrite, after washing as described, can be used directly.

**Triptycene.** In a 100-ml round-bottomed flask equipped with an upright condenser place 20 ml of 1,2-dichloroethane (*flammable!*), 0.01 mole (1.8 g) of anthracene, 0.02 mole (1.5 ml, 1.2 g) of propylene oxide (*flammable!*), and the yield of benzenediazonium-2-carboxylate hydrochloride prepared above. Heat the mixture under reflux on a steam bath for 1 hr. Arrange the apparatus for downward distillation, add 20 ml of diglyme and remove the low boiling materials by distillation until a head temperature of 150° is reached.

To remove unreacted anthracene, cool the solution slightly, add 0.01 mole (1 g) of maleic anhydride and heat under reflux for 15 min. Cool the dark mixture to room temperature and add a solution of 3.0 g of potassium hydroxide in 30 ml of 2:1 methanol-water. Allow the mixture to stand for 15 min, collect the insoluble residue of triptycene by suction filtration and wash it with methanol-water (4:1). A yield of 1.0–1.4 g of nearly pure off-white triptycene, mp 252–254°, is obtained. The product can be recrystallized from methylcyclohexane or toluene.

## Questions

**1.** Write the structure of the Diels-Alder adduct formed by combination of the following reactants: (a) isoprene and acrylonitrile; (b) 2, 5-dimethylfuran and maleic anhydride; (c) 2,3-dimethoxybutadiene and 1,4-naphthoquinone; (d) 1-methylcyclopentadiene and N-phenylmaleimide; (e) ethyl propiolate ($HC{\equiv}C{-}CO_2Et$) and 2-ethoxybutadiene.

**2.** The structure of the adducts formed in parts (a), (b), and (e) of the preceding question can be established by conversion to known benzene derivatives. How is this accomplished?

**3.** Write the structure of the cyclic dimers formed by the following compounds through self-addition, in a reaction of the Diels-Alder type: (a) isoprene; (b) cyclopentadiene; (c) acrolein ($CH_2{=}CH{-}CH{=}O$).

**4.** Write equations for the reactions used in manufacturing the insecticides Chlordan and Aldrin, starting from cyclopentadiene.

**5.** What product would result from the Diels-Alder reaction of benzyne with furan; predict its behavior on treatment with highly acidic reagents.

EXPERIMENT

# 56

# Polymerization

## Synthetic Polymers

Polymerization is the process of forming a single large molecule by combining a number of identical or similar small molecules.[1] Simple organic molecules can polymerize in several different ways: by direct **addition** reactions of compounds containing one or more reactive multiple bonds, such as $CH_2{=}CH{-}R$, $HC{\equiv}CH$, $H_2C{=}O$, $R{-}N{=}C{=}O$, $CH_2{=}CH{-}CH{=}CH_2$, etc.; by **ring opening** of cyclic structures such as epoxides, lactams, lactones, and cyclic anhydrides or imides; by **condensation** reactions of polyfunctional molecules (glycols, diamines, dibasic acids, amino acids, etc.) to form polyesters, polyamides, polyethers and the like, through elimination of water, alcohol, or ammonia.

A polymer may be described as a large molecule (macromolecule) that can be formed by these processes of multiple combination, or a macromolecule that can be degraded to simple molecules by the reverse process (depolymerization). The simple molecules from which the polymers are built up are called monomers. Molecules derived from two, three, or four units of the monomer are called dimers, trimers, tetramers, etc. The prefix *oligo-*, meaning few or small, is used occasionally for polymers of low molecular weight containing only a small number of monomer units (oligomers, oligosaccharides, etc.).

Naturally occurring polymers of high molecular weight have for cen-

[1] Wallace Carothers, *Chem. Revs.*, **8**, 353 (1931). General reference books in this field are: Ellis, *Chemistry of Synthetic Resins*, Reinhold Publishing Corp., New York (1935); Meyer, *Natural and Synthetic High Polymers*, Interscience Publishers, Inc., New York (1950); Marvel, *Introduction to the Organic Chemistry of High Polymers*, John Wiley and Sons, Inc., New York (1959).

turies played an important role in everyday life—as foodstuffs (starches and proteins) and for providing shelter, fuel, clothing, and a variety of necessities and luxuries. A list of such polymers would include wood, cotton, paper, silk, wool, leather, rubber, lacquers and finishes, and a host of others. Chemical studies of native polymers have given a good working knowledge of their chemical constitution. This information has permitted modifications to be made in the properties of natural polymers by chemical transformations (rayon, celanese, cellophane, lacquers and finishes, from cellulose) and laid the foundation for the synthesis of new polymeric materials from simple organic molecules. Examples of purely synthetic polymers are: Bakelite resins from phenols and aldehydes, alkyd resins from glycols and dibasic organic acids, glass-like polymers (Butacite and Lucite) from unsaturated esters, rubber-like elastomers (Neoprene, Koroseal, Styrene-butadiene and Butyl rubber) from acetylene and other unsaturated compounds, and silk-like fibers (Nylon) and bristles from diamines and dibasic organic acids.

**Addition Polymers.** The simplest type of addition polymer is that formed by self-combination of a symmetrical molecule such as ethylene or tetrafluoroethylene. The process may be initiated by means of free radical or ionic catalysts. The resulting polymeric molecules are long chains made up of a repeating series of identical structural units, each unit corresponding to one molecule of the monomer. Such linear polymers are usually mixtures containing chains of different lengths and, in free radical polymerizations, highly branched chains are formed by a chain-transfer mechanism. This occurs because a growing free radical chain can abstract a hydrogen atom from a chain already formed, thereby creating an active center that can start the growth of a new chain.

The terminal valences of the chain are saturated by univalent groups such as hydrogen, hydroxyl, acetoxyl, halogen, etc., or an organic fragment derived from the polymerization catalyst. The polymers from ethylene and from tetrafluoroethylene may be represented by the following formulas:

$$-CH_2-CH_2-CH_2-CH_2-CH_2-CH_2-$$

Linear Polyethylene

$$-CF_2-CF_2-CF_2-CF_2-CF_2-CF_2-$$

Teflon

Polymerized ethylene is a waxlike solid having a higher melting point than any known paraffin wax and possessing valuable dielectric properties. Polymerized tetrafluoroethylene, known as Teflon, is a very tough hornlike material of high melting point, showing remarkable inertness toward all chemical reagents. It has found an important application in the form of gaskets for equipment used in handling hot corrosive chemicals and containers for corrosive liquids. Because it is insoluble in all solvents, a difficult milling operation is necessary to convert the polymer into useful shapes and forms.

The development of new ionic polymerization catalysts has resulted in great advances in modifying the polymerization of alkenes and 1,3-dienes. Polyethylene made at low pressures by means of coördination catalysts, such as a complex of triethylaluminum and titanium tetrachloride, is a true linear polymer free of branched chains. It is distinctly crystalline and has a higher melting point than the branched polymer obtained with free radical catalysts. In the polymerization of propylene, the use of ionic catalysts on appropriate solid supports permits stereochemical control of the process. With crystalline catalyst supports a highly crystalline, *isotactic* polypropylene can be obtained, that has a relatively high melting point and can be drawn into strong fibers. With amorphous supports a randomly oriented, *atactic* polypropylene is produced, that is soft and rubber-like.

The simplest unsymmetrical monomers may be illustrated by the general formula $CH_2{=}CH{-}X$ to represent compounds such as vinyl chloride ($X = Cl$), vinyl acetate ($X = O{-}CO{-}CH_3$), vinyl cyanide or acrylonitrile ($X = CN$), methyl acrylate ($X = CO_2CH_3$), styrene ($X = -C_6H_5$), etc. Three different types of polymeric structure may result from the self-combination of these unsymmetrical compounds, owing to different arrangements of the structural units in the chain:

$$\begin{array}{llllllllll} -CH_2 & CH- & CH_2 & CH- & CH_2 & CH- & CH_2 & CH- & CH_2 & CH- \\ & \;| & & \;| & & \;| & & \;| & & \;| \\ & X & & X & & X & & X & & X \end{array}$$

Type I (X groups in 1,3-positions)

$$\begin{array}{llllllllll} -CH & CH_2- & CH_2 & CH- & CH & CH_2- & CH_2 & CH- & CH & CH_2- \\ \;\;| & & & \;| & \;| & & & \;| & \;| & \\ \;\;X & & & X & X & & & X & X & \end{array}$$

Type II (X groups in 1,2-positions)

$$\begin{array}{llllllllll} -CH & CH_2- & CH_2 & CH- & CH_2 & CH- & CH & CH_2- & CH & CH_2- \\ \;\;| & & & \;| & & \;| & \;| & & \;| & \\ \;\;X & & & X & & X & X & & X & \end{array}$$

Type III (X groups in random positions)

The structure of the linear polymer formed from a specific compound, under a particular set of conditions, can be formulated definitely only after an examination of its properties. The structure of commercial vinyl acetate polymer has been shown definitely to be of Type I, but the constitution of many other polymers of this group is still in doubt.

Polystyrene is one of the few water-clear plastics and on account of its cheapness is widely used. Analogous to Teflon is the polymer of chlorotrifluoroethylene ($CF_2{=}CFCl$). It is very inert to chemical action and has an advantage over Teflon in being thermoplastic and therefore readily molded into a variety of useful shapes.

Examples of other types of unsymmetrical monomers are vinylidene chloride, $CH_2{=}CCl_2$; isobutylene, $CH_2{=}C(CH_3)_2$; vinyl fluoride, $CH_2{=}CHF$; methyl methacrylate, $CH_2{=}C(CH_3)CO_2CH_3$. The commercial polymer from vinylidene chloride forms fibers that are made into special upholstery fabrics (used in trains and buses). These fabrics can be cleaned readily but are not suitable for wearing apparel as the material softens at too low a temperature to permit ironing. The polymerization of isobutylene at low temperatures gives an elastomer known as Vistanex. Vinylidene chloride polymer and Vistanex are more resistant to penetration of water vapor than any other polymers now known.

Methyl methacrylate forms a beautiful crystal-clear polymer (Lucite, Plexiglas) that has found many applications. During World War II, it was employed for the manufacture of shatter-resistant windows for the turrets of bombers. When polymerized in aqueous emulsion the polymer is obtained in the form of small beads that are convenient for molding operations. Methyl methacrylate polymer, like many addition polymers (rubber, polystyrene, polymethylene glycol), can be degraded to the parent monomer by cracking at high temperatures.

Elastomers are formed from 1,3-dienes and their derivatives, in which the presence of a conjugated system makes possible a 1,4-addition process. Natural rubber is a macromolecular 1,4-polymer of 2-methyl-1,3-butadiene (isoprene) in which all of the units are combined uniformly at the 1,4-positions (head-to-tail fashion) and have a *cis* configuration ($CH_3$:H) about the double bond. The terminal valences of the chain are satisfied by hydroxyl groups.

$$\begin{array}{ccccccccc} HO\cdots CH_2 & & CH_2{-}CH_2 & & CH_2{-}CH_2 & & CH_2\cdots OH \\ & \diagdown\; C{=}C \;\diagup & & \diagdown\; C{=}C \;\diagup & & \diagdown\; C{=}C \;\diagup & \\ CH_3 & & H \quad CH_3 & & H \quad CH_3 & & H \end{array}$$

Synthetic polyisoprene made with free radical catalysts (organic peroxides, redox systems) differs markedly from natural rubber: some 1,2-addition occurs, giving rise to pendent vinyl or isopropenyl groups; the units are combined in random fashion, so that the methyl groups are not present at regular intervals along the chain; and the configuration of the structural units is mainly *trans* instead of *cis*. By means of the new coördination catalysts, and anionic catalysts such as *n*-butyl-lithium, it has been possible to obtain synthetic *cis*-1,4-polyisoprene that is essentially the same as natural rubber. Butadiene is polymerized by coördination catalysts in a similar manner, to provide *cis*-1,4-polybutadiene. This polymer exhibits better abrasive resistance, toughness and heat resistance than *cis*-polyisoprene and has great potential commercial value. In fabrication it is usually blended with *cis*-polyisoprene, natural rubber, or more often conventional butadiene-styrene rubber (BSR, see below).

On heating in the presence of an inhibitor (hydroquinone, *t*-butylcatechol) isoprene and 1,3-butadiene undergo a much simpler polymerization and give mainly cyclic dimers instead of rubbery polymers of high molecular weight. It is noteworthy that many natural products of plant origin (camphor

$$\begin{array}{l} HC{=}CH_2 \\ | \\ CH_3C{=}CH_2 \end{array} + \begin{array}{l} CH_3 \\ | \\ CH{-}C{=}CH_2 \\ \| \\ CH_2 \end{array} \longrightarrow \begin{array}{l} \quad CH_2 \qquad CH_3 \\ HC \qquad CH{-}C{=}CH_2 \\ \| \qquad\quad | \\ CH_3C \qquad CH_2 \\ \quad CH_2 \end{array}$$

*dl*-Limonene, a cyclic dimer of isoprene

and other terpenes, carotene, lycopene, etc.) contain a carbon skeleton made up of isoprene units. An intermediate in the biosynthesis of isoprenoid compounds is mevalonic acid (3,5-dihydroxy-3-methylpentanoic acid).

Another substance analogous to isoprene is 2-chloro-1,3-butadiene (chloroprene), which polymerizes to a rubber-like material known as Neoprene. The latter has an advantage over natural rubber in being more inert to hydrocarbon solvents and also retains its elastic properties over a wider temperature range.

Examples of addition polymerization by ring opening of cyclic monomers are the formation of polyethylene glycols from ethylene oxide, polybutylene glycols from tetrahydrofuran, Nylon 4 from pyrrolidone, and Nylon 6 from ε-caprolactam.

**Copolymers.** Many industrially important products are obtained by addition polymerization of a mixture of two different monomers, such as vinyl chloride and vinyl acetate. In this way, long chains are formed in which some of the units are derived from one monomer and some from the other. The properties of the product depend on the chemical nature and proportions of the two monomers and on the average molecular weight of the polymer. One of the most common copolymers is butadiene-styrene rubber (GRS or BSR), which is prepared in aqueous emulsion at low temperature (cold rubber) with a hydroperoxide-redox catalyst system.

Copolymerization of isobutylene with a small amount of butadiene, at low temperatures, results in an elastomer known as Butyl rubber. It is valuable for the manufacture of inner tubes and also tubeless tires since it is very impervious to air and other gases. Other commercial copolymers are produced from maleic anhydride and styrene, and vinyl chloride and acrylonitrile.

Elastomers of an unusual type, containing urethan linkages (polyurethans), are manufactured from polymeric glycols of low molecular weight (1800–3000) and diisocyanates, in a stepwise process. For the preparation of a foam rubber, a low molecular weight polyester from adipic acid and excess ethylene glycol, having terminal hydroxyl groups, is treated with a diisocyanate in sufficient amount to form a diurethan that will have one free isocyanate group at each end of the chain. This intermediate is treated with a carefully controlled amount of water, and an amine catalyst, so that some of the isocyanate groups are converted to carbamic acid groups. The latter decompose with evolution of carbon dioxide (foaming) and formation of a primary amine. The newly formed amino groups react with residual isocyanate groups to link the small polymer chains into a macromolecule by means of urea linkages. The polyurethan elastomers have excellent resistance to abrasion and to air oxidation.

Lycra, which has superior properties for the manufacture of elastic fabrics, is an elastomer of the polyurethan type. Polybutylene glycol (from tetrahydrofuran), having a molecular weight of 2500–3000, is treated with sufficient tolylene-2,4-diisocyanate to form the diurethan, in which one unreacted isocyanate function remains at each end of the chain. This inter-

O=C=N−Tolyl−NH−CO−O−R−O−CO−NH−Tolyl−N=C=O

−Tolyl−NH−CO−NH−NH−CO−NH−Tolyl−NH−CO−O−R−O−CO−NH−

mediate is then converted to a polymer of high molecular weight by reaction with hydrazine, a bifunctional amine. A fragment of the final polymeric structure is shown above.

**Condensation Polymers.** Polyfunctional molecules (glycols, dibasic acids, hydroxy acids, amino acids, hydroxy aldehydes, etc.) can undergo reactions that give rise to cyclic structures, usually 5-or 6-membered rings, or to linear polymers. With but few exceptions, the esterification of a dihydric alcohol (glycol) and a dibasic acid furnishes linear polyesters.

The polymeric ester derived from terephthalic acid and ethylene glycol has found numerous industrial applications, such as the manufacture of textile fibers and fabrics (Dacron or Terylene), transparent sheets and coatings of high strength and flexibility (Mylar), and photographic film (Cronar). The polyester is manufactured by an ester exchange reaction from dimethyl terephthalate and ethylene glycol. Since an excess of the glycol is used, the

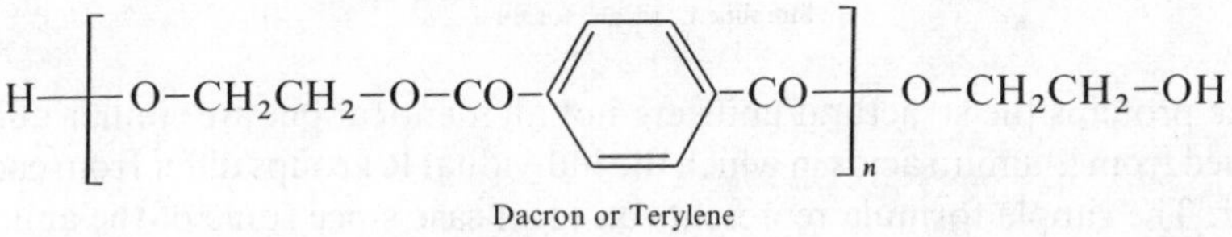

$$H{-}\left[O{-}CH_2CH_2{-}O{-}CO{-}C_6H_4{-}CO\right]_n{-}O{-}CH_2CH_2{-}OH$$

Dacron or Terylene

polymer has alcohol groups at the ends of the polymer chain. The physical and chemical characteristics of the polymeric material are influenced markedly by the degree of polymerization (number of structural units in the polymer) and the nature of the terminal groups. Polyesters containing a free hydroxyl and a free carboxyl group at the terminals can be obtained by intermolecular esterification of hydroxy acids of the type $HO-(CH_2)_n-CO_2H$, where $n$ is 5 or greater.

An important commercial example of polyesterification is the formation of synthetic resins and plastics (Alkyd polymers) from phthalic anhydride and polyhydric alcohols, particularly glycerol. They are especially useful for surface coatings. By controlling the relative amounts of the anhydride and alcohol, and the conditions of reaction, various polymers of different properties are obtained. A group of these has been given the name *glyptals*, derived from the words glycerol and phthalic. Under mild conditions (150°) only the primary alcohol groups are esterified and the secondary alcohol group remains free; the structural unit of the resulting polymers is the following:

$$—CO-C_6H_4-CO-O-CH_2-CHOH-CH_2-O—$$

These are relatively soft materials and are soluble in a number of organic solvents. Under more drastic conditions (200–210°), and with a larger proportion of phthalic anhydride, the secondary alcohol groups are esterified and the simple chains become crosslinked, three-dimensional polymers of much higher molecular weight. These are relatively infusible, brittle materials that are insoluble in most solvents. In commercial practice the properties of the polymers are varied by mixing aliphatic acids or anhydrides with the phthalic anhydride. Like simple esters, the various polyesters can be saponified by alkalies to regenerate the parent acid and alcohol.

Other types of polymeric materials are polyacetals, from polyfunctional glycols and aldehydes, or from hydroxyaldehydes (e.g., carbohydrates); polyamides, from polyfunctional acids and polyamines or from amino acids (e.g., proteins). In all of these the constitution of the polymer is represented as a long chain formed by combination of a large number of identical or similar structural units.

$$H_2N-\underset{}{\overset{R'}{\overset{|}{C}}}H-CO-\left[NH-\overset{R}{\overset{|}{C}}H-CO\right]_n-NH-\overset{R''}{\overset{|}{C}}H-CO_2H$$

Polyamide ("polypeptide") of protein type, free amino at one terminal and free carboxyl group at the other. Simplified "Ideal" formula

In the proteins the structural units are not all identical but are similar units derived from α-amino acids in which the individual R groups differ from each other. The simple formula represents an ideal case since some of the amino acid units in native proteins contain more than one amino or carboxyl group (lysine, arginine, aspartic acid, glutamic acid).

Wool and silk are examples of natural polypeptide fibers. Wool-like fibers have been manufactured from the proteins obtained from peanuts, from casein of milk, and from the protein zein of corn.

# 56a Addition Polymers

## Methyl Methacrylate Polymer

**Depolymerization by Heat.** Place 20 g of commercial granulated methyl methacrylate polymer (Lucite or Plexiglas)[2] in a small pyrex distilling flask provided with a tight cork and fitted to a condenser set for distillation. Heat the sides of the flask gently with a small luminous flame (3–4 cm) that is rotated constantly.

The polymer gradually softens and at 300–325° undergoes rapid depolymerization to the monomer, methyl methacrylate,[2] which distills into the receiver. Continue the distillation *cautiously* until only a small black residue (3–4 g) remains in the distilling flask. Place the distillate, amounting to 14–15 g, in a small flask and redistill; record the boiling range. Perform tests immediately; do not store the monomer.

**Polymerization.** Place small samples of the monomer in three narrow test tubes. To one add nothing, to another add a trace of hydroquinone (inhibitor), and to the third add a *very small* amount of catalyst (benzoyl peroxide). Make sure that no trace of hydroquinone is introduced into the third tube. *Cork the tubes firmly* and expose to sunlight for several days. Observe the physical change of the material from time to time.

➧**CAUTION:** Under some conditions benzoyl peroxide can be dangerously explosive. It should be manipulated carefully and in very small quantities. It must not be stored in a glass-stoppered bottle, and not compressed by grinding in a mortar, or handled with an iron spatula.

*Questions follow Experiment 56b.*

# 56b Condensation Polymers

## An Alkyd Polyester[3]

**Formation of a Glyptal Resin (Brittle Type).** In a 200-ml beaker place 0.06 mole (5.5 g; 4.5 ml) of glycerol and 0.075 mole (11 g) of pulverized phthalic

[2] Methyl methacrylate is the methyl ester of α-methylacrylic acid (2-methylpropenoic acid). The granular form of the polymer is obtained by polymerizing the ester in an emulsified form; it may be purchased in small quantities from the Eastman Kodak Co., or in larger quantities from the manufacturers, E. I. du Pont de Nemours and Co., Wilmington, Del., or the Röhm and Haas Co., Philadelphia, Pa.

[3] An interesting preparation of the polyamide Nylon 6–10 by interfacial polymerization of 1,6-hexanediamine (hexamethylenediamine) and sebacoyl chloride ($Cl-CO(CH_2)_8CO-Cl$) has been described by Morgan and Kwolek, *J. Chem. Educ.*, **36**, 182, 530 (1959).

anhydride. Mix the materials thoroughly with a glass stirring rod and heat gently (150–175°) on a wire gauze over a small flame, preferably under a hood. Take care to avoid overheating, which causes decomposition and charring of the product. As esterification proceeds water distills from the reaction mixture. Increase the heating gradually (200–210°) until the mass finally forms large bubbles that cause it to puff up to a large volume. At this point remove the flame and allow the polyester to cool. When cold, break up the brittle, clinker-like mass and remove it from the beaker as completely as possible. Pulverize the polymer in a mortar to obtain a fine powder.

Test the solubility of this polyester in the following solvents, using about 0.2 g of polymer per 4–5 ml of solvent and warming to the boiling point (*caution with flammable solvents*): acetone, carbon tetrachloride, toluene, ethanol, and any commercial solvents or plasticizers that may be available (cellosolves, carbitols, di-*n*-butyl phthalate, etc.).

A soft, soluble polyester may be obtained by heating one part by weight of glycerol with two parts of phthalic anhydride to 150–160° and discontinuing the heating when the first condensation is complete (see page 460).

**Degradation of the Polymer.** In a small round-bottomed flask place about 5 g of the polyester and 25 ml of aqueous sodium hydroxide solution (10 percent). Boil the mixture gently under reflux until the saponification is complete. What organic compounds are present in the solution? Outline a procedure for isolating and identifying them.

## Questions

**1.** Write equations showing how methacrylic acid, $CH_2{=}C(CH_3)CO_2H$, could be synthesized from acetone.

**2.** Write equations, including the conditions of reaction, showing how ethylene oxide may be converted into: diethylene glycol, dioxane, methyl cellosolve, triethanolamine.

**3.** Write structural formulas for the terpenes, limonene and geraniol, and show how they may be derived from isoprene units.

**4.** Write a structural formula for the polymer Nylon 6-6 obtained by heating hexamethylenediamine $[NH_2(CH_2)_6NH_2]$ and adipic acid $[HO_2C(CH_2)_4CO_2H]$. Indicate clearly the structural unit of the polymer.

**5.** Write formulas showing the structural units that would be formed by condensation of the following compounds: *n*-butyraldehyde and 1,5-pentanediol (polyacetal formation); succinaldehyde and the Grignard reagent from 1,5-dibromopentane (followed by treatment with dilute acid).

# Suggestions for Supplementary Experiments

It is desirable that several laboratory manuals be available for the benefit of students who are interested in organic syntheses and are qualified to perform supplementary experiments of an advanced type. The following are particularly useful: *Organic Syntheses*,[1] Collective Volumes **I–IV**; *Organic Reactions*; Fieser, *Experiments in Organic Chemistry* (3rd edition) or *Organic Experiments* (2nd edition); Lucas and Pressman, *Principles and Practice in Organic Chemistry*.

Qualified students are encouraged to use the well-known German laboratory manual, Gattermann-Wieland, *Die Praxis des organischen Chemikers* (obtainable from Stechert-Hafner Inc., 31 East Tenth Street, New York, N.Y. 10003). Students who read French may consult the laboratory book of Vavon, Doulou, and Lozac'h, *Manipulations de Chimie Organique* (Masson and Co., Paris, 1946), or Velluz, *Substances Naturelles de Synthèses* (Masson and Co., Paris, 1957).

Although the following list gives references to specific laboratory manuals, the student should be encouraged to consult comprehensive reference works and original articles in the scientific journals.[2]

In general, the experiments listed below are comparable in scope to those in the last third of this manual, and an effort has been made to choose those illustrating important general reactions and typical laboratory operations. In a number of cases, experiments are suggested in which preparations described in this manual serve as starting materials.

A few special notes are necessary in connection with the use of *Organic Syn-*

[1] The individual volumes of Organic Syntheses are published annually. The collective volumes are compilations and revisions of preparations appearing in the individual volumes: Collective Volume **I** (volumes **1–9**); **II** (**10–19**); **III** (**20–29**); **IV** (**30–39**). The annual volumes **44** (1964) and **49** (1969) have cumulative indexes for volumes **40–44** and **45–49**, respectively.

[2] For a list of selected reference works and scientific journals consult the following section of this manual, "Literature of Organic Chemistry."

*theses*. The directions given there call for the use of relatively large quantities of materials and frequently for mechanical stirring. For a laboratory preparation the student will generally wish to use only one-fifth or one-tenth of the quantities mentioned. Care must be taken to decrease proportionately *all* the reagents used. With smaller quantities the time required for various operations can usually be reduced considerably (but not proportionately) and stirring or shaking by hand will suffice if mechanical stirrers are not available.[3]

In the following list the citations in *Organic Syntheses* refer to the Collective Volume **I** (2nd edition, 1941), Volume **II** (1943), Volume **III** (1955), and Volume **IV** (1963). The other manuals cited are: Gattermann-Wieland, *Die Praxis des organischen Chemikers*, 40th edition. (Walter de Gruyter and Co., Berlin, 1961); Fieser, *Experiments in Organic Chemistry*, 3rd edition, revised (D. C. Heath and Co., Boston, 1957) and Fieser, *Organic Experiments*, 2nd edition (Raytheon Education Company, Lexington, Mass., 1968)[4] differentiated from the former by the entry Fieser*.

2,4-Dihydroxyacetophenone (Resacetophenone) from resorcinol and acetonitrile (Expt 13): Gattermann-Wieland, p. 298, *see also*, Organic Syntheses, **III**, p. 761.

Salicylaldehyde from phenol and chloroform: Gattermann-Wieland, p. 206.

Catechol from salicylaldehyde: Organic Syntheses, **I**, p. 149 (use one-tenth quantities).

Coumarone (Benzofuran) from salicylaldehyde and chloroacetic acid: Organic Syntheses, **46**, p. 28 (use one-fifth or one-tenth quantities).

dl-Mandelic Acid from benzaldehyde: Organic Syntheses, **I**, p. 336 (use one-tenth quantities); Gattermann-Wieland, p. 198; Fieser, p. 97, Fieser*, p. 111.

d(-)-Mandelic Acid (resolution of racemic mandelic acid): Gattermann-Wieland, p. 199.

Nicotinic Acid (Pyridine-3-Carboxylic Acid) from nicotine: Organic Syntheses, **I**, p. 385 (use one-tenth quantities); from $\beta$-picoline; Velluz, **VIII**, p. 6 (use one-twentieth quantities).

Dibenzoylmethane from benzalacetophenone (Expt 36): Organic Syntheses, **I**, p. 205 (use about one-seventh quantities); *see also*, Organic Syntheses, **III**, p. 251.

dl-$\alpha$-Phenylethylamine from acetophenone (Expt 17c): Organic Syntheses, **II**, p. 503 (use one-fifth or one-tenth quantities).

Resolution of dl-Phenylethylamine: Organic Syntheses, **II**, p. 506 (use all of material from preceding preparation).

Triphenylchloromethane from benzene and carbon tetrachloride: Gattermann-Wieland, p. 297; from triphenylcarbinol (Expt 41) and acetyl chloride (Expt 11a) Organic Syntheses, **III**, p. 841 (use one-twentieth quantities).

[3] A simple and cheap mechanical stirrer, driven by compressed air, is described on page 170.

[4] This recent Fieser manual contains new experiments not given in the *Experiments in Organic Chemistry*, but the chapters 43–50 (Accessories for Reactions, Reagents, etc.) have been omitted.

HEXAPHENYLETHANE (TRIPHENYLMETHYL) from triphenylchloromethane: Gattermann-Wieland, p. 306.

4-AMINOTETRAPHENYLMETHANE from triphenylcarbinol and aniline hydrochloride: Organic Syntheses, **IV**, p. 45 (use one-tenth quantities). The amino group may be replaced by hydrogen (procedure similar to Expt 29) to give tetraphenylmethane; see Ullmann and Münzheimer, *Ber.*, **36**, 309, 407 (1903).

2-BROMO-4-METHYLBENZALDEHYDE from 2-bromo-4-methylaniline (Expt 29b) and formaldoxime: Organic Syntheses, **46**, p. 13 (use one-half quantities). 2-Chloro-4-methylaniline or *p*-bromoaniline (Expt 24) may be converted to the corresponding aldehyde by a similar procedure.

IODOBENZENE from aniline: Organic Syntheses, **II**, p. 351 (use one-tenth quantities); Gattermann-Wieland, p. 246.

IODOBENZENE DICHLORIDE, IODOSOBENZENE and IODOXYBENZENE from iodobenzene: Organic Syntheses, **III**, pp. 482–487; Gattermann-Wieland, p. 246.

TETRACYCLONE (TETRAPHENYLCYCLOPENTADIENONE) from benzil (Expt 37b) and dibenzylketone: Organic Syntheses, **III**, p. 806 (use one-tenth quantities); Fieser*, p. 295.

1,2,3,4-TETRAPHENYLNAPHTHALENE from *o*-iodobenzoic acid and tetracyclone, *via* diphenyliodonium-2-carboxylate and benzyne: Organic Syntheses, **46**, p. 107; Fieser*, p. 303.

4-METHYL-6-HYDROXYPYRIMIDINE from thiourea and ethyl acetoacetate: Organic Syntheses, **IV**, p. 638 (use one-tenth quantities).

ε-CAPROLACTAM and ε-AMINOCAPROIC ACID from cyclohexanone oxime (Expt 9B): Organic Syntheses, **II**, pp. 77, 371; **IV**, p. 39 (use one-tenth quantities).

2-METHYL-1,4-NAPHTHOQUINONE from 2-methylnaphthalene: Velluz, **V**, p. 64; Fieser, p. 209, Fieser*, p. 246.

QUINIZARIN from phthalic anhydride and hydroquinone: Gattermann-Wieland, p. 299; *see also*, Organic Syntheses, **I**, p. 476.

MALACHITE GREEN from benzaldehyde and dimethylaniline: Gattermann-Wieland, p. 279.

MARTIUS YELLOW and its reduction products, from α-naphthol: Fieser, pp. 234–238, Fieser*, p. 272.

ADIPIC ACID from cyclohexanol: Organic Syntheses, **I**, p. 18 (use one-tenth or one-twentieth quantities); Fieser, pp. 94–96, Fieser*, p. 108.

DL-ALANINE from acetaldehyde: Organic Syntheses, **I**, p. 21 (use one-tenth quantities); Gattermann-Wieland, p. 200.

DL-β-PHENYLALANINE from benzaldehyde and hippuric acid (Expt 51b): Organic Syntheses, **II**, p. 489; from benzaldehyde and acetylglycine, Organic Syntheses, **II**, p. 491.

DL-PENICILLAMINE HYDROCHLORIDE from *dl*-Valine in four steps: Biochemical Preparations, **3**, pp. 111–113 (John Wiley and Sons, Inc., 1953).

D- and L-PENICILLAMINE from isopropylidene-*dl*-penicillamine in four steps: Biochemical Preparations, **3**, pp. 114–118.

# Literature of Organic Chemistry

Scientific publications dealing with organic chemistry and related fields have been increasing at an extremely rapid rate. Many journals and new annual reviews and surveys in specialized areas (organometallic chemistry, heterocyclic chemistry, chromatography, and others) have become available in recent years, mainly for the benefit of advanced students and research workers. The following short account is intended for beginners.[1]

It is convenient to group literature sources into a number of broad categories based upon the nature of the subject matter covered and the particular purpose served by the publication. The primary sources of information are the research journals, in which the results of original scholarly investigations are described and new theoretical developments are presented. Secondary sources include abstract journals, encyclopedias and handbooks, surveys and reviews, advanced treatises and textbooks.

Means of access to the literature are numerous and well organized. Time and effort will be saved by learning the particular function and usefulness of the various kinds of chemical publications that are available.

Advanced textbooks, treatises and monographs, and review articles are appropriate starting points for obtaining references that will furnish more complete coverage of a specific *subject*. This approach is especially recommended for students who are not experienced in making systematic literature searches.

The literature search for information about a specific *organic compound* will follow a pattern different from that used in seeking coverage of a specific topic or reaction. For a particular compound one uses chemical dictionaries and handbooks, and the subject and formula indexes of an abstract journal. For full documentation it is standard practice to use Beilstein's *Handbuch der organischen Chemie* and the

[1] A useful survey has been published recently by Professor J. E. H. Hancock, of Reed College in three articles: *J. Chem. Educ.*, **45**, 193–199, 260–266, 336–339 (1968).

indexes of *Chemical Abstracts*. The number of known organic compounds is approaching the three million mark—at the rate of about eight hundred new compounds per day. Nevertheless an ardent student after some practice and with patience, in a good library, can usually assemble in an hour or two extensive information about methods of preparation, physical constants, and chemical reactions of a particular compound.

The small selection of books and periodicals listed below is intended merely to give representative examples of the different kinds of publications and to indicate their uses. For a comprehensive survey of chemical publications in all fields, with discussion of literature searches, see one of the following:

Crane, Patterson, and Marr: *A Guide to the Literature of Chemistry*, John Wiley and Sons, Inc., New York, 2nd edition, 1957.

Gould (editor): *Searching the Chemical Literature*, Advances in Chemistry Series #30, American Chemical Society, Washington, D.C., 1961.

Mellon: *Chemical Publications—Their Nature and Use*, McGraw-Hill Book Co., New York, 4th edition, 1965.

# Textbooks and Reference Books

## General Organic Chemistry

Fieser and Fieser: *Advanced Organic Chemistry*, Reinhold Publishing Corp., New York, 1961 (1158 pp.); also, *Topics in Organic Chemistry*, 1963–.

Fuson: *Reactions of Organic Compounds*, John Wiley and Sons, Inc., New York, 1962 (765 pp.).

Noller: *Chemistry of Organic Compounds*, W.B. Saunders Co., Philadelphia, 3rd edition, 1965 (1115 pp.).

Roberts and Caserio, *Basic Principles of Organic Chemistry*, W. A. Benjamin, Inc., New York, 1964 (1315 pp.); also, *Supplement for Basic Principles of Organic Chemistry*, 1964 (435 pp.).

Wheland: *Advanced Organic Chemistry*, John Wiley and Sons, Inc., New York, 3rd edition, 1960 (871 pp.).

Coffey (editor): Rodd's *Chemistry of Carbon Compounds. A modern comprehensive treatise*, American Elsevier Publishing Co., New York, 2nd edition, 1964–. In five volumes, each in several parts: **I**, Aliphatic Compounds; **II**, Alicyclic Compounds; **III**, Aromatic Compounds; **IV**, Heterocyclic Compounds; **V**, Miscellaneous, and General Index. The new edition has not been completed so it is necessary to consult the first edition for the later parts. This modern textbook of factual organic chemistry is the most extensive general reference book in the English language that covers the whole field in a systematic way.

## Theoretical and Physical Organic Chemistry

Hermans: *Introduction to Theoretical Organic Chemistry*, edited and revised by R. E. Reeves, American Elsevier Publishing Co., New York, 1954 (507 pp.).

Cram and Hammond: *Organic Chemistry*, McGraw-Hill Book Co., New York, 2nd edition, 1964 (846 pp.).

Hine: *Physical Organic Chemistry*, McGraw-Hill Book Co., New York, 2nd edition 1962 (552 pp.).

Wiberg, *Physical Organic Chemistry*, John Wiley and Sons, Inc., New York, 1964 (591 pp.).

Gould: *Mechanism and Structure in Organic Chemistry*, Holt, Rinehart and Winston, Inc., New York, 1959 (790 pp.).

Ingold: *Structure and Mechanism in Organic Chemistry*, Cornell University Press, Ithaca, N.Y., 1953 (828 pp.).

Liberles: *Introduction to Theoretical Organic Chemistry*, The Macmillan Co., New York, 1968 (896 pp.).

Ferguson: *The Modern Structural Theory of Organic Chemistry*, Prentice-Hall, Inc., Englewood Cliffs, N.J., 1963 (600 pp.).

Velluz (editor): *Principes de Synthèse Organique* (Introduction to the Mechanism of Organic Reactions), Masson et Cie., Paris, 1957 (598 pp.).

Breslow: *Organic Reaction Mechanisms. An Introduction*, W. A. Benjamin, Inc., New York, 1966 (232 pp.). This is a volume in *The Organic Chemistry Monograph Series* (paperbacks) edited by Professor Breslow.

Roberts, *Notes on Molecular Orbital Calculations*, W. A. Benjamin, Inc., New York, 1961 (156 pp.).

Streitwieser, *Molecular Orbital Theory for Organic Chemists*, John Wiley and Sons, Inc., New York, 1961 (489 pp.).

## Synthetic Methods and Techniques

Fieser, *Experiments in Organic Chemistry*, D. C. Heath and Co., Boston (now Raytheon Education Company, Lexington, Mass.), 3rd edition, revised, 1957 (360 pp.). This manual has a number of interesting reaction sequences and an extremely informative chapter listing more than three hundred reagents used in organic laboratory work, including methods of preparation and references to articles describing the reactions and specific uses of the reagents. Also, *Organic Experiments*, D. C. Heath and Co., Boston, 2nd edition, 1968 (342 pp.). This is a new version of the foregoing manual in which many interesting experiments have been added—but the chapters in Part 2 of the earlier manual, on accessories for reactions, solvents, reagents, etc., have been deleted.

Fieser and Fieser: *Reagents for Organic Synthesis*, John Wiley and Sons, Inc., New York, 1967 (1457 pp.). This remarkable collection of concise essays on reagents is replete with features that render it extremely useful. Conventional practices that are faulty or ill-advised are pointed out. The appendix material gives names and addresses for suppliers of chemicals, solvents, and apparatus; reagents listed according to types; and author and subject indexes.

Gattermann-Wieland, *Die Praxis des organischen Chemikers*, Walter de Gruyter and Co., Berlin, 40th edition, revised by Theodor Wieland, 1961 (411 pp.). This famous German manual presents a wide range of preparations, accompanied by stimulating discussions of the methods and reactions.

Lucas and Pressman, *Principles and Practice in Organic Chemistry*, John Wiley and Sons, Inc., New York, 1949 (557 pp.). This text has good discussions of physicochemical principles and carefully tested experimental procedures, with extensive explanatory notes.

Hickinbottom, *Reactions of Organic Compounds*, Longmans, Green and Co., Inc., New York, 3rd edition, 1957 (608 pp.). This book of modest size gives a comprehensive coverage of the reactions of all important groups of organic compounds from the point of view of laboratory procedures, with an abundance of literature references. Laboratory directions are presented in a concise form and usually must be supplemented by consulting the sources.

*Organic Syntheses*, John Wiley and Sons, Inc., New York (1921– ). An annual publication of tested synthetic laboratory procedures. There are four Collective Volumes: **I** (2nd edition, 1941), covers annual volumes **1–9**; **II** (1943), volumes **10–19**; **III** (1955), volumes **20–29**; **IV** (1963), volumes **30–39**. In addition to formula and name indexes, the collective volumes have indexes of types of reaction, types of compounds, preparation or purification of solvents and reagents, and illustrations of special apparatus. The annual volumes **44** (1964) and **49** (1969) have cumulative indexes for volumes **40–44** and **45–49**, respectively.

*Organic Reactions*, John Wiley and Sons, Inc., New York (1942– ). The volumes of this series are collections of five to twelve chapters, each of which deals with a general reaction of wide applicability, or a definite segment of a reaction of broad scope, and includes several detailed procedures illustrating variations of the method. Each new volume contains a cumulative author and chapter-title index covering the entire series and a subject index for the individual volume. New volumes are published at intervals of one or two years; volume **17** appeared in 1969.

*Organikum. Organisch-chemisches Grundpraktikum*, VEB Deutscher Verlag der Wissenschaften, Berlin (East Germany), 7th edition, 1967 (696 pp.). This is an excellent compilation of general procedures for all types of organic preparations.

Theilheimer, *Synthetic Methods of Organic Chemistry*, S. Karger, Basel and New York (1948– ). This annual publication gives a concise and comprehensive survey of synthetic procedures described in the current literature. Entries are classified according to the type of reaction involved and the material is well indexed. Cumulative indexes have appeared in volumes **5** (1951), **10** (1956), **15** (1961), **20** (1966). Beginning with Volume **8** (1954) there is a section entitled *Trends in Synthetic Organic Chemistry* that calls attention to highlights of general interest.

Velluz (editor), Mathieu, Allais, and Valls, *Cahiers de Synthèse Organique*, Masson et Cie., Paris (1957–1962). The volumes of this series consist of chapters devoted to synthetic methods for achieving a particular type of transformation. There are ten volumes comprising twenty-one chapters. Examples of chapter titles are: Acylation, and Bifunctional Condensations (volume **IV**), Rearrangement and

Migration (volume **VI**), Formation of mono-Heterocycles (volume **IX**). The coverage is thorough.

Velluz (editor), Mathieu, Petit and Poirier, *Substances Naturelles de Synthèse*, Masson et Cie., Paris (1951–1954). In addition to a fine assembly of stepwise syntheses of natural products, this series presents discussions of the types of reactions and the techniques involved. There are collective indexes in volumes **5** and **10**.

Wiberg, *Laboratory Technique in Organic Chemistry*, McGraw-Hill Book Co., Inc., New York, 1960 (262 pp.).

Weissberger and Perry (editors): *Technique of Organic Chemistry*, Interscience Publishers, Inc., New York. This is a comprehensive modern treatise on laboratory manipulations and apparatus. Examples of individual volumes are: **I**, Physical Methods in Organic Chemistry (3rd edition, 1961, in four parts); **VIII**, Investigation of Rates and Mechanisms of Reaction (2nd edition, 1961, in two parts); **XII**, Thin-Layer Chromatography (1967); **XIII**, Gas Chromatography (1968).

*Biochemical Preparations*, John Wiley and Sons, Inc., New York (1949– ). This series includes syntheses of compounds of biochemical interest and enzymatic transformations. Volume **12** appeared in 1968.

# Identification and Analysis of Organic Compounds

Cheronis and Entrikin: *Semimicro Qualitative Organic Analysis*, Interscience Publishers, Inc., New York, 2nd edition, 1957 (774 pp.); also, a student's text by the same authors, *Identification of Organic Compounds*, Interscience Publishers, New York, 1963 (477 pp.).

Shriner, Fuson, and Curtin: *The Systematic Identification of Organic Compounds*, John Wiley and Sons, Inc., New York, 5th edition, 1964 (458 pp.).

Rappoport (editor): *Handbook of Tables for Identification of Organic Compounds*, Chemical Publishing Company, Cleveland, 3rd edition, 1966–1967.

Siggia: *Quantitative Analysis via Functional Groups*, John Wiley and Sons, Inc., New York, 3rd edition, 1963 (697 pp.).

Fritz and Hammond: *Quantitative Organic Analysis*, John Wiley and Sons, Inc., New York, 1957 (303 pp.).

# Monographs and Reference Books on Selected Topics

## General

Pauling: *The Nature of the Chemical Bond*, Cornell University Press, Ithaca, N.Y., 3rd edition, 1960 (644 pp.).

Sidgwick: *The Organic Chemistry of Nitrogen*, 3rd edition, revised by Millar and Springall, Oxford University Press, New York, 1966 (909 pp.).

Eliel: *Stereochemistry of Carbon Compounds*, McGraw-Hill Book Co., New York, 1962 (486 pp.).

Kharasch and Reinmuth, *Grignard Reactions of Nonmetallic Substances*, Prentice-Hall, Inc., Englewood Cliffs, N.J., 1954 (1384 pp.).

Walling: *Free Radicals in Solution*, John Wiley and Sons, Inc., New York, 1957 (631 pp.).

Patai (general editor): *The Chemistry of Functional Groups*, Interscience Publishers, New York. A series of volumes on individual functional groups: *The Chemistry of Alkenes*, 1964 (1315 pp.); *The Chemistry of the Carbonyl Group*, 1966 (1027 pp.); *The Chemistry of the Ether Linkage*, 1967 (785 pp.).

Olah (editor): *Friedel-Crafts and Related Reactions*, Interscience Publishers, Inc., New York, 1963–1965; in four volumes, with cumulative index in Volume **4**.

Blomquist (editor): *Organic Chemistry. A Series of Monographs*, Academic Press, New York. Examples of volumes in this series are: **1**, Kirmse, *Carbene Chemistry*, 1964; **3**, Hanack, *Conformation Theory*, 1965; **7**, A. W. Johnson, *Ylid Chemistry*, 1966.

Rinehart (editor): *Foundations of Modern Chemistry Series*, Prentice-Hall, Inc., Englewood Cliffs, N.J., 1965– . Examples of volumes in this series of paperbacks are: Stock, *Aromatic Substitution Reactions*; DePuy, *Molecular Reactions and Photochemistry*; Stewart, *The Investigation of Organic Reactions*.

Farber (editor): *Great Chemists*, Interscience Publishers, Inc., New York, 1961 (489 pp.). This fine collection of biographies is notable for its inclusion of a number of modern organic chemists who have died within the past few decades. Excellent biographies of American scientists are found in the *Biographical Memoirs* of the National Academy of Sciences (U.S.A.), published by the Columbia University Press, New York. Volume 40 (1969) has a cumulative name index for volumes 1–40.

Surrey: *Name Reactions in Organic Chemistry*, Academic Press, New York, 2nd edition, 1961 (278 pp.). About 110 name reactions are included, with brief biographical sketches of their discoverers.

Krauch and Kunz: *Organic Name Reactions*, translation of the 2nd German edition, with addendum, by John M. Harkin, John Wiley and Sons, Inc., New York, 1964 (620 pp.).

## Physical Methods[2]

Schwarz (editor): *Physical Methods in Organic Chemistry*, Holden-Day, Inc., San Francisco, 1964 (350 pp.).

Gillam and Stern: *An Introduction to Electronic Absorption Spectra in Organic Chemistry*, E. Arnold, London, 2nd edition, 1958 (326 pp.).

[2] See also various volumes in Perry-Weissberger, *Technique of Organic Chemistry*, cited on page 470.

Silverstein and Basler: *Spectrometric Identification of Organic Compounds*, John Wiley and Sons, Inc., New York, 2nd edition, 1967 (256 pp.).

Jaffé and Orchin: *Theory and Applications of Ultraviolet Spectroscopy*, John Wiley and Sons, Inc., New York, 1962 (624 pp.).

Bellamy: *The Infrared Spectra of Complex Molecules*, Methuen, London, and John Wiley and Sons, Inc., New York, 2nd edition, 1958 (425 pp.). This book is supplemented by *Advances in Infrared Group Frequencies*, Methuen, London, and Barnes and Noble, Inc., New York, 1968 (304 pp.).

Conley: *Infrared Spectroscopy*, Allyn and Bacon, Boston, 1966 (293 pp.).

Meloan: *Elementary Infrared Spectroscopy*, The Macmillan Co., New York, 1963 (193 pp.).

Nakanishi: *Infrared Absorption Spectroscopy. Practical*, Holden-Day, Inc., San Francisco, 1962 (233 pp.).

Roberts: *Nuclear Magnetic Resonance, Applications to Organic Chemistry*, McGraw-Hill Book Co., New York, 1959 (118 pp.).

Mathieson (editor): *Nuclear Magnetic Resonance for Organic Chemists*, Academic Press, New York, 1966 (287 pp.).

Reed: *Applications of Mass Spectrometry to Organic Chemistry*, Academic Press, New York, 1966 (256 pp.).

Dal Nogare and Juvet: *Gas-Liquid Chromatography*, Interscience Publishers, Inc., New York, 1962 (450 pp.).

Keulemans, *Gas Chromatography*, Reinhold Publishing Corp., New York, 2nd edition, edited by Verver, 1959 (234 pp.).

Stahl (editor): *Thin-Layer Chromatography. A Laboratory Handbook*, Springer Verlag, New York, 1965 (553 pp.).

Truter: *Thin Film Chromatography*, Interscience Publishers, Inc., New York, 1963 (205 pp.).

## Handbooks and Chemical Dictionaries

*Handbook of Chemistry and Physics*, Chemical Rubber Publishing Co., Cleveland, Ohio. New editions are published frequently.

Lange's *Handbook of Chemistry*, Handbook Publishers, Inc., Sandusky, Ohio.

SOCMA Handbook,[3] *Commerical Organic Chemical Names*, Chemical Abstracts Service, American Chemical Society, Columbus, Ohio, 1965.

*The Merck Index of Chemicals and Drugs*, Merck and Co., Inc., Rahway, N.J., 8th edition, 1968 (1714 pp.). This encyclopedia gives a concise summary of the physical and biological properties of about 10,000 chemical compounds, with numerous structural formulas and references to the literature. It has a cross index of synonyms and trademarked names with nearly 30,000 entries. There is a very useful section on Organic Name Reactions, in which each reaction is described and references to original and review articles are given. A table of

[3] SOCMA means Synthetic Organic Chemicals Manufacturers Association.

physical data on radioactive isotopes and information about their use in medicine are included.

Heilbron's *Dictionary of Organic Compounds*, Oxford University Press, New York, 4th edition, edited by Pollock and Stevens. Five volumes, with annual supplements, 1965– . This comprehensive reference work gives for each compound a brief summary of physical and chemical properties, and a few important literature references.

Kirk-Othmer: *Encyclopedia of Chemical Technology*, Interscience Publishers, Inc., New York, 2nd edition, in fifteen volumes, 1963–1968.

*Encyclopedia of Science and Technology. An International Reference Work*, McGraw-Hill Book Co., New York, 2nd edition, in fifteen volumes (including index), 1965–1966. Supplements are published as McGraw-Hill *Yearbook of Science and Technology.*

## The Compleat Encyclopedia

Beilstein's *Handbuch der organischen Chemie*, Verlag of Julius Springer, Berlin, 4th edition, with supplements, edited by Prager, Jacobson, and Richter. This impressive compendium surveys the literature for *all organic compounds of known structure*[4] published during specified periods. The main series (Hauptwerk), volumes **1–27**, covers the literature to 1910; the first supplement (Erstes Ergänzungswerk), 1910–1919; the second supplement (Zweites Ergänzungswerk), 1920–1929; these comprise more than 60,000 pages. The third supplement (Drittes Ergänzungswerk), now in the course of publication, covers the literature from 1930–1949, but current volumes have references to publications much later than 1949. The supplements have volume numbers corresponding to those of the main series and a special (centered) page numbering referring to the main series, such as H, 132, corresponding to page 132 of Hauptwerk.[5]

[4] In the original plan of the 4th edition of Beilstein, natural products of complex and uncertain structure were to be included but this part of the compilation was discontinued after two volumes had been published (in 1938): Volume **30**, dealing with rubber, gutta-percha, balata, and carotenoids; Volume **31**, covering monosaccharides and oligosaccharides. Great progress has been made during the past twenty years in structure determination of natural products and this field is covered by review and survey periodicals such as those listed in the following section.

[5] Abbreviations used for the various series are: H, for main series (Hauptwerk); E I or I, for first supplement; E II or II, for second supplement. Positional isomers, such as *o*-, *m*-, and *p*-disubstitution products, are not entered as separate names but are included in the page numbers following the index name. Thus, the three dibromobenzenes are entered in the Index Volumes, **28** and **29**, in a simplified form: Dibrombenzol **5**, 210, 211, I 116, II 162, 163. The location in the third supplement (actually III 564–569) can be found easily by using the centered paging H210, 211.

Volume **28** of the second supplementary series, published in 1955, is a cumulative name index (General-Sachregister) covering the main series and the first and second supplements. Volume **29** of the second supplementary series, published in 1956–1957, is a cumulative formula index (General-Formelregister) of similar scope. These indexes facilitate access to the information in Beilstein but it is necessary to have some knowledge of German nomenclature and the organization of the handbook to use this encyclopedia effectively. For coverage of the literature since the second supplementary series of Beilstein (ending in 1929) one may use the cumulative indexes of *Chemical Abstracts* (**III**, 1927–1936; **IV**, 1937–1946; **V**, 1947–1956; **VI**, 1957–1961; **VII**, 1962–1965) and subsequent annual and semiannual indexes. For compounds covered in volumes of the third supplement that have already appeared one may begin with Collective Index **V**. The *Chemical Abstracts* collective Formula Index covering the period 1920–1946 is also useful.

In Beilstein all compounds are regarded either as index compounds or as derivatives of index compounds. There are twenty-eight arbitrary classes of index compounds, comprising the fundamental parent systems (stammkerne) and twenty-seven functional groups, which are listed below. Carbon-nitrogen compounds of the type $R_2C{=}NH$ and $R{-}C{\equiv}N$ (also amides and amidines) are not treated as independent functional groups but are considered to be functional derivatives of the compounds $R_2C{=}O$ and $R{-}CO_2H$. Compounds such as $R{-}CHOH{-}Cl$ and $R{-}CHOH{-}SO_3Na$ (a bisulfite adduct) are listed under the index compound $R{-}CH{=}O$.

The entries in Beilstein under a particular compound are arranged under the headings: name (and synonyms); molecular formula and structural formula; occurrence; methods of formation; methods of preparation; physical properties; chemical reactions, classified according to the reagent involved; uses; detection and quantitative determination; addition compounds and salts; functional derivatives; substitution products (comprising halogen, nitro, nitroso, and azido derivatives) of the main compound and its functional derivatives; replacement derivatives, formed by replacement of oxygen by sulfur or selenium. The coverage is thorough and the style of presentation is concise. Lists of abbreviations used in the text and literature references are given at the beginning of each volume.[6] Beilstein can be used with a limited vocabulary of technical German terms and occasional reference to a good German-English dictionary.

There are three main Divisions in the Beilstein classification:

Volumes **1–4** Open-chain Compounds (Acyclische Verbindungen)
Volumes **5–16** Carbocyclic Compounds (Isocyclische Verbindungen)
Volumes **17–27** Heterocyclic Compounds (Heterocyclische Verbindungen)

[6] The abbreviations commonly encountered are: V. (Vorkommen) = occurrence in nature or in technical products; B. (Bildung) = methods of formation; D. (Darstellung) = preferred preparative methods; F. (Frierpunkt) = melting point; Kp. (Kochpunkt) = boiling point; bezw. (beziehungsweise) = or, respectively; Tl., Tle., Tln. (Teil, Teile, Teilen) = part(s), portion(s); verd. (verdünnt) = dilute(d). The abbreviations used for chemical periodicals are more condensed than those used in *Chemical Abstracts*.

Within each of these Divisions the sequence of entries is organized in a highly systematic manner. Parent systems and functional groups are presented in the following order:

1........Parent Hydrocarbon or Heterocycle (Stammkerne)
2........Hydroxy(Oxy) Compounds: alcohols, phenols, etc.
3........Carbonyl(Oxo) Compounds: aldehydes, ketones, etc.
4........Carboxylic Acids: $R-CO_2H$
5–6......Sulfinic and Sulfonic Acids: $R-SO_2H$ and $R-SO_3H$
7........Selenium analogs: $R-SeO_2H$ and $R-SeO_3H$
8........Amines, including quaternary ammonium compounds
9–10.....Hydroxylamines and Hydrazines: $R-NH-OH$ and $R-NHNH_2$, etc.
11–13....Azo, Diazo, and Diazonium Compounds
14–15....Azoxy Compounds and Isonitramines
16–22....Functions with 3 and more nitrogen atoms, excluding azides
23.......Carbon bonded to elements of Group V (except nitrogen)
24–28....Carbon bonded to elements of Groups IV, III, II, I, VIII.

Compounds with two or more like functional groups are placed in the first subdivision following the monofunctional compounds of the same type. When two or more *unlike* functional groups are present, the compound is placed in a subdivision of the class of compounds that occurs *latest* in the list given above.

The principle of latest placement is an arbitrary one, devised to avoid having the entries pile up in the early sections of the handbook. It works out as shown in the example below, for the treatment of subdivisions of the Acylic Carboxylic Acids, in volumes **2** and **3**.

## Carboxylic Acids

*General:* nomenclature, classification of derivatives

MONOCARBOXYLIC ACIDS (in order of increasing unsaturation)

Acids $C_nH_{2n}O_2$ (in order of increasing molecular weight)

Name of the specific acid; occurrence; properties; etc.

*Salts, Functional Derivatives:* esters; anhydrides, acyl peroxides, acyl halides; ammonia derivatives (simple amides, imino esters, nitriles, amidines); hydroxamic acids; hydrazides.

*Substitution Products:* halo-, nitro-, nitroso-, azido-.

*Replacement Derivatives:* thio acids, thio amides, etc.

Acids $C_nH_{2n-2}$, $C_nH_{2n-4}O_2$, etc.

DICARBOXYLIC ACIDS AND POLYCARBOXYLIC ACIDS

Acids $C_nH_{2n-4}O_4$, $C_nH_{2n-6}O_4$, etc.; $C_nH_{2n-4}O_6$, $C_nH_{2n-6}O_6$, etc.

HYDROXY(OXY)-CARBOXYLIC ACIDS: $HO-CO_2H$, $HOCH_2-CO_2H$, etc.

OXO-CARBOXYLIC ACIDS: $H-CO-CO_2H$, $CH_3-CO-CH_2-CO_2H$, etc.

HYDROXY(OXY)-OXO-CARBOXYLIC ACIDS: $HOCH_2-CO-CO_2H$, etc.

Carboxylic acids containing a functional group such as $-SO_3H$, $-NH_2$, $-N{=}N-$, or $-HgCl$, are placed under a later class of compounds. Lactones and cyclic acid anhydrides are listed under oxygen heterocycles, lactams and cyclic imides under nitrogen heterocycles.

Learning to use Beilstein's *Handbuch* easily and effectively requires some exercise of patience and ingenuity—but it is a challenge and a rewarding experience.[7] This remarkable enterprise of Beilstein, which started modestly in 1880–1881 as a two-volume edition of about 2200 pages, has become a bulwark of organic chemistry throughout the world.

## Reviews and Surveys: Books and Periodicals

These publications afford an excellent means of locating information on a wide variety of topics. The articles survey specific subjects and usually are written by scientists actively working in the fields covered.

It is customary to use a standard form of abbreviation in citing references to the periodical literature. The abbreviations used by *Chemical Abstracts*, and adopted by many other American publications, are indicated in the following lists by the letters printed in boldface type.[8] The usual form of citation for a journal reference gives the names of the authors, abbreviated name of the journal, volume number (if any), page number, and year of publication. For books the citation should give the name of the author or editor, name and location of the publisher, number of the edition (except for the first edition), and date of publication. Typical examples will be found among the references cited in this manual.

Journal of **Chem**ical **Educ**ation
**Chem**ical **Rev**iews
**Quarterly Reviews**
**Ann**ual **Reports** on the **Prog**ress of **Chem**istry.
**Org**anic **Reactions**
**Adv**ances in **Org**anic **Chem**istry: Methods and Results
Zeitschrift für **Angew**andte **Chem**ie **Internat**ional **Ed**ition
**Fortschr**itte der **Chem**ie **Org**anischer **Naturstoffe** (Progress in the Chemistry of Organic Natural Products)
**Advances** in **Carbohydrate Chem**istry

[7] Guides to the intricacies of Beilstein, in English, are: Huntress, *A Brief Introduction to the Use of Beilstein's Handbuch der Organischen Chemie*, John Wiley and Sons, Inc., New York, 1938 (44 pp.); Runquist, *A Programmed Guide to Beilstein's Handbuch*, Burgess Publishing Co., Minneapolis, 1966 (53 pp.). See also, Hancock, *J. Chem. Educ.*, **45**, 336 (1968).

[8] For other journals and their abbreviations, see the *List of Periodicals Abstracted by Chemical Abstracts*, Chemical Abstracts Service, American Chemical Society, Columbus, Ohio, 1956 and later supplements. An interesting new periodical (July, 1969) is: *Organic Preparations and Procedures. An International Journal for Rapid Communication.*

**Adv**ances in **Chromatog**raphy
**Adv**ances in **Phys**ical **Org**anic **Chem**istry
**Adv**ances in **Heterocycl**ic **Chem**istry
**Ann**ual **Rev**iew of **Phys**ical **Chem**istry
**Ann**ual **Rev**iew of **Biochem**istry

## Research Journals

The brief list given below includes the principal journals that publish articles in organic chemistry and related fields. The standard form of abbreviation used by Chemical Abstracts is indicated by the letters in boldface type.

**J**ournal of the **Amer**ican **Chem**ical **Soc**iety
**J**ournal of **Org**anic **Chem**istry
**J**ournal of **Biol**ogical **Chem**istry
**J**ournal of the **Chem**ical **Soc**iety (British)
**Tetrahedron** (International)
**Can**adian **J**ournal of **Chem**istry
**Ann**alen der Chemie (Liebig's Annalen) (German)
**Chem**ische **Ber**ichte (formerly **Ber**ichte der deutschen chemischen Gesellschaft)
**Helv**etica **Chim**ica **Acta** (Swiss)
**Ann**ales de **chimie** (**Paris**)
**Bull**etin de la **soc**iété **chim**ique de **France**
**Bull**etin des **soc**ietes **chim**iques **Belges** (Belgian)
**Rec**ueil des **trav**aux **chim**iques des Pays-Bas (Dutch)
**Gaz**etta **chim**ica **ital**iana
**Bull**etin of the **Chem**ical **Soc**iety of **Japan**
**J**ournal of **Gen**eral **Chem**istry (U.S.S.R.) (Russian)

There are many additional journals of a general character and also many journals devoted to specific fields, such as heterocyclic chemistry, organometallic chemistry, medicinal chemistry, polymers, and others.[8] Numerous less well-known journals are published in Sweden, Czechoslovakia, Jugoslavia, and other countries.

## Abstract Journals

These journals publish concise summaries of articles appearing in the research journals and other scientific periodicals, give listings of reviews, surveys and patents, and announcements of new books. Two of the most important journals of this type are *Chemisches Zentralblatt* and *Chemical Abstracts*.

*Chemisches Zentralblatt* (from 1856 in its present form). Fewer periodicals are abstracted than in Chemical Abstracts but the coverage of organic chemistry is quite good. There are collective indexes of authors, subjects, formulas, and patents, that are published at frequent intervals.

*Chemical Abstracts* (from 1907– ). Now published in separate sections (Organic Chemistry, Physical Chemistry, etc.). Author and Subject indexes were published annually through 1960. Currently the author, subject and formula indexes appear semiannually. There is a collective formula index covering 1920–1946; subsequent formula indexes are included in regular collective indexes. There are five collective indexes covering periods of ten years: **I**, 1907–1916; **II**, 1917–1926; **III**, 1927–1936; **IV**, 1937–1946; **V**, 1947–1956. Thereafter the periods are five years: **VI**, 1957–1961; **VII**, 1962–1966.[9]

Effective use of the subject indexes of Chemical Abstracts requires a working knowledge of the system used for naming and indexing organic compounds. This is described fully in the introduction to the subject index of volume **56** (Jan–June 1962).[10] This survey presents a comprehensive explanation of inorganic and organic chemical nomenclature for systematic indexing, a classified bibliography, and the following appendixes (lists): Miscellaneous Chemical Prefixes, Inorganic Groups and Radicals, Anions, Organic Groups and Radicals (by name and structure), Organic Suffixes. For the naming and numbering of positions of cyclic structures, *The Ring Index*[11] is extremely helpful.

Organic compounds are indexed on the basis of *index compounds* and the names of substituents follow, in alphabetical order (principle of inversion). Thus, 2-iodo-3,3-dimethyl-1-pentanol is listed in the form, 1-Pentanol, 2-iodo-3,3-dimethyl-. A few of the basic general rules are summarized below, merely to indicate some of the principles of the system:

The *chief function* is expressed within the main part of the name (index compound) and not as a substituent. Thus, hexanol is used instead of hydroxyhexane; naphthoic acid, instead of carboxynaphthalene.

Common or trivial names of long established usage are employed rather than the less familiar systematic names. Examples are: Acetylene instead of Ethyne; Palmitic acid instead of Hexadecanoic acid. There are many exceptions to this rule; substitution products or homologs may appear under the systematic index names. This is troublesome.

For compounds containing two or more different functions, the *chief function* is determined by an arbitrary order of precedence (not based on the relative importance of the functions): -onium compound, acid (carboxylic, sulfonic and others), acid halide, amide, imide, amidine, aldehyde, nitrile, isocyanide, ketone, alcohol, phenol, thiol, amine, imine, ether, sulfoxide, sulfone. Thus, acetonyltrimethylammonium

[9] The current issues of Chemical Abstracts have a *Keyword Index* with entries from both titles and texts of the abstracts in each *complete* issue regardless of the particular section (organic, physical, etc.) in which they are published.

[10] Published separately as a reprint entitled: *The Naming and Indexing of Chemical Compounds by Chemical Abstracts*; available from Chemical Abstracts Service, American Chemical Society, Columbus, Ohio.

[11] Patterson, Capell, and Walker, *The Ring Index*, American Chemical Society Monograph Series, Reinhold Publishing Corp., New York, 2nd edition, 1960; also, three supplements, 1963–1965.

iodide is indexed under Ammonium compounds, substituted. Acetonyltrimethyl-iodide. The compound $HOCH_2{-}CH{=}C(CH_3){-}CO{-}CH_3$ is indexed as 3-Penten-2-one, 5-hydroxy-3-methyl-.

For compounds containing two or more functions of the same kind, where feasible, both functions are included in the index compound: thus, 2,4-Pentanediol; 1,5-Naphthalenedicarboxylic acid. The fundamental chain is selected to contain the maximum numbers of such functions, although it may not be the longest chain present.

Some functions (as nitro and nitroso) are designated *always* as prefixes, which will be inverted in the index; other functions (as double and triple bonds) are indicated only by suffixes. Many functions can be denoted either by a prefix or a suffix (as hydroxy- or -ol, for alcohols and phenols)—the choice will depend upon the rules for choosing the proper name for the index compound; thus, aminopropanol; hydroxypropionaldehyde.

For complex molecules (and some simple ones) selection of the proper index compound may be difficult. It is helpful in learning the system to consult the comprehensive guide[10] and prepare a concise digest of the principal rules. One can then obtain experience by writing names for specific compounds and checking the results against actual index entries in one of the collective subject indexes.

# Appendix

# Materials and Reagents

The amounts listed below for each experiment are the quantities of materials and reagents needed for ten students. Wherever an experiment specifies a variable quantity, the maximum amount has been taken. It is necessary in ordering materials to add to the quantities listed, on the average about 10 percent, to take care of normal losses in handling and distributing. Frequently, repetition of experiments is required, which may increase the figures still more. Desk reagents[1] are included in the list except for concd hydrochloric acid, concd sulfuric acid, and 10 percent aqueous sodium hydroxide, *which are not shown for quantities less than 100 ml* (10 ml per student).

In the following list the solutions indicated are aqueous solutions unless another solvent is specified.

## Exercises on Laboratory Operations

### 1. Simple Distillation

**(A)** Carbon tetrachloride: 250 ml; 365 g
**(B)** Carbon tetrachloride–toluene mixture (equal volumes): 250 ml (185 g carbon tetrachloride, 110 g toluene)
**(C)** Distillation mixture: 250 ml

[1] It will usually be found convenient to place certain common materials and reagents on the desk divider and the side-shelves. A list of appropriate **Desk Reagents** is given in Table 1, page 5. **Side-Shelf Reagents and Materials** are listed at the end of this section after the materials for Experiment 56.

## 2. Fractional Distillation

Carbon tetrachloride–toluene mixture (equal volumes): 1.2 liters (950 g carbon tetrachloride, 520 g toluene)

*or* Benzene–toluene mixture (equal volumes): 1.2 liters (530 g benzene, 520 g toluene)

*or* Benzene–acetic acid mixture (equal volumes): 1.2 liters (530 g benzene, 630 g glacial acetic acid)

*or* Methanol–water mixture (equal volumes): 1.2 liters (475 g methanol, 600 g water)

## 3. Vacuum Distillation

**(A)** Benzaldehyde (practical grade): 580 ml; 600 g
Sodium carbonate solution (10 percent): 1 liter
Anhydrous magnesium sulfate: 100 g (see footnote 11)
Hydroquinone *or* catechol (anti-oxidant): 5 g
**(B)** Ethyl acetoacetate (practical grade): 600 ml; 600 g

## 4. Steam Distillation

**(A)** *o*-Chlorotoluene *or* bromobenzene *or* *p*-chlorotoluene *or* furfural: 280 ml
**(B)** *p*-Dichlorobenzene and salicylic acid: 50 g of each

## 5. Determination of Melting Points

Benzoic acid and 2-naphthol: 10 g of each
*or* Cinnamic acid and urea: 10 g of each
Paraffin oil *or* glycerol (*or* other bath liquid): 1 liter; 920 g

## 6. Crystallization and Sublimation

**(A)** Impure acetanilide:[2] 100 g
Decolorizing carbon: 20 g
**(B)** Impure naphthalene:[3] 100 g
Decolorizing carbon: 5 g
Ethanol (95 percent): 500 ml
**(C)** Impure *p*-dichlorobenzene:[4] 10 g

[2] Satisfactory material for this purpose may be prepared by mixing thoroughly 100 g of ordinary acetanilide, 20 g of ordinary sugar (sucrose), and 4–5 g of sawdust that has been impregnated with 4–5 g of aniline.

[3] Satisfactory material for this purpose may be prepared by mixing thoroughly 100 g of ordinary naphthalene, 5 g of commercial biphenyl (or acetanilide), and 4–5 g of sawdust.

[4] Suitable material for this purpose may be prepared by mixing together thoroughly 10 g of *p*-dichlorobenzene and 1 g of methyl orange (or charcoal).

## 7. Simple and Multiple Extraction

**(A)** and **(B)** Glacial acetic acid *or* propionic acid: 50 ml; 53 g
Ether: 3.5 liters
Standard sodium hydroxide solution (about 0.3 *N*): 1 liter
**(C)** Dry tea: 300 g
Calcium carbonate: 150 g
Chloroform: 750 ml; 1.12 kg
Benzene: 100 ml
Ligroin, bp 60–90: 200 ml; 140 g

## 8. Chromatography

**(A)** Dye-mixture:[5] 5 ml
Alumina: 50 g (Merck Alumina 71707 is suitable)
Celite: 50 g
**(B)** Spinach *or* green leaves *or* carrot tops: 10 g
Acetone: 50 ml; 40 g
Petroleum ether: 50 ml; 35 g
Developing solvent (8 volumes of petroleum ether or pentane, 2 volumes of acetone)
Thin layer plates (Eastman Chromagram Sheet, Type 6060 or 6061 is suitable).
**(C)** Sand: 100 g
Silica gel: 250 g (Fisher 60–200 mesh, grade 950 is suitable)
Other reagents as in part **(B)**

# Identification of Organic Compounds

## 9a. Qualitative Detection of Elements

Sugar (sucrose): 20 g
Cupric oxide: 20 g
Quicklime (calcium oxide), halogen-free: 60 g
Copper wire: 3 meters
Carbon tetrachloride: 15 ml; 25 g
Sulfanilic acid: 5 g
*p*-Dichlorobenzene: 5 g
Metallic sodium (under kerosene or xylene): 20 g
Lead acetate solution (10 percent): 10 ml
Sodium nitroprusside solution (2 percent): 10 ml
Silver nitrate solution (5 percent): 80 ml

[5] A suitable dye solution may be prepared by dissolving 0.1 g each of crystal violet, auramine hydrochloride, and malachite green in 100 ml of water, or commercial food color, 1 g.

Lime water (saturated calcium hydroxide solution): 50 ml
Potassium fluoride solution (10 percent): 10 ml
Ferrous sulfate solution (10 percent): 10 ml
Asbestos board

### 9b. Solubility Tests

Acetone: 50 ml; 40 g
Sucrose: 50 g
Benzoic acid: 50 g
Benzene sulfonamide: 50 g
Dimethylaniline: 50 ml, 48 g
Cyclohexanol: 50 ml, 48 g
Benzene: 50 ml
Acetanilide: 50 g
Solubility unknowns: 5 g samples for each student

# Preparation and Reactions of Typical Organic Compounds

### Experiment 1: Hydrocarbons

*n*-Heptane *or* *n*-pentane:[6] 40 ml; 28 g
1-Pentene *or* cyclohexene (from Expt 2): 40 ml; 28 g
Benzene: 40 ml
1-Hexyne:[7] 10 ml; 10 g
Bromine in carbon tetrachloride (2 percent): 280 ml
Potassium permanganate solution (0.5 percent): 120 ml
Sodium carbonate solution (10 percent): 120 ml
Ammoniacal cuprous chloride solution: 50 ml
*or* the following reagents: copper sulfate crystals ($CuSO_4 \cdot 5H_2O$) (20 g), sodium chloride (5 g), sodium bisulfite (5 g)
Light source (200–300 watts)

[6] Ligroin (bp 60–90°) or petroleum ether may be used instead of the individual hydrocarbons, but the commercial material must be tested for unsaturation. If unsaturated hydrocarbons are present they may be removed by shaking the material with half its volume of cold concentrated sulfuric acid, and then with small portions of fuming sulfuric acid (7 percent or stronger), until a fresh portion of acid no longer becomes colored; finally, the material is washed with water, dried over anhydrous calcium chloride, and filtered from the drying agent.

[7] 1-Hexyne is available from Farchan Research Laboratories, 4702 East 355th Street, Willoughby, Ohio, 44094, at $9 per 100 g.

## 2a. Methylpentenes

4-Methyl-2-pentanol (practical grade): 190 ml; 153 g
Concd sulfuric acid: 215 ml; 395 g
Anhydrous calcium chloride: 20 g
Ice
Test reagents for unsaturation:
    Potassium permanganate solution (0.5 percent): 30 ml
    Bromine in carbon tetrachloride (2 percent): 40 ml

## 2b. 2-Pentene

3-Pentanol (practical grade) *or* 2-pentanol: 215 ml; 175 g
Concd sulfuric acid: 215 ml; 395 g
Anhydrous calcium chloride: 20 g
Ice
Test reagents as in 2a

## 2c. Cyclohexene

Cyclohexanol (practical grade): 210 ml; 200 g
Sodium carbonate solution (10 percent): 20 ml
Anhydrous calcium chloride: 40 g
Ice
Test reagents as in 2a

## 3a. Ethanol

Sucrose: 800 g
Pasteur's salts solution: 800 ml
    Potassium phosphate: 2 g
    Calcium phosphate: 0.2 g
    Magnesium sulfate: 0.2 g
    Ammonium tartrate: 10 g
Compressed yeast: 5 cakes
Kerosene or Paraffin oil: 75 ml; 60 g
Barium hydroxide solution (saturated): 100 ml
Anhydrous potassium carbonate: 200 g
Iodine solution in aqueous potassium iodide (as in 3b): 100 ml

## 3b. Test Reactions of Alcohols[8]

Iodine solution, in aqueous potassium iodide: 250 ml
  Iodine: 30 g
  Potassium iodide: 60 g
Glacial acetic acid: 20 ml; 22 g
Metallic sodium (under kerosene or xylene): 3 g
Sodium dichromate solution (1 percent): 150 ml
Lucas' reagent (zinc chloride–hydrochloric acid): 175 ml
  Anhydrous zinc chloride (fused): 170 g
  Concd hydrochloric acid: 115 ml; 135 g
Alcohols for testing and for "unknowns":
  Ethanol (95 percent): 30–40 ml
  *n*-Propyl alcohol: 30–40 ml
  *s*-Butyl alcohol: 30–40 ml
  *t*-Butyl alcohol: 30–40 ml
  Isopentyl alcohol: 10–20 ml
Additional alcohols for testing and for "unknowns": 15–25 ml of each
  Isopropyl alcohol, Isobutyl alcohol, *n*-Butyl alcohol, *n*-Pentyl alcohol, *s*-Pentyl alcohols (2- or 3-pentanol), *t*-Pentyl alcohol
For optional experiment: 1-naphthyl isocyanate (6 g), ligroin (50 ml), sodium acetate, fused (2 g)

## 4. 2-Methyl-2-pentanol

Magnesium turnings: 80 g
*n*-Propyl bromide: 300 ml; 405 g
Ether (absolute): 1.7 liters
Iodine: 1 g
Methyl iodide *or* 1,2-dibromoethane: 5 g (for initiating sluggish reactions)
Acetone (anhydrous): 250 ml; 200 g
Anhydrous potassium carbonate *or* Anhydrous magnesium sulfate: 250 g
Ether (ordinary): 1 liter
Potassium permanganate solution (0.5 percent): 30 ml
Bromine in carbon tetrachloride (2 percent): 30 ml
Ice

## 5a. *n*-Butyl Bromide

*n*-Butyl alcohol: 460 ml; 370 g
Sodium bromide ($NaBr \cdot 2H_2O$): 840 g

[8] The quantities of alcohols and reagents will vary according to the number of test reactions and of "unknowns" assigned to the class. The amounts listed will permit rather extensive testing. Technical or practical grades of alcohols may be used for the Lucas test, provided the student is forewarned.

Concd sulfuric acid: 900 ml; 1670 g
Sodium bicarbonate solution (saturated): 300 ml
Anhydrous calcium chloride: 60 g

## 5b. *s*-Butyl Chloride

*s*-Butyl alcohol (practical grade): 310 ml; 250 g
Anhydrous zinc chloride (fused); 900 g
Concd hydrochloric acid: 550 ml; 630 g
Concd sulfuric acid: 250 ml; 460 g
Anhydrous calcium chloride: 60 g

## 5c. *t*-Butyl Chloride

*t*-Butyl alcohol: 240 ml; 185 g
Concd hydrochloric acid: 700 ml; 830 g
Sodium bicarbonate solution (5 percent): 300 ml
Anhydrous calcium chloride: 60 g

## 5d. Ethyl Iodide

Phosphorus (red): 25 g
Ethanol (95 percent): 300 ml; 250 g
Iodine: 255 g
Sodium hydroxide solution (10 percent): 100 ml
Anhydrous calcium chloride: 30 g
Ice

## 6. Methyl *n*-Butyl Ether

Anhydrous sodium methoxide: 160 g
*n*-Butyl bromide (from Expt 5a): 215 ml; 275 g
Methanol: 150 ml
Calcium chloride solution (25 percent): 500 ml
Anhydrous calcium chloride: 40 g

## 7a. Acetaldehyde

Paraldehyde: 200 ml; 200 g
Sulfamic acid *or* *p*-toluenesulfonic acid: 20 g
Anhydrous calcium chloride: 10 g
Ice

### 7b. Reactions of Aldehydes

Formalin (40 percent formaldehyde solution): 20 ml
Acetaldehyde (from above): 25 ml; 20 g
Benzaldehyde (optional): 25 ml; 26 g
Acetone or cyclohexanone (from Expt 8): 25 ml; 24 g
Silver nitrate solution (5 percent): 60 ml
Ammonia solution (2 percent): 75 ml
Schiff's fuchsin-aldehyde reagent (0.5 g *p*-rosaniline hydrochloride per liter): 30 ml
Methone (optional): 5 g
Diethylamine (or piperidine): 1 g

### 8a. Cyclohexanone

Cyclohexanol (practical grade): 210 ml; 200 g
Sodium dichromate crystals ($2H_2O$): 210 g
Concd sulfuric acid: 170 ml; 315 g
Sodium chloride (ordinary salt): 250 g
Methylene chloride: 150 ml; 200 g *or* pentane: 150 ml
Anhydrous magnesium sulfate: 60 g

### 8b. Diethyl Ketone

3-Pentanol (practical grade): 325 ml; 265 g
Sodium dichromate crystals ($2H_2O$): 300 g
Concd sulfuric acid: 220 ml; 400 g
Sodium carbonate: 5 g
Sodium chloride (ordinary salt): 100 g
Anhydrous potassium carbonate: 50 g

### 9. Reactions of the Carbonyl Group in Aldehydes and Ketones[9]

**(A)** *n*-Butyraldehyde (*or* benzaldehyde): 50 ml; 40 g
Acetone: 50 ml; 40 g
Ethanol (95 percent): 750 ml
Sodium bisulfite solution (saturated): 200 ml
**(B)** Acetone: 80 ml; 65 g
Hydroxylamine hydrochloride: 50 g
Sodium hydroxide (sticks *or* pellets): 30 g

[9] The quantities of aldehydes and ketones, and of reagents, will vary according to the number of tests assigned to the class. The amounts listed here cover the specific examples given in the Manual; if optional tests are to be done, larger quantities of the various reagents will be needed.

Optional oxime test: furfural: 50 ml; 55 g
*or* Cyclohexanone (from Expt 8a): 50 g; 53 ml
Hydroxylamine hydrochloride: 50 g
Sodium acetate crystals ($3H_2O$): 75 g
**(C)** *n*-Butyraldehyde: 10 ml; 10 g
Semicarbazide hydrochloride: 10 g
Sodium acetate crystals ($3H_2O$): 15 g
Optional semicarbazone tests: acetone, furfural, cyclohexanone
Ice (for parts **A**, **B**, and **C**)
**(D)** Furfural: 5 ml; 6 g
Cyclohexanone (from Expt 8a): 5 ml; 6 g
Phenylhydrazine hydrochloride:[10] 20 g
Sodium acetate crystals ($3H_2O$): 30 g
*or* Phenylhydrazine (base):[10] 20 ml; 22 g
**(E)** Cyclohexanone: 10 g, 11 ml
2,4-Dinitrophenylhydrazine: 10 g
Ethanol (95 percent): 750 ml
Optional 2,4-dinitrophenylhydrazone tests: 2,4-dinitrophenylhydrazine (20 g), *n*-butyraldehyde, acetone, etc.
**(F)** Acetone: 5 ml; 4 g
*n*-Butyraldehyde: 5 ml; 4 g
Sodium hydroxide solution (10 percent): 50 ml
Iodine: 10 g
Potassium iodide: 20 g
**(G)** Carbonyl compounds for unknowns (Table 19): 5-ml samples

## 10a. *n*-Butyl Acetate

**(A)** *n*-Butyl alcohol: 150 ml; 110 g
Glacial acetic acid: 170 ml; 180 g
Sodium carbonate solution (saturated): 500 ml
Anhydrous magnesium sulfate:[11] 60 g *or* Drierite: 100 g
**(B)** *n*-Butyl acetate (from part **A**): 50 ml; 45 g
Sodium hydroxide solution (10 percent): 250 ml

[10] Phenylhydrazine hydrochloride may be prepared from phenylhydrazine (base) by the following method: Dissolve 20 ml (22 g) of the base in 250 ml of 95 percent ethanol and add slowly, with shaking, a mixture of 25 ml of concentrated hydrochloric acid and 50 ml of ethanol. Chill thoroughly and filter the hydrochloride with suction. Wash the crystals with two 20-ml portions of cold ethanol, and finally with ether. The yield is 24–28 g. Note that phenylhydrazine (base) is poisonous and must be handled carefully.

[11] Anhydrous magnesium sulfate may be prepared readily by heating the crystalline hydrate, Epsom salts ($MgSO_4 \cdot 7H_2O$), to dryness in a porcelain dish or casserole over a wire gauze, and stirring with a porcelain spatula or glass rod until a dry powder is obtained. The loss in weight is about 50 percent.

Sodium chloride (ordinary salt): 100 g
**(C)** Potassium hydroxide (pellets): 30 g
Ethanol (95 percent): 500 ml
*n*-Butyl acetate (from part **A**): 10 ml; 9 g
Standardized hydrochloric acid (0.5 *N*): 500 ml
**(D)** Sodium hydroxide (sticks or pellets): 10 g
Unknown organic acid ($C_2$–$C_5$): 30 g
**(E)** Unknown ester: 1–2 ml
Hydroxylamine hydrochloride solution in ethanol (5 percent): 10 ml
Ferric chloride solution in water (3 percent): 2 ml

## 10b. Methyl Benzoate

Benzoic acid: 122 g
Methanol: 400 ml; 320 g
Methylene chloride: 750 ml; 1 kg
Sodium carbonate solution (5 percent): 500 ml
Anhydrous magnesium sulfate:[11] 100 g

## 11a. Acetyl Chloride

Glacial acetic acid: 230 ml; 240 g
Phosphorus trichloride: 130 ml; 210 g
Aniline: 10 ml; 10 g
Ethanol (95 percent): 25 ml; 20 g
Sodium chloride solution (saturated): 20 ml
Ice

## 11b. Acetic Anhydride

Anhydrous sodium acetate: 350 g
Acetyl chloride (from part **A**): 180 ml; 200 g
Ethanol (95 percent): 20 ml
Aniline: 10 ml; 10 g

## 12. Amines

Aliphatic amines for tests: examples such as *n*-propylamine, isobutylamine, cyclohexylamine, diethylamine; aqueous solutions of methylamine, dimethylamine, ethylamine, etc.
Aniline: 100 ml; 100 g
Methylaniline: 100 ml; 100 g
Dimethylaniline: 100 ml; 100 g
Acetic anhydride: 93 ml; 100 g

Benzenesulfonyl chloride: 30 ml; 42 g
Methyl iodide *or* Ethyl iodide: 15 ml; 35 g
Benzoyl chloride: 20 ml; 24 g
Ethanol for crystallizations: 100 ml
Samples of amines for identification (e.g., *t*-butylamine, piperidine, *o*-toluidine, *p*-toluidine, *m*-toluidine, *o*-anisidine)
Medicine droppers, or graduated 1-ml pipettes are desirable for this experiment

## 13a. Capryl Chloride and Capramide

Capric acid: 20 g
Thionyl chloride: 20 ml; 35 g
Aqueous ammonia (concd): 200 ml
Methylene chloride: 150 ml; 200 g *or* ethanol-free ether: 150 ml

## 13b. Acetamide

Ammonium acetate: 500 g
Glacial acetic acid: 400 ml; 420 g
Benzene: 30 ml
Ethyl acetate: 10 ml; 9 g
Yellow mercuric oxide: 5 g
For preparation of Acetonitrile (optional): Acetamide (180 g), Phosphorus pentoxide (300 g), Anhydrous potassium carbonate (250 g)

## 14a. Wöhler's Synthesis of Urea

Potassium cyanate: 200 g
Ammonium sulfate: 260 g
Methanol: 700 ml; 560 g
Sodium nitrite: 5 g
Sodium hypochlorite solution ("Clorox"): 50 ml
Copper sulfate solution (dilute): 5 ml
For preparation of Dulcin (optional): *p*-Phenetidine (14 ml; 14 g), Glacial acetic acid (20 ml; 21 g), Potassium cyanate (16 g) *or* Sodium cyanate (13 g)
For Eicosane-Urea Complex (optional): Eicosane (100 ml of 50 percent solution of technical eicosane in xylene), Urea (200 g), Methanol (1.2 liters)

## 14b. Urea Nitrate and Urea from Urine

Urine: 5 liters
Concd nitric acid: 300 ml; 430 g
Ethanol (95 percent): 2.5 liters; 2 kg
Potassium permanganate solution (0.5 percent): 25 ml

Decolorizing carbon: 75 g
Sodium hypochlorite solution ("Clorox"): 10 ml
Sodium nitrite: 5 g
Barium carbonate: 120 g

## 15. *n*-Valeronitrile and *n*-Valeric Acid

Sodium cyanide (95 percent): 270 g
Potassium iodide: 15 g
*n*-Butyl bromide (from Expt 5a): 475 ml; 620 g
2-Ethoxyethanol (Cellosolve): 700 ml
Concd sulfuric acid: 200 ml; 370 g
Sodium bicarbonate solution (5 percent): 200 ml
Anhydrous calcium chloride: 60 g
*n*-Valeronitrile (from first step): 210 g
Sodium hydroxide (sticks *or* pellets): 200 g
Concd sulfuric acid: 170 ml; 315 g
Benzene: 1 liter
Anhydrous magnesium sulfate: 60 g
Reagents for preparation of *n*-Valeranilide (optional): Thionyl chloride (20 g; 15 ml), Benzene (300 ml), Aniline (20 ml; 20 g), Dilute hydrochloric acid (5 percent) (100 ml), Sodium hydroxide solution (5 percent) (100 ml)

## 16. Carbohydrates—Sugars

Glucose (40 g), Fructose, Sucrose, Lactose, Maltose (10 g each for tests)
Fehling's solutions, No. 1 and No. 2: 150 ml of each
To prepare one liter of each of these solutions the following amounts of reagents are required:
Copper sulfate crystals ($5H_2O$) (35 g), Sodium potassium tartrate (Rochelle salts) (170 g), Sodium hydroxide (sticks) (50 g) *or* Benedict's solution
To prepare one liter of this solution the following amounts of reagents are required:
Sodium citrate (173 g), Sodium carbonate, anhydrous (100 g), Copper sulfate crystals ($5H_2O$) (17 g)
Phenylhydrazine hydrochloride: 50 g
Sodium acetate crystals ($3H_2O$): 75 g
*or* Phenylhydrazine (base): 50 ml; 55 g
Sodium bisulfite solution (saturated): 10 ml
Fused sodium acetate: 15 g
Acetic anhydride: 140 ml; 150 g
Methanol: 150 ml
Benzoyl chloride: 20 ml; 25 g

Sodium hydroxide solution (10 percent): 150 ml
Ethanol (95 percent): 200 ml

## 17a. Ethyl Benzene

Ethyl bromide: 310 ml; 440 g
Benzene: 1.75 liters
Anhydrous aluminum chloride: 130 g
Anhydrous calcium chloride: 150 g
Concd hydrochloric acid: 200 ml; 240 g
Ice

## 17b. *t*-Butylbenzene

Benzene: 600 ml
*t*-Butyl chloride (from Expt 5c): 270 ml; 230 g
Aluminum, thin sheets or foil: 10 g
Mercuric chloride solution (5 percent): 100 ml
Ethanol (95 percent): 100 ml
Anhydrous calcium chloride: 120 g
Methanol: 200 ml
Asbestos paper

## 17c. Acetophenone

Acetic anhydride: 200 ml; 210 g
Benzene, thiophene-free: 1.05 liters
Anhydrous aluminum chloride: 670 g
Concd hydrochloric acid: 1.2 liters; 1.4 kg
Ether (ordinary): 500 ml
Sodium hydroxide solution (10 percent): 400 ml
Anhydrous calcium chloride: 200 g
Ice
For oxidation to Benzoic acid (optional): Sodium hypochlorite solution ("Clorox"): 400 ml

## 17d. 4-Acetylbiphenyl

1,2-Dichloroethane: 1.2 liters; 1.5 kg
Anhydrous aluminum chloride: 160 g
Acetyl chloride: 80 ml; 90 g
Biphenyl: 160 g
Anhydrous magnesium sulfate: 100 g
Ethanol (95 percent): 1 liter

Sodium hydroxide solution (5 percent): 500 ml
Concd hydrochloric acid: 400 ml; 475 g
Ice

## 18a. Nitrobenzene

Benzene: 270 ml
Concd nitric acid: 250 ml; 350 g
Concd sulfuric acid: 300 ml; 550 g
Anhydrous calcium chloride: 100 g

## 18b. *m*-Dinitrobenzene

Nitrobenzene: 30 ml; 36 g
Concd nitric acid: 50 ml; 70 g
Ethanol (95 percent): 350 ml

## 18c. *p*-Bromonitrobenzene

Bromobenzene: 55 ml; 80 g
Concd nitric acid: 100 ml; 142 g
Concd sulfuric acid: 100 ml; 185 g
Ethanol (95 percent): 600 ml

## 18d. Methyl *m*-Nitrobenzoate

Methyl benzoate (from Expt 10b): 63 ml; 70 g
Concd nitric acid: 50 ml; 70 g
Concd sulfuric acid: 195 ml; 340 g
Methanol: 100 ml

## 19. Bromobenzene and *p*-Dibromobenzene

Benzene: 270 ml; 240 g
Iron wire (or tacks): 30 g
*or* Anhydrous aluminum chloride: 5 g
Bromine: 130 ml; 400 g
Ethanol (95 percent): 350 ml
Anhydrous calcium chloride: 60 g
Decolorizing carbon: 20 g

## 20a. Sodium Benzenesulfonate

Benzene: 140 ml
Fuming sulfuric äcid (7 percent): 300 ml; 570 g

Sodium bicarbonate: 180 g
Sodium chloride (ordinary salt): 500 g
S-Benzylthiouronium chloride: 10 g
*or* Benzyl chloride (16 ml; 25 g), Thiourea (15 g), Ethanol (95 percent) (60 ml)
Phosphorus oxychloride: 50 ml; 85 g

## 20b. Sodium *p*-Bromobenzenesulfonate

Bromobenzene: 55 ml; 80 g
Fuming sulfuric acid (7 percent): 110 ml; 200 g
Sodium chloride (ordinary salt): 200 g

## 20c. Sulfanilic Acid

Aniline: 140 ml; 140 g
Decolorizing carbon: 30 g
Concd sulfuric acid: 220 ml; 400 g

## 21a. *p*-Nitrobenzoic Acid

*p*-Nitrotoluene: 90 g
Sodium dichromate crystals ($2H_2O$): 270 g
Concd sulfuric acid: 480 ml; 900 g
Sodium hydroxide solution (5 percent): 800 ml
Decolorizing carbon: 30 g

## 21b. *o*-Nitrobenzoic Acid

*o*-Nitrotoluene: 30 ml; 35 g
Sodium bisulfite: 5 g
Concd hydrochloric acid: 100 ml; 120 g
Decolorizing carbon: 5 g
Potassium permanganate: 90 g

## 22a. Aniline

Nitrobenzene: 210 ml; 250 g
Tin, mossy or granulated: 480 g
Concd hydrochloric acid: 1 liter; 1.2 kg
Sodium hydroxide (sticks *or* pellets): 750 g
Benzene: 500 ml *or* methylene chloride: 500 ml, 670 g
Sodium chloride: 1 kg
Anhydrous magnesium sulfate: 100 g

## 22b. Acetanilide

**(A)** Aniline (from 22a): 46 ml; 47 g
Concd hydrochloric acid: 42 ml; 47 g
Decolorizing charcoal: 30 g
Acetic anhydride: 60 ml; 65 g
Sodium acetate: 80 g
Ice
**(B)** Aniline (from 22a): 46 ml; 47 g
Acetic anhydride: 60 ml; 65 g
Decolorizing charcoal: 20 g
**(C)** Aniline (from 22a): 182 ml; 186 g
Glacial acetic acid: 230 ml; 240 g
Decolorizing charcoal: 20 g
Ice

## 23. *p*-Nitroacetanilide and *p*-Nitroaniline

Acetanilide: 135 g
Glacial acetic acid: 150 ml; 160 g
Concd nitric acid: 70 ml; 100 g
Concd sulfuric acid: 300 ml; 560 g
Ice
*p*-Nitroacetanilide (from first part): 100 g
Concd hydrochloric acid: 350 ml; 420 g
Aqueous ammonia (concd): 400 ml; 360 g

## 24. *p*-Bromoacetanilide and *p*-Bromoaniline

Acetanilide: 135 g
Glacial acetic acid: 500 ml; 530 g
Bromine: 53 ml; 170 g
Decolorizing carbon: 30 g
Ethanol (95 percent): 1.2 liters
Sodium bisulfite solution (saturated): 50 ml
*p*-Bromoacetanilide (from first part): 100 g
Ethanol (95 percent): 200 ml
Potassium hydroxide (sticks *or* pellets): 50 g
Ice
For optional experiment: glacial acetic acid (32 g), sodium cyanate (6.5 g) *or* potassium cyanate (8 g), ethanol (95 percent) (150 ml)

## 25. *p*-Aminobenzenesulfonamide (Sulfanilamide)

Acetanilide (dried thoroughly): 70 g
Chlorosulfonic acid: 200 ml; 360 g
Aqueous ammonia (concd): 250 ml; 200 g
Decolorizing carbon: 20 g
Ice

## 26. N-Phenylsydnone

Chloroacetic acid: 95 g
Aqueous sodium hydroxide (10 percent): 400 ml
Aniline: 100 ml; 105 g
Methylene chloride: 100 ml; 135 g
Sodium nitrite: 46 g
Acetic anhydride: 500 ml; 540 g
Ice

## 27. Hydrazobenzene–Benzidine–Azobenzene

Nitrobenzene: 150 ml; 185 g
Ethanol (95 percent): 2.2 liters
Sodium hydroxide (flakes *or* sticks): 100 g
Zinc dust (good quality): 400 g
Concd hydrochloric acid: 650 ml; 770 g
Ether (ordinary): 900 ml
Sodium dichromate crystals ($2H_2O$): 30 g
Glacial acetic acid: 50 ml; 53 g
Dilute acetic acid (10 percent): 150 ml
Ice

## 28. *o*-Chlorotoluene (*or p*-Chlorotoluene)

*o*-Toluidine (*or p*-toluidine): 215 ml; 215 g
Concd hydrochloric acid: 1.35 liters; 1.7 kg
Copper sulfate crystals ($5H_2O$):[12] 600 g
Sodium chloride (ordinary salt): 180 g
Sodium bisulfite: 140 g

[12] Cuprous chloride (200 g) may be used instead of the copper sulfate, sodium chloride, and sodium bisulfite.

Sodium nitrite: 140 g
Sodium hydroxide (sticks *or* pellets): 90 g
Concd sulfuric acid: 350 ml; 650 g
Starch-iodide test paper
Anhydrous calcium chloride: 40 g
Ice

## 29a. *m*-Chlorotoluene

*p*-Toluidine: 535 g
Glacial acetic acid: 1.6 liters; 1.68 kg
Acetic anhydride: 520 ml; 560 g
Concd hydrochloride acid: 2.7 liters; 3.4 kg
Sodium chlorate: 220 g
Sodium bisulfite: 70 g
Ethanol (95 percent): 3.8 liters
Sodium hydroxide (sticks *or* pellets): 200 g
Concd sulfuric acid: 900 ml; 1.7 kg
Sodium nitrite: 280 g
Powdered copper (copper bronze): 90 g
Anhydrous calcium chloride: 40 g
Starch-iodide test paper
Ice

## 29b. *m*-Bromotoluene

*p*-Toluidine: 535 g
Glacial acetic acid: 1.6 liters; 1.6 kg
Acetic anhydride: 520 ml; 560 g
Bromine: 270 ml; 820 g
Sodium bisulfite: 70 g
Ethanol (95 percent): 4 liters
Concd hydrochloric acid: 1.3 liters; 1.6 kg
Sodium hydroxide (flakes *or* sticks): 250 g
Concd sulfuric acid: 900 ml; 1.7 kg
Sodium nitrite: 280 g
Powdered copper (copper bronze): 90 g
Anhydrous calcium chloride: 40 g
Starch-iodide test paper
Ice

## 30. *m*-Nitroaniline and *m*-Nitrophenol

Sodium sulfide crystals ($9H_2O$): 500 g
*or* Sodium hydrosulfide ($2H_2O$): 190 g and methanol: 1.5 liters
Anhydrous sodium bicarbonate: 170 g
*m*-Dinitrobenzene: 200 g
Methanol: 1.8 liters
*m*-Nitroaniline (from first part): 140 g
Concd sulfuric acid: 870 ml; 1.6 kg
Sodium nitrite: 70 g
Concd hydrochloric acid: 500 ml; 600 g
Starch-iodide test paper
Ice

## 31a. Methyl Orange

Sulfanilic acid (anhydrous *or* monohydrate): 110 g (*or* 115 g)
Sodium nitrite: 35 g
Glacial acetic acid: 30 ml; 32 g
Dimethylaniline: 63 ml; 60 g
Sodium carbonate solution (5 percent): 600 ml
Sodium hydroxide (sticks *or* pellets): 70 g
Sodium chloride (ordinary salt): 200 g

## 31b. Para Red

Sodium hydroxide: 10 g
Trisodium phosphate crystals ($12H_2O$): 95 g
2-Naphthol: 14 g
*p*-Nitroaniline: 14 g (diazotized as in Expt 32)
Ethanol (95 percent): 100 ml
Ice

## 32. *p*-Nitrostilbene

*p*-Nitroaniline (from Expt 23): 70 g
Sodium nitrite: 40 g
Concd hydrochloric acid: 150 ml; 160 g
Cinnamic acid: 75 g
Acetone: 750 ml; 560 g
Cupric chloride (dihydrate): 25 g

Sodium acetate (trihydrate): 250 g
Sodium bicarbonate solution (5 percent): 1.5 liters
Methylene chloride: 1 liter; 1340 g
Decolorizing carbon: 5 g
Ice

## 33. *p*-Methoxystilbene

Triethyl phosphite: 90 ml; 83 g
Benzyl chloride: 58 ml; 63 g
Sodium methoxide: 28 g (see footnote 6, page 348)
*p*-Methoxybenzaldehyde (anisaldehyde): 60 ml; 68 g
Dimethylformamide: 400 ml; 390 g
Ethanol (95 percent): 500 ml

## 34a. Reactions of Phenols

Phenol, aqueous solution (5 percent) (parts **A–D**): 70 ml
Aqueous solutions (about 5 percent) of benzyl alcohol, oxalic acid, acetic acid: 10 ml of each
Ferric chloride solution (3 percent): 5 ml
Small amounts of salicylic acid, 2-naphthol, catechol, acetone (*or* allyl alcohol), anisole
Potassium permanganate solution (0.5 percent): 50 ml
Bromine water (3 percent): 350 ml
Sodium bisulfite solution (5 percent): 200 ml
Ethanol (95 percent): 225 ml
**(E)** Phenol, aqueous solution (5 percent): 200 ml
Sodium nitrite: 9 g
Decolorizing carbon: 10 g
Ice
**(F)** Phenol, aqueous solution (5 percent): 200 ml
Benzoyl chloride: 10 ml; 12 g
Sodium hydroxide solution (10 percent): 200 ml
Methanol: 200 ml

## 34b. 2-Naphthyl Acetate

2-Naphthol ($\beta$-naphthol): 10 g
Sodium hydroxide solution (10 percent): 50 ml
Acetic anhydride: 20 ml; 22 g
Ice

## 34c. Aspirin

Salicylic acid: 14 g
Acetic anhydride: 35 ml; 38 g
Phosphoric acid, syrupy (85 percent): 5 ml
Ether: 250 ml
Petroleum ether (bp 30–60): 250 ml; 170 g
Methanol: 10 ml
Ferric chloride solution (3 percent): 10 ml

## 35. Benzyl Alcohol and Benzoic Acid

Benzaldehyde (freshly distilled): 200 ml; 210 g
Potassium hydroxide (pellets *or* sticks): 180 g
Ether (ordinary): 800 ml
Sodium bisulfite solution (20 percent): 100 ml
Anhydrous magnesium sulfate: 50 g
Concd hydrochloric acid: 400 ml; 475 g
Ice
For optional experiments: 1-naphthyl isocyanate (5 g), ligroin (50 ml); *p*-nitrobenzoic acid, from Expt 21a (5 g), thionyl chloride (33 g), pyridine (12 g); standardized sodium hydroxide solution, about 0.1 *N* (200 ml)

## 36. Benzalacetophenone

Acetophenone (from Expt 17c): 60 ml; 60 g
Benzaldehyde (freshly distilled): 50 ml; 53 g
Ethanol (95 percent): 700 ml
Sodium hydroxide solution (10 percent): 250 ml
For optional experiments: carbon tetrachloride (40 ml), 20 percent bromine in carbon tetrachloride (40 ml); ethanol (330 ml), 30 percent hydrogen peroxide solution (15 ml); glacial acetic acid (100 ml), acetone (50 ml)

## 37. Benzoin and Benzil

Benzaldehyde: 150 ml; 160 g
Ethanol (95 percent): 500 ml
Sodium cyanide: 15 g
For optional experiments with benzoin: glacial acetic acid (10 ml); acetic anhydride (10 ml); ethanol (100 ml), 35 percent aqueous hydroxylamine hydrochloride solution (10 ml), 30 percent sodium hydroxide solution (20 ml)
Ammonium nitrate: 40 g
Cupric acetate solution (2 percent): 50 ml

Glacial acetic acid: 250 ml; 265 g

For optional experiments with benzil: ethanol (50 ml), 25 percent aqueous hydrazine hydrate solution (10 ml); ethanol (50 ml), 35 percent aqueous hydroxylamine hydrochloride solution (10 ml), 30 percent sodium hydroxide solution (20 ml); ethanol (200 ml), semicarbazide hydrochloride (5 g), sodium acetate crystals (15 g); urea (5 g), ethanol (150 ml), 30 percent sodium hydroxide solution (30 ml)

## 38. Benzilic Acid

**(A)** Benzil (recrystallized; from Expt 37): 50 g
Potassium hydroxide (sticks): 50 g
Decolorizing carbon: 5 g
Ethanol (95 percent): 200 ml
Benzene: 300 ml
**(B)** Benzoin (from Expt 37): 65 g
Sodium hydroxide (sticks *or* pellets): 70 g
Sodium bromate: 15 g
*or* Potassium bromate: 17 g
Benzene: 1.2 liters
For optional experiments: glacial acetic acid (10 ml), acetic anhydride (10 ml); methanol (50 ml); glacial acetic acid (100 ml), red phosphorus (6 g), iodine (2 g), sodium bisulfite (10 g), ethanol (210 ml)

## 39. Cinnamic Acid (*or* Furylacrylic Acid)

Potassium acetate (freshly fused): 60 g
Acetic anhydride: 150 ml; 160 g
Benzaldehyde (freshly distilled): 100 ml; 105 g
*or* Furfural (freshly distilled): 80 ml; 93 g
Decolorizing carbon: 40 g
For optional experiments: carbon tetrachloride (410 ml), 20 percent bromine in carbon tetrachloride (160 ml), ethanol (200 ml), potassium hydroxide (20 g), barium chloride dihydrate (10 g)

## 40. Benzohydrol

Benzophenone: 55 g
2-Propanol: 250 ml; 200 g
Sodium borohydride: 6 g
Sodium hydroxide solution (10 percent): 300 ml
Methylene chloride: 80 ml; 110 g

## 41. Triphenylcarbinol

Magnesium turnings: 24 g
Ether (absolute): 1 liter; 720 g
Iodine: 1 g
Bromobenzene: 110 ml; 165 g
Methyl benzoate (from Expt 19): 65 ml; 70 g
*or* Ethyl benzoate: 80 ml; 80 g
Ether (ordinary): 1 liter
Ethanol (95 percent): 250 ml
Sodium bisulfite solution (saturated): 50 ml
Decolorizing carbon: 5 g
Ice

## 42. Benzopinacol and Benzopinacolone

**(A)** Benzophenone: 91 g
2-Propanol: 1 liter; 785 g
Glacial acetic acid: 1 ml
**(B)** Benzophenone: 55 g
Magnesium powder: 15 g
Benzene: 400 ml
Iodine: 45 g
Ether (absolute): 100 ml
Ether (ordinary): 100 ml
Ethanol (95 percent): 100 ml
Benzopinacol (from procedure **A** or **B**): 30 g
Glacial acetic acid: 150 ml; 160 g
Iodine: 1 g
Ethanol (95 percent): 150 ml

## 43. Benzoquinone

Hydroquinone: 55 g
Sodium dichromate crystals ($2H_2O$): 70 g
For optional experiments: acetic anhydride (30 ml), ethanol (50 ml)

## 44a. Anthraquinone

*o*-Benzoylbenzoic acid (anhydrous): 50 g
Concd sulfuric acid: 250 ml; 470 g
Aqueous ammonia (concd): 100 ml; 90 g
Acetone: 100 ml; 80 g
Ice
For optional experiments: sodium hydrosulfite dihydrate (120 g), sodium hydroxide (50 g)

## 44b. Phenanthrenequinone

Phenanthrene (technical 90 percent): 60 g
Glacial acetic acid: 900 ml; 960 g
Concd sulfuric acid: 600 ml; 1.1 kg
Sodium dichromate crystals ($2H_2O$): 360 g
Ethanol (95 percent): 750 ml
Sodium bisulfite: 300 g
Ice

## 45. Di-*p*-chlorophenyltrichloroethane (DDT)

Chloral hydrate: 200 g
Chlorobenzene: 260 ml; 280 g
Concd sulfuric acid: 1 liter; 1.85 kg
Fuming sulfuric acid (15 percent oleum): 420 ml; 800 g
*or* 20 percent oleum: 350 ml; 600 g
Methylene chloride: 600 ml; 820 g
Anhydrous magnesium sulfate: 60 g
Ethanol (95 percent): 2 liters

## 46. 2-Acetylcyclohexanone

Pyrrolidine: 92 ml; 78 g
Cyclohexanone (from Expt 8a): 98 ml; 98 g
*p*-Toluenesulfonic acid: 1 g
Toluene: 600 ml; 520 g
Acetic anhydride: 104 ml; 112 g
Hydrochloric acid (5 percent): 250 ml
Anhydrous calcium chloride: 25 g

## 47. Reactions of Ethyl Acetoacetate

Ethyl acetoacetate: 75 ml; 75 g
Ethanol (95 percent): 160 ml
Ferric chloride solution (3 percent): 10 ml
Small amount of salicylic acid or of a 5 percent aqueous solution of phenol
Bromine in carbon tetrachloride (2 percent): 10 ml
Semicarbazide hydrochloride: 20 g
Sodium acetate crystals ($3H_2O$): 30 g
Copper sulfate crystals ($5H_2O$): 20 g

## 48. Ethyl *n*-Butylacetoacetate and 2-Heptanone

Ethanol (absolute): 2 liters
Methanol (absolute): 200 ml
 and Magnesium turnings: 24 g
*or* Metallic sodium: 15 g
 and Diethyl phthalate: 60 g
Kerosene *or* xylene: 250 ml
Metallic sodium (clean): 69 g
Ethyl acetoacetate (pure): 380 ml; 390 g
*n*-Butyl bromide (pure; from Expt 5a): 360 ml; 460 g
Anhydrous magnesium sulfate: 100 g
Ice
Sodium hydroxide solution (5 percent): 2.5 liters
Ethyl *n*-butylacetoacetate (from first part): 380 g
Concd sulfuric acid: 150 ml; 280 g
Sodium hydroxide (sticks *or* pellets); 100 g
Calcium chloride solution (35–40 percent): 300 ml
Anhydrous magnesium sulfate *or* Drierite: 50 g
For optional preparations of derivatives:
 Semicarbazide hydrochloride: 12 g
 Sodium acetate crystals ($3H_2O$): 18 g
 2,4-Dinitrophenylhydrazine: 10 g
 Ethanol (95 percent): 1 liter

## 49. Ethyl-*n*-Butylmalonate and *n*-Caproic Acid

Ethanol (absolute): 3 liters
Methanol (absolute): 200 ml
 and Magnesium turnings: 24 g
*or* Metallic sodium: 23 g
 and Diethyl phthalate: 90 g
Kerosene *or* xylene: 250 ml
Metallic sodium (clean): 115 g
Ethyl malonate (pure): 800 ml; 880 g
*n*-Butyl bromide (pure; from Expt 5a): 540 ml; 685 g
Anhydrous magnesium sulfate: 300 g
Ethyl *n*-butylmalonate (from first part): 750 g
Potassium hydroxide (sticks *or* pellets): 750 g
Concd sulfuric acid: 830 ml; 1.5 kg
Benzene: 4.5 liters
For optional experiment: Thionyl chloride (20 g), benzene (300 ml), aniline (20 g)
Ice

## 50a. Fluorescein and Eosin

Phthalic anhydride: 75 g
Resorcinol: 120 g
Zinc chloride (fused): 60 g
Concd hydrochloric acid: 150 ml; 180 g
Acetic anhydride: 300 ml; 315 g
Glacial acetic acid: 350 ml; 375 g
Bromine: 120 g; 38 ml
Sodium hydroxide solution (10 percent): 200 ml
Sodium bisulfite solution (saturated): 75 ml
Anhydrous sodium carbonate: 10 g
Ethanol (95 percent): 300 ml

## 50b. Cresol Red

*o*-Sulfobenzoic anhydride: 10 g
*o*-Cresol: 16 g
Zinc chloride: 10 g
Sodium hydroxide solution (10 percent): 500 ml
Concd hydrochloric acid: 120 ml; 150 g

## 51a. Hippuric Acid from Urine

Sodium benzoate *or* Ammonium benzoate: 50 g
Ammonium sulfate: 2 kg
Decolorizing carbon: 20 g

## 51b. Hippuric Acid from Glycine

Glycine: 75 g
Benzoyl chloride: 130 ml; 155 g
Sodium hydroxide (10 percent): 1.1 liters
Ether (ordinary): 500 ml
Decolorizing carbon: 20 g
Ice
For optional azlactone preparation: Hippuric acid (36 g), Benzaldehyde (20 ml; 21 g), Sodium acetate, fused (16 g), Acetic anhydride (60 ml; 65 g)

## 52. Quinoline

Aniline: 190 ml; 190 g
Nitrobenzene: 125 ml; 150 g
Glycerol: 750 g
Boric acid: 120 g

Concd sulfuric acid: 350 ml; 650 g
Ferrous sulfate crystals ($7H_2O$): 70 g
Sodium hydroxide (flakes *or* sticks): 850 g
Concd hydrochloric acid: 150 ml; 180 g
*p*-Toluenesulfonyl chloride: 40 g
Benzene: 1.8 liters
Anhydrous magnesium sulfate: 100 g
Ice

## 53. Fumaric Acid from Furfural

Sodium chlorate: 450 g
Furfural: 175 ml; 200 g
Vanadium pentoxide catalyst: 3 g
*or* Ammonium vanadate, cp: 6 g

## 54. Camphor from Camphene

Camphene: 300 g
Glacial acetic acid: 750 ml; 800 g
Sodium carbonate solution (10 percent): 500 ml
Potassium hydroxide (sticks *or* pellets): 100 g
Ethanol (95 percent): 500 ml
Acetone: 250 ml; 200 g
Chromium trioxide: 135 g
Concd sulfuric acid: 125 ml; 230 g
Ice

## 55a. Dihydroxytriptycene

Anthracene: 18 g
*p*-Benzoquinone (from Expt 43): 11 g
Xylene: 100 ml; 86 g
Potassium hydroxide (sticks *or* pellets): 5 g
Ethanol (95 percent): 500 ml

## 55b. Triptycene

Anthranilic acid: 14 g
*n*-Butyl nitrite: 23 ml; 21 g *or* Isopentyl nitrite: 26 ml; 23 g
Ethanol (95 percent): 150 ml
Ethyl ether: 350 ml
1,2-Dichloroethane: 200 ml; 250 g
Anthracene: 18 g

Propylene oxide: 15 ml; 12 g
Diglyme (diethylene glycol dimethyl ether): 200 ml
Maleic anhydride: 10 g
Potassium hydroxide: 30 g
Methanol: 400 ml

### 56. Polymerization

Glycerol: 100 g
Phthalic anhydride: 150 g
Solvents (acetone, carbon tetrachloride, toluene, ethanol; and others such as cellosolves, carbitols, dioxane, etc.): 50 ml of each
Sodium hydroxide solution (10 percent): 250 ml
Granular methyl methacrylate polymer (Lucite): 200 g
Hydroquinone: 1 g
Benzoyl peroxide:[13] 1 g

## Side-Shelf Reagents and Materials

Anhydrous calcium chloride
Anhydrous magnesium sulfate
Anhydrous sodium carbonate
Anhydrous sodium bicarbonate
Common salt
Sodium bisulfite
Sodium bicarbonate solution (saturated)
Sodium chloride
Sodium hydroxide, pellets
Stopcock grease
Boiling chips
Glass wool
Cotton
Decolorizing carbon
Glycerol
Litmus paper, red and blue

[13] Benzoyl peroxide should be handled carefully and should not be stored in quantities over 20–30 g, and not placed in a glass stoppered container. It should be stored in a dark glass bottle fitted with a clean cork.

# Apparatus and Equipment

Lists A, B, and C indicate items of glassware, ironware, and other supplies, that may be provided as standard equipment in the student's laboratory desk for a full year's laboratory course. For shorter courses some items may be reduced in quantity and variety of sizes or omitted entirely. Expendable (nonreturnable) supplies issued from the stockroom are given in list D. Supplementary apparatus for limited or special purposes, that may be issued to students for a short period of usage, is shown in list E.

## A. Standard Taper Jointed Glassware[14, 15]

1 Adapter, Claisen, three ₸ joints
1 Adapter, thermometer, one ₸ joint
1 Adapter, three-way, three ₸ joints
1 Adapter, vacuum, two ₸ joints
2 Condensers, Liebig-West, 25 cm, two ₸ joints

[14] The ₸ 19/22 joint is recommended as the best single joint size for apparatus covering the volume range of 25–500 ml. Kits of ₸ jointed glassware are available from several manufacturers (Ace, Corning, Kontes) in a variety of joint sizes and components.

[15] In lieu of the taper jointed equipment in list A the following standard glassware (preferably Kimax or Pyrex) is suitable:
2 Adapters, curved, 25 × 180 mm
2 Condensers, Liebig-West, 20—30-cm
1 Condenser tube (air cooled)
5 Distilling flasks, 25—250-ml
5 Flasks, round-bottomed, 100—500-ml
1 Fractionating column
1 Separatory funnel, cylindrical, 50—100-ml

1 Funnel, separatory and addition, 125 ml, ₮ joints with Teflon plug
1 Flask, pear-shaped, 25 ml, ₮ joint
1 Flask, pear-shaped, 50 ml, ₮ joint
1 Flask, round-bottomed, 100 ml, ₮ joint
1 Flask, round-bottomed, 250 ml, ₮ joint
1 Flask, round-bottomed, 500 ml, ₮ joint
2 Stoppers, hollow, pennyhead, ₮ joint

## B. Plain Glassware and Porcelain

2 Beakers, 100-ml, graduated*
2 Beakers, 250-ml, graduated*
1 Beaker, 600-ml, graduated*
1 Bottle, wide-mouth, screw cap, 8 oz
3 Bottles, narrow-mouth, 500-ml
2 Centrifuge tubes, 15-ml
1 Drying tube, calcium chloride, 100-mm
1 Filter plate, No. 4, 25-mm
2 Flasks, Erlenmeyer, 25-ml, graduated*
3 Flasks, Erlenmeyer, 50-ml, graduated*
3 Flasks, Erlenmeyer, 125-ml, graduated*
3 Flasks, Erlenmeyer, 250-ml, graduated*
1 Flask, filtering, 125-ml
1 Flask, filtering, 500-ml
1 Funnel, Büchner, No. 2, 75-mm inside diam.
1 Funnel, conical, short stem, 100-mm
1 Funnel, Hirsch, No. 00, 47-mm plate
1 Funnel, powder, 60-mm
1 Funnel, separatory, with stopper, 500-ml
1 Graduate cylinder, 10-ml
1 Graduated cylinder, 100-ml
10 Test tubes, $10 \times 75$-mm
1 Test tube, $16 \times 150$-mm
1 Test tube, side-arm, $16 \times 150$-mm
1 Thermometer, $-10°$ to $+360°$
1 Thermometer, $-10^0$ to $+260°$
2 Wash bottles, polyethylene, 250- and 500-ml
1 Watch glass, 125-mm

* Beakers and Erlenmeyer flasks are available with imprinted graduations which are useful for estimating volumes of liquids.

## C. Hardware and Accessories

1 Burner, micro
1 Burner, Tirrill-type
1 Burner tip, wing top
2 Clamps, burette, single
2 Clamps, condenser
2 Clamp holders
1 Clamp, screw, Hofmann
1 File, triangular
1 Filter adapter, Neoprene, No. 1
1 Filter adapter, Neoprene, No. 2
1 Filter adapter, Neoprene, No. 3
2 Pans, enameled, 5- and 9-inch
2 Rings, iron, 4-inch
3 Ring stands, 24-inch
1 Spatula, nickel
1 Spatula, laboratory spoon
1 Steam bath, with rings
1 Test tube rack
1 Test tube clamp
18 ft Rubber tubing, 3–4 ft lengths
3 ft Rubber tubing, heavy wall
2 Wire gauzes, asbestos center

## D. Expendable Supplies

Corks, assorted sizes
Filter papers, assorted sizes
Glass stirring rods
Glass tubing
Indicator papers
Labels, gummed
Marking pencils, for glass
Matches
Medicine droppers and bulbs
Melting point tubes
Packing material for columns

Rubber stoppers, assorted
Sample vials
Soap and scouring powder
Sponges, cellulose
Test tube and beaker brushes
Towels, cotton or cheesecloth

## E. Supplementary Apparatus and Optional Equipment

Burettes and pipettes
Cork borers and sharpener
Cork (suberite) rings
Desiccators
Drying ovens, electric
Evaporating dishes, porcelain
Flasks, round bottomed, 1- and 2-liter
Hot plates, electric
Ice crushing machine
Manometers
Mechanical stirrers and motors
Microscope, low power
Mortars and pestles
Vacuum pumps
Water pumps (for filtrations)

# Tables of Physical Data

## Vapor Pressures of Organic Substances and of Water at 30° to 120°C

(In millimeters of mercury)

| T, °C | WATER | BENZENE | BROMO-BENZENE | *para*-DIBROMO BENZENE | *para*-DICHLORO-BENZENE | CARBON TETRA-CHLORIDE | TOLUENE | ETHANOL |
|---|---|---|---|---|---|---|---|---|
| 30 | 31.8 | 118 | 6 | .. | ... | 143 | 37 | 78 |
| 40 | 55.3 | 181 | 10 | .. | 5 | 216 | 60 | 135 |
| 50 | 92.5 | 269 | 17 | .. | 9 | 317 | 93 | 223 |
| 60 | 149.2 | 389 | 28 | .. | 15 | 451 | 138 | 353 |
| 70 | 233.8 | 547 | 43 | 4 | 24 | 622 | 203 | 543 |
| 80 | 355.5 | 754 | 66 | 7 | 37 | 843 | 288 | 813 |
| 90 | 526.0 | 1016 | 98 | 12 | 56 | 1122 | 402 | 1187 |
| 100 | 760.0 | 1344 | 141 | 18 | 83 | 1463 | 557 | 1695 |
| 110 | 1074. | 1748 | 199 | 28 | 119 | 1884 | 747 | 2364 |
| 120 | 1489. | 2238 | 275 | 42 | 167 | .... | 965 | .... |

## Vapor Pressures of Organic Substances at 100°C

(In millimeters of mercury)

| Substance | mm Hg | Substance | mm Hg |
|---|---|---|---|
| Acetophenone | 26.5 | *p*-Dichlorobenzene | 83 |
| Aniline | 45.7 | Dimethylaniline | 37.9 |
| Benzaldehyde | 60.5 | Ethylbenzene | 307 |
| Benzoic acid | 1.8 | Ethyl benzoate | 17.4 |
| Bromobenzene | 141 | Furfural | 110 |
| Biphenyl | 9 | Nitrobenzene | 20.8 |
| *n*-Butyl acetate | 325 | *p*-Nitrobenzoic acid | 0.006 |
| *n*-Butyl alcohol | 390 | *p*-Nitrotoluene | 7.9 |
| *o*-Chlorotoluene | 133 | Salicylic acid | 0.86 |
| *p*-Chlorotoluene | 123 | Toluene | 557 |
| *o*-Cresol | 13.6 | Triphenylcarbinol | 0.01 |
| *p*-Dibromobenzene | 18 | *n*-Valeric acid | 28.2 |

NOTE: The vapor pressures of salts, such as aniline hydrochloride and sodium acetate, are generally so low at 100°C that they can be considered to be nil.

## Density and Vapor Pressure of Water 0° to 35°C

| T, °C | VAPOR PRESSURE MM MERCURY | DENSITY $d_{4^\circ}^{t^\circ}$ | T, °C | VAPOR PRESSURE MM MERCURY | DENSITY $d_{4^\circ}^{t^\circ}$ | T, °C | VAPOR PRESSURE MM MERCURY | DENSITY $d_{4^\circ}^{t^\circ}$ |
|---|---|---|---|---|---|---|---|---|
| 0 | 4.58 | 0.99987 | 12 | 10.48 | 0.99952 | 24 | 22.18 | 0.99733 |
| 1 | 4.92 | 0.99993 | 13 | 11.19 | 0.99940 | 25 | 23.54 | 0.99708 |
| 2 | 5.29 | 0.99997 | 14 | 11.94 | 0.99927 | 26 | 24.99 | 0.99682 |
| 3 | 5.68 | 0.99999 | 15 | 12.73 | 0.99913 | 27 | 26.50 | 0.99655 |
| 4 | 6.09 | 1.00000 | 16 | 13.56 | 0.99897 | 28 | 28.10 | 0.99627 |
| 5 | 6.53 | 0.99999 | 17 | 14.45 | 0.99880 | 29 | 29.78 | 0.99597 |
| 6 | 7.00 | 0.99997 | 18 | 15.38 | 0.99862 | 30 | 31.55 | 0.99568 |
| 7 | 7.49 | 0.99993 | 19 | 16.37 | 0.99843 | 31 | 33.42 | 0.99537 |
| 8 | 8.02 | 0.99988 | 20 | 17.41 | 0.99823 | 32 | 35.37 | 0.99505 |
| 9 | 8.58 | 0.99981 | 21 | 18.50 | 0.99802 | 33 | 37.43 | 0.99473 |
| 10 | 9.18 | 0.99973 | 22 | 19.66 | 0.99780 | 34 | 39.59 | 0.99440 |
| 11 | 9.81 | 0.99963 | 23 | 20.88 | 0.99757 | 35 | 41.85 | 0.99406 |

## Aqueous Ethanol (Ethyl Alcohol)

| DENSITY $d_{4^\circ}^{20^\circ}$ | PERCENT $C_2H_5OH$ BY WEIGHT | PERCENT $C_2H_5OH$ BY VOLUME | GRAMS $C_2H_5OH$ PER 100 ML | DENSITY $d_{4^\circ}^{20^\circ}$ | PERCENT $C_2H_5OH$ BY WEIGHT | PERCENT $C_2H_5OH$ BY VOLUME | GRAMS $C_2H_5OH$ PER 100 ML |
|---|---|---|---|---|---|---|---|
| 0.98938 | 5 | 6.2 | 4.9 | 0.85564 | 75 | 81.3 | 64.2 |
| 0.98187 | 10 | 12.4 | 9.8 | 0.84344 | 80 | 85.5 | 67.5 |
| 0.97514 | 15 | 18.5 | 14.6 | 0.83095 | 85 | 89.5 | 70.6 |
| 0.96864 | 20 | 24.5 | 19.4 | 0.81797 | 90 | 93.3 | 73.6 |
| 0.96168 | 25 | 30.4 | 24.0 | 0.81529 | 91 | 94.0 | 74.2 |
| 0.95382 | 30 | 36.2 | 28.6 | 0.81257 | 92 | 94.7 | 74.8 |
| 0.94494 | 35 | 41.8 | 33.1 | 0.80983 | 93 | 95.4 | 75.4 |
| 0.93518 | 40 | 47.3 | 37.4 | 0.80705 | 94 | 96.1 | 75.9 |
| 0.92472 | 45 | 52.7 | 41.6 | 0.80424 | 95 | 96.8 | 76.4 |
| 0.91384 | 50 | 57.8 | 45.7 | 0.80138 | 96 | 97.5 | 76.9 |
| 0.90258 | 55 | 62.8 | 49.6 | 0.79846 | 97 | 98.1 | 77.4 |
| 0.89113 | 60 | 67.7 | 53.5 | 0.79547 | 98 | 98.8 | 77.9 |
| 0.87948 | 65 | 72.4 | 57.1 | 0.79243 | 99 | 99.4 | 78.4 |
| 0.86766 | 70 | 76.9 | 60.7 | 0.78934 | 100 | 100 | 78.9 |

## Aqueous Hydrochloric Acid

| DENSITY $d_{4^\circ}^{15^\circ}$ | PERCENT HCl BY WEIGHT | GRAMS HCl PER 100 ML | DENSITY $d_{4^\circ}^{15^\circ}$ | PERCENT HCl BY WEIGHT | GRAMS HCl PER 100 ML | DENSITY $d_{4^\circ}^{15^\circ}$ | PERCENT HCl BY WEIGHT | GRAMS HCl PER 100 ML |
|---|---|---|---|---|---|---|---|---|
| 1.010 | 2.14 | 2.2 | 1.075 | 15.16 | 16.3 | 1.140 | 27.66 | 31.5 |
| 1.015 | 3.12 | 3.2 | 1.080 | 16.15 | 17.4 | 1.145 | 28.61 | 32.8 |
| 1.020 | 4.13 | 4.2 | 1.085 | 17.13 | 18.6 | 1.150 | 29.57 | 34.0 |
| 1.025 | 5.15 | 5.3 | 1.090 | 18.11 | 19.7 | 1.155 | 30.55 | 35.3 |
| 1.030 | 6.15 | 6.4 | 1.095 | 19.06 | 20.9 | 1.160 | 31.52 | 36.6 |
| 1.035 | 7.15 | 7.4 | 1.100 | 20.01 | 22.0 | 1.165 | 32.49 | 37.9 |
| 1.040 | 8.16 | 8.5 | 1.105 | 20.97 | 23.2 | 1.170 | 33.46 | 39.2 |
| 1.045 | 9.16 | 9.6 | 1.110 | 21.92 | 24.3 | 1.175 | 34.42 | 40.4 |
| 1.050 | 10.17 | 10.7 | 1.115 | 22.86 | 25.5 | 1.180 | 35.39 | 41.8 |
| 1.055 | 11.18 | 11.8 | 1.120 | 23.82 | 26.7 | 1.185 | 36.31 | 43.0 |
| 1.060 | 12.19 | 12.9 | 1.125 | 24.78 | 27.8 | 1.190 | 37.23 | 44.3 |
| 1.065 | 13.19 | 14.1 | 1.130 | 25.75 | 29.1 | 1.195 | 38.16 | 45.6 |
| 1.070 | 14.17 | 15.2 | 1.135 | 26.70 | 30.3 | 1.200 | 39.11 | 46.9 |

## Aqueous Nitric Acid

| PERCENT $HNO_3$ BY WEIGHT | DENSITY $d_{4^\circ}^{20^\circ}$ | GRAMS $HNO_3$ PER 100 ML | PERCENT $HNO_3$ BY WEIGHT | DENSITY $d_{4^\circ}^{20^\circ}$ | GRAMS $HNO_3$ PER 100 ML | PERCENT $HNO_3$ BY WEIGHT | DENSITY $d_{4^\circ}^{20^\circ}$ | GRAMS $HNO_3$ PER 100 ML |
|---|---|---|---|---|---|---|---|---|
| 5 | 1.0256 | 5.1 | 55 | 1.3393 | 73.7 | 91 | 1.4850 | 135.1 |
| 10 | 1.0543 | 10.5 | 60 | 1.3667 | 82.0 | 92 | 1.4873 | 136.8 |
| 15 | 1.0842 | 16.3 | 65 | 1.3913 | 90.4 | 93 | 1.4892 | 138.5 |
| 20 | 1.1150 | 22.3 | 70 | 1.4134 | 98.9 | 94 | 1.4912 | 140.2 |
| 25 | 1.1469 | 28.7 | 75 | 1.4337 | 107.5 | 95 | 1.4932 | 141.9 |
| 30 | 1.1800 | 35.4 | 80 | 1.4521 | 116.2 | 96 | 1.4952 | 143.5 |
| 35 | 1.2140 | 42.5 | 85 | 1.4686 | 124.8 | 97 | 1.4974 | 145.2 |
| 40 | 1.2463 | 49.9 | 90 | 1.4826 | 133.5 | 98 | 1.5008 | 147.1 |
| 45 | 1.2783 | 57.5 | 95 | 1.4932 | 141.9 | 99 | 1.5056 | 149.0 |
| 50 | 1.3100 | 65.5 | 100 | 1.5129 | 151.3 | 100 | 1.5129 | 151.3 |

## Aqueous Sulfuric Acid

| PERCENT $H_2SO_4$ BY WEIGHT | DENSITY $d_{4^\circ}^{20^\circ}$ | GRAMS $H_2SO_4$ PER 100 ML | PERCENT $H^2SO_4$ BY WEIGHT | DENSITY $d_{4^\circ}^{20^\circ}$ | GRAMS $H_2SO_4$ PER 100 ML | PERCENT $H_2SO_4$ BY WEIGHT | DENSITY $d_{4^\circ}^{20^\circ}$ | GRAMS $H_2SO_4$ PER 100 ML |
|---|---|---|---|---|---|---|---|---|
| 5 | 1.0317 | 5.2 | 55 | 1.4453 | 79.5 | 91 | 1.8195 | 165.6 |
| 10 | 1.0661 | 10.7 | 60 | 1.4983 | 89.6 | 92 | 1.8240 | 167.8 |
| 15 | 1.1020 | 16.5 | 65 | 1.5533 | 101.0 | 93 | 1.8279 | 170.0 |
| 20 | 1.1394 | 22.8 | 70 | 1.6105 | 112.7 | 94 | 1.8312 | 172.1 |
| 25 | 1.1783 | 29.5 | 75 | 1.6692 | 125.2 | 95 | 1.8337 | 174.2 |
| 30 | 1.2185 | 36.6 | 80 | 1.7272 | 138.2 | 96 | 1.8355 | 176.3 |
| 35 | 1.2599 | 43.8 | 85 | 1.7786 | 151.2 | 97 | 1.8364 | 178.1 |
| 40 | 1.3028 | 52.1 | 90 | 1.8144 | 164.3 | 98 | 1.8361 | 179.9 |
| 45 | 1.3476 | 60.6 | 95 | 1.8337 | 174.2 | 99 | 1.8342 | 181.6 |
| 50 | 1.3951 | 69.6 | 100 | 1.8305 | 183.0 | 100 | 1.8305 | 183.0 |

## Aqueous Acids and Bases

| *Aqueous Sodium Hydroxide* | | | *Aqueous Ammonia* | | | *Fuming Sulfuric Acid* | | |
|---|---|---|---|---|---|---|---|---|
| DENSITY $d_{4^\circ}^{20^\circ}$ | PERCENT NaOH BY WEIGHT | GRAMS NaOH PER 100 ML | DENSITY $d_{4^\circ}^{20^\circ}$ | PERCENT $NH_3$ BY WEIGHT | GRAMS $NH_3$ PER 100 ML | SPECIFIC GRAVITY $d_{20^\circ}^{20^\circ}$ | PERCENT FREE $SO_3$ BY WEIGHT | GRAMS FREE $SO_3$ PER 100 ML |
| 1.0207 | 2 | 2.0 | 0.9939 | 1 | 1.0 | 1.860 | 1.54 | 2.8 |
| 1.0428 | 4 | 4.2 | 0.9895 | 2 | 2.0 | 1.865 | 2.66 | 5.0 |
| 1.0648 | 6 | 6.4 | 0.9811 | 4 | 3.9 | 1.870 | 4.28 | 8.0 |
| 1.0869 | 8 | 8.7 | 0.9730 | 6 | 5.8 | 1.875 | 5.44 | 10.2 |
| 1.1089 | 10 | 11.1 | 0.9651 | 8 | 7.7 | 1.880 | 6.42 | 12.1 |
| 1.1309 | 12 | 13.6 | 0.9575 | 10 | 9.6 | 1.885 | 7.29 | 13.7 |
| 1.1530 | 14 | 16.1 | 0.9501 | 12 | 11.4 | 1.890 | 8.16 | 15.4 |
| 1.1751 | 16 | 18.8 | 0.9430 | 14 | 13.2 | 1.895 | 9.43 | 17.7 |
| 1.1972 | 18 | 21.5 | 0.9362 | 16 | 15.0 | 1.900 | 10.07 | 19.1 |
| 1.2191 | 20 | 24.4 | 0.9295 | 18 | 16.8 | 1.905 | 10.56 | 20.1 |
| 1.2738 | 25 | 31.8 | 0.9229 | 20 | 18.5 | 1.910 | 11.43 | 21.8 |
| 1.3279 | 30 | 39.8 | 0.9164 | 22 | 20.2 | 1.915 | 13.33 | 25.5 |
| 1.3798 | 35 | 48.3 | 0.9101 | 24 | 21.8 | 1.920 | 15.95 | 30.6 |
| 1.4300 | 40 | 57.2 | 0.9040 | 26 | 23.5 | 1.925 | 18.67 | 35.9 |
| 1.4779 | 45 | 66.5 | 0.8980 | 28 | 25.1 | 1.930 | 21.34 | 41.2 |
| 1.5253 | 50 | 76.3 | 0.8920 | 30 | 26.8 | 1.935 | 25.65 | 49.6 |

| *Aqueous Acetic Acid* | | | *Aqueous Hydrobromic Acid* | | | *Aqueous Sodium Carbonate* | | |
|---|---|---|---|---|---|---|---|---|
| PERCENT Acetic Acid BY WEIGHT | DENSITY $d_{4^\circ}^{20^\circ}$ | GRAMS Acetic Acid PER 100 ML | PERCENT HBr BY WEIGHT | DENSITY $d_{4^\circ}^{20^\circ}$ | GRAMS HBr PER 100 ML | PERCENT $Na_2CO_3$ BY WEIGHT | DENSITY $d_{4^\circ}^{20^\circ}$ | GRAMS $Na_2CO_3$ PER 100 ML |
| 10 | 1.0125 | 10.1 | 10 | 1.0723 | 10.7 | 2 | 1.0190 | 2.04 |
| 20 | 1.0263 | 20.5 | 20 | 1.1579 | 23.2 | 4 | 1.0398 | 4.16 |
| 30 | 1.0384 | 31.1 | 30 | 1.2580 | 37.7 | 6 | 1.0606 | 6.36 |
| 40 | 1.0488 | 42.0 | 35 | 1.3150 | 46.0 | 8 | 1.0816 | 8.65 |
| 50 | 1.0575 | 52.9 | 40 | 1.3772 | 56.1 | 10 | 1.1029 | 11.03 |
| 60 | 1.0642 | 63.8 | 45 | 1.4446 | 65.0 | 12 | 1.1244 | 13.50 |
| 70 | 1.0685 | 74.8 | 50 | 1.5173 | 75.8 | 14 | 1.1463 | 16.05 |
| 80 | 1.0700 | 85.6 | 55 | 1.5953 | 87.7 | 16 | 1.1682 | 18.50 |
| 90 | 1.0661 | 96.0 | 60 | 1.6787 | 100.7 | 18 | 1.1905 | 21.33 |
| 100 | 1.0498 | 105.0 | 65 | 1.7675 | 114.9 | 20 | 1.2132 | 24.26 |

## Atomic Weights (1969)

| Element | Symbol | Weight | Element | Symbol | Weight |
|---|---|---|---|---|---|
| Aluminum | Al | 26.98 | Manganese | Mn | 54.94 |
| Antimony | Sb | 121.75 | Mercury | Hg | 200.59 |
| Arsenic | As | 74.92 | Nickel | Ni | 58.71 |
| Barium | Ba | 137.34 | Nitrogen | N | 14.01 |
| Boron | B | 10.81 | Oxygen | O | 16.00 |
| Bromine | Br | 79.90 | Phosphorus | P | 30.97 |
| Calcium | Ca | 40.08 | Platinum | Pt | 195.09 |
| Carbon | C | 12.01 | Potassium | K | 39.10 |
| Chlorine | Cl | 35.45 | Silicon | Si | 28.09 |
| Chromium | Cr | 52.00 | Silver | Ag | 107.87 |
| Copper | Cu | 63.54 | Sodium | Na | 22.99 |
| Fluorine | F | 19.00 | Sulfur | S | 32.06 |
| Hydrogen | H | 1.01 | Tin | Sn | 118.69 |
| Iodine | I | 126.90 | Titanium | Ti | 47.90 |
| Iron | Fe | 55.85 | Vanadium | V | 50.94 |
| Magnesium | Mg | 24.31 | Zinc | Zn | 65.37 |

# Treatment of Burns and Injuries

## Treatment of Severe Burns

Extensive burns, in which the burned area amounts to as much as three square feet, require special first aid to avoid a fatal outcome. *Summon medical aid at once.*

Large burns may be fatal as a result of the burn itself or of shock. Take precautions to avoid shock when giving first aid for any burn over 30–40 square inches in area. As soon as the fire is extinguished, carry or conduct the patient to a quiet room, lay him on a cot, and cover the unburned area with blankets. Keep the patient warm and quiet, and give water to drink.

## In Case of Accidents

Always call or notify a laboratory instructor as soon as possible.

### Fire

***Burning Reagents.*** Immediately extinguish any gas burners in the vicinity. Fire extinguishers are available in various parts of the laboratory. The newer type of extinguisher charged with carbon dioxide under pressure is usually more satisfactory than a liquid (carbon tetrachloride) extinguisher. If the latter is used, direct the liquid to the base of the flame. It is important to note that toxic vapors may be formed when carbon tetrachloride extinguishers are used, and are a source of danger in a confined space.

For burning oil use powdered solid sodium bicarbonate. Dry chemical extinguishers, filled with dry powder under air pressure, are suitable for burning oil and flammable liquids.

***Burning Clothing.*** Avoid running (which fans the flame) and take great care not to inhale the flame. Rolling on the floor is often the quickest and best method for extinguishing a fire on one's own clothing.

Smother the fire as quickly as possible using wet towels, laboratory coats, heavy (fire) blankets, or carbon dioxide extinguisher. *Do not use carbon tetrachloride (pyrene).*

***Treatment of Small Burns.*** Wash with sterile gauze, soap and water. Cover with cod liver oil ointment (or vaseline) and sterile gauze pad. Fasten dressing in place with bandage or adhesive tape. In small second or third degree burns in which blisters have formed or broken, or in which deep burns are encountered, a physician should be consulted.

***Extensive Burns.*** These require special treatment to avoid serious or fatal outcome—*summon medical treatment at once.* Combat the effects of shock by keeping the patient warm and quiet.

## Injuries and Chemical Burns

***Reagents in the Eye.*** Wash *immediately* with a large amount of water, using the ordinary sink hose, eye-wash fountain or eye-wash bottle—*do not touch the eye.* If any severe discomfort remains after this treatment, the patient should see a physician.

***Reagents on the Skin.*** *Acids*—Wash *immediately* with a large amount of water, then soak the burned part in water for at least 3 hr. Cover the burned area with cod liver ointment (or vaseline) and a dressing.

*Alkalies*—Wash *immediately* with a large amount of water, then soak the burned part in water for at least 3 hr before applying boric acid ointment and a dressing.

*Bromine*—Wash *immediately* with a large amount of water, then soak the burned area in 10% sodium thiosulfate, or cover with a *wet* sodium thiosulfate dressing, for at least 3 hr. Cod liver oil ointment and a dressing may be applied later.

*Organic Substances*—Most organic substances can be removed from the skin by washing immediately with ordinary ethanol, followed by washing with soap and warm water. If the skin is burned (as by phenol), soak the injured part in water for at least 3 hr, then apply a cod liver oil or boric acid ointment and a dressing.

***Cuts.*** Wash the wound with sterile gauze, soap and water. Cover with a sterile dressing and keep dry.

# Index

Numbers in **boldface** refer to laboratory preparations